ANALYTICAL CHEMISTRY SYMPOSIA SERIES — volume 22

ion-selective electrodes, 4

Fourth Symposium held at Mátrafüred, Hungary,
8—12 October, 1984

editor

Prof. E. Pungor, Ph. D., D. Sc.

Member of the Hungarian Academy of Sciences

associate editor

I. Buzás, Ph. D., C. Sc.

ELSEVIER
Amsterdam - Oxford - New York - Tokyo 1985

ORGANIZING COMMITTEE OF THE SYMPOSIUM

President: Prof. E. PUNGOR, Ph. D., D. Sc., Member of the Hungarian Academy of Sciences; *Secretary:* K. TÓTH, Ph. D., C. Sc.: *Members:* G. FARSANG, Ph. D., C. Sc., E. LINDNER, Ph. D., C. Sc., G. HORVAI, Ph. D., C. Sc., G. NAGY, Ph. D., C. Sc.; *Organizers:* E. G. HARSÁNYI, Ph. D., A. HRABÉCZY-PÁLL, Ph. D.

Joint edition published by
Elsevier Science Publishers, Amsterdam, The Netherlands and
Akadémiai Kiadó, The Publishing House of the Hungarian Academy of Sciences, Budapest, Hungary

The distribution of this book is being handled by the following publishers
for the U.S.A. and Canada
Elsevier Science Publishing Company, Inc.
52 Vanderbilt Avenue
New York, New York 10017, U.S.A.

for the East European Countries, Democratic People's Republic of Korea, People's Republic of Mongolia, Republic of Cuba and Socialist Republic of Vietnam
Akadémiai Kiadó, The Publishing House of the
Hungarian Academy of Sciences, Budapest

for all remaining areas
Elsevier Science Publishers
P.O. Box 211, 1000 AE Amsterdam, The Netherlands

Library of Congress Cataloging in Publication Data
Main entry under title:

Ion-selective electrodes, 4.

(Analytical chemistry symposia series; v. 22)
Includes index.
1. Ion-selective electrodes—Congresses. I. Pungor,
E. (Ernő) II. Buzás, I. III. Series.
QD571.I585 1985 541.3'724 85-12916
ISBN 0-444-99553-6 (U.S.)

ISBN 0-444-99553-6 (Vol. 22)
ISBN 0-444-41786-9 (Series)

Printed in Hungary

ANALYTICAL CHEMISTRY SYMPOSIA SERIES

Volume 1 Recent Developments in Chromatography and Electrophoresis. Proceedings of the 9th International Symposium on Chromatography and Electrophoresis, Riva del Garda, May 15–17, 1978
edited by A. Frigerio and L. Renoz

Volume 2 Electroanalysis in Hygiene, Environmental, Clinical and Pharmaceutical Chemistry. Proceedings of a Conference, organised by the Electroanalytical Group of the Chemical Society, London, held at Chelsea College, University of London, April 17–20, 1979
edited by W. F. Smyth

Volume 3 Recent Developments in Chromatography and Electrophoresis, 10. Proceedings of the 10th International Symposium on Chromatography and Electrophoresis, June 19–20, 1979
edited by A. Frigerio and M. McCamish

Volume 4 Recent Developments in Mass Spectrometry in Biochemistry and Medicine, 6. Proceedigs of the 6th International Symposium on Mass Spectrometry in Biochemistry and Medicine, Venice, June 21–22, 1979
edited by A. Frigerio and M. McCamish

Volume 5 Biochemical and Biological Applications of Isotachophoresis. Proceedings of the First International Symposium, Baconfoy, May 4–5, 1979
edited by A. Adam and C. Schots

Volume 6 Analytical Isotachophoresis. Proceedings of the 2nd International Symposium on Isotachophoresis, Eindhoven, September 9–11, 1980
edited by F. M. Everaerts

Volume 7 Recent Developments in Mass Spectrometry in Biochemistry, Medicine and Environmental Research, 7. Proceedings of the 7th International Symposium on Mass Spectrometry in Biochemistry, Medicine and Environmental Research, Milan, June 16–18, 1980
edited by A. Frigerio

Volume 8 Ion-Selective Electrodes, 3. Proceedings of the Third Symposium, Mátrafüred, Hungary, October 13–15, 1980
edited by E. Pungor

Volume 9 Affinity Chromatography and Related Techniques. Theoretical Aspects/Industrial and Biomedical Applications. Proceedings of the 4th International Symposium, Veldhoven, The Netherlands, June 22–26, 1981
edited by T. C. J. Gribnau, J. Visser and R. J. F. Nivard

Volume 10 Advances in Steroid Analysis. Proceedings of the Symposium on the Analysis of Steroids, Eger, Hungary, May 20–22, 1981
edited by S. Görög

Volume 11 Stable Isotopes. Proceedings of the 4th International Conference, Jülich, March 23–26, 1981
edited by H.-L. Schmidt, H. Förstel and K. Heinzinger

Volume 12 Recent Developments in Mass Spectrometry in Biochemistry, Medicine and Environmental Research, 8. Proceedings of the 8th International Symposium on Mass Spectrometry in Biochemistry, Medicine and Environmental Research, Venice, June 18–19, 1981
edited by A. Frigerio

Volume 13 Chromatography in Biochemistry, Medicine and Environmental Research, 1. Proceedings of the 1st International Symposium of Chromatography in Biochemistry, Medicine and Environmental Research, Venice, June 16–17, 1981
edited by A. Frigerio

Volume 14 Chromatography and Mass Spectrometry in Biomedical Sciences, 2. Proceedings of the International Conference on Chromatography and Mass Spectrometry in Biomedical Sciences, Bordighera, Italy, June 20–23, 1982
edited by A. Frigerio

Volume 15 Computer Applications in Chemistry. Proceedings of the 6th International Conference on Computers in Chemical Research and Education (ICCCRE), held in Washington, DC, July 11–16, 1982
edited by S. R. Heller and R. Potenzone, Jr.

Volume 16 New Approaches in Liquid Chromatography. Proceedings of the 2nd Annual American–Eastern European Symposium on Advances in Liquid Chromatography, Szeged, Hungary, June 16–18, 1982
edited by H. Kalász

Volume 17 Chemical Sensors. Proceedings of the International Meeting on Chemical Sensors, held in Fukuoka, Japan, September 19–22, 1983
edited by T. Seiyama, K. Fueki, J. Shiokawa and S. Suzuki

Volume 18 Modern Trends in Analytical Chemistry. Part A: Electrochemical Detection in Flow Analysis; Proceedings of the Scientific Symposium, Mátrafüred, Hungary, October 17–20, 1982. Part B: Pattern Recognition in Analytical Chemistry; Proceedings of the Scientific Symposiun, Mátrafüred, Hungary, October 20–22, 1982
edited by E. Pungor, G. E. Veress and I. Buzás

Volume 19 Analytical Spectroscopy. Proceedings of the 26th Conference on Analytical Chemistry in Energy Technology, Knoxville, TN, October 11–13, 1983
edited by W. S. Lyon

Volume 20 Topics in Forensic and Analytical Toxicology. Proceedings of the Annual Meeting of the International Association of Forensic Toxicologists, Munich, August 21–25, 1983
edited by R. A. A. Maes

Volume 21 Chromatography and Mass Spectrometry in Nutrition Science and Food Safety. Proceedings of the International Symposiun on Chromatography and Mass Spectrometry in Nutrition Science and Food Safety, Montreux, June 19–22, 1983
edited by A. Frigerio

Volume 22 Ion-Selective Electrodes, 4. Proceedings of the Fourth Symposium, Mátrafüred, Hungary, October 8–12, 1984
edited by E. Pungor and I. Buzás

PREFACE

The Fourth Scientific Session on Ion-Selective Electrodes has raised a number of important issues in the field. Emphasis was primarily on theoretical questions but practical applications of ion-selective sensors were also discussed.

Organizing such a series of symposia is most worthwhile for it promotes the advancement of this branch of science by offering the participants thought-provoking ideas for further research.

In publishing the papers given at the symposium as well as the discussions, a complete appraisal of the state-of-the-art of ion-selective electrodes can be made.

I am convinced that this book will be useful for experts working in this field enhancing and updating their knowledge on these sensors and thereby raising problems to be studied in the future, the solution of which we hope to discuss during the Fifth Scientific Session on Ion-Selective Electrodes in 1988.

Ernő Pungor

LIST OF PARTICIPANTS

Bagó-Szmodits, E. (Mrs)
Hungary

Bartalits, L. (Mrs)
Hungary

Becker, P.
Switzerland

Boksay, Z.
Hungary

Buck, R.P.
USA

Buøen, S. (Mrs)
Norway

Busch, Hejlskov, I. (Mrs)
Denmark

Bychkov, E.
USSR

Čakrt, M.
Czechoslovakia

Cammann, K.
GFR

Camões, M.F. (Mrs)
Portugal

Csősz, M. (Mrs)
Hungary

Djikanović, M. (Mrs)
Yugoslavia

Durst, R.A.
USA

Ebel, H.
Austria

Ebel, M. (Mrs)
Austria

Eckert, W.
GFR

Egry, R.
Hungary

Elter, D.
Hungary

Ernyei, L.
Hungary

Farsang, G.
Hungary

Fehér, Zs. (Mrs)
Hungary

Fišl, Julija (Mrs)
Yugoslavia

Fodor, M.
Hungary

Frevert, T.
GFR

Fucskó, J.
Hungary

Gaál, F.
Yugoslavia

Gábor-Klatsmányi, P. (Mrs)
Hungary

Garay, T.
Hungary

Gráf-Harsányi, E. (Mrs)
Hungary

Gratzl, M.
Hungary

Gyenge, R. (Mrs)
Hungary

Hangos-Mahr, M. (Mrs)
Hungary

Hödrejärv, H. (Mrs)
USSR

Hoffmann, W.
GDR

Hopîrtean, E. (Mrs)
Romania

Hopkala, H. (Mrs)
Poland

Horvai, G.
Hungary

Hrabéczy-Páll, A. (Mrs)
Hungary

Ishibashi, N.
Japan

Janata, J.
USA

Jänchen, M.
GDR

Jovanović, M.
Yugoslavia

Jovanović, V. (Mrs)
Yugoslavia

Kellner, R.
Austria

Kimura, K.
Japan

Kolev, S.D.
Bulgaria

Kónya, G.
Hungary

Kormos, F. (Mrs)
Romania

Kucharkowski, R.
GDR

László, Á. (Mrs)
Hungary

Levy, P.F.
USA

Marsoner, H.
Austria

Márton, L. (Mrs)
Hungary

Milch, G.
Hungary

Morf, W.
Switzerland

Müller, H.
GDR

Nagy, G.
Hungary

Nagy, K.
Norway

Neshkova, M. (Mrs)
Bulgaria

Noszticzius, Z.
Hungary

Nowak, P.
Poland

Oesch, U.
Switzerland

Papp-Sziklay, Zs. (Mrs)
Hungary

Papp, J. (Mrs)
Hungary

Pastor, T.
Yugoslavia

Pékli, M. (Mrs)
Hungary

Radovanović, M. (Mrs)
Yugoslavia

Redinha, J.S.
Portugal

Reichenbach, K.
GDR

Rohonczy-Boksay, E. (Mrs)
Hungary

Seago, J.L.
USA

Simon, W.
Switzerland

Stefănigă, E. (Mrs)
Romania

Stella, R.
Italy

Szücs, Z. (Mrs)
Hungary

Tarczali, J.
Hungary

Tarjányi, K. (Mrs)
Hungary

Timár-Horváth, V. (Mrs)
Hungary

Thomas, J.D.R.
UK

Tomcsányi, L.
Hungary

Végh, G.
Hungary

Veress, E. (Mrs)
Romania

Vlasov, J.
USSR

Werner, B. (Mrs)
GDR

Wuhrmann, H.
Switzerland

Young, C.C.
USA

Zhang, Zong-Rang
China

Zhukov, A.
USSR

CONTENTS

DISCUSSION LECTURES

ROUND TABLE DISCUSSION ON THE PROBLEMS OF STANDARDS IN WORK WITH ION-SELECTIVE ELECTRODES

PLENARY LECTURES

CHEMICAL SENSOR MECHANISMS
A PROGRESS REPORT

RICHARD P. BUCK

Department of Chemistry
University of North Carolina
Chapel Hill, North Carolina 27514, USA

ABSTRACT

The mechanisms of the chemical sensors in this report are based on electrochemical effects. These are physical phenomena that obey electrostatic laws of classical physics, applicable to charged species (ions and electrons), and dipolar species (solvents, plasticizers and membrane components). Models of charge separations at interfaces are followed by models for calculating potential differences at equilibrium and kinetically-controlled interfaces. Conditions for generation of potential differences within homogeneous and non-homogeneous phases are also enumerated.

Five physical systems are used as examples, from a longer list of interface types, for which the general models are applicable. These include selectivity theory of solid state sensors, a mapping of potentials involved in the "suspension effect", a general treatment of liquid ion exchanger membrane potentials (at zero current) and two cases of biosensors based on unusual electrostatic effects.

AIMS AND TYPES OF THEORY

Chemical sensors in this review are those based on electrochemical effects. Because of the proliferation of devices for the monitoring of charged species, neutral species in solution or in the gas phase, and because of the advent of non-faradaic, capacitively-coupled as well as faradaic devices, the number of common features is difficult to enumerate (1-3). The number of phases and interfaces, differing by one, is not crucial. However, the presence of non-equilibrium interfaces is very important. For potentiometric, zero-current measurements of activities, cells with reversible, permselective interfaces lead to salt activity determinations. One or more non-equilibrium interfaces, of the junction type, provide the opportunity to measure single-ion activities, within the uncertainties of the junction potentials (4-7). Non-equilibrium cells with imposed voltages or currents(amperometric and chronopotentiometric) allow determination of single ion and neutral species concentrations in the transient state or steady state (8).

The phases fall into physical categories of ionic, electronic and mixed conductors. The conductivities are frequently in the semiconductor range, but metals are used, at least, in reference roles, and insulators are found in some newer devices based on field effects (9-11).

The devices are aimed at measuring activities (concentrations) of species. This end is possible because of the characteristic system function (input-output function) that relates current, voltage, activities and time. The currents and voltages are measured through the lead wires going to the device. The designs often use applied voltages or currents fixed in time; or they may use programmed, time-dependent applied voltages or currents as inputs. The measured result, the output, is most generally a time-varying current or voltage through the system. The output relates to the species activities via calibration, or occasionally by calculation, using predetermined parameters. When the measured output varies with time, the system is said to be in a transient, non-equilibrium state. When the output is constant (steady in time) the system is in a steady state or at equilibrium (8).

The aim of theory, at its least demanding level, has been to relate input and output, and to describe the relation between output and activities of species. Subsequently, intermediate level theory sought to relate output to physical-chemical processes at the phenomenological level. This means that models (mechanisms) were constructed. Based on the analytical outcome, one could predict what processes and factors needed to be modified to improve output. Improvements included better sensitivity to low activities of species, better selectivity among sensed species, and better (usually faster) time responses and avoidance of unwanted time-dependent outputs. The beauty of intermediate level theory was that the practical observables could be predicted, or at least explained. The results of these theories were at the same level as thermodynamic predictions, so one could always test the theories for consistency with thermodynamics (4-7).

Today, transport properties of ionic conductors, principally the "fast" ionic conductors, have become fashionable topics for statistical mechanicians. The deepest level of theory focuses on local ionic and atomic properties to predict the stepwise mechanisms of charge motion. Phenomenological effects, based on large numbers of ions, are taken as "knowns" or observables, to be interpreted at the most fundamental levels (12,13).

MODELS FOR CHARGE SEPARATION

Models of electrochemical sensors make use of electrostatic laws as first principles (4,5,14). All potential differences arise from separation of charge.

1. Separation of free charges (ions and electron) are permitted by energy differences of the species in different environments. This fact is the origin of charge separations at interfaces. The principle is applied generally for reactive metal/metal ion interfaces, inert metal/redox couple interfaces, and all kinds of electrolyte interfaces, among them electrolyte/salt, electrolyte/semiconductor, and electrolyte/electrolyte.
2. Separation of bound charges to form dipoles in a given phase arises from different energies of electrons in molecules. Dipoles of solvent molecules determine the liquid phase dielectric constants, and dipoles aligned by ions determine, in part, the energies in 1. Alignment of dipoles at interfaces creates local potential differences that determine electrical energies of ions in the interior bulk solution.
3. Separation of free charges by virtue of different energy environments in two phases can lead to dipole-like behavior. Partitioning of single kinds of ions from one phase to another produces diffuse layers in one or both phases. When one phase is metallic, the diffuse electron charge is classically treated as a surface sheet; however, modern theory and data show that this notion is over simplified (15,16). The two opposing layers of diffuse charge behave like a sheet of oriented dipoles.
4. Separation of free charges by virtue of different energy environments in a single phase, can lead to dipole-like behavior. A single salt solution with preferential surface adsorption (or absorption) of cations relative to anions behaves like an oriented dipole sheet.
5. Separation of free charges in a uniform-concentration phase occurs only when an external field (or other external force) is applied to make energies of ions different in different regions of the uniform phase. This separation of charge is found in experiments such as conductivity measurements and in some kinds of transference studies. Separation of free charges also occurs in non-uniform concentration phases by diffusion of ions with differing mobilities. This phenomenon is basic to the "junction potential".

Separation of charges at interfaces are driven by thermal energy and are opposed by local fields and a restoring force that arises from the violation of electroneutrality. Rates of charge separation are determined by the energy barrier inhibiting charge motion, relative to the driving energy kT. The time scale for the spontaneous, intrinsic process is the dielectric constant divided by the specific conductivity.

MODELS FOR POTENTIAL DIFFERENCES AT EQUILIBRIUM INTERFACES

The next step in modeling is calculation of potential differences (pds) from separated charges and from dipoles, usually dipole layers of the various types mentioned above (4-9,14,17-20). Isolated charges, well-separated in space, at long distances from a phase are taken to have zero energy, as a reference condition. The work done in moving charges near to a phase is the energy at any point. The electrostatic energy per unit of charge moved, is the local potential, ϕ, referred to infinity as zero. The work done in moving charges <u>into</u> a phase is made up of several components of which one is the electrostatic potential difference. Absolute calculation of total energies and potential requires calculation of at least four energy components in addition to the electrostatic term:

1. Calculation of energies inside charge free phases, treated as though the phases were structureless dielectrics.
2. Calculation of surface dipole potential contributions.
3. Calculation of specific ionic interactions or electronic interactions with molecular environment (local dielectric effects such as alignments, relaxation and saturation),
4. Calculation of bonding as ions or electrons react with local molecules of solvent, ionic components and neutral carriers.

The details of the procedures depend on whether one is considering ions or electrons (the statistics), and whether the phase is monatomic, molecular, ionic or a mixture. The procedure also depends on whether the receiving phase is liquid or solid; whether it is a metal, semiconductor (ionic or electronic), or an insulator; and whether it is mainly bulk or mainly surface. Certain similarities in procedure and theoretical concepts to exist:

1. The energies of free neutrals, ions and electrons in phases are composed of a "chemical" term, a concentration (activity) term, and electrostatic potential terms. These sums comprise the electrochemical potential which, for electrons, is called the Fermi Energy. Other work terms such as pressure-volume, or gravitation can be included into the simple sum, as needed.
2. At equilibrium, the electrochemical potentials of all rapidly, reversibly, exchanging species are equal. Fluxes of species and exchange current densities are equal and opposite; no net current flows. For several charged species at equilibrium, the individual exchange current densities can be different. Establishment of the equality of fluxes when phases are contacted happens because of the rapidly generated space charge and dipole alignment. Potential differences take the Nernst, Nicolsky or Goldman forms

depending on the number, and sign or exchanging charged species.
Reversible interfaces may be perturbed with an applied voltage; current flows and near-surface concentrations are changed from the bulk values. Nevertheless, the exchange fluxes and currents take on new equilibrium values corresponding to the new, often time-dependent surface concentrations. This is the origin of the observation that the interfacial pd is given by the equilibrium Nernst Equation even though current is flowing and the system is formally not in equilibrium.

3. Poisson's equation and Gauss' Law are generally applicable, but are rarely needed for the calculation of measurable net pds between phases. Space charge, created at interfaces, is a result of energy imbalance of free charges. Alignment of permanent dipoles at interfaces presumably follows from the local field established, in part, by the space charge. Unless charge-potential relationships are sought, or potential-distance profiles are needed, Poisson's Equation and Gauss' Law are peripheral; net pds are found directly from energy considerations.
4. Bulk phases are electrically neutral (quasi-electroneutral) for electrolyte concentrations with corresponding Debye screening lengths that are small compared with bulk phase dimensions. The generated space charge reside mainly at interfaces.
5. Systems with rapid surface exchange rates respond to peturbations by voltage concentration or charge changes, by laws of mass transport. Responses are mass transport-controlled.

MODELS FOR POTENTIAL DIFFERENCES AT KINETICALLY CONTROLLED INTERFACES

The kinetics of pure ion and electron crossings of interfaces obey Butler-Volmer form equations because rate constants reflect potential-dependent barriers; concentrations near an interface are not uniform, but are potential-dependent because of potential profiles adequately expressed through Gouy-Chapman theory (8). Surface barrier-controlled transport determines the smaller-than-reversible-limit fluxes and exchange currents. At equilibrium, slow, surface-barrier controlled fluxes have equal values. However these values and corresponding exchange current densities are much smaller than the rapid, reversible values. Kinetically-limited charge transport at interfaces is not exclusively caused by barriers to simple ion or electron transfers. Rates of ion transfers into liquid ion exchangers phases are most frequently limited by the homogeneous reaction rates of ions with aqueous ion exchangers at low concentration. The subsequent transport of ion pairs into the organic phase can be rapid or slow. Similarly, formation of ion-carrier complexes in advance of interfacial ion transport may be rate limiting, although rates of ion-carrier formation at, or in, the organic phase cannot be ruled out. Together these are examples of rate limiting pre- or post-transfer reactions analogous to CE and EC mechanisms in electron transfer kinetics.

When ions of both sign transfer, as in salt transfers, one or both ions may be kinetically limited (21). The exchange currents for each ion need not be equal, at equilibrium, because electroneutrality of the phases will not be violated. However, at non-equilibrium salt extracting interface fluxes of transfering ions of opposite charges must be equal to each other, to avoid electroneutrality violation. Potential differences for this type of kinetically slow interface may differ from the equilibrium value depending on the flux values of the transporting ions. This added potential difference is an overpotential that mainly reflects potential-dependence of the rate constants.

When surface rates are very small, mass transport in an electrolyte or in a membrane can be faster than the interfacial crossing. In that limiting case, stirring or flow does not affect pds in principle. However, there is a wide range between very slow (totally irreversible) and rapid, reversible cases in which surface transport and bulk mass transport are coupled. Both must be considered in deducing the measured pd (8).

MODELS FOR BULK PHASE TRANSPORT

Non-equilibrium transport of ions and electrons in bulk phases (uniform and non-uniform concentrations) is treated macroscopically. The systems are presumed to be thoroughly damped, even on the microsecond time scale (7-9,17). Resulting force-velocity equations (Nernst-Planck, Stefan-Maxwell and Onsager Irreversible Thermodynamic) give fluxes as a linear sum of diffusion, migration (drift) and interaction (cross term) forces. Proportionality factors between forces and velocities are mobilities that depend on ionic strength (relaxation and electrophoretic effects). Transport in multiple phases, typically laminates, and in grossly heterogeneous phases, may require additional theory, such as "percolation" theory, or application of regular theory in each part of the system.

Transport of salts is accompanied by dipole-level separation of charge, but quasi-electroneutrality is not violated since the net charge is zero in any volume large compared with inter-ion distances. A result of the need to maintain electroneutrality in bulk phases is the idea of "electroneutrality of diffusion". Spontaneous diffusion-migration of fast and slow ions creates local fields and potentials (so-called "diffusion potentials") such that fluxes of opposite-sign species beecome equal. Fast ions are slowed and slow ions are speeded. This principle is basic to the derivation of "junction potentials" where salts diffuse from high concentrations to low. Maintenance of electroneutrality is also a guiding principle in counter-direction transport, at junctions of all kinds, and in membranes with fixed, mobile or no sites.

APPLICABLE SYSTEMS

Closely related theories using the principles outlined above have been developed for many interfacial systems (11). These have been studied and continued to be studied, in the contexts of sensors, modified metal electrodes, modified semiconductors and photovoltaic and photogalvanic devices. There are further applications to the continuing work on batteries and to corrosion that involve many of the same interfaces and bulk phases. A short table of interfaces with their applications shows the extent of the field for theoretical analyses (Table 1). Combining interfaces and phases into electrodes (half-cells) and whole cells, makes a set of combinations that are too numerous to be mentioned here.

As far as sensors are concerned, response theory has progressed to cover many conditions: potentiometry (no current) amperometry (fixed applied cell voltage), chronopotentiometry (constant cell current), cyclic voltammetry (triangular voltage-time) applications, and AC methods (small amplitude applied cell voltages). There are numerous other techniques with less generality and less usefulness in sensor applications, i.e., chronoamperometry.

SOLID STATE ELECTRODES

Progress in predicting solids for use as ion-sensing membranes, and metal-contact electrodes (1-3) does not prove existing theories, but does not deny validity because there have been no surprises. All solids that have been uncovered are ionic or mixed conductors, and they show rapid ion exchange when Nernstian responses are encountered, as judged from time-responses. Effects of complexing agents on sensed ion activities, identification of side reactions such as air oxidation, leaching or hydrolysis of some solid electrodes, and interpretation of surface species and surface reactions via surface analysis methods, confirm expectations of theories of solid state electrode sensing mechanisms. These very general theories predict location of the potential-determining processes. They also indicate factors that should affect responses at equilibrium, steady state and transient time regions. The extensive studies reported in recent reviews may not prove the models or the equations rigorously, but they are consistent with theoretical predictions.

Previous mechanistic topics include metathetic surface reactions and redox reactions among species being determined and the compounds of the electrode surface. Since the earliest papers, these topics have been avoided because of the difficulty in describing them thermodynamically and kinetically. In addition, the modern analyses lead to response functions that do not always look like the Nicolsky form. It becomes difficult to appreciate the factors determining slopes and selectivities.

As an example, consider the AgCl crystal membrane in the presence of mixed electrolytes of soluble Cl^-, Br^- and/or I^-. From thermodynamics, the membrane potential is

$$\Delta\phi_{membrane} = \frac{RT}{F} \ln \frac{Ag^+(test)}{\overline{Ag^+}(memb.\ test)} \cdot \frac{\overline{Ag^+}(memb.\ ref.)}{Ag^+(ref.)} \quad (1)$$

where bars are silver ion activities inside the membrane, and silver ions are presumed to be potential-determining. For a pure AgCl membrane, it is logical that membrane phase activities are uniform and constant. However, upon exposure to Br^- or I^- the membrane converts methathetically to AgBr or AgI. It is not clear that $\overline{Ag^+}$ in AgBr or AgI is the same activity as in AgCl. But this assumption is made:

$$\text{Activity of salt } (\overline{Ag^+})(\overline{Cl^-}) = (\overline{Ag^+})(\overline{Br^-}) = 1 \text{ pure phase} \quad (2)$$

Activities of mixed salts or island phase

$$(\overline{Ag^+})(\overline{Cl^-}+\overline{Br^-}) = 1 \quad (3)$$

From mass balance:

$$\frac{(Ag^+)(Cl^-)}{K_{AgCl}\overline{\gamma}_{Cl^-}} + \frac{(Ag^+)(Br^-)}{K_{AgBr}\overline{\gamma}_{Br^-}} = 1 \quad (4)$$

and

$$\Delta\phi_{membrane} = -\frac{RT}{F}\ln\left(Cl^- + \frac{K_{AgCl}\overline{\gamma}_{Cl^-}}{K_{AgBr}\overline{\gamma}_{Br^-}}Br^-\right) - \frac{RT}{F}\ln\frac{Ag^+(ref.)}{K_{AgCl}\overline{\gamma}_{Cl^-}} \quad (5)$$

The equilibrium selectivity coefficient is:

$$k_{Cl^-,Br^-} = \frac{K_{AgCl}}{K_{AgBr}} \quad (6)$$

This result arises from assumptions that are not thermodynamic! The result is peculiar for other reasons: If solid activities are 1, the response should be two intersecting straight lines for variable Cl^- and constant Br^- or I^-. Since no mixed solids form, then the membrane should change sharply from all AgCl to all AgBr or AgI at the equilibrium activity ratio. Introduction of mixed phases may help in the analysis because the mixed phases with less than unit activity, say s and 1-s, should give lower, rather than greater solubilities in the mixture region. However, formation of miscible, mixed solid phases is not a general property of all halide salts; those with different crystal structures are not expected to form mixed phases. A possibility that surface phases have different (higher) activities and greater than normal K_{sp} values, has not been included.

One possible conclusion is that curvature response region for mixtures of soluble anions is a non-equilibrium effect and should be treated by kinetic methods. A corrolary is that the potentiometric selectivity coefficient given just above, is an artifact. If the interference processes are kinetically-controlled, then it is problematic whether Cl^- interference at AgBr operates by the same processes and rates as Br^- attack on AgCl, for example.

There have been several attempts to treat experimental data. One is simply to ignore explanations and to curve-fit potentiometric responses (22-24). Another approach of Hulanicki and Lewenstam (25-27), Lindner, Toth and Pungor (28) and Morf (29) is to treat the change of phases by diffusion-controlled kinetics. The idea admits that the electrode surface activities determine the measured responses, but the bulk solution activities are not the same as surface values until equilibrium is reached. Then, one expresses the difference between surface and bulk concentrations of anions by balancing fluxes of entering and leaving ions at zero current. The result is a time-dependent expression for the salt activities AgCl (1-s) and AgBr (s), for example. From this analysis, the apparent, time-dependent selectivity coefficient is deduced. The limiting expressions are:

$$k^{o}_{Cl^-,Br^-} = \frac{D_{Br^-}}{D_{Cl^-}} \quad \text{at t about zero (all bulk AgCl)} \tag{7}$$

$$k^{t}_{Cl^-,Br^-} = \frac{k^{\infty}_{Cl^-,Br^-} k^{o}_{Cl^-,Br^-}}{k^{\infty}_{Cl^-Br^-}(1-s) + k^{o}_{Cl^-,Br^-}(s)} \tag{8}$$

$$k^{\infty}_{Cl^-,Br^-} = \frac{K_{AgCl}}{K_{AgBr}} \qquad t = \infty \quad \text{(all AgBr)} \tag{9}$$

Experimental potential decay curves have been reported by Sandifer (30,31) and by Rhodes and Buck (32), in addition to some of the above. When the responses are monotonic with one inflection, the Morf theory with reasonable, but adjustable, parameters suffices, as in Fig. 1. However, the data are not always so simple, in Fig. 2 and Fig. 3. It is possible that a model using only transport in the bathing phase is not sufficient, and that diffusion in pores and crevasses may need consideration.

Closely related problems involve the mixed-valence sulfide sensor CuS/Cu_2S, and the venerable glass electrode.

Hepel has shown (33) the reactive nature of Cu(I)/Cu(II) surfaces in the presence of Cu^{2+} response regions (slope 29 mV/decade activity) and Cu^{+} response regions (slope 59 mV/decade activity). Once again the species in bulk are not sensed, but the species at the electrode surface are followed. He finds that Cu^{2+} replaces Cu^{+} from the surface to convert Cu_2S to CuS, and vice versa, depending on ligand stability.

Although not an identical situation, because glasses are pure ion exchangers, surface layers on glasses cannot be ignored. To explain slower time responses than those predicted by mass transport in bathing solutions, a gel layer with rather low site densities, such as that provided by silica gel, seems indicated. This is not a new idea, but time responses were not considered criteria for gel layers, even though dehydration experiments, and direct surface thickness measurements, showed the films. Thermodynamic, i.e. equilibrium, theories do not require consideration of passive films, and three-layer transport problem analysis is difficult. However, transient-overshoot potential responses with highly ion-exchange-favored, but low mobility ions, can only be analyzed assuming low site-density films that permit separation of ion exchange and mobility in space and in time. Homogeneous membrane theories do not permit this, because single ion exchange constants and mobilities are coupled in time, via the Teorell-Meyer-Sievers potential sum, to arrive at the total membrane pd.

FIXED SITE MEMBRANES, ION EXCHANGE RESINS AND THE "SUSPENSION EFFECT"

Fixed site ion exchanger membranes and corresponding resin beads are electrically neutral in the dry state. In contact with water, beads swell and the mobile counter ions (opposite sign of charge from the sites) find themselves in a low energy environment. Some, albeit few, counter ions are lost to the bathing phase to form space charge and an interfacial potential difference (positive aqueous phase and negative resin phase). The space charge region is very close to the surface, so the interstitial water pH is unaffected. When many resin beads touch, the space charge overlaps; the ensemble of beads defines a conducting phase. The net, uncompensated charge resides at the interface with the supernatant. The slurry "phase" interior is internally charge-compensated. An illustration is given in Fig. 4.

The significance of the overlapping charge is that the supernatant phase is different in potential (the inner Galvani potential) from the interstitial water (slurry) phase. Both are at the same pH as demonstrated by filtering the interstitial fluid. Constancy of pH is also shown by use of two matched glass electrodes, one in each phase. However, placement of the reference electrode, of the junction type, is crucial to the measured results: These experiments are analyzed electrostatically in Fig. 5.

Two matched glass electrodes in the same phase, or opposite phases, read zero pd. This result, demanded by thermodynamics, means that the interface pd between supernatant and slurry also exists between the glass electrode surface and the slurry. Consequently, matched electrodes read zero pd in principle, and nearly zero experimentally. Two matched junction-type reference electrodes read zero pd in supernatant, and have been reported to read approximately zero in the slurry. We find considerable variability in the slurry measurements, although the measured pds settle down to a near zero value on standing. This variability, that depends on reference placement in the slurry, gives a hint that the junction potential magnitude depends on the extent of bead contact vs interstitial fluid contact . Transference numbers of the junction potential-determining ions ought to vary with position in space charge regions vs electroneutral interstitial solution regions. There is considerable experimental evidence to support this hypothesis, although the original effort by Jenny et. al (34) overestimated the importance of the slurry-phase junction potential variability. One can diminish the variability effect by wrapping the reference electrodes in inert, ion-conducting dialysis membranes. Our results suggest that greater contact area between reference electrode tip and slurry makes the non-equilibrium junction potential more uniform and reproducible.

The actual "suspension effect" is the name given to the apparent measured pH change occurring when the glass and reference electrodes are both in supernatant, or both in the slurry. The lower half of Fig. 5 shows these cases, and two other combinations of the electrodes in opposite phases. There are only two different pd results: one gives the pH of the supernatant and interstitial fluid when the reference electrode is in the supernatant; the other gives a lower pH value for supernatant and interstitial fluid when the reference electrode is in the slurry. The mV difference, corresponding to the pH difference, is surprisingly close to the measured interfacial pd between supernatant and slurry using two reference electrodes, one is each phase!

Recently Brezinsky (35) made a study of the suspension effect by repeating and extending Jenny's et. al results. Although the notion is not new that the suspension acts as a charged phase, Brezinsky made a strong case. He analyzed the interfacial pd and charge, and emphasized its dominating control over the effect. This argument ran counter to that of Jenny et. al, who emphasized the reference electrode junction potential variability in the slurry.

The electrostatic analysis supports the supernatant/slurry ionic interfacial pd as the principal contributor to the suspension effect. Furthermore, the two reference electrode measurements (one electrode in each phase) gives the quantity $\Delta\phi_{slurry} + \Delta E_J$, because the junction with free ion flow "swamps" any interfacial components at the junction tip. Swamping occurs when the inner filling electrolyte contains KCl at typical concentrations large compared with the proton space charge in the vicinity. It is experimentally found that some "artifactual" junction potential at the slurry interface remains, depending on fortuitous electrode placement. The ΔE_J is then not precisely zero.

A second result of the electrostatic analysis is that

$$\Delta\phi_{slurry} = \frac{\Delta\mu^{o}}{F} + \frac{RT}{F} \ln \frac{H^{+}\ (beads)}{H^{+}\ (solution)} \quad . \tag{10}$$

The standard state difference can be near zero since the beads contain mostly water. With electrodes in supernatant,

$$\Delta\phi_{measured} = \Delta\phi^{o}_{glass} - \frac{RT}{F} pH_{su} - E_J \tag{11}$$

where su = supernatant. With the reference electrode in the slurry,

$$\Delta\phi_{measured} = \Delta\phi^{o}_{glass} - \frac{RT}{F} pH_{su} - \Delta\phi_{slurry} - E_J \tag{12}$$

$$\simeq \Delta\phi^{o}_{glass} - \frac{RT}{F} pH_{beads} - E_J - \frac{\Delta\mu^{o}}{F}$$

Brezinsky hypothesized that the suspension effect pH seemed remarkably like the pH anticipated for the beads!

When inert salts are added to the system, i.e. KCl, the aqueous phase becomes quite acidic as the beads take up K^+ and release H^+. The phase pd is then dependent on the activities of K^+ in, and out, of the bead. For high concentrations of added KCl, the slurry phase pd approaches zero and the suspension effect disappears. However, if one starts with water and beads already prepared in the K^+ form, and no excess of added KCl, the suspension effect reappears.

LIQUID SALT/ION EXCHANGER MEMBRANES

In 1970 Higuchi, et al. (36) and Liteanu (37) introduced liquid membrane electrodes responsive to organic and inorganic ions. These electrodes were neither straightforward ion exchanger- nor neutral carrier-based. Although the equilibrium principle was available (equality of electrochemical potential of each ion that reversibly equilibrates across an immiscible liquid/liquid interface), the elementary theory and consequences were not explored until recently (38). To develop an interfacial pd at a liquid interface, two ions M^+, X^- that partition are required. This condition is necessary but not sufficient, because the pd produced is independent of salt concentration in each phase. To develop a pd dependent solely on an ion activity, say M^+, three ions are required -- M^+, X^- and Y^-, of which Y^- is typically very oil soluble; X^- is water soluble and M^+ is soluble in both phases. The salt MY is typically an organic ion pair that may be isolated and dissolved in an organic solvent or prepared by extraction. The anion is typically picrate, tetraphenylborate, triphenylstilbenylborate or tetrabiphenylborate. The salt MX, where $X^- = Cl^-$ preferably, is the sample whose M^+ activity is to be measured at variable values in the aqueous phase. When MX is varied, the interfacial pd is S-shaped (mV vs log[MX]), Fig. 6 (upper curve).

Each ion M^+, X^- and Y^- will generally have different energies in water and in an organic phase: an ester such as dioctyladipate (low dielectric constant) or nitrobenzene (high dielectric constant). The partition free energy

$$\Delta G = \bar{\mu}^o - \mu^o = -RT\ln K_i \qquad (13)$$

is a measure of the intrinsic ionic oil-solubility, where K_i is the single-ion partition coefficient. For oil soluble (hydrophobic) ions ΔG is more negative and K_i is larger than for water-soluble (hydrophilic) ions. For a typical two phase water/organic system containing equilibrated salts MX and MY, some MX and MY will be present in each phase. The concentrations in each depend on the salt partition coefficients $K_M K_X$ and $K_M K_Y$. MX is predominately in water while MY is predominately in the organic phase when $K_X << K_Y$. When MX activity is varied an S-shaped curve is found such that

$$\Delta\phi = \frac{RT}{F} \ln \frac{K_M a_M}{\bar{a}_M} \qquad (14)$$

(Nernstian response) over a wide activity range. Bars signify organic phase activities. However, at very low MX activities, the pd becomes insensitive to decreasing M^+ activities and levels off at a value

$$\Delta\phi = \frac{RT}{2F} \ln (K_M/K_Y), \qquad (15)$$

while at very high MX activities (generally only seen when $X^- = I^-$, NO_3^-, ClO_4^-) the pd again levels off since Donnan Exclusion by Y^- is violated in the organic phase and

$$\Delta\phi = \frac{RT}{2F} \ln (K_M/K_X) \quad . \qquad (16)$$

Electrodes for anion Y^- can be envisioned along the same lines: select cation M^+ to be hydrophobic and N^+ to be hydrophilic, such as Na^+, K^+, etc. Dissolve MY in the membrane while NY is the aqueous salt whose anion activity is to be determined. Note that a single electrode can respond in Nernstian fashion to anions or cations! This is shown in Fig. 6 (lower), where activity N^+Y^- is plotted. The lower limit of detectability of Y^- is again given by (15). Breakdown of Donnan Exclusion at high added NY activities produces a limit potential

$$\Delta\phi = \frac{RT}{2F} \ln \left(\frac{K_N}{K_Y}\right) \tag{17}$$

analogous to eq. (16).

Although this theory has not been succinctly reported until recently, intuitive applications have been applied earlier. Ruzicka, et al. (39) increased the oil solubility of phosphate ester anions to increase the sensitivity of the Ca^{2+} electrode. Gavach, et al. (40) and Birch, et al. (41) made detergent sensors for cations and anions.

In a recent paper (42) we reported the construction and performance characteristics of membrane electrodes for succinylcholine, hexamethonium and decamethonium, based on their ion pair complexes with triphenylstilbenylborate. The electrodes show near-Nernstian responses over a large range of concentrations and very low detection limits. These electrodes are not affected by pH in the range 2-10. Their selectivity relative to a number of inorganic ions, amino acids, neurotransmitters, drugs and various drug-excipients is outstanding (detection limit of 0.3 microgm/ml).

As an example of an anion-sensitive electrode, the construction and performance characteristics of an ion-selective membrane electrode sensitive to phenytoin-drug, based on its ion-pair complex with tricaprylylmethylammonium, in PVC matrix has been described (43). The electrode shows near-Nernstian response over 10^{-1}-10^{-4} mol/l range and a detection limit of 1.5×10^{-5} mol/l. The selectivities to a number of inorganic and organic anions are reported. OH^- interference, in the linear range of calibration curve, is negligible up to pH 11.

THEORY FOR BIOSENSOR APPLICATIONS WITH BLOCKED AND NEARLY BLOCKED INTERFACES

Properties of blocked interfaces have long interested electrochemists because these interfaces represent examples of the electrified surface state. The thermodynamic treatment was done long ago by J. W. Gibbs, noted in this context, for the Gibbs Adsorption Isotherm. Further analysis by Lippmann showed the possibility of determining charge densities of interfaces and the presence of an extra degree of freedom possessed by blocked vs classical, reversible interfaces (8). Blocked interfaces are those which pass no charge and are ideally polarizable by application of an external voltage. Their electrical properties depend on charge, in addition to voltage, concentration and time. Any current that appears to flow on application of voltage is transient, capacitive in character, and develops space charge or other forms of surface excess species; accumulation and depletion are general effects called "adsorption" when either specific contact or space charge regions are predominantly developed near an electrode surface. This rather abstract topic became important in sensor research with the advent of CHEMFETS or Chemically-Sensitive Field Effect Devices (Transistors), and was introduced by Janata and Buck at the first meeting devoted to solid state sensors (6,44), and later amplified by Janata and Huber (10).

The relevant blocked interfaces in sensor work are electrolyte/metal, electrolyte/membrane (where the electrolyte is a large polyelectrolyte or protein that cannot pass into the membrane), electrolyte/insulator, electrolyte/insulator/metal and electrolyte/insulator/semiconductor. For biosensor applications wide voltage range, blocked interfaces are not required. Probably, less than one volt range of pds would be adequate and, in that case, inert metals in their double layer region may suffice.

There are three potential/charge effects at blocked electrolyte/metal interfaces and blocked membrane interfaces that are easily visualized. These are effects that, hopefully, can be exploited for sensors of activity or activity changes for very low concentrations of species in solution.

1. Esin-Markov Effect
2. Charge cancellation/charge generation (adsorption potentials)
3. Frumkin Effects on rates of pilot ions that are irreversibly crossing incompletely-blocked interfaces

The first is described by the Esin-Markov coefficient that relates the space charge potential changes to the solution activity changes at a charged metal surface. Thermodynamics provides a general relation for the potential of a cell with a blocked interface electrode vs a constant reference electrode potential. This effect is non-discriminating of species, and all salts in an electrolyte mixture are involved. Consequently, changes in blocked electrode potentials with changing solution activities arenot expected to be a generally applicable analytical principle.

Charge cancellation at a blocked interface is, at first sight, likely to be a more specific event. For example, consider a metal with a bound, charged dye, a surfactant, or an antigen in equilibrium with an electrolyte. There will be some interfacial pd corresponding to the inner layer and diffuse layer potential components, as determined by the ionic concentration and dipole effects. Upon addition of a specific counter-charge reagent, the bound layer will be discharged partially, completely, or even charged with opposite sign. The measured potential should then reflect this charge cancellation reaction. The probable requirements are:

1. Fully blocked interface
2. Establishment of charged interface using a closed circuit
3. Measurement of potential difference at nearly open circuit using an electrometer
4. Measurement of potential change upon charge cancellation
5. Large, favorable equilibrium constant for the charge cancellation reaction

The sensitivity could be quite large. For a change in surface concentration $\triangle c/cm^2$, the charge change is

$$\triangle q = \triangle c/cm^2 \times n \text{ (charge/species)} \times F \qquad (18)$$

For picomole/cm^2 changes and a blood electrolyte with the Gouy-Chapman capacitance value 86.6 μcoul/volt cm^2, the expected voltage change is:

$$\frac{d\phi_2}{dc} = 1.11 \times n \ \ \text{mV/picomole/cm}^2 \qquad (19)$$

By using integral capacitance data, determined experimentally in terms of actual, total rational electrode potential, one finds a lower capacitance value. Using a more realistic integral capacitance of 10-20 μF/cm^2, then the result 5 - 10 mV/picomole/cm^2 is found.

These experiments are not easy to perform. There are several severe problems that are sources of error or failure:

1. Charge leakage to exterior
2. Charge leakage to interior:
   a. via ion penetration of incompletely blocked interface, a non-zero ion exchange current
   b. via electron exchange at the incompletely blocked interface, a non-zero electron exchange current
3. Unfavorable charge cancellation equilibrium constant

In a series of papers beginning in 1978 (45) and most recently, 1981 (46), Yamamoto et al. have tested these hypotheses using trypsin and aprotinin alternately bound to TiO_2 and in solution. About the same time, Boitieux (47) tried very similar experiments. At the 10^{-6} - 10^{-7} mol/l level of trypsin and aprotinin, potential shifts of a few millivolts were found upon addition of the dilute solution of one to a bound form of the second species. The pH-dependence was measured and maximum effect was detected at pH 9.5 where the binding reaction constant shows, by independent measurement, the greatest value.

Another group of effects that lead to pds at electrolyte/membrane interfaces are associated with Frumkin. The basic notion is that leaky membranes admit salts from bathing solutions as illustrated in Fig. 7. The dissolving process for kinetically slow ion transfers establishes a pd; and the salt becomes a kind of pilot salt. When the membrane is subsequently modified by a surface reaction with a large molecule such as a protein, the rates of pilot ion transfers are changed and the new pd is a monitor of the protein surface raction.

Buck treated this subtle effect that appears with imperfectly permselective materials. He showed that it arises when there is a failure of co-ion exclusion accompanied by slow interfacial kinetics of ion crossings (21). Whenever the site density of ion exchanging materials is low in comparison with the external electrolyte concentration multiplied by the salt-extraction coefficient, it is probably that ions of both signs will enter a membrane. Thus for organic liquids and polymers used as film materials, it is probable that salts from the bathing electrolyte will dissolve in the organic phase. This effect has been observed and verified by Collins and Janata (48), and it is likely that electrodes made from site-free materials for binding haptens, enzymes, and antibodies may be plagued by salt encroachment from bathing electrolytes.

An important result of electrochemical kinetics is that the equilibrium interfacial potential difference is independent of rates, but the individual forward and backward fluxes can be quite different. For example, a saturated solution of an ionic substance can receive cations and anions at different rates because their simultaneous removal rates balance the generation rates. However, away from saturation equilibrium, rates of dissolving or rates of precipatation must produce equal fluxes of ionic charge in each phase. Otherwise, electroneutrality will be violated.

Consider, in Fig. 7, a liquid-membrane film without ion exchange sites and exhibiting co-ion exclusion failure. An increase in NaCl activity at x = 0 will be accompanied by a distribution of $\underline{NaCl}$ at x = 0'. There will be a net equal flux of Na^+ and Cl^- from left to right across the interface, and a lower than equilibrium concentration of NaCl will result as the salt diffuses from x = 0' to x = d. During the filling time, C(0',t) for each ion will be low relative to the equilibrium value. The back reaction

$$\overline{Na}^+ + \overline{Cl}^- \rightarrow Na^+ + Cl^- \qquad (20)$$

is slower than normal. The partial currents are shown in

Fig. 8 for the initial state and the transient state. Initially, the forward and backward partial currents are the exchange currents, and they are equal for each species at $\triangle\phi^{eq}$ (dotted vertical line). Later, the forward fluxes will dominate. In fact, if the backward rates are small enough to be neglected, the interfacial potential difference will shift to the value shown by the potential of equal fluxes of Na^+ and Cl^- [solid line marked $\triangle\phi$(total irreversibility)]. As the film fills with NaCl, the system reaches equilibrium and the interfacial potential difference will return to a value near the initial value before the activity step was made. For this illustration, i_-^o for Cl^- was taken to be larger than i_+^o for Na^+. If the reverse were true, then $\triangle\phi$(totally irreversible) would have shifted positively (left) rather than negatively (right).

THEORY FOR BIOSENSOR APPLICATIONS WITH NEARLY REVERSIBLE INTERFACES

The last item in the previous section offered a mechanism for potential generation via pilot salts whose interfacial transfer rates were slow enough to be only quasi-reversible or irreversible. In that event, the pd developed was interfacial. However it is well known that diffusion potentials occur in reversible interface membranes with low site densities as a result of Donnan Exclusion Failure. When salts penetrate the membrane interface and pass by diffusion-migration from one side to the other, electroneutrality of diffusion generates local interior fields and diffusion potentials, as outlined in an earlier section. The most recent theoretical analysis was applied to liquid membranes (49-51) and earlier references to fixed-site membranes are contained in these papers. The diffusion potential is a component of the total membrane potential and is directly measureable by potentiometry.

One can visualize two, or perhaps, more ways that Donnan Failure, or the reverse, Permselectivity Generation, can be developed in membranes or at membrane interfaces.

1. Cancellation of fixed sites monitored by permeable pilot ions using membrane potential (Donnan Breakdown) measurements
2. Generation of fixed sites monitored by permeable pilot ions using membrane potential measurements

These processes are monitored by the diffusion potential of a pilot salt (or salts), and they are generated by very low activities of charged species, such as antibodies. Aizawa, Kato and Suzuki (52) have reported a blood typing membrane electrode containing membrane-bound blood group substances. In the presence of added antibody (agglutin) in serum, the membrane potential is changed significantly by the agglutination reaction that seems to occur mainly at the surface. The interesting mode of operation: use of a fixed, non-unity ratio of NaCl bathing electrolyte, suggests that the measured membrane potential change, before and after agglutination, is a diffusion potential change due to NaCl transport through the membrane. The model is being elaborated by other workers in Suzuki's group in terms of surface adsorption, charge cancellation and changes in the diffusion potential according to the Donnan Exclusion Failure theory. It is too early to say whether the system is better treated by this model or by the surface rate-potential dependence model given in the previous section. It is probable that the pilot ion mechanism via diffusion or interface control, is involved in the immunosensor of Solsky and Rechnitz (53).

REFERENCES

1. G. H. Fricke, Anal. Chem., 52, (1980) 259R.
2. M. E. Meyerhoff and Y. M. Fraticelli, Anal. Chem. 54, (1982) 27R.
3. M. A. Arnold and M. E. Meyerhoff, Anal. Chem., 54, (1984) 20R.
4. R. P. Buck, Sens. and Act., 1, (1980) 197.
5. R. P. Buck, "Theory and Principles of Ion Selective Electrodes" in "Ion-Selective Electrodes in Analytical Chemistry " Vol. 1, H. Freiser (ed.), Plenum Press, N.Y., 1978, pp 1-141.
6. R. P. Buck, "Potential Generating Processes at Interfaces: From Electrolyte/Metal and Electrolyte/Membrane to Electrolyte/Semiconductor" in "Theory, Design and Biomedical Applications of Solid State Chemical Sensors", P. W. Cheung, D. G. Fleming, W. H. Ko and M. R. Neuman (eds.), CRC Press, W. Palm Beach, FL., 1978, pp 3-39.
7. W. E. Morf, "Principles of Ion-Selective Electrodes and of Membrane Transport", Elsevier Sci. Pub. Co., N.Y., 1981.
8. A. J. Bard and L. R. Faulkner, "Electrochemical Methods, Fundamentals and Applications", John Wiley and Sons, Inc., N.Y., 1980.
9. R. P. Buck, Sens. and Act., 1, (1981) 137.
10. J. Janata and R. J. Huber, Ion-Sel. Elect. Revs., 1, (1979) 31.
11. J. Zemel and P. Bergveld (eds.), "Chemically Sensitive Electronic Devices, Principles and Applications", Elsevier Sequoia SA, Lausanne, 1981.
12. P. Hagenmuller and W. VanGool (eds.), "Solid Electrolytes: General Principles, Characterization, Materials and Applications", Academic Press, N.Y., 1978.
13. J. B. Bates and G. C. Farrington (eds.), "Fast Ionic Transport in Solids", North-Holland Pub. Co., Amsterdam, Neth., 1981.
14. R. deLevie, "The Structure of the Charged Interface", in Ref. 11, pp 79-110.
15. A. Kornyshev, W. Schmickler and M. Vorotyntsev, Phys. Rev., B25, (1982) 5244.
16. W. Schmickler, J. Electroanal. Chem., 150, (1983) 19.
17. R. P. Buck, J. Membr. Sci., 17, (1984) 1.
18. R. P. Buck, "Structure of the Double Layer and Rates of Ion Crossings at 'Single' Immiscible Liquid/Liquid

Interfaces", Proc. NATO A.S.I. Conf., July 2-13, 1984, Reidel Pub. Co., Dordrecht, Neth. (in press 1984).
19. G. A. Bootsma, N. F. deRooij and A. van Selfhout, "The Solid/Liquid Interface", Ref. 11, pp 111-125.
20. I. Lauks, "Polarizable Interfaces", in Ref. 11, pp 261-288.
21. R. P. Buck, IEEE Trans. Elec. Dev., ED-29 (1982) 108.
22. D. Midgley, Anal. Chem., 49, (1977) 1211.
23. D. Midgley, Anal. Chim. Acta., 87, (1976) 7.
24. D. Midgley, Anal. Chim. Acta., 87, (1976) 19.
25. A. Hulanicki and A. Lewenstam, Talanta, 23, (1976) 661.
26. A. Hulanicki and A. Lewenstam, Talanta, 24, (1977) 171.
27. A. Hulanicki and A. Lewenstam, Anal. Chem., 53, (1981) 1401.
28. E. Lindner, K. Tóth and E. Pungor, Anal. Chem., 54, (1982) 202.
29. W. E. Morf, Anal. Chem., 55, (1983) 1165.
30. J. R. Sandifer, Anal. Chem., 53 (1981) 312.
31. J. R. Sandifer, Anal. Chem., 53, (1981) 1164.
32. R. K. Rhodes and R. P. Buck, Anal. Chim. Acta, 113, (1980) 67.
33. T. Hepel, Anal. Chim. Acta., 142, (1982) 217.
34. H. Jenny, T. R. Nielsen, N. T. Coleman and D. E. Williams, Science, 112, (1950) 164.
35. D. P. Brezinsky, Talanta, 30, (1983) 347.
36. T. Higuchi, C. R. Illian and J. L. Tossounian, Anal. Chem., 42 (1970) 1674.
37. C. Liteanu and E. Hopirtean, Talanta, 17, (1970) 1067.
38. O. R. Melroy and R. P. Buck, J. Electroanal. Chem., 143, (1983) 23.
39. J. Ruzicka, E. H. Hansen and J. C. Tjell, Anal. Chim. Acta., 67, (1973) 155.
40. C. Gavach and P. Seta, Anal. Chim. Acta., 50, (1970) 407.
41. B. J. Birch and D. E. Clarke, Anal. Chim. Acta., 67, (1973) 387.
42. V. V. Cosofret and R. P. Buck, Anal. Chim. Acta. (scheduled 1985).
43. V. V. Cosofret and R. P. Buck, J. Pharmaceut. and Biomed. Anal. (scheduled 1985).
44. J. Janata, "Thermodynamics of Chemically Sensitive Field Effect Transistors" in "Theory, Design and Biomedical Applications of Solid State Chemical Sensors", P. W. Cheung, D. G. Fleming, W. H. Ko and M. R. Neuman (eds.), CRC Press, W. Palm Beach, FL, 1978, pp 41-51.
45. N. Yamamoto, Y. Nagasawa, M. Sawai, T. Sudo and H. Tsubomura, J. Immunol. Methods, 22 (1978) 309.
46. N. Yamamoto, S. Shuto and H. Tsubomura, Appl. Biochem. Biotech., 6, (1981) 319.
47. J. L. Boitieux, G. Desmet and D. Thomas, Clin. Chem., 25, (1979) 318.
48. S. G. Collins and J. Janata, Anal. Chim. Acta., 136, (1982) 93.

49. R. P. Buck, F. S. Stover and D. E. Mathis, J. Electroanal. Chem., 82 (1977) 345.
50. F. S. Stover and R. P. Buck, J. Electroanal. Chem., 94 (1978) 59.
51. R. P. Buck, F. S. Stover and D. E. Mathis, J. Electroanal. Chem., 100 (1979) 63.
52. M. Aizawa, S. Kato and S. Suzuki, J. Memb. Sci., 2 (1977) 125.
53. R. L. Solsky and G. A. Rechnitz, Anal. Chim. Acta., 123 (1981) 135.

Table 1

Electrostatic Interfaces Susceptible to General Theory

Equilibrium Interfaces	Device or Application
Metal/Gas	Field Effect Sensors/Classical Redox Electrodes
Metal/Electrolyte	Classical Electrodes/ Batteries
Metal/Membrane	Modified Electrodes/Field Effect Devices
Metal/Electronic Semiconductor	Diode Devices/Contacts
Metal/Salt (Ionic Semiconductor or Fast Conductor)	ISEs/Classical Electrodes/ Batteries
Metal Insulator	Field Effect Application in Electronics and Sensors
Electronic Semiconductor/ Electrolyte	Photovoltaic/Photogalvanic Devices
Electronic Semiconductor/Membrane	Stabilized Interfaces of Photosensitive Devices
Insulator/Electrolyte; Insulator/Membrane	Field Effect Devices and Sensors
Membrane/Electrolyte	ISEs/Mass Transport-Stabilized Interfaces for Devices
Membrane/Membrane	Layered (Non-homogeneous) Sensors, Protected Sensors
Membrane/Salt	Coated Wire ISEs/Field Effect Sensors
Electrolyte/Electrolyte	ISEs, Junctions

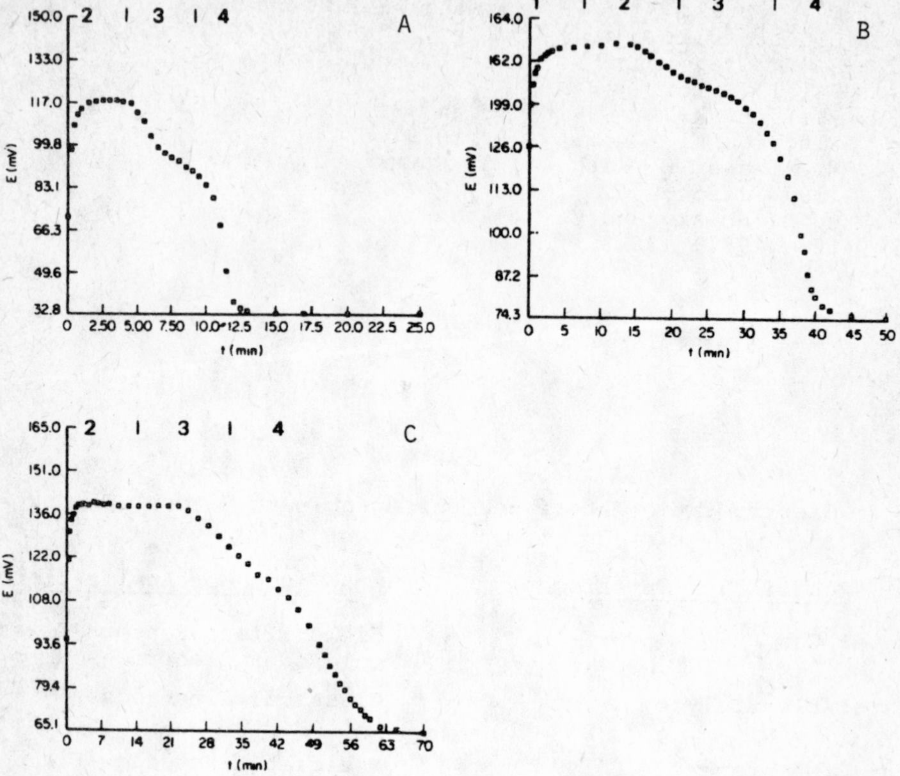

Fig. 1. Potential-time transients resulting for activity step solutions with different initial halide concentrations: (A) to 4.43×10^{-4} M in KBr and 1.47×10^{-3} M in KCl (electrode thickness 1.2×10^{-4} cm); (B) to 1.47×10^{-4} M in KBr and 1.57×10^{-4} M in KCl (electrode thickness 1.2×10^{-4} cm); (C) to 2.96×10^{-4} M in KBr and 2.96×10^{-4} M in KCl (electrode thickness 2.4×10^{-4} cm). [R. K. Rhodes and R. P. Buck, Anal. Chim. Acta., 113 (1980) 67.]

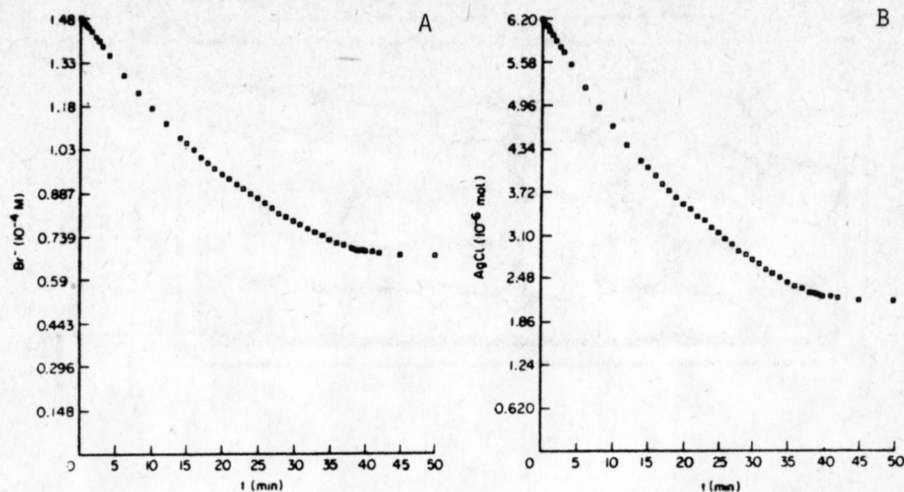

Fig. 2. (A) Bromide activity decay in solution versus time and (B) silver chloride depletion versus time, for a 1.2 x 10^{-4} cm thick electrode in solution initially 1.47 x 10^{-4} M in KBr and 1.57 x 10^{-4} M in KCl. [R. K. Rhodes and R. P. Buck, Anal. Chim. Acta., 113 (1980) 67.]

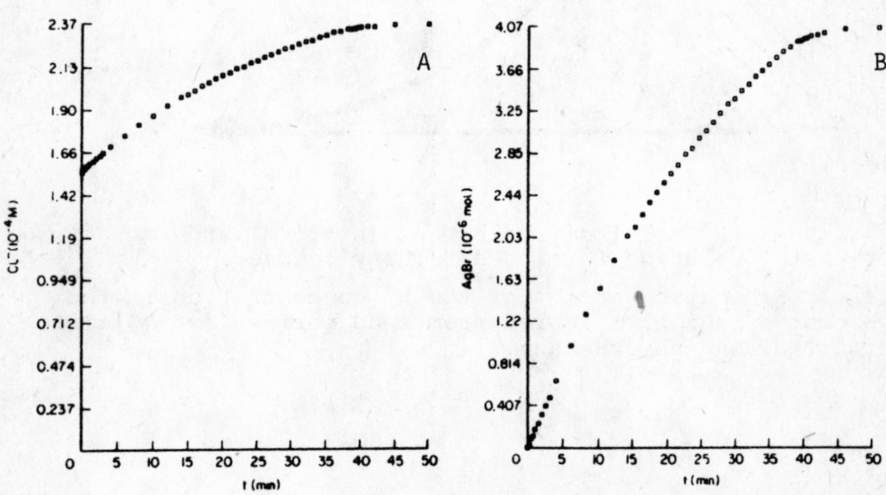

Fig. 3. (A) Chloride activity enhancement in solution versus time and (B) silver bromide formation versus time, for a 1.2 x 10^{-4} cm thick electrode in solution initially 1.47 x 10^{-4} M in KBr and 1.57 x 10^{-4} in KCl. [R. K. Rhodes and R. P. Buck, Anal. Chim. Acta., 113 (1980) 67.]

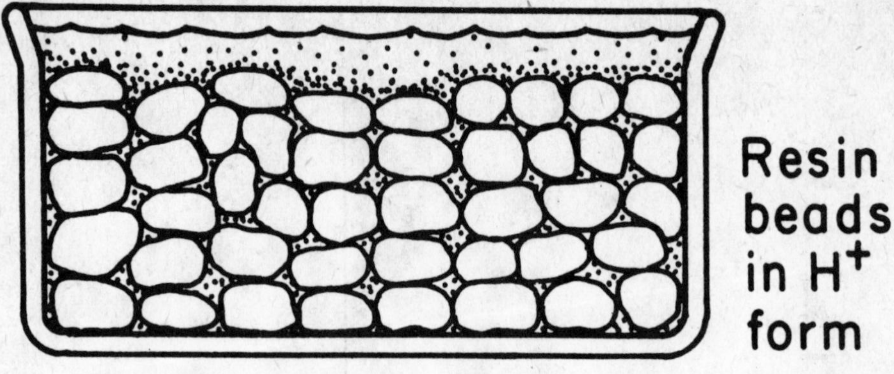

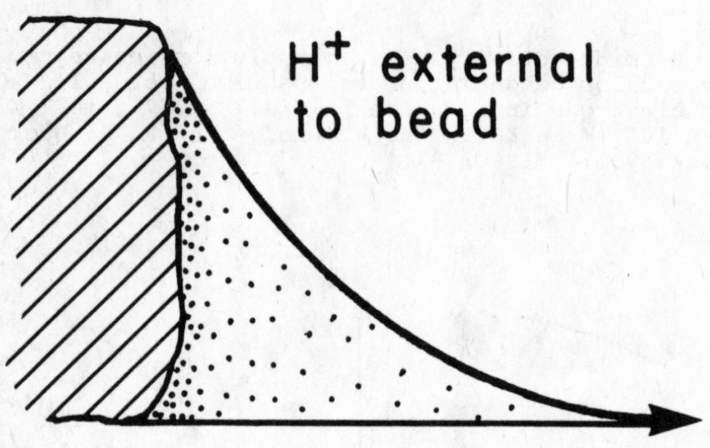

Fig. 4 (Upper). Model of resin bead (H^+-form) and aqueous, interstitial solution forming a "slurry" phase.

(Lower). Schematic version of the H^+ concentration in the space charge region, measured from bead surface ($x = 0$) into interstitial <u>and</u> supernatant.

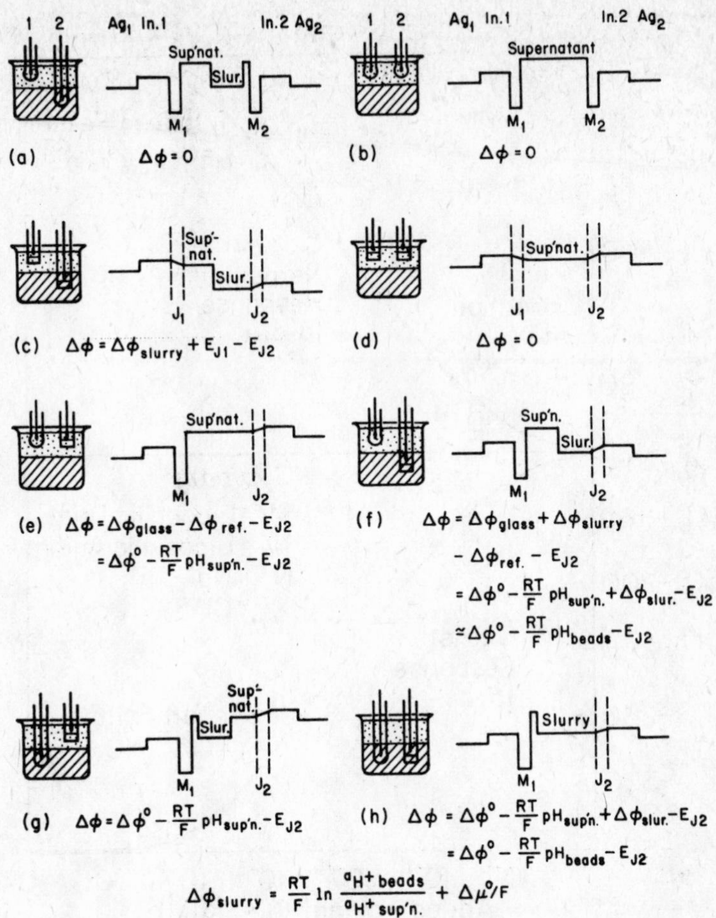

Fig. 5. Illustration of the "Suspension Effect": Cells containing glass pH electrodes and junction reference electrodes in all significant configurations - upper phase: supernatant; lower phase: slurry. Figs. a,b,d potential profiles (matched electrodes) read net zero pd. Fig. c measures crucial "Suspension Effect" pd. Figs. e,f give same pd. Figs. g,h give same pd (approx.). Figs. g,h pd differs from Figs. e,f pd by result in Fig. c.

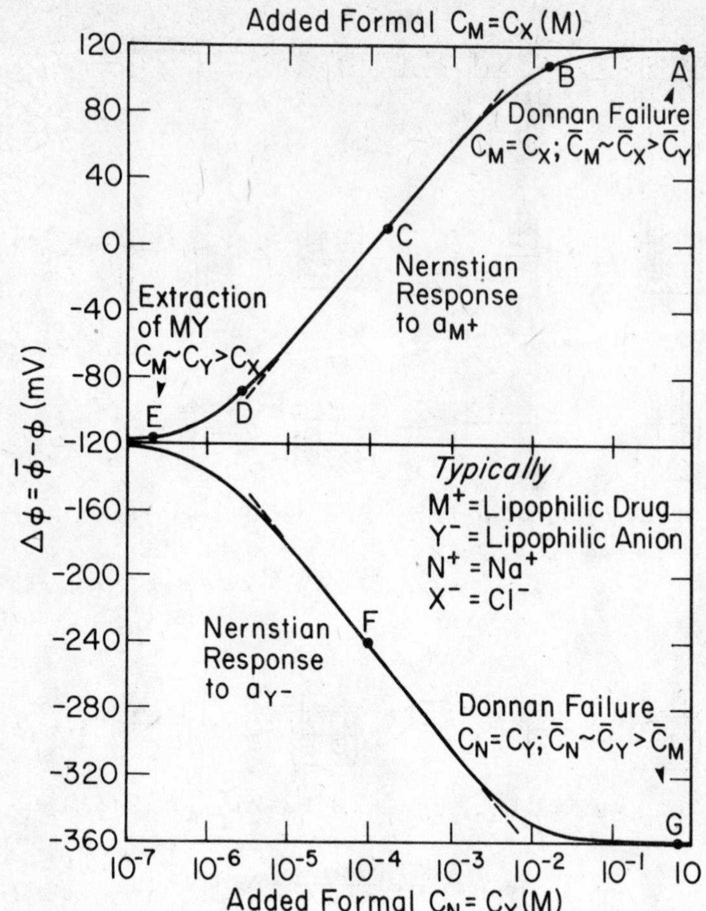

Fig. 6. Bi-function Membrane Response. (Upper). An example of the potential "windows", computation for two salts MX, mainly water-soluble and MY, mainly oil-soluble. Concentration of MY = 10^{-2} M; single ion partition coefficients $K_Y = 10^6$, $K_M = 10^2$, $K_X = 10^{-2}$. The computed potential window is 336 mV using for E eqn. (15) and for A eqn. (16). C_M and C_X are added concentrations.

(Lower). Computation for two salts NY, mainly water-soluble and MY, mainly oil-soluble. Parameters as above with $K_N = 10^{-6}$. C_N and C_Y are added concentrations. Computed potential window is again 336 mV; using eqn. (17) for G. Points C and F are in the Nernstian region, while points B and D require detailed balance calculations.

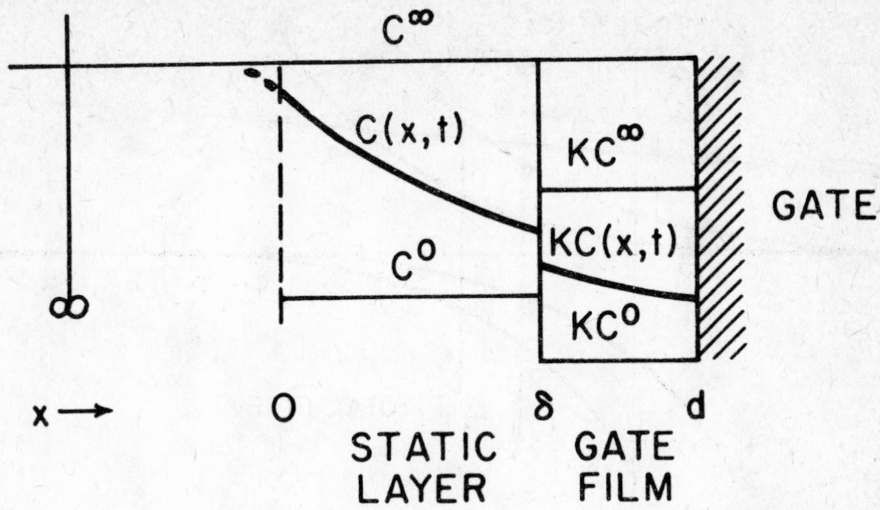

Fig. 7. Physical model for one-dimensional mass transport to a CSSD gate. Stirring occurs for $x < 0$. The Nernst layer approximation asserts perfect stirring at $x = 0$ so that $C(0,t) = C^{\infty}$. Positive flux moves from left to right. Two-region model for thick, permeable gate at $x = \delta$. Salt-partition equilibrium for site-free films, and single-ion-partition equilibrium for permselective (high site density) films are assumed as equilibrium boundary conditions. Impenetrable wall is $x = d$. Mass transport and kinetics of ion crossings at $x = \delta$ determine potential drift in this theory.

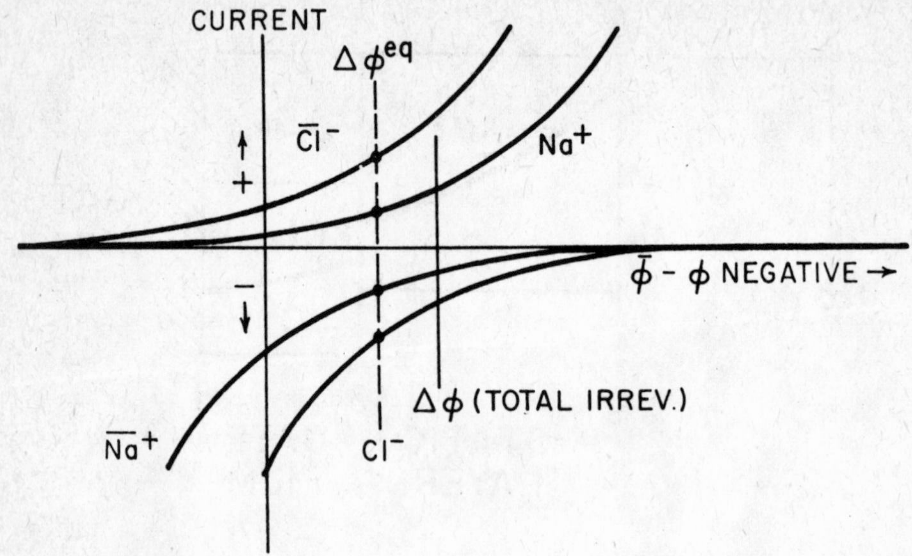

Fig. 8. A schematic drawing of single-ion irreversible current-voltage curves for ion crossings. Positive current is plus charge moving from left to right (from aqueous to organic phase) in Fig. 7, or negative charge moving from right to left. The equilibrium interfacial potential difference term is $\bar{\phi}-\phi$. The totally irreversible case corresponds to no back reaction of ions from the organic phase.

QUESTIONS AND COMMENTS

Participants of the discussion: Z.Boksay, R.P.Buck, J.Janata, P.F.Levy, Z.Noszticzius and E.Pungor

Question:

You presented a very interesting experiment where some interfacial potential was measured. You could measure a potential difference through a special interfacial layer where there were ion exchanger beads and an electrolyte. Sometimes it is said that a Galvani-type potential difference, that is, the potential difference between points inside two phases cannot be measured. Could you please comment on that ? Is it a Galvani--type potential difference which you could measure, or is it a Volta-type difference ?

Answer:

The measurement of a single interfacial Galvani potential is not theoretically possible, but the nearest thing to it is to measure the single interfacial potential vs. two reference electrodes whose junction potentials can be assumed to have some value reasonably calculated by the Henderson equation, with some corrections for medium effects if necessary. So the measurement I proposed was to be a Galvani potential difference between the inside of the phase containing the beads, that is, the slurry phase, and the supernatant. And the error in that is the difference in the two junction potentials, the one in the supernatant and the one in the slurry phase. That difference could be important and I know that the actual values are somewhat sensitive to the position of the reference electrode in the slury phase. I have also found that if you wrap the electrodes with dialysis membranes, they settle down to a more nearly constant value, which, I believe, is accurately representing the Galvani potential difference.

Question:

Are you using the two reference electrodes as two probes of the inner potentials or is the difference which you measure indeed the difference due to the liquid junction potential ?

The second question is: If you take an average lake in North Carolina and you stick one reference electrode in the mud and the other one in the water, can you make electricity out of that arrangement ?

Answer:

I am sure that if I put two glass electrodes in the two phases, I measure zero because I cannot get any work out of the system. Also I know that if I torture the water out of the region between the beads, the electrode measures separately the same pH as in the supernatant. However, when I have two non-equilibrium reference electrodes in the two phases, there is always the possibility that the entire effect is due to a difference in the junction potentials in the two phases. We have shown that putting the reference electrode in the slurry and the glass electrode in either phase we get the same answer. This, I believe, says that there is some kind of interfacial potential and that we can move the reference electrode around and find very much the same effect. The argument in favour of a total junction effect says that the only way we can get that is to have the entire space charge at the tip of the junction. I think that the entire amount of junction potential you can get from that effect is not as great as this interfacial potential.

Question:

You indicated that the slurry effect was also applicable to blood. What is the effect of this phenomenon on measurements in whole blood ?

Answer:

If I said that I overstated the case. In connection with the suspension effect I can say that it is possible that whole blood measurements are in error because of charges on blood cells, and the present discrepancy between actual potassium or chloride activities in whole blood and what you measure when you spin it down, is only partially explained by the excluded volume effect; part of it may be due to the charge on the blood cells. But this was suggested by Brzezinsky. We have not yet explored the connection between the suspension effect using Dowex resin beads and the suspension effect in whole blood. But there ought to be a connection, as there ought to be connection with measurements in mud or in alumina suspensions or in any kind of charged phase, slurry, where the particles are so close together that you can actually spread the space charge out over them all.

Question:

Concerning the silver iodide-silver bromide problem. What is the structure of the mixed phase, if we consider e.g. an ISE made of silver iodide, and the interfering ion is bromide ? Silver iodide has a sphalerite structure, whereas silver bromide has a structure similar to sodium chloride.

Comment:

Silver iodide may have other structures, too, e.g. wurtzite structure.

Answer:

I do not think that the problem here is that of mixed phases. Let us consider silver chloride - which is stable at a given temperature, and silver bromide, which is also stable and they have different solubility products. When you make a thin film of silver chloride, and expose it to bromide or iodide, at a concentration less than corresponding to the solubility product, nothing happens. But when you exceed the solubility product, the whole membrane converts, starting at the surface and working all the way through. And you see the complete change from one phase to the other and the structure of the phase, as far as I know, may influence the kinetics, but not the thermodynamics. So one would think that the response, if it is due to the silver ion activity ratio, which is my first assumption, should give two intersecting straight lines. And my point was that if you see a curvature, that means some kind of slow kinetics, either diffusion to the surface or diffusion within the crystal.

Question:

Do you think that silver iodide is covered by silver bromide or not ? How do you calculate this ?

Answer:

I start with pure silver iodide and calculate the equilibrium silver, and I do similarly for the bromide and if $[Br^-]/[I^-]$ exceeds the ratio of the solubility products, then the whole phase should slowly change from one to the other. And they may change through a series of mixtures. But I would not think it is an equilibrium, as I think that the solubility products for mixtures are smaller than the solubility products for the pure phases.

4th Symposium on Ion-Selective Electrodes
Mátrafüred, 1984

POLARIZATION STUDIES WITH ION-SELECTIVE ELECTRODES

K. CAMMANN* and G.A. RECHNITZ

Department of Chemistry
University of Delaware
Newark, Delaware 19716, USA

ABSTRACT

Polarization studies with ion-selective electrodes (I.S.E.'s) and/or membranes allow the determination of a key kinetic parameter, the exchange current density, at those non-metallic interfaces. A high exchange current density of the potential-determining (primary) ions is essential for an ideal Nernstian sensor. However, for an accurate prediction of the selectivity coefficient, more information concerning the course of the individual current-voltage characteristic (C-V curve) for all participating ions is needed. In principle, the new techniques of voltammetry at the interface of two immiscible electrolytes can give this further information. The exchange current densities at I.S.E.'s determined thus far, lie more or less in the $\mu A \cdot cm^{-2}$ range if the typical concentrations of the primary ions are considered. This is too low to assume a thermodynamic equilibrium at the interface in all cases. Therefore, the classical theoretical approach has to be revised; the interface can under certain conditions be polarized by the ion-transfer of interfering ions. Parallel electrochemical processes necessarily lead to a mixed potential which can best be illustrated in the form of an Evans-diagram. This theoretical approach treats all kinds of I.S.E. membranes in the same manner and allows an easy

---

*on sabbatical leave from: Abteilung für Analytische Chemie
Universität Ulm, 79 Ulm, F.R.G.

explanation for all known properties of an I.S.E., e.g. selectivity, time response, sub- and super-Nernstian slope, transient response to interfering ions (also with respect to time), potential, overshoot, etc. Applying this theoretical approach (with the recent findings of the polarizability of the interface) to biomembranes, the active transport of ions together with its complex dependence from the composition of the electrolyte solution can be understood.

INTRODUCTION

In ion-selective electrode potentiometry, an ion-selective change in the absolute potential difference across the phase boundary, i.e., ion-selective membrane (or surface)/sample solution, is measured. Consequently, this demands a corresponding ion-selective change in the charge distribution of that specific ion across that interface. This is exactly the process which gives rise to the electric field variations one measures by determining different membrane potentials. The challenge of finding a useful selective membrane is therefore oriented toward preventing interfering ions from participating in this interface charging process.

The potential-determining ion is consequently that which has the greatest contribution to the change of the charge distribution. Since, in general, the timescale for this process is rather short (millisecond-range), those ions which are the fastest in crossing the phase boundary are likely to determine the membrane potential in this time range. Simultaneously, this newly established potential is working as an overpotential for all interfering charge transfer reactions which are possible at this particular interface. This creates small currents carried by interfering ions the direction of which depends on the particular sign of the overpotential. In the beginning these small current densities may be in the range of the exchange current density of the primary potential-determining ion and thus create a response proportional to $t^{-1/2}$, a situation similiar to chronoamperometry. Depending on the direction of these overvoltage-induced currents caused by interfering ions (including counterions), there may also be a potential overshoot in the e.m.f. - time course after an activity change of the measured ion. A knowledge of the exchange current densities of the different ions at the interface of I.S.E membranes is of fundamental importance concerning a reliable prediction for its proper functioning as an ideal Nernstian sensor. The exchange current density can be obtained via polarization studies in an analogous manner as in the case of metallic electrodes [1]. This similarity of the electrochemical kinetics is further corroborated by the whole field of electrolysis at the "interface of two immiscible electrolyte solutions" (I.T.I.E.S.) [2]. Thus, parallel electrochemical reactions, i.e., more than one kind of ion transfer simultaneously at those interfaces, must also lead to a mixed potential as in the case of corroding metallic electrodes. It has been shown that a corrosion-like theoretical approach results in a convincingly simple and very illustrative model, which could serve as an alternative model for those researchers thinking more in pictures than in mathematical equations [3].

COMPARISON OF THE THREE-SEGMENTED-POTENTIAL WITH THE MIXED-POTENTIAL MODEL

Historically, potentiometry is based on the Nernst equation. In its derivation, two important conditions must be met and verified if applied: a) thermodynamic equilibrium of the ions under study, b) the reversibility of this equilibrium. Reversibility is often interpreted as a case of no kinetic hindrance of the ions crossing the interface region, i.e., rapid ion exchange. An exact measure of the latter is the exchange current density. In the case of a rapid ion exchange (large exchange current density), it is very likely that the thermodynamic equilibrium is achieved within the measuring time. Therefore, the determination of the exchange current density at ion-selective membranes is very important. Reversibility also means that any small deviation of the electrode potential from its equilibrium value should immediately result in a finite, fast, and single electrochemical reaction which changes direction only according to the sign of the applied overpotential. By this definition, any mixed potential which gives rise to different electrochemical reactions, depending on the sign of the additional overpotential (externally applied) is excluded from being described by the Nernst equation. Despite this, many text books still apply the Nernst equations to such cases as MnO_4^-/Mn^{2+} or $Cr_2O_7^{2-}/2\ Cr^{3+}$ in acid media. Contrary to the general opinion, thermodynamically reversible electrode reactions are rare, and the proof of the latter at the molecular level is virtually impossible [4].

Most good I.S.E.'s obey the Nernst equation rather well. Therefore, almost all theoretical approaches assume that for the primary ion, the thermodynamic equilibrium at the phase boundary of interest is established. What remains to be described by a proper theory are the selectivity and some artifacts of certain membranes showing considerable sub-Nernstian, and sometime even super-Nernstian, behavior. The excellent monography by Morf [5] summarizes all theoretical approaches and goes further in proposing a logically closed theoretical model, based on the three-segmented-potential approach, which allows the explanation of nearly all the different properties of the different types of I.S.E.'s. To our knowledge, no further theoretical approach differing in major positions from the last model has been undertaken. However, the approaches using electronic equivalent circuits in describing the behavior of I.S.E. membrane [6,7] have been continued and seem to have increased recently, possibly due to the commercial availability of sophisticated instrumentation which permits the automatic recording of complex impedance spectra [8]. This technique is extremely valuable for the reliable determination of exchange current densities at I.S.E. membranes [9,10]. However, it should be noted that the differentiation between the exchange current density at a mixed potential point and at the point of real thermodynamic equilibrium is difficult. With the necessary linearization (use of small amplitude pertubations), the mostly asymmetric behavior of the current-voltage curve near a mixed-potential steady-state is difficult to verify. Furthermore, in the case of extreme asymmetry, and/or at extremely low frequencies, the frequency spectrum may be distorted, rendering any interpretation impossible.

On the other hand, the segmented potential model, which assumes a thermodynamic equilibrium at the phase boundary and describes deviations from Nernstian behavior with an additional diffusion potential inside a homogeneous membrane phase, uses the Nernst-Planck flux equation for the calculation of the latter. In relating ion fluxes to the corresponding driving forces, a linearization is also performed in this model. The same is true if, in a non-equilibrium thermodynamics treatment (irreversible thermodynamics), the cross coefficients of the friction coefficients are omitted as Buck [11] pointed out recently. One should note that in a case of activation control of an ion transfer across a phase boundary, the analogy of the friction coefficient depends exponentially on the driving force, (i.e., overpotential). For example, in the case of a cation flux creating a large positive potential difference across a phase boundary or a layer segment resulting in an anion flux which is no longer strictly proportional to the driving force, there will be no potential-independent ion mobility.

Mirkin et al. [12] bridged the gap between the three-segmented potential model and the mixed potential model suggested by one of the authors [3,13] in expressing the mixed potential steady-state situation mathematically. As a matter of necessity, there was formally no difference between the equations obtained for the mixed potential model and the Nikolsky-type equations for the classical approach. However, the physical significance of the entities which describe the selectivity coefficient were different. In the segmented potential model, the mobilities within the membrane phase play a key role. In the mixed potential model, the mobilities across the phase boundary, i.e., the ion transfer rates, are essential, because the potential differences are created only there. This is also corroborated by the fast response times of thick membranes and by studies on glass and PVC-membranes [14,15]. Again, the bridge between the two models is that the transfer rates depend on the composition of the membrane phase or surface. There is a further apparent merging of the two approaches since recent studies show evidence that our usual treatment of phase boundaries has to be changed -a sharp boundary with corresponding steep changes in the physical constants is likely to be non-existent [16,17]. In addition, the behavior of an electrode depends on surface structure rather than on bulk structure [18] as it can be seen also by the ion-selectivity of extremely thin surface layers at ISFET's. Both models suffer from the fact that the physico-chemical entities which determine the selectivity coefficient are so far rather impossible to measure in an independent way,rendering both theoretical approaches hypothetical and thereby open to criticism by theory of science.

However, this may change in the future for the mixed potential model: as a consequence of all the uncertainties mentioned above, an approach which takes into account all possible complications at the specific phase boundary has distinct advantages. Depending on the long-range forces in the two adjacent phases, a new additional interface may be formed and the properties of that monolayer or ion-pair compound may be responsible for the parameters of the electrochemical reactions [18]. The prevailing influence of the chemical interactions on the transfer Gibbs energy at I.T.I.E.S. has been demonstrated [19]. Determining the current-voltage (C-V) characteristic of every major electrochemical reaction, which is likely to occur at the interface under study (including the effect of surface-active substances, etc.) would be very valuable, since this would permit one to find the mixed potential by the superposition of C-V curves.

Progress in the field of voltammetry at I.T.I.E.S. could make it possible to verify the mixed potential model of I.S.E.'s. In principle, PVC-based phases should be accessible to this technique, too. Thus the properties of an ion-selective membrane could be fairly well predicted if the individual C-V curves of the main ions could be obtained and the key parameters which determine the latter could be found.

THE MIXED POTENTIAL MODEL AND VOLTAMMETRY AT I.T.I.E.S.'S

The mixed potential approach [3,13,20] depends upon whether the electrochemical kinetics developed thus far for metallic electrodes [21] are also applicable to I.S.E. membranes and I.T.I.E.S.'s. The principal similarity was noted as early as 1973 [13] without awareness of Guastella's [22] and Gavach's [23,24] research in the field of I.T.I.E.S. In 1975, the first exchange current density, j_o, determinations on commercial I.S.E.'s were performed by the authors [1] demonstrating definitely that the techniques developed for metallic electrodes could also be used in this new field. Surprisingly, these exchange current densities, determined with an extremely simple galvanostatic current step method (using the resistance measuring mode of an electrometer-multimeter), agree very well with recent redeterminations using the new and sophisticated technique of complex impedance spectrum analysis [10]. In the case of the galvanostatic current step technique, the uncertainties include: the amount of the Frumkin- and double layer charging correction, measuring problems of determining the change of a normally small charge transfer resistance in series with a very large (and sometimes slowly changing) ohmic (bulk) membrane resistance. These uncertainties were considered by carefully speaking only of "apparent exchange current density estimates" [1,3,20,26,28]. Armstrong et al. [10,25] in two recent publications on exchange current determinations at valinomycin-based potassium-selective membranes using the most sophisticated measuring equipment (Solartron 1174 frequency response analyzer coupled to a Solartron 1186 electrochemical interface with a four-electrode cell, all controlled by a microcomputer with data storage and handling), obtained greater precision than could previously be achieved. The main results were: firstly, in case of a liquid membrane, the exchange current densities lie in the $\mu A \cdot cm^{-2}$ range; secondly, the selectivity coefficient $K^{pot}_{K,Na}$ could be approximately expressed by the exchange current density ratio $j_{o,Na}/j_{o,K}$, as it was also reported earlier by the authors [1,26,27]. However, in case of a PVC membrane, a ratio of $j_{o,Na}/j_{o,K} = 1/60$ was found although this corresponded to a selectivity coefficient of $K^{pot}_{K,Na} \approx 10^{-3}$. The authors [25] interpreted this behavior as if the selectivity does not arise wholly at the membrane/solution interface, but rather, in part from processes deeper in the membrane than the thickness of the electrical double-layer (0.5 nm).

Our finding of a correlation between the exchange current density ratio and the corresponding selectivity coefficient was an empirical observation which, nevertheless, points strongly toward a kinetic interpretation. With a proper understanding of the mixed potential model, these findings can be interpreted as the result of a very special situation at the interface. Analysis of the relation between the different exchange current densities and the corresponding selectivity coefficients leads to the conclusion that a time dependence of the latter is possible, if not likely, because when an I.S.E. membrane is initially brought in contact with the sample solution, the relative contributions of every ion-transfer to the total change of the charge distribution across the interface expresses the selectivity at that time. Being aware of this charging process (see Fig. 11 of [28], one can easily deduce the special condition for $q_I/q_M \approx j_{o,I}/j_{o,M} \approx K^{pot}_{M-I}$ assuming the same spatial distribution pattern of both ions. However, on a longer time scale, as a result of this primary charging process, a mixed potential is formed in most cases, i.e., whenever the Nernst equation is not fully obeyed. Here, the particular selectivity coefficient depends upon the exact course of the partial (C-V) curves leading to the mixed potential as Koryta showed convincingly [29]. Only in case of more or less linear (C-V)-curves, can the above mentioned proportionality between the exchange current density ratio and the selectivity coefficient be theoretically rationalized.

In the meantime, all the research in the field of voltammetry at I.T.I.E.S.'s unambiguously demonstrates the similiarity of analogous charge transfer processes at metallic electrodes. Of course, there are differences which have to be taken into account:

a) There might be two diffuse double layers on either side of the interface influencing the overpotential dependence of the ion-transfer processes.

b) At high positive or negative galvanipotentials across the interface, interfacial transport of counterions must be considered (salt partition).

c) Partition equilibrium is a non-linear ionic process that can lead to insoluble equations [30].

d) In certain cases where one side of the interface shows a very low diffusional behavior (e.g., PVC membranes with certain carriers and plasticizers) apparently no concentration polarization can be observed [15,27,31-33].

Summarizing the differences between I.S.E.'s and I.T.I.E.S.'s and metallic/ electrolyte interfaces, other researchers come to similiar conclusions [2,9,29]. Points b) and c) can best be explained by the mixed potential model; here the failure of the Donnan exclusion principle and the non-linearity of the salt partition is demonstrated in a simple illustrative way (Fig. 1). In cases described in d), a steady-state between the overpotential and diffusion (concentration polarization) may be established, e.g., a concentration change at the surface of one interface leads to a new equilibrium potential reducing the overpotential and thus reducing the transfer rate until it fits the diffusion rate. In this way, the constant interfering ion flux, necessary for a stable mixed potential, can be explained. Recently the importance of point a) was demonstrated by showing in certain cases of PVC-based I.S.E. membranes [27,31] that the apparent exchange current density decreased with increasing concentration of the primary and/or interfering ions in the solution.

Despite such differences, the published results of

exchange current density ranges at I.S.E.'s [1,3,20,26-28,31,34-36] should nevertheless have an impact upon the classical theoretical treatment. In all those approaches, the interface is assumed to be in total thermodynamic equilibrium (Donnan potential term). In order to achieve this, a rapid, reversible ion-exchange is necessary, i.e., an exchange current density $j_{o,M} > 0.1\ A \cdot cm^{-2}$ for the primary ion [9]. Values of $j_{o,M} \ll 10^{-2} A \cdot cm^{-2}$ suggest a slow or kinetically limited, irreversible ion-exchange [9], where it is likely that in case of interfering parallel ion transfer processes the thermodynamic equilibrium at the interface is not fully established. Most exchange current densities published so far for ion-selective membranes lie more or less in the $10^{-6} A \cdot cm^{-2}$ range considering the typical analytical concentration range. Any flux of interfering ions in the range of only 10^{-11} mol/s corresponding (for univalent ions) to a current density of about 100 $nA \cdot cm^{-2}$ across the interface, will shift the equilibrium potential of the primary ions by as much as ca. ±3 mV depending upon the direction (assuming a linear C-V relationship near the equilibrium potential, i.e., overpotential < 20 mV/z).

In the mixed potential model, in most cases well above the detection limit of the I.S.E., a negative overpotential for the primary ion is most likely, leading to slightly sub-Nernstian behavior, which also depends upon the counterions present. Thus, this typical behavior of real I.S.E.'s is explained without additional diffusion potential within the membrane phase. Super-Nernstian behavior is a special case related to wrong conditioning, among other factors. For a full description of the behavior of an I.S.E. with the mixed potential model, a knowledge of the exchange current density of the different charge transfer processes alone is not sufficient. What is further needed is the equilibrium potential difference of each interfering charge transfer process and its dependence on the overpotential, i.e., the (C-V) characteristic. Since one cannot assert a priori that all overpotential occurs exclusively across the space charge in the diffuse double layer or exclusively across the compact layer [37], because of all uncertainties concerning the surface state as mentioned above and also because of "discreteness-of-charge" effects [38], an empirical approach must be taken. In this respect, the new techniques of voltammetry at I.T.I.E.S.'s should gain fundamental importance in the field of I.S.E's, as will be shown below.

Studies with I.T.I.E.S.'s have already lead to tables of standard Gibbs free energies $\Delta G^{o \to w}_{tr,i}$ and corresponding standard equilibrium potential differences $\Delta\phi^{o}{}_{eq,i}$ for simple ion transfer processes across water/organic solvent interfaces [39-42]. According to this, the individual equilibrium potentials of the cations and anions can differ as much as 0.5 volt. In the case of a useful I.S.E. membrane, the exchange current density of the primary ions is always much higher than the exchange current densities of interfering co- and counterions at the same interface. Therefore, any mixed potential will always lie close to the equilibrium potential of the primary ions. However, the latter behaves like an externally applied overpotential on the ion transfer of all interfering ions. According to the exponential relationship in electrode kinetics, an overpotential of about 0.5 V may increase the rate in which the counterions cross the phase boundary. For example, assuming an exchange current density of only 10^{-9} $A \cdot cm^{-2}$ for counterions and an apparent transfer coefficient $\alpha = 0.5$, an overpotential of + 0.4 V would result in a directed current flow of about 2 $\mu A \cdot cm^{-2}$ of those ions across the interface (from the aqueous phase into the membrane phase). This internally generated polarization current in turn will shift the thermodynamic equilibrium potential $\Delta\phi_{eq,M}$ of the measured ions according to their exchange current density by $\eta = 0.0257 \cdot 2/j_{o,M}$ volts (univalent ion, 25°C, j_o in $\mu A \cdot cm^{-2}$) [4].

In Figure 1, the influence of a counterion flux on the equilibrium potential $\Delta\phi_{eq,M+}$ of measured cations (M^+) as an example is shown. For the M^+-ions, a realistic exchange current density of $j_{o,M^+} = 10^{-6}$ $A \cdot cm^{-2}$ is assumed and drawn to scale in the form of an Evans diagram [43]. In the case of a negative overpotential-η (anion interference), the C-V curve for the primary ion is drawn linear within about 20 mV (z=1). The slope η/i is related to the exchange current density by

$$-\eta/i = \frac{R \cdot T}{z \cdot F \cdot j_o}$$

where R,T,z,F = constants with their usual meaning [44]. Since the exchange current density is concentration dependent ($j_{o,M+} \propto C^{\alpha}$, with the apparent charge transfer coefficient α equal to an assumed value of 0.5), the slope decreases with decreasing concentration of the primary ions in the solution, rendering the interface increasingly polarizable. This demonstrates the concentration dependence of K^{pot}. Lacking at the time any further information on the shape of the positive overpotential-current curve for the corresponding anion charge transfer, e.g., j_{o,I^-}, $\Delta\phi_{eq,I^-}$, α or diffusion controlled in the example given, the mixed potentials (intersection of the two C-V curves) are chosen in such a way that the known e.m.f.-concentration behavior of the interface under study is obtained. Deviations from ideal Nernstian behavior are thus explained by mixed potentials due to a parallel charge transfer process by interfering ions. At a first glance this looks like a logical circle. However, such self-consistent approaches are very customary in science, e.g., assuming single ion activities in solid state matrices, level of vacancies or interstitial concentrations, which are not independently measurable, etc. It is an additional advantage of the mixed potential model that the necessary entities can also be obtained by independent measurements, e.g., voltammetry at I.T.I.E.S.'s or tracer techniques [45] can verify those entities.

In Figure 1, the flux of anions from the aqueous into the membrane phase is regarded as a negative current exactly balancing the positive current of the flux of primary ions in the same direction. It is not possible to differentiate, with the aid of a C-V curve, between an anion flux into the membrane and a possible flux of interfering cations, I^+, contained in the membrane in the opposite direction. The latter will definitely take place if those cations are confronted with the positive overpotential created for them by the primary ions. Thus, in any case, the instantaneously created potential difference across any interface governed by the charge separation and distribution of the primary ions immediately after contacting the I.S.E. membrane with the sample solution, will act as an overpotential to all other ion transfer equilibria (the time constant for this charging process is: $\tau = \frac{RT}{z \cdot F \cdot j_o} C_{dl}$, with C_{dl} = double layer capacitance [4]). As in the case of the analogous situation in the field of chronoamperometry, corresponding currents of interfering ions will start flowing with a direction depending on the sign of the overpotential working on each individual ion transfer equilibrium. The magnitude of those partial currents of interfering ions can be estimated using the Cottrell equation. Thus, according to this, more than one partial-current-time relationship influencing the mixed potential time response is possible. This is reflected in more than one time constant for the e.m.f. vs. time response for the primary ion, e.g., as it is obtained from the above mentioned complex impedance spectra, which previously were difficult to understand. In the same manner the dependence of the response times of I.S.E.'s and on the concentrations of the primary ions and interfering ions can be explained, the influence of the latter being greater with ions which interfere to a larger extent. It was shown above that only 10^{-11} mol/s of interfering ions need to be transported across the interface to account for the small deviation from the equilibrium potential of the primary ions. It seems impossible to prevent the presence of impurities at the phase boundary at such a low level, especially since sufficient H^+ and OH^- ions are always available to act as interfering ions. Furthermore, the transient response in those cases in which the interfering ion concentration is suddenly changed can be understood.

If the net (C-V)-curve for all possible interfering ion transfer processes is parallel to the voltage axis, e.g., in the case of diffusion or steady-state control, a super-

Nernstian behavior of the e.m.f. vs. concentration relationship will be obtained in cases where the interfering ion flux is not strongly concentration dependent ($I^+_{membrane} \rightarrow I^+_{solution}$). Aside from the power of such an illustrative model, other artifacts connected with such rare cases can easily be explained. In that case, if it is a diffusion controlled interfering anion flux into a cation-selective membrane which gives rise to this anomalous behavior, a sensitivity to rate of stirring will be seen. This sensitivity will be greatest when both intersecting (C-V) curves meet at a low angle, rendering any reproducible e.m.f. measurement impossible. In the former case, the super-Nernstian response is due to a parallel (C-V)-curve of an interfering cation leaving the membrane phase. A false conditioning procedure was applied in this case; the so-called interfering cation is thermodynamically prefered by the membrane phase.

It is well-known that certain liquid and PVC membranes show a cation selectivity, e.g., slope of ca. 40 mV, even without any special ion carrier or ion-exchanger present. The reason for this behavior lies in the smaller exchange current densities for anions in cases of a negatively charged surface. Also Frumkin [46] reported earlier that the rate of discharge of ions is drastically reduced if the charge of the metal surface is identical in sign with the charge of the ion.

With the mixed potential model, the non-linearity of salt-extraction can be explained and with the knowledge of the determining parameters ($\Delta\phi_{eq.}$ difference between cation and anion, $j_{o,+-}$, and α), even it can be calculated.

CONCLUSIONS

Optimizing the charge density at the surface of an I.S.E. membrane (addition of so-called moderators) seems to be a straightforward way of influencing the kinetics of unwanted charge transfer reactions. The published values of exchange current densities at I.S.E. membranes determined with different methods and by different groups are only in the range of $j_o < 10^{-3}$ A·cm^{-2} and should have an impact also on the theoretical treatment in the biomembrane field, insofar as the thermodynamic equilibrium at an interface may not always be fully established. Studies with the reliable impedance spectrum analysis technique demonstrate that also in the field of lipid bilayer research, the charge transfer resistance is higher than expected and nearly impossible to separate from the bulk resistance especially if the errors in the low frequency range are considered [48,49]. With awareness of the similarities between I.S.E.'s and biological membranes, the mixed potential approach allows a straightforward explanation of the active ion transport across biomembranes and its dependence on the concentration of participating ions [50].

REFERENCES

1. K. Cammann, G. A. Rechnitz, Anal. Chem., 48/ 1976/ 856.
2. J. Koryta, Ion-Selective Electrode Rev., Vol. 5 / 1983 / 131.
3. K. Cammann, A Mixed-Potential I.S.E. Theory, in: Conference on Ion-Selective Electrodes 1977, ed. E. Pungor, Akadémiai Kiadó, Budapest 1978.
4. H. A. Laitinen, W. E. Harris, Chemical Analysis, 2nd ed., McGraw-Hill, Inc. New York 1975.
5. W. E. Morf, The Principles of Ion-Selective Electrodes and of Membrane Transport, Elsevier, Amsterdam-Oxford-New York 1981.
6. R. P. Buck, I. Krull, J. Electroanal. Chem. and Interfacial Electrochemistry, 18 / 1968 / 387.
7. M. J. D. Brand, G. A. Rechnitz, Anal. Chem. 41 / 1969 / 1185, 1788, and 42 / 1970 / 478.
8. C. Gabrielli, Identification of Electrochemical Processes by Frequency Response Analysis, Solartron/ Schlumberger Publication, Farnborough, U.K. 1980.
9. R. P. Buck, The Impedance Method Applied to the Investigation of In-Selective Electrodes, in: Ion-Selective Electrode Rev., Vol. 4 / 1982 / 3.
10. R. D. Armstrong, A. K. Covington, G. P. Evans, T. Handyside, Electrochimica Acta, Vol. 29, No. 8 / 1984 / 1127.
11. R. P. Buck, Journal of Membrane Science, 17 / 1984 / 1.
12. M. A. Iljuschenko, W. A. Mirkin, Chabarschysny Vestnik, Kasakstan Academy of Science, 4 / 1981 / 41; translated into English in [50].
13. K. Cammann, Das Arbeiten mit ionenselektiven Elektroden, 1st ed., Springer-Verlag, Berlin-Heidelberg-New York 1973.
14. F. G. K. Baucke, Non-Crystalline Solids, 14 / 1974 / 13.

15. J. R. Luch, T. Higuchi, L. A. Sternson, Anal. Chem., 54 / 1982 / 1583.
16. A. A. Kornyshev, M. A. Vorotyntsev, J. Electroanal. Chem., 167 / 1984 / 1.
17. Yu. I. Kharkats, H. Nielsen, J. Ulstrup, ibid., 169 / 1984 / 47.
18. A. T. Hubbard et al., ibid., 168 / 1984 / 43.
19. Z. Koczorowski, I. Paleska, G. Geblewicz, ibid., 164 / 1984 / 201.
20. K. Cammann, Dissertation, Ludwig-Maximilians-Universitat, Munchen 1975.
21. K. J. Vetter, Electrochemical Kinetics, Academic Press, New York 1967.
22. J. Guastalla, Nature 227 / 1970 / 485.
23. C. Gavach, T. Mlodnicka, J. Guastalla, C. R. Acad. Sci., Ser. C, 266 / 1968 / 1196.
24. C. Gavach, F. Henry, J. Electroanal. Chem., 54 / 1975 / 361.
25. R. D. Armstrong, A. K. Covington, G. P. Evans, ibid., 159 / 1983 / 33.
26. K. Cammann, Anal. Chem., 50 / 1978 / 936.
27. C. D. Crawley, G. A. Rechnitz, unpublished results.
28. K. Cammann, Working with Ion-Selective Electrodes, English tranl. by A. H. Schroeder, Springer-Verlag, Berlin-Heidelberg-New York 1977.
29. J. Koryta, Electrolysis at the Interface of Two Immiscible Electrolyte Solutions and its Analytical Aspects, in: Ion-Selective Electrodes. Third Symposium held at Matrafüred, Hungary 1980, ed. E. Pungor, Elsevier, Amsterdam-Oxford-New York 1981.
30. O. R. Melroy, Thesis, University of North Carolina at Chapel Hill, Dept. of Chemistry 1982.
31. U. Reymers, Diplomarbeit, Universtat Ulm 1982.
32. F. Honold, Diplomarbeit, Universitat Ulm 1983.
33. M. M. Abu Samrah, R. A. Bitar, A. M. Zihlif, A. M. Y. Jaber, Applied Physics Communication, 3 (3) / 1983 / 225.
34. P. Nowak, A. Pomianowski, paper on the fourth scientific session on Ion-Selective Electrodes, Matrafüred, Hungary, Oct. 8-12 / 1984.
35. M. S. Turaeva, O. O. Lyalin, ibid.
36. O. O. Lyalin, M. S. Turaeva, ibid.
37. O. R. Melroy, R. P. Buck, J. Electroanal. Chem. 136 / 1982 / 19.
38. W. R. Fawcett, S. Levine, Electroanalytical Chemistry and Interfacial Electrochemistry, 43 / 1973 / 175.
39. P. Vanýsek, M. Behrend, J. Electroanal. Chem., 130 / 1981 / 287.
40. L. Q. Hung, ibid., 149 / 1983 / 1.
41. T. Solomon, H. Almenu, B. Hundhammer, ibid., 169 / 1984 / 303.
42. Y. Marcus, Pure Appl. Chem., 55 / 1983 / 977.
43. U. R. Evans, The Corrosion and Oxidation of Metals - Scientific Principles and Practical Application, St. Martin's Press, Inc., New York 1960.

44. A. J. Bard, L. R. Faulkner, Electrochemical Methods - Fundamentals and Applications, J. Wiley & Sons, New York 1980.
45. C. Feuillade, Electrochimica Acta, 14 / 1969 / 317.
46. A. N. Frumkin, Ref. J. Electrochem. Soc., 107 / 1960 / 461.
47. B. B. Damaskin, L. I. Krishtalik, Elektrokhimiya, 20 / 1984 / 291.
48. R. De Levie, D. Vukadin, Electroanalytical Chemistry and Interfacial Electrochemistry, 62 / 1975 / 95.
49. C. Gabrielli, M. Keddam, J. F. Lizee, J. Electroanal. Chem., 163 / 1984 / 419.
50. K. Cammann, Ion-Selective Bulk Membranes as Models for Biomembranes, chapter in "Membranes", Springer-Verlag, Berlin-Heidelberg-New York, in press.

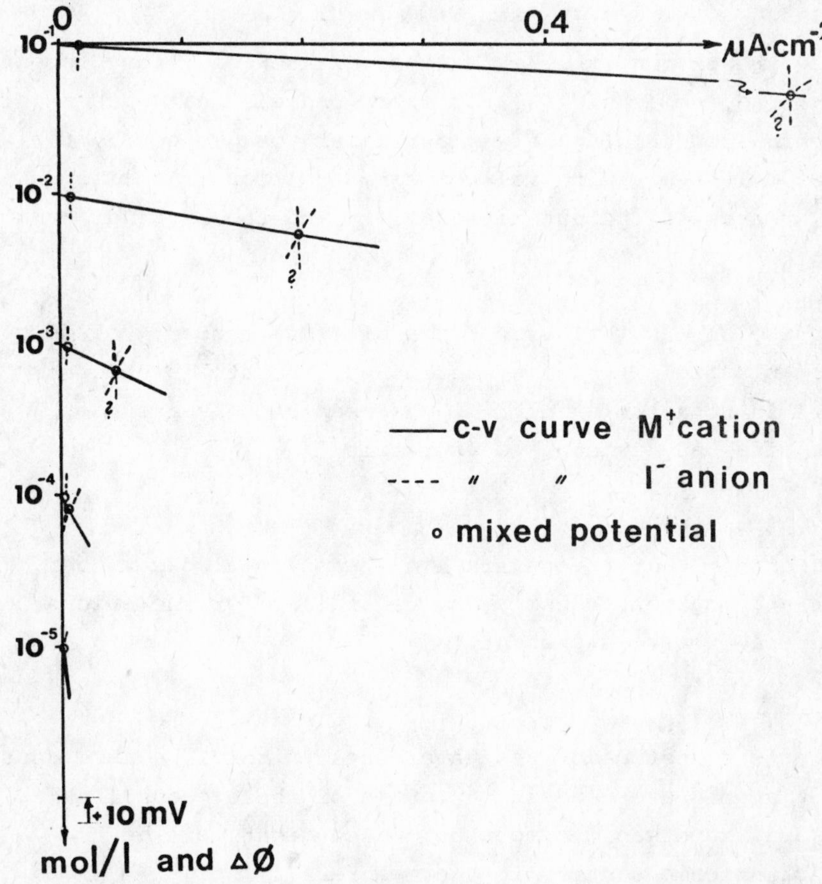

Fig. 1. Evans-diagram of a cation-selective membrane (assumptions: jo = 10^{-6} A cm^{-2} in a 0.1 mol/l solution, α = 0.5). The straight lines are the linear part of the c-v curves relating to negative overpotentials for the cation (in scale). The broken lines are the important sections from the c-v curves of interfering ions at positive overpotentials. The intersection (circles) is the mixed potential. A Nernstian and a sub-Nernstian behavior is shown.

QUESTIONS AND COMMENTS

Participants of the discussion: R.P.Buck, K.Camman, J.Janata, E.Pungor, J.S.Redinha and W.Simon

Question:

You presented a number of curves showing how the electrode potential is influenced by the amount of organic compound added to the membrane phase. What happens if you use silicone rubber containing the active component, but no additive ?

Answer:

I have never done this experiment. I know of a very recent result found in Rechnitz's laboratory, namely that with a pvc membrane containing only plasticizer you get nearly the same selectivity as with valinomycin, addition of valinomycin only increases the potential stability and the exchange current.

Question:

My question was that if you use silicone rubber, you do not need any plasticizer, and no tetraphenyl borate is necessary as there is no anion effect. It has been shown years ago. Can you offer any explanation for this ?

Answer:

It might be that anions are hindered by silicone rubber, the exchange current density may be smaller for the anions if silicone rubber is present.

Comment:

Silicone rubber membranes have sites in them, because cross--linking agents are used anyway. Some of these cross-linking agents are tin-organic compounds, which might well be the source of anionic centres in the membrane.

Question:

You presented the example of the interference by hydroxyl ions with the lanthanum fluoride electrode. According to your theory, we can interpret this in terms of the exchange current densities of the primary and interfering ion. On a molecular scale, one has to admit that both species are potential-determining ions, otherwise it is meaningless of speaking about selectivity coefficients and selectivities. If I refuse to employ the term selectivity coefficient when $K > 1$, I refuse to accept the term for ions that do not undergo an exchange between the solution and membrane phase.

However, let us assume that the interfering ion, hydroxyl is specifically adsorbed at the interface, consequently that it is in the internal Helmholtz plane. If it is so, the adsorbed ion will affect the surface charge and at the same time the transport of the primary ion which is the only potential determining ion. Thus we have no completition between the ions, we have only put something in the way of the fluoride ions and the kinetics will be responsible for what we observe. Hence, one has to be very careful when interpreting the selectivity. I would like to hear your opinion, whether you agree with this or not.

Answer:

Yes, I agree.

Comment:

You pointed out that there is a theoretical problem connected with the counter-transport system you were describing, and you referred to Mitchell's theory. I do not see any problem there. You have shown a counter-transport system in which the driving force was the proton flux, and it was thermodynamically all right.

Answer:

I was just thinking why people use his theory as hypothesis, even if it is proved to be correct. He is probably not well accepted by the bio-science people.

Question:

Do you consider the theory of mixed potentials to be applicable to all ion-selective electrodes, even when they are in the reversible domain, or just to systems where the exchange of ions of positive or negative charge is kinetically limited and the rate constant is dependent on voltage ?

Answer:

I considered all cases. Even for reversible behaviour the charge transfer resistance or the exchange current density of the interfering ion are so small that you cannot detect it. But it might be that you still have a flow of the interfering ion in one or the other direction and that this flow may also slowly change the membrane composition and may give rise to a drift. And maybe you can explain this by a constant amount of interfering ions being transported out of the membrane. Or maybe you have interfering ions in the solution, lipophilic ones, which will be slowly enriched in the membrane, and will have their own exchange current density, and the electrode deteriorates, does not give Nernstian response any longer.

Comment:

I would like to add that mixed potential is exactly the same as corrosion potential. If you take a piece of aluminium which is dissolving violently with evolution of hydrogen, you are still establishing a corrosion potential, and the important thing is that there are two processes which have approximately equal exchange current densities, one is the hydrogen ion going in and out, the other one is the aluminium going in and out of the interface. And the same situation helds for ISEs. It does not matter what the magnitude is, it is the relative magnitude of the two competing processes which is important for ISEs.

Answer:

I agree with this suggestion. I would like to raise a question to the audience, connected to the silver halide detectors. Why is the selectivity coefficient different for a silver halide single crystal electrode and a silver halide--silver sulphide mixed precipitate electrode ?

Comment:

If you are measuring the selectivity coefficient of the silver iodide electrode for bromide or chloride, you always get the same value, whether you use the mixed membrane or pure membrane. How did you make the measurements ?

Answer:

It is in the manufacturer's literature. Radiometer presents the selectivity coefficients of the silver chloride electrode to bromide as 3, and iodide as 5. And from the solubility product data we calculate much higher data.

Comment:

These values are not selectivity coefficients, because selectivity coefficients are always smaller than 1. So, there is no use of discussing things that do not exist.

Remark:

I think that the examples shown as reversible cases, which were intended to be treated as mixed potentials, were all examples where there was something in the solution which was not in the membrane at the beginning, and which was going into the membrane with a high exchange current density. But that is a non-equilibrium process. And I think the theory outlined is wonderful and I work on it, too, but I think it is limited to systems which are out of equilibrium, either systems with high exchange currents or low exchange currents.

4th Symposium on Ion-Selective Electrodes
Mátrafüred, 1984

SELECTIVITY PROPERTIES OF ION-SELECTIVE LIQUID MEMBRANE ELECTRODES AND THEIR APPLICATION TO FLOW ANALYSIS

NOBUHIKO ISHIBASHI, TOSHIHIKO IMATO,
MASARU YAMAUCHI, MASAHIRO KATAHIRA
and AKINORI JYO*

Faculty of Engineering, Kyushu University,
Hakozaki, Higashiku, Fukuoka, 812, Japan

*Faculty of Engineering, Kumamoto University,
Kurokami, Kumamoto, 860, Japan

ABSTRACT

The condition necessary to evaluate the selectivity coefficients intrinsic to the membrane media of the liquid ion exchange membrane electrodes is clarified, based on the theoretical treatment on the concentration variation at the interface between an electrode membrane and a sample solution. Observed intrinsic selectivity coefficients between many organic as well as inorganic ions are diagramatically shown on some typical membrane solvents. Selectivity of neutral carrier type liquid membrane electrodes shows dependence on concentration of the carriers, characteristic to the compositions (molar ratio) of the metal-carrier complexes. Enhancement of selectivity utilizing the concentration dependence is shown for potassium ion sensitive electrodes based on naphtho-15-crown-5 and its analogues. Synergisitic effect is shown to be useful for selectivity enhancement of the neutral carrier-based lithium ion selective electrode. Sensitivities of ion selective electrode detectors in flow analysis are discussed in connection with their selectivity. Less selective electrodes are illustrated to be more sensitive.

INTRODUCTION

Selectivity of liquid ion exchange membrane electrodes has been evaluated from electrode potentials E_M and ionic activities in the bulk of samples, in which primary and foreign ions i and j or the foreign ion j alone is contained, under the assumption that the electrode potentials obey the following Nikolsky equation:

$$E_M = \text{const.} + (RT/z_iF)\ \ln\ (a_i' + K_{i,j}^{pot}\ a_j') \qquad (1)$$

where symbols have their usual meanings. It is known that the selectivity coefficient theoretically should be constant for a given pair of ions i and j and for a given membrane. However, observed selectivity coefficients are not constant and vary with methods and conditions of their evaluation [1-7]. It is well known today that the observed variation of selectivity is mainly due to the deviation of ionic concentration at the interface between the membrane surface and the bulk of the sample solution, which is caused by ion-exchange between the membrane and the sample solution. The relation between selectivity and a concentration at the interface has been theoretically treated by several investigators. Our theoretical treatment was introduced at the previous Symposium by Dr. W. E. Morf as the Jyo-Ishibashi theory [8,9]. So, we present here the "intrinsic selectivity coefficient" diagrams which are supported by our theory on the membrane potential of the liquid membrane electrodes.

For neutral carrier type liquid membrane electrodes, the effects of the concentration of a neutral carrier in the membrane and of the membrane media such as the membrane solvents and of synergists will be mainly discussed on the selectivity of the electrodes.

Application of the ion-selective electrdes for flow analysis will be discussed from the view-point of influence of selectivity on sensitivity.

Apparent selectivity and ion-exchange at the membrane-solution interface.

Intrinsic selectivity diagram.

When a liquid membrane of an i ion-sensitive electrode is in contact with a sample containing a foreign ion j, the ion exchange reaction (2) takes place between the membrane and the sample solution:

$$i(\text{membrane}) + j(\text{solution}) \rightleftharpoons i(\text{solution}) + j(\text{membrane}) \qquad (2)$$

This leads to formation of a diffusion layer at the vicinity of the sample-membrane interface and the sample, and the concentration polarization occurs at the interface. Then an ionic composition in the bulk of the sample is not always equal to that on the membrane surface which may essentially govern the membrane potential. Concentrations of ions i and j on the membrane surface of the sample side, $C_i'(0)$ and $C_j'(0)$, are expressed by equation (3) under the following assumptions. The assumptions are (1) electrochemical equilibria are set up at the interfaces, (2) electrolytes in the aqueous phase dissociate completely, while those in the organic phase are in two extreme cases, i.e., associate strongly or dissociate completely. For the case of strong association, association constants have the same values, irrespective of counter ions species, (3) a co-ion in the aqueous phase is excluded by the membrane, and an ion exchange site S is confined within the membrane, (4) diffusion coefficients or mobilities of species are equal in respective phases, and (5) activities are equal to concentrations.

$$C_i'(0) = \frac{(C-C_j')K_{i,j} - \sqrt{[(C_j'-C)K_{i,j}]^2+4K_{i,j}CC_j'}}{2(K_{i,j}-1)}$$

$$C_j'(0) = \frac{(C_j'-C)K_{i,j} + \sqrt{[(C_j'-C)K_{i,j}]^2 + 4K_{i,j}CC_j'}}{2(K_{i,j}-1)} \qquad (3)$$

where $C=C_S^*/U$ and $U = (D/D^*)(\delta^*/\delta)$, and where C_S^* is the concentration of ion exchanger site in the membrane, $K_{i,j}$ is the equilibrium constant of the ion exchange reaction (2). The membrane potential is expressed by equation (4), based on electrochemical equilibria at the membrane-solution interfaces and on assumption (4) which leads to neglect of the diffusion potential.

$$E_M = (RT/z_iF) \ln [(C_i'(0) + K_{i,j}C_j'(0))/C_i'']$$
$$= \text{const.} + (RT/z_iF) \ln [C_i'(0) + K_{i,j}C_j'(0)] \quad (4)$$

Detailed examinations on the effects of concentration of the ion j in the sample bulk and the site concentration C_S^* and the constant $K_{i,j}$ upon the membrane potential were performed using equations (3) and (4). The results showed that only the case, in which $K_{i,j}$>>1 and C_j' is much larger than C_S^*, gives the membrane potential dependent on $K_{i,j}$ and C_j', as expressed by equation (5). In other cases, the membrane potential was proved to vary complicately with variations of $C_i'(0)$ and $C_j'(0)$.

$$E_M = \text{const.} + (RT/z_iF) \ln K_{i,j}C_j' \quad (5)$$

The ion exchange constant $K_{i,j}$ is related to the partition coefficients k_i and k_j or extraction coefficients (K^{ex}) in partitions to the electrode membrane as follows:

$$K_{i,j} = k_j/k_i \quad \text{(for the case of complete dissociation in the membrane)} \quad (6)$$

$$K_{i,j} = K_j^{ex}/K_i^{ex} \quad \text{(for the case of strong association)} \quad (7)$$

The Nikolsky equation is simplified to the next form (8) for the sample solution containing a foreign ion j alone,

$$E_M = \text{const.} + (RT/z_iF) \ln K_{i,j}^{pot} \; C_j' \quad (8)$$

and one can obtain the $K_{i,j}^{pot}$ value, comparing the two observed membrane potentials for the sample solution containing the j ion and for the sample solution containing the primary ion i, according to the separate solution method. However, it is known that physico-chemically significant and constant values of selectivity coefficients are obtainable only in the case where the potentials are measured under the condition described in derivation of equation (5), i.e., $K_{i,j}$>>1 and

$C_j^{'} > C_S^{*}$. The "constant" selectivity coefficients thus obtained correspond to the equilibrium constants of ion exchange $K_{i,j}$, which are intrinsic values determined by the medium of the membrane (solvent and ion exchanger species) and of course, the ionic species i and j. When no specific interaction occurs between an ion exchanger and counter ions, selectivity coefficients may be governed by the membrane solvent used. Ion exchange site species and its concentration exert no effect on the selectivity. The following relation for intrinsic selectivity coefficients is easily understood from equations (6) and (7).

$$K_{1,n}^{pot} = K_{1,2}^{pot} \times K_{2,3}^{pot} \times \ldots\ldots \times K_{n-1,n}^{pot} \qquad (9)$$

where $K_{n-1,n}^{pot}$ represents the selectivity coefficient between the (n-1)th and (n)th ions.

Using the relation (9), the selectivity diagrams have been constructed for liquid ion exchanger membranes and neutral carrier type liquid membranes [4,5,7], which are showed in Figs. 1, 2 and 3.

<u>Selectivity of neutral carrier liquid membranes. Effect of neutral carrier concentration.</u>

Selectivity behaviors of neutral carrier liquid membrane electrodes have extensively investigated by many researchers, especially by Professor W. Simon's group[10]. Here, we present our recent works on effects of concentration of neutral carriers on the selectivity and of some synergists added to the membrane [11]. Figure 4 shows selectivity coefficients of the dibenzo-18-crown-6-containing liquid membrane electrodes with different membrane solvents for some alkali metal ions and one monovalent organic cation. Variations of selectivity coefficients of the organic cation for all of three solvent membranes and of the cesium and rubidium ions for 1,2-dichloroethane and chloroform membranes can be theoretically explained by taking into account of formations of complexes with different molar ratio [7,8,10].

Theoretical treatment on potentials and selectivity of the neutral carrier liquid membrane electrodes gives the following expression on the selectivity coefficient of the j ion over the reference ion, Na^+, for the membrane system of the j-ion/membrane/Na^+.

$$K^{pot}_{Na,j} = \frac{u^*_j k_j + u^*_{jS} k_{jS} K^f_{jS} (C^{*tot}_S / k_S) + u^*_{jS_2} k_{jS_2} K_{jS} K_{jS_2} (C^{*tot}_S / k_S)^2}{u^*_{NaS} k_{NaS} K^f_{NaS} (C^{*tot}_S / k_S)} \qquad (10)$$

where u^* is a mobility in the membrane, k is a partition coefficient, K^f is a formation constant of the metal-carrier complex, C^{*tot}_S is total concentration of S in the membrane, S is the neutral carrier species, jS, jS_2, and NaS are complexes of metal ions with the carrier. In equation 10, the sodium ion is assumed to form only the 1:1 complex NaS, which is distributed to the membrane phase. Equation 10 predicts the following concentration dependence of the selectivity.

A) For the case that the j ion forms no complex,

$$\log K^{pot}_{Na,j} = \text{const.} - \log C^{*tot}_S \qquad (11)$$

B) For the case that the j ion forms the 1:1 (molar ratio of metal to carrier) complex,

$$\log K^{pot}_{Na,j} = \text{const.} \qquad (12)$$

C) For the case that the j ion forms the 1:2 complex,

$$\log K^{pot}_{Na,j} = \text{const.} + \log C^{*tot}_S \qquad (13)$$

From relations A, B and C, the change of selectivity of the $(CH_3)_4N^+$ ion indicates that the ion forms no complex with dibenzo-18-crown-6. The selectivity behavior of the cesium and rubidium ions implies that the ions tend to form the 1:2 complexes, to some extent, as well as the 1:1 complexes, whereas the potassium ion forms the complex of the same molar composition as that of the sodium ion complex. The concentration dependence of selectivity as shown in Fig. 4 may be used to enhance the ion selectivity of the electrodes and the selective transport property of the membrane. Figure 5 shows the selectivity variation of the naphtho-15-crown-5 containing nitrobenzene membrane electrode for alkali metal ions. It is known that this crown ether tends to form the 1:2 complex with the potassium or rubidium ions and the 1:1 complex with the sodium ion. Increase of this crown ether content in the nitrobenzene membrane brings a great enhancement of the selectivity for the potassium ion. The high selectivity shown in Figs. 5 and 6 is due to the high degree of partition of the 1:2 complex to the membrane phase, since mobilities of various species are not significantly different in the liquid membrane phase. The partition of relevant chemical species between 1,2-dichloroethane and aqueous phases was examined by solvent extraction. Extraction parameters measured are listed in Table 1 with the related equilibrium constants. The extraction constant $K^{ex}_{MS_2A}$ of the potassium 1:2 complex is much higher than K^{ex}_{MSA} of the sodium 1:1 complex, as expected. However, the excellent selectivity of the naphtho-15-crown-5 based membrane could not fully explained only by the higher partition of the 1:2 complex. The complex cations are both in the forms of free cations and associates with the counter anion in the membrane. The 1:1 complex is in a state of higher degree of association, compared with the 1:2 complex, as shown in Table 1. The higher dissociative character of the 1:2 complex may also contribute to enhancement of the selectivity.

Selectivities of the naphtho-12-crown-4 and dibenzo-14-crown-4 based liquid membranes are also greatly dependent on total concentrations of respective crown ethers in the liquid membranes. Observed results are shown in Figs. 7 and 8. It is said that naphtho-12-crown-4 forms the 1:2 complex with sodium ion, while dibenzo-14-crown-4 forms the 1:1 complex with lithium ion. Selectivity variations shown in Figs. 7 and 8 will be explained from difference in the molar ratio of the metal to neutral carrier in the complexes.

Enhancement of selectivity of neutral carrier liquid membrane electrodes by adduct formation and modification of carriers.[11]

In solvent extraction, it is well known that the distribution of species to an organic phase is increased, to large extent, by an addition of chemical reagents having a synergistic action. We tried to use this synergistic effect to enhance selectivity of the lithium ion sensitive liquid membrane electrode. Four organophosphorus compounds, trioctyl phosphine oxide (TOPO), butyldibutyl phosphinate (BDBP), dioctyloctyl phosphonate (DOOP) and trioctyl phosphate (TOP), were used as the synergistic reagents. They were added into the nitrobenzene membrane containing dibenzo-14-crown-4 as a neutral carrier and sodium dipicrylaminate. Observed effects of added organophosphorus reagents are listed in Table 2. Enhancement effects are larger in a sequence of the basicities of the reagents, i.e., TOPO>BDBP>DOOP>TOP. The extent of adduct formation was evaluated by separately conducted solvent extraction experiments. Reactions in extraction are expressed by the next equations.

$$M^{+}(w) + A^{-}(w) + mS(o) \rightleftharpoons MS_{m}A(o), \qquad K^{ex}_{MSmA} \qquad (14)$$

$$M^{+}(w) + A^{-}(w) + mB(o) \rightleftharpoons MB_{m}A(o), \qquad K^{ex}_{MBmA} \qquad (15)$$

$$M^{+}(w) + A^{-}(w) + S(o) + mB(o) \rightleftharpoons MSB_{m}A, \qquad K^{ex}_{MSBmA} \qquad (16)$$

Association constants between the MSB_{m}^{+} ion and A^{-} ion in the organic phase were obtained and the results are listed in Table 3. From analysis of solvent extraction experiments, the values of m were found to be 2 for the lithium ion and 1 for the sodium and potassium ions. From Table 3, we can see that extractabilities of the lithium ion and the sodium ion by dibenzo-14-crown-4 alone are in the same level, but the extractability of the lithium ion can be enhanced, to high degree, by an addition of DOOP. The reason is ascribed to additive coordination of 2 moles of DOOP to one mole of the lithium-crown ether complex.

The sodium ion selectivity of the 12-crown-4 based liquid membrane electrode is increased by an introduction of a long alkyl chain into the 12-crown-4 moiety. C_6H_{13}-12-crown-4 and $C_{12}H_{25}$-12-crown-4 were synthesized and added to the electrode membrane. The measured selectivity coefficients $K^{pot}_{Na,K}$ were 0.33, 0.045 and 0.030 for naphtho-12-crown-4-based-, C_6H_{13}-12-crown-4-based- and $C_{12}H_{25}$-12-crown-4 based- liquid membrane electrodes, respectively. This increase of the selectivity for sodium ion probably comes from increased tendency for formation the 1:2 (Na: crown ether) complex by self-interaction of the long alkyl chain.

Sensitivity of the ion selective electrode detector for flow analysis and its selectivity.

In the flow analysis, it is also advisable to use the electrodes with high selectivity for an objective ion when selective detection is required. However, less selective electrodes are sometimes very useful for a quite simple solution which may contain only one interferent at a low levels, like effluents from HPLC column. A small damage of the membrane surface suffered from a passage of an interferent may be recovered by the primary ion of low concentration which previously is added into a flowing solution. Disturbance of the streaming potential for potential measurement may be avoided by adding an insensitive electrolyte of a high concentration into a flowing solution. We named the solution containing the primary ion at low level and an insensitive electrolyte at high level as "Baseline Supporting Electrolyte Solution (BLSS) [12]". The solution containing 10^{-5} M $NaNO_3$ - 0.5 M Na_2SO_4 is one of the good BLSS for the nitrate ion selective electrode and it gives very stable potential when it is added to effluent from HPLC column by the post column manner. In Fig. 9, chromatographic peaks of four oxyacids, IO_3^-, BrO_3^-, NO_3^- and ClO_3^- are shown. These are detected with the oleophilic anion exchange resin membrane electrode [13]. The ion exchange site is a benzyl-trioctyl ammonium group ion, and the resin can hold an organic solvent like nitrobenzene even in aqueous surroundings. Therefore, this membrane electrode can be assumed essentially one of liquid membrane electrodes. Figure 10 shows effects of three kind of solvents imbibed in the membrane. o-nitrophenyl octyl ether shows the most highest peaks and decan-1-ol makes the membrane less sensitive. Figure 11 shows effect of the ionic forms on the sensitivity of the detector. The chloride form membrane is more sensitive than the nitrate form membrane. These relative sensitiveness can be understood based on the selective diagrams shown in Fig. 2.

REFERENCES

1. J. W. Ross, in Ion-Selective Electrodes (R. A. Durst, ed.), National Bureau of Standards Special Publication 314, Washington, 1964.
2. A. Hulanicki and R. Lewandowski, Chemia Analityczna, 19/53/1974.
3. N. Yoshida and N. Ishibashi, Chem. Lett., 1974/493.
4. A. Jyo, H. Torikai and N. Ishibashi, Bull. Chem. Soc. Jpn., 47/2862/1974.
5. A. Jyo, H. Mihara and N. Ishibashi, Denki Kagaku, 44/268/1976.
6. N. Yoshida and N. Ishibashi, Bull. Chem. Soc. Jpn., 50 /3189/1977.
7. A. Jyo, H. Seto and N. Ishibashi, Nippon Kagaku Kaishi, /1980/1423.
8. W. E. Morf, The Principles of Ion-Selective Electrodes and of Membrane Transport, Akademiai Kiado, Budapest, 1980.
9. W. E. Morf, 3rd Symposium on Ion-Selective Electrodes (E. Pungor ed.), Akademiai Kiado, Budapest 1982.
10. W. E. Morf and W. Simon, in Ion-Selective Electrodes in Analytical Chemistry vol. 1 (E. Freiser ed.), Plenum Press, New York, 1978.

11. T. Imato, M. Katahira and N. Ishibashi, Anal. Chim. Acta, in press.
12. N. Ishibashi, A. Jyo and T. Imato, Proceeding of the International Meeting on Chemical Sensors, (T. Seiyama ed.), Kodansha/Elsevier, Tokyo, 1983.
13. T. Imato, A. Jyo and N. Ishibashi, Anal. Chem., 52 /1893/1980.

Table 1.* Extraction and complexing parameters for alkali metal - picrate-naphtho-15-crown-5 system at 15 °C. Solvent is 1,2-dichloroethane.

M^+	Na^+	K^+
K^f_{MS}	0.35	0.81
$k_{MS}k_A$	3.0×10^2	(3.0×10^2)
K^{org}_{f1}	——	(1.8×10^5)
K^{org}_{f2}	——	(2.3×10^3)
K^{ass}_{MSA}	1.3×10^5	(1.3×10^5)
$K^{ass}_{MS_2A}$	——	1.7×10^3
K^{ex}_{MSA}	7.2×10^3	(1.8×10^4)
$K^{ex}_{MS_2A}$	——	4.1×10^7

$k_S = 1.8 \times 10^3$. Values in parentheses are estimated ones.

* see Scheme 1.

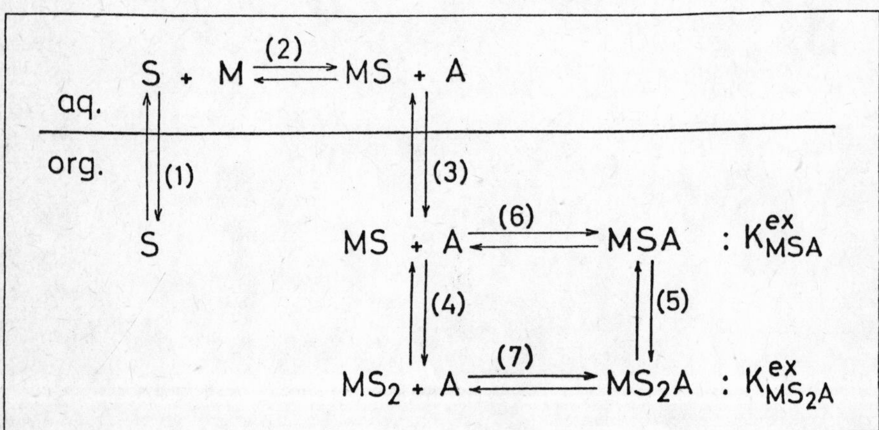

Scheme 1. Extraction system and extraction parameters.*
(1): k_S, (2): K^f_{MS}, (3): $k_{MS}k_A$, (4): K^{org}_{f1}.
(5): K^{org}_{f2}, (6): K^{ass}_{MSA}, (7): K^{ass}_{MS2A}.

* see Table 1

Table 2. Extraction parameters for alkali metal picrate - dibenzo-14-crown-4 - synergist ternary system at 25 °C. Solvent is benzene.

	K^{ex}_{MSA}	$K^{ex}_{MB_2A}$	$K^{ex}_{MSB_mA}$	$K^{f}_{MSB_mA}$
Li^+	38.0	53.3	8.3×10^3	2.2×10^2
Na^+	35.0	1.7	3.4×10^2	9.7
K^+	2.7	0.4	34.7	12.9

Li^+ : m = 2, Na^+ : m = 1, K^+ : m = 1. B: DOOP, S: dibenzo-14-crown-4.

Table 3. Synergistic effect of organophosphorus compounds on selectivity of neutral carrier type liquid membrane electrode. Solvent of membrane is nitrobenzene.

Synergist	Concn. of synergist (M)	$k^{pot}_{Li,Na}$	$k^{pot}_{Li,K}$	$k^{pot}_{Li,Rb}$	$k^{pot}_{Li,Cs}$
none		1.0×10^{-1}	2.5×10^{-1}	3.0×10^{-1}	6.0×10^{-1}
TOPO	1.0×10^{-3}	6.3×10^{-2}	1.3×10^{-1}	1.7×10^{-1}	3.0×10^{-1}
	5.0×10^{-3}	3.6×10^{-2}	6.2×10^{-2}	7.5×10^{-2}	1.4×10^{-1}
	1.0×10^{-2}	2.7×10^{-2}	4.4×10^{-2}	4.8×10^{-2}	8.2×10^{-1}
BDBP	1.0×10^{-2}	3.9×10^{-2}	7.6×10^{-2}	9.7×10^{-2}	1.9×10^{-1}
DOOP	1.0×10^{-2}	6.6×10^{-2}	1.3×10^{-1}	1.5×10^{-1}	3.1×10^{-1}
TOP	1.0×10^{-2}	6.6×10^{-2}	1.3×10^{-1}	1.6×10^{-1}	3.8×10^{-1}

Membrane composition: 1.0×10^{-2} M dibenzo-14-crown-4,
1.0×10^{-4} M sodium dipicrylaminate.

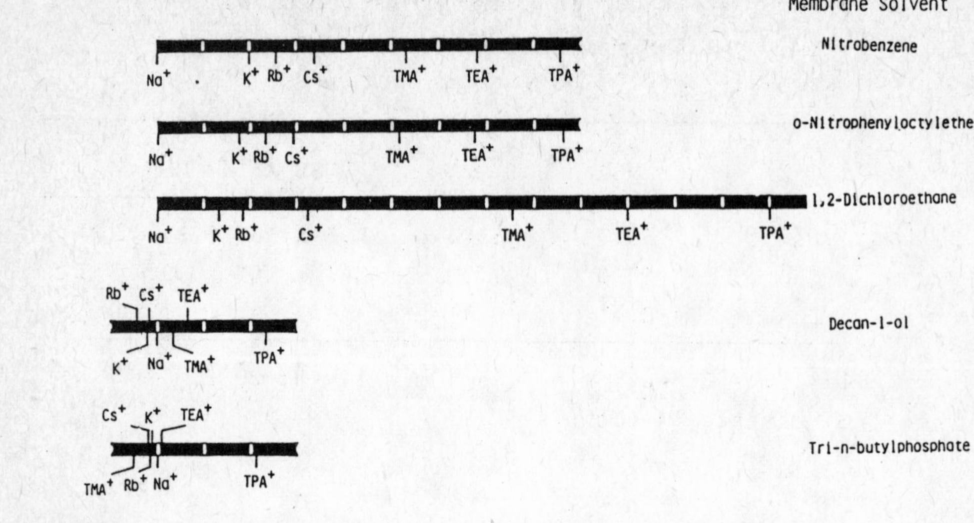

Figure 1. Selectivity diagram for liquid cation exchanger membrane. Reference ion: Na^+

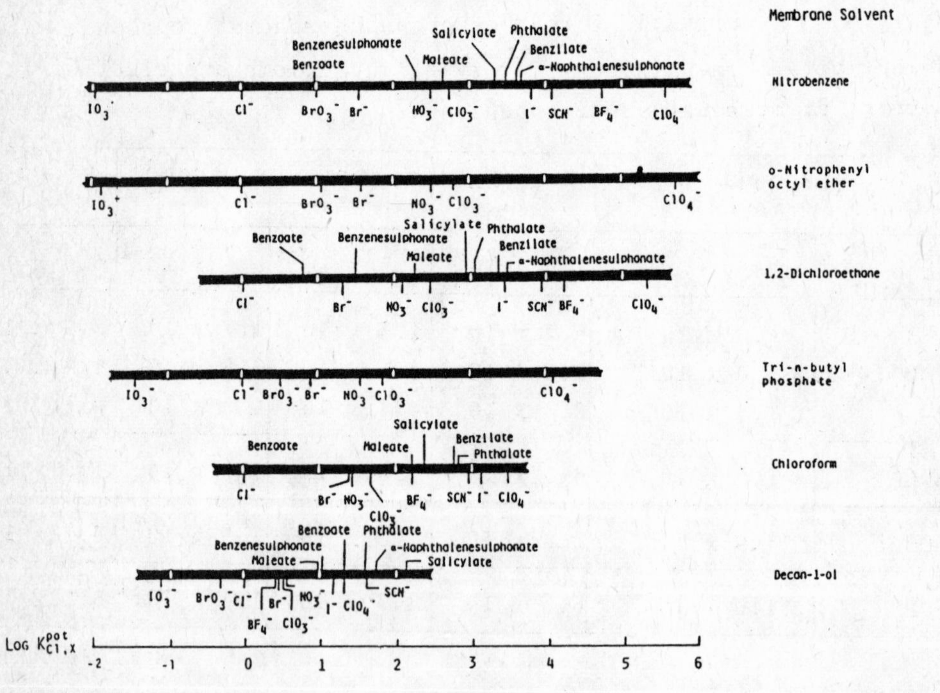

Figure 2. Selectivity diagram for liquid anion exchanger membrane. Reference ion: Cl^-.

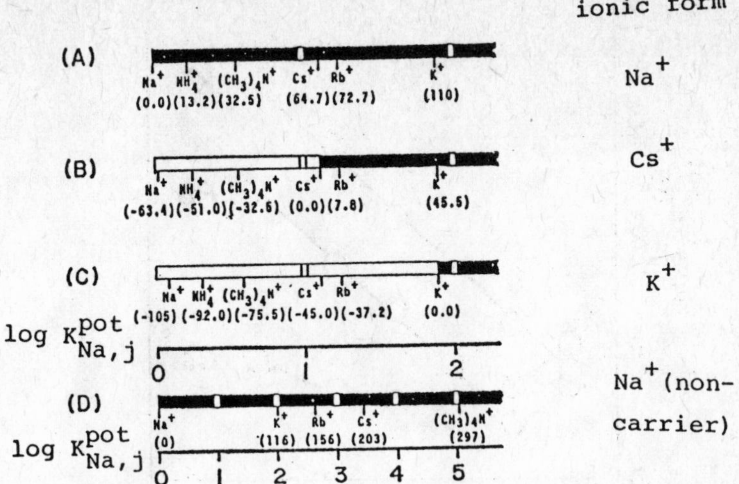

Figure3. Additivity of membrane potential and selectivity diagram for neutral carrier liquid membrane. neutral carrier: dibenzo-18-crown-6, solvent: nitrobenzene

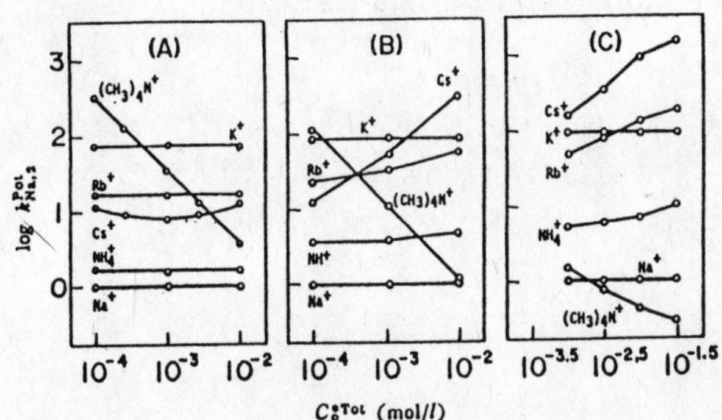

Figure 4. Dependence of selectivity on the concentration of dibenzo-18-crown-6 in the liquid membrane. solvent of the membrane ; (A): nitrobenzene, (B): 1,2-dichloroethane, (C): chloroform.

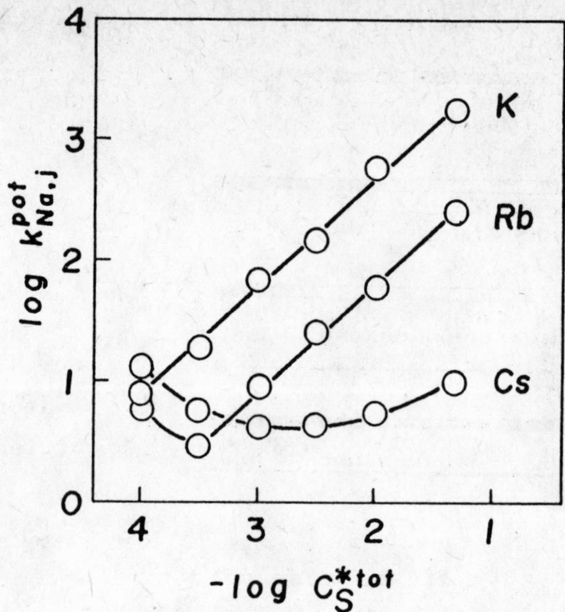

Figure 5. Dependence of selectivity of the naphtho-15-crown-5 containing nitrobenzene membrane on the concentration of the carrier.

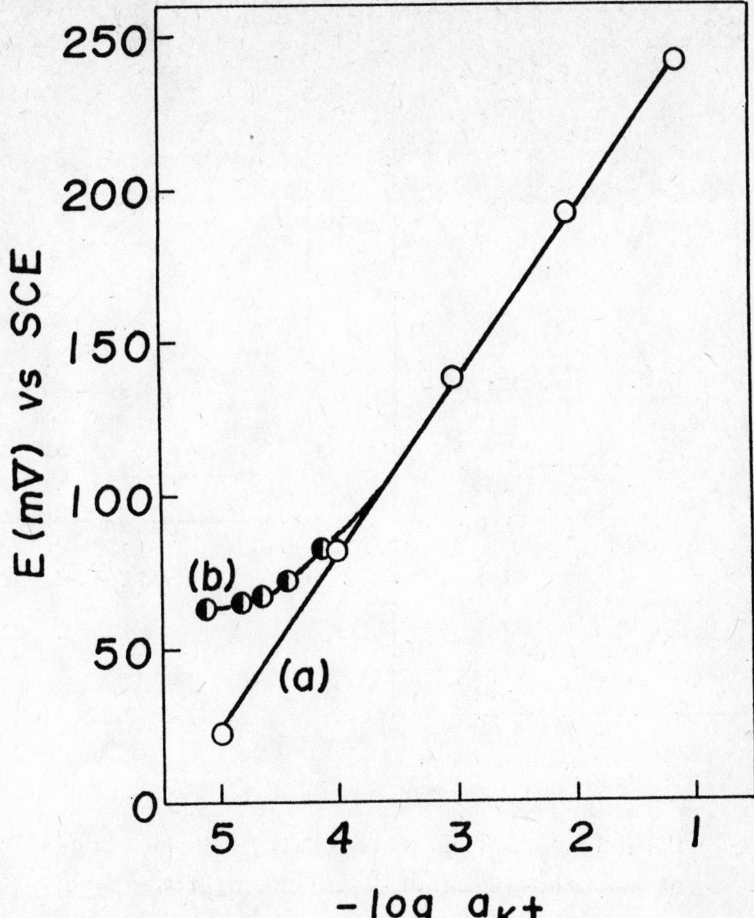

Figure 6. Calibration curves for the naphtho-15-crown-5 based PVC membrane electrode. Composition of the membrane; PVC: 0.4g, o-nitrophenyl octyl ether: 1.0 g, naphtho-15-crown-5: 0.08 g. Calibration curves; (a) potassium ion only, (b) in the presence of 0.1 M sodium ion.

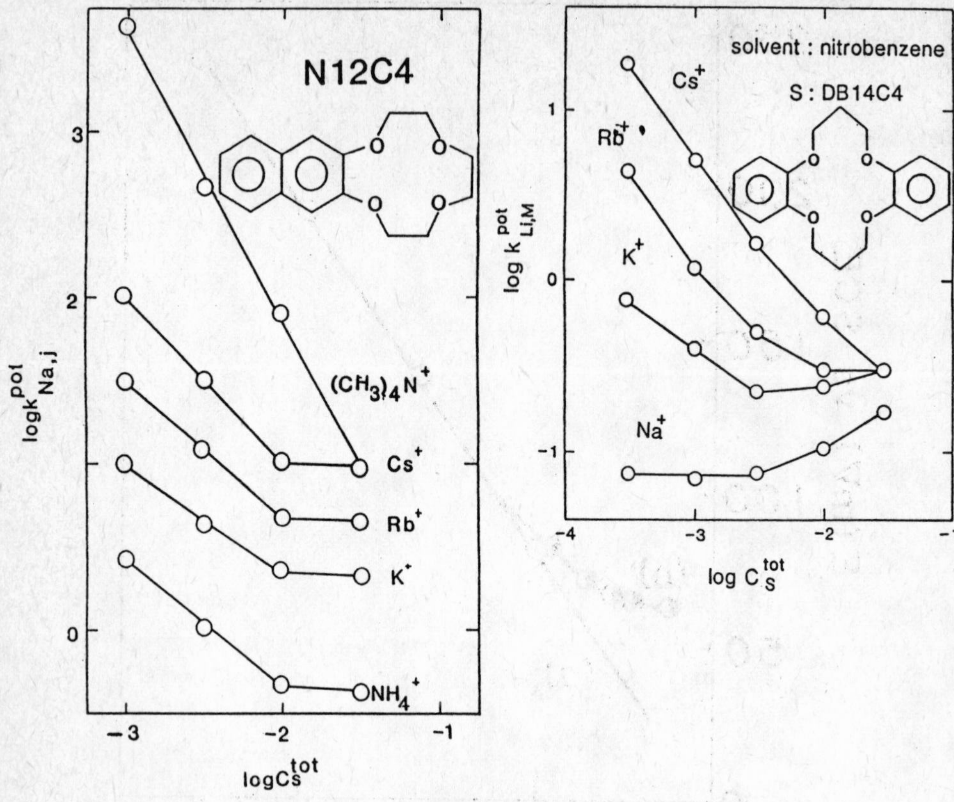

Figure 7. Dependence of the selectivity on the concentration of naphtho-12-crown-4 in the nitrobenzene.

Figure 8. Dependence of the selectivity on the concentration of dibenzo-14-crown-4 in the nitrobenzene membrane.

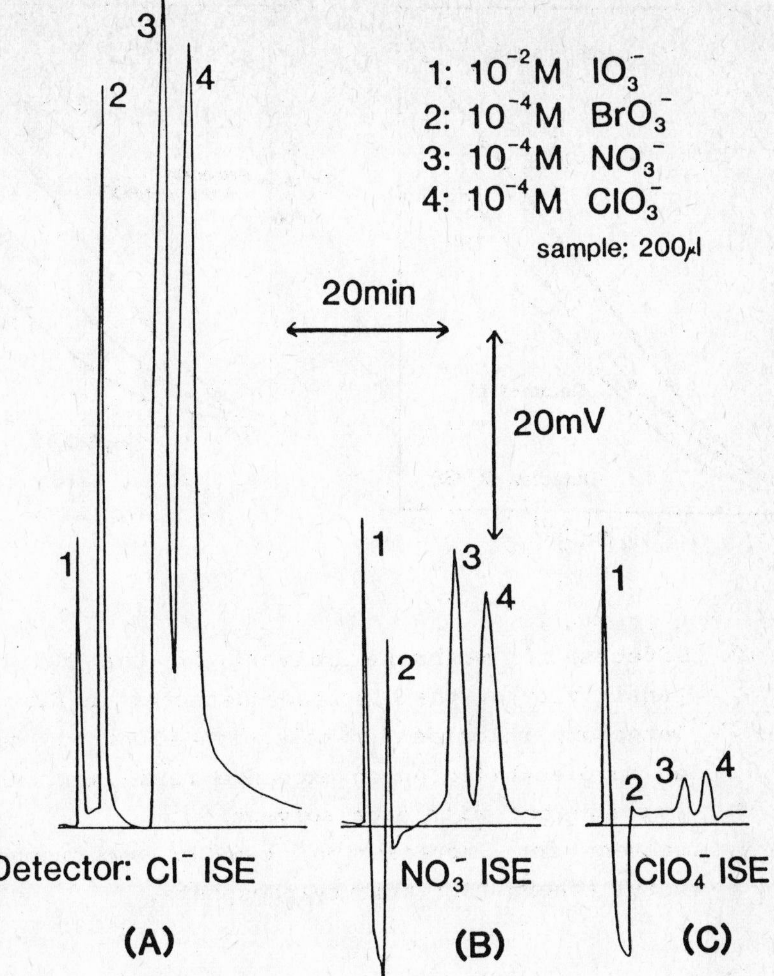

Figure 9. Chromatograms for oxyacid anions.
(A) Detector: chloride ion-selective electrode. carrier: 0.0010 M Na_2SO_4 + 1.0 x 10^{-5} NaCl (1.0 ml/min)
(B) Detector: nitrate ion-selective electrode. carrier: 0.0010 M Na_2SO_4 + 1.0 x 10^{-5} $NaNO_3$ (1.0 ml/min)
(C) Detector: perchlorate ion-selective electrode. carrier: 0.0010 M Na_2SO_4 (1.0 ml/min).
The sensitive membrane electrode is an oleophilic anion exchange resin membrane impregnated with o-nitrophenyl octyl ether. Separation column: Zipax SAX (4mm i.d. x 300mm).

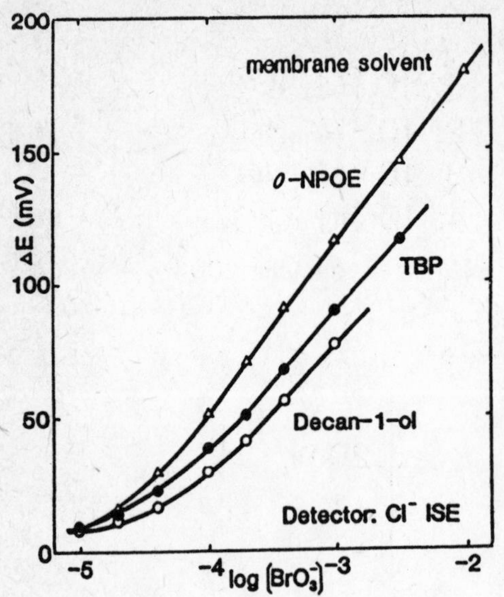

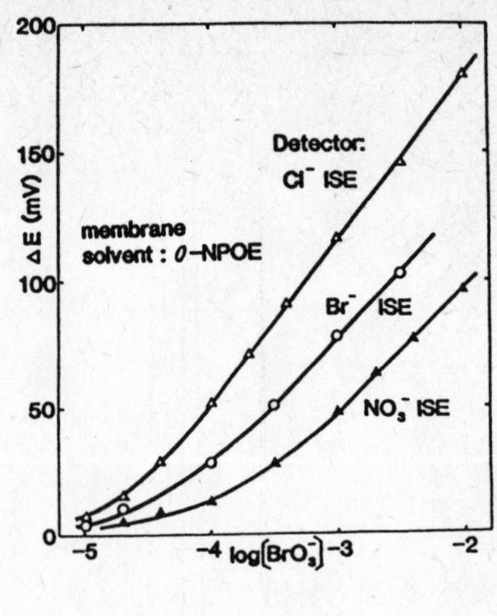

Figure 10. Effects of membrane solvent on the relative sensitivity of the electrode detector in flow analysis. Detector: chloride ion-selective electrode based on an oleophilic anion exchange resin membrane impregnated with each solvent.
solvent for impregantion; o-NPOE: o-nitrophenyl octyl ether, TBP: tributylphosphate.

Figure 11. Effects of ionic forms of the oleophilic anion exchange resin membrane on sensitivity of the detector in flow analysis.
Detector: o-nitrophenyl octyl ether impregnated resin membrane in each ionic form.

QUESTIONS AND COMMENTS

Participants of the discussion: P.Becker, R.P.Buck, K.Cammann, G.Horvai, N.Ishibashi, E.Pungor, W.Simon and J.D.R.Thomas

Question:

Your measurements were made at 25°C. Manufacturers usually claim a temperature range of 0 to 50°C. How does the selectivity change in this temperature range?

Answer:

If the temperature rises, the selectivity coefficient may approach unity because the selectivity depends on the unequal distribution of ionic species between the organic, membrane phase and the aqueous phase. As the temperature rises, the different character of ionic species is reduced.

It is also necessary that the membrane medium is suited to high temperature. We have to find a solvent of high viscosity and low vapour pressure. But actually, we have not measured the temperature dependence of the selectivity coefficient.

Comment:

There is probably a maximum in selectivity at some temperature, because equilibrium constants, such as acid-base equilibrium constants have a maximum.

Comment:

We have not made a systematic study of selectivity coefficient changes with temperature, but we have from time to time measured selectivity coefficients for the calcium ion-selective electrode at 25°C and also at 37°C. I cannot recall in it any definite trend, in other words the differences were not sufficiently great to attract attention to any big variations.

Comment:

It may also be that the activation energy differences level out.

Comment:

Based on thermodynamic considerations, I would not expect great changes in the selectivity coefficient with the temperature.

Comment:

Nevertheless, I have to repeat that the equilibrium constants have a maximum at some temperature, which depends on the functionality of the groups which undergo the equilibrium reaction. But at very low temperature, one might be faced with another problem, and this is that the exchange reaction will be so slow that the membrane will not respond to the primary ion any more. In other words, the ligand exchange reaction is so slow that the ionophor will trap the primary ion and it will not act as e.g. a cation selective system any more, but will become an anion-selective system.

Question:

You spoke about a maleate electrode and showed figures which indicated that the electrode responds better to other anions. Why do you call this electrode a maleate electrode ?

Answer:

The electrode responds better to maleate than to some other anions such as chloride or acetate. But you are right, the electrode response was found to be even better for perchlorate than to maleate.

Question:

Why was the slope of the perchlorate electrode super-Nernstian ?

Answer:

The slope depended on the concentration at the interface.

Question:

You have shown us a micro flow-through cell with a very small diameter. Could you please tell us something about the construction of the electrode you used ?

Answer:

We used a pvc tube with a diameter of 0.25 mm. Then we cut a piece of the tube with a razor, then put a thin steel wire into the tube, and applied on it the pvc containing the ion exchanger.

4th Symposium on Ion-Selective Electrodes
Mátrafüred, 1984

MEMBRANE TECHNOLOGICAL OPTIMIZATION OF ION-SELECTIVE ELECTRODES BASED ON SOLVENT POLYMERIC MEMBRANES FOR CLINICAL APPLICATIONS

U. OESCH, P. ANKER, D. AMMANN and W. SIMON

Swiss Federal Institute of Technology,
Department of Organic Chemistry,
CH-8092 Zürich, Switzerland

ABSTRACT

Requirements for a reliable monitoring of ion-activities with ion-selective solvent polymeric membranes for clinical application in flow-through analyzers are discussed and compared to currently available membrane systems.

INTRODUCTION

Ion-selective membrane electrodes are now widely used for the direct potentiometric determination of ion-activities or ion-concentrations in biological samples [1,2]. Particularly, the feasibility of their use in continuous as well as in situ applications imposes a strong competition on the currently established methods like flame photometry. Under a variety of membrane types, solvent polymeric membranes have proved to be especially suited for clinical analysis since they can easily be manufactured in different sizes and shapes and are less affected by the presence of biological substrates (e.g. proteins) [3]. Today, most of the competitive ion-selective electrodes based on solvent polymeric membranes utilise a neutral carrier as the ion-selective component. This is underlined by the strong research activities by several laboratories around the world to create highly selective neutral carriers [3-20] and the increasingly wide-spread use of carrier membranes in commercial clinical analyzers. The intent of this contribution is to claim some selected requirements on such membrane systems and to compare these demands with the performance of currently available neutral carrier membrane electrodes.

REQUIREMENTS FOR ION-SELECTIVE ELECTRODES FOR CLINICAL APPLICATIONS

The two major applications of clinical ion-measurement are in blood and urine analysis. In the following the discussion will be focused on these. The in a certain respect well-defined conditions are reflected e.g. by the distinct physiological concentration ranges (Table 1). The general demands on potentiometric ion-measurements by membrane electrodes may therefore be reduced, or enhanced, due to their distinct environments within these particular samples. Thus, a meaningful discussion about electrode characteristics inevitably should deal rather with a well-balanced optimization of all the properties relevant for the application than with the superiority of a single property. Relevant properties reflecting the reliability and quality of sensors are selectivity, stability, response time and life time.

Selectivity. Evidently, a sufficient selectivity in respect to other ions in the sample is a prerequisite. The extent of the necessary selectivity depends on the prevailing activities of the measuring and the interfering ion. Using the Nicolsky-Eisenman formalism and assuming representative physiological concentration ranges (Table 1) the required selectivity factors can be calculated with Equation (1).

$$K^{Pot}_{IJ,max} = \frac{a_{I,min}}{(a_{J,max})^{z_I/z_J}} \cdot \frac{p_{IJ}}{100} \qquad (1)$$

with

$K^{Pot}_{IJ,max}$: highest tolerable value of the selectivity factor

$a_{I,min}$: lowest expected activity of the measuring ion I^{z_I}

$a_{J,max}$: highest expected activity of the interfering ion J^{z_J}

p_{IJ} : highest tolerable error in the activity a_I due to interference of a_J (in %)

The required $\log K_{IJ}^{Pot}$-values given in Tables 2 - 7 are calculated for $p_{IJ} = 1$, i.e. for a maximally tolerable error of 1% (worst case). These values can be compared with ion-selectivities of available solvent polymeric membrane electrodes. However the comparison might be impeded since measured selectivity factors depend on the method as well as on the conditions of their experimental determination [22].

Stability. In view of clinically meaningful results the performance of the electrode cell assembly should allow an adequate subdivision of the small emf ranges corresponding to the physiological normal concentration ranges (Table 1). Consequently this imposes a high stability on the emf response. It is advised to discuss the stability of the sensor in a flowing system in terms of drift and residual standard deviation of the response in one given sample and in terms of reproducibility between two alternating samples. Ideally a system should be without drift. The residual standard deviation is then at least limited by the electrical system (e.g. 0.012 mV [1]). Often reported stability data do not differentiate between drift and residual standard deviation. Instead they are given as a standard deviation of collected sequential data without correcting for eventual drifts and therefore are dependent on the time period they are collected. On the other hand the reproducibility usually is drift independent since it is given as the standard deviation of the emf difference of two consecutively measured solutions.

Since commercial analyzers are normally working with high sampling rates and with a one-point calibration after each sample, drift problems are minimized. But in cases of long term continuous recording (e.g. monitoring in an extracorporeal circulation) the exact knowledge of drift is of utmost importance.

The required standard deviations claimed in Tables 2 - 7 represent values which would allow a five-fold subdivision (95% confidence limit) of the physiological normal range.

Response time. The required response time of the membrane electrode cell assembly has to be compatible with the analysis time, which is in the order of $\sim$30 s for commercial analyzers. There are many different processes that contribute to the dynamic response behaviour [23]. Since some of these processes (e.g. diffusion of sample ions through the stagnant layer adhering to the membrane surface) are not intrinsic to the membrane itself, it is difficult to properly compare reported response times as characteristics of membrane systems.

Life time. Beside mechanical defects, electrical leakage pathways, surface contaminations and membrane poisoning the life time of a liquid membrane electrode is generally limited due to a gradual extraction of membrane components (carriers, plasticizers) into the samples. On one side a loss of the plasticizer down to a level of less than 30% plasticizer content results in an extremly high electrical resistance of plasticized PVC membranes [24]. On the other side a reduction of the carrier concentration in the membrane causes a breakdown of the ion-selectivity as well as of the electrode function [24]. The extent of such a reduction which can be tolerated depends on the sensor properties required for the particular application considered. The time until the system reaches this critical state is considered the life time of the sensor.

Equation (2) gives a theoretical description of the life time of a solvent polymeric membrane electrode in a continuous flow-through system of a typical commercial clinical flow analyzer (Figure 1).

$$t^{lim} = \frac{V_m \delta_s K}{A D_s} \left(\frac{(q-1)^2 \ln q}{(q-1)^2 + \frac{\pi V_m D_s}{A D_m \delta_s K} \ln q} \right) \qquad (2)$$

with $$q = c_m^o / c_m^{lim} \qquad (3)$$

$$\delta_s = \frac{\pi r D_s^{1/2} \ell^{1/2}}{u^{1/2}} \qquad (4)$$

where t^{lim} is the life time (s), A the sensing electrode area (cm^2), V_m the membrane volume (cm^3), ℓ the sensor length in the direction of the flow (cm), u the flow of the sample stream ($cm^3\ s^{-1}$), r the channel radius (cm), K the partition coefficient of the considered component between membrane and sample, D_s and D_m the diffusion coefficient of the considered component in the sample and in the membrane, respectively ($cm^2\ s^{-1}$) and c_m^o and c_m^{lim} the initial and limiting concentration of the considered component in the membrane ($mol\ cm^{-3}$). Equation (2) is derived from and combines the two kinetic processes which govern the extraction of a component from the membrane into a continuously streaming sample. They are (a) diffusion through the stagnant Nernst diffusion layer in the sample and adhering on the membrane surface and (b) diffusion through the bulk of the membrane towards the membrane surface (see [25] for the mathematical derivation). Furthermore Equation (2) cares about the generation of a concentration gradient within the membrane if the extraction is limited by process (b) and considers c_m^{lim} only as the required concentration at the surface of the membrane for determining the life time. Equation (2) obviously shows that an enhancement of t^{lim} can be achieved, although practically only to a limited extent, by a larger V_m, a smaller A and a larger c_m^o. Logically these parameters together have a reservoir effect. But more serious are the parameters K and c_m^{lim}. With large K c_m^{lim} has barely an impact while with small values of K c_m^{lim} is of utmost importance. Particularly severe demands are imposed on these two parameters in a clinical application. On one hand a crucial deterioration of the selectivity can only be prevented by maintaining a certain concentration within the membrane. The required concentration can be within a small range only [26] or set by a minimal value [24]. On the other hand the lipophilic character of the sample (whole blood, serum, plasma) favors a substantial and fast extraction of the membrane components. Therefore, for prevention an extremely high lipophilicity P of the membrane components should be realized. The correlation between lipophilicity P_{TLC} of a component (lipophilicity P as determined by thin layer chromatography [26]) and its partition coefficient K in the membrane/sample system can be assessed by Equations (5) and (6) (see [25]).

<u>blood</u>: $\log K = 0.48 + 0.33 \log P_{TLC}$ (DOS/Serum) (5)

<u>urine</u>: $\log K = 1.42 + 0.80 \log P_{TLC}$ (o-NPOE/Water) (6)

The required values for the lipophilicities P_{TLC} of the carriers and the plasticizers mentioned in Tables 2 - 7 are estimated by Equations (2) and (5) or (6), respectively, to guarantee a continuous use life time of at least one month. For this assessment a geometric arrangement of the membrane electrode has been taken as it is typically found in a commercial flow analyzer (Figure 1a). The parameters set for this assessment are $r = 0.5$ mm, $A = 1\ mm^2$, $V = 4\ \mu l$, $u = 18\ ml\,h^{-1}$, $\ell = 1$ mm, $D_s = 3\cdot10^{-6}\ cm^2\,s^{-1}$ [25], $D_m = 2\cdot10^{-9}\ cm^2\,s^{-1}$ [25], $q = 1000$ for carriers and $q = 0.28$ for plasticizers. The values obtained in this way represent a worst case situation since the assumption of a continuous use is only fulfilled in few applications like an in situ monitoring in intensive care units. In conventional commerical analyzers the stream of samples is usually interrupted after each sample by calibration and/or washing solution which on one hand do not require high values for P_{TLC}. On the other hand they allow to recover an eventual depletion of components in the vicinity of the membrane surface by a resupply from the interiour of the membrane. The life time will obviously be extended in such cases.

CURRENTLY AVAILABLE ION-SELECTIVE ELECTRODES FOR CLINICAL APPLICATIONS

Tables 2 - 7 allow comparisons between the requirements as outlined above and the characteristics of currently available membrane systems. Since the concentration ranges for urine are only available as a range of extreme concentrations the calculation of the required selectivities for Na^+- and K^+-sensors using Equation (1) yields stringent demands. Equivalent calculations for H^+- and Li^+-electrodes additionally seem to be obsolete since H^+- and Li^+-activities in urine hardly are of physiological interest. Furthermore, the tabulated figures for calcium and magnesium represent only total concentrations whereas the ratio of their ionized to bound forms is unknown. All experimental stability values reported in Tables 2 - 7 are standard deviations of data collected over a period of 5 h and thus comprising drift as well as residual standard deviation. All of the experimental values for $\log P_{TLC}$ given in Tables 2 - 7 have been determined in our laboratory using a recently introduced, very convenient method [26].

H^+-selective electrodes. As shown in Table 2 pH glass membrane electrodes have an outstanding electromotive behaviour although their mounting into a miniaturized electrode arrangement is technically not simple. In addition glass membrane electrodes are more likely to be fouled by biological samples [29] than solvent polymeric membrane electrodes. Satisfactory electrodes of the latter type have been realized with carrier 1 (Figure 2) as H^+-selective component.

Li^+-selective electrodes. Due to the rather low physiological Li^+-activities (Table 1) the selectivities of the electrodes described to date are insufficient for direct Li^+-measurements. In the widely applied therapy of maniacal depressive psychosis, the Li^+-activity is elevated to the millimolar range. Therefore the required selectivities given in Table 3 are calculated for a therapeutic Li^+ concentration level of 1 mM. The two best Li^+ sensors reported so far clearly fail by their relatively low Li^+/Na^+-selectivity.

Na^+-selective electrodes. Na^+-selective glass membrane electrodes exhibiting high selectivities (Table 4) have found wide spread use in clinical analyzers. However they suffer from the same drawbacks as mentioned above for H^+-selective glass membrane electrodes. The feasibility of a selective Na^+-determination in blood as well as in urine samples is determined mainly by the Na^+/K^+-selectivity. Electrodes based on carrier 4 and 6 marginally satisfy the required selectivity for a urine application while all listed liquid membrane electrodes match sufficiently an application in blood. However the lipophilicities of carriers 5 and 6 are far beyond the value for a comforting long-term application in blood.

K^+-selective electrodes. Table 5 clearly demonstrates that valinomycin based silicone rubber as well as PVC membrane electrodes exhibit a so far unmatched superior performance in any respect considered. In spite of it, many efforts have been invested to replace the natural carrier-antibiotic by synthetic carriers. Electrodes based on the synthetic bis-(crown)ethers 8 - 10 approximate the K^+/Na^+-selectivity of membrane electrodes with valinomycin. Unfortunately their lipophilicities as known so far are poor. However a more lipophilic derivative of 10 showing a comparable performance has been realized very recently [39]. In addition, the listed P_{TLC}-values underline that evidently more consideration should be given to the lipophilicities of the PVC plasticizers used.

Mg^{2+}-selective electrodes. Up to now no membrane electrode for extracellular Mg^{2+}-activity measurement has been developed with great success. Among the few electrodes described the sensor characterized in Table 6 approaches the required selectivities best. Anyhow, according to Table 6 this sensor still is far from a suitable clinical application.

Ca^{2+}-selective electrodes. All the electrodes presented in Table 7 satisfy the selectivity requirements. The outstanding stability and life time characteristics of the membrane electrode based on carrier 12 could be achieved only by accepting a drastically reduced but still sufficient selectivity especially due to the use of a non-polar plasticizer. Using the same carrier in membranes with the polar but slimly lipophilic plasticizer o-NPOE the Ca^{2+}/Na^{+}- and Ca^{2+}/K^{+}-selectivities are nearly by a factor of 1000 enhanced [3] and well above the corresponding values of membranes with 13 or 14.

Cl^{-}-selective electrodes. Presently Cl^{-}-selective solvent polymeric membranes used for determination in blood are exclusively based on classical ion-exchangers. Independent on their membrane composition and on the structure of the ion-exchanger salt such electrodes exhibit typically a Cl^{-}/HCO_3^{-}-selectivity of barely 10 [43]. An estimation of the required selectivity using Equation 1 and Table 1 would lead to a value of 25. Since lipophilic anions are unavoidably preferred over Cl^{-} by such ion-exchanger electrodes careful considerations have to be given to the interference by SCN^{-}, NO_3^{-} and Br^{-}. For example the required $\log K^{Pot}_{ClSCN}$-value amounts to 1.7 which is hardly achieved by currently available solvent polymeric membrane electrodes [44].

HCO_3^{-}-selective electrodes. No HCO_3^{-}-selective carrier molecule has been described so far which would be applicable as a component in a solvent polymeric membrane electrode of clinical relevance. On the basis of classical ion-exchanger sites there have been claims for both HCO_3^{-}- [45]

and CO_3^{2-}-selective [46] liquid membrane electrodes. So far the two similar membrane systems have found only rarely a clinical application. Another type of HCO_3^--selective electrodes on the basis of gas-permeable, H^+-selective polymeric membranes have been proposed [47,48]. In employing electrodes of this kind attention has to be given to interferences of neutral, acidic components in the samples such as acetic acid and lactic acid rather than the inorganic anions listed in Table 1 [49].

ACKNOWLEDGEMENT

This work was partly supported by the Swiss National Science Foundation.

REFERENCES

[1] P.C. Meier, D. Ammann, W.E. Morf, and W. Simon, in Medical and Biological Applications of Electrochemical Devices, J. Koryta (Ed.), John Wiley & Sons, Chichester, New York, Brisbane, Toronto, 1980, p. 13.

[2] J.G. Schindler and M.M. Schindler, Bioelektrochemische Membranelektroden, de Gruyter, Berlin, New York, 1983.

[3] D. Ammann, W.E. Morf, P. Anker, P.C. Meier, E. Pretsch, and W. Simon, Ion-Sel. El. Rev. 5, 3 (1983).

[4] K. Kimura, A. Ishikawa, H. Tamura, and T. Shono, J. Chem. Soc. Perkin Trans. II, 1984, 447.

[5] L. Toeke, B. Agai, I. Bitter, E. Pungor, K. Toth, E. Lindner, M. Horvath, and J. Havas, PCT Int. Appl. WO 8300149, 1983.

[6] I.J. Borowitz, J.D. Readio, and V.S. Li, Tetrahedron 40, 1009 (1984).

[7] J. Petranek and O. Ryba, Coll. Czech. Chem. Commun. 48, 1944 (1983).

[8] Sh.K. Norov, A.K. Tashmukhamedova, and N.Zh. Saifullina, Zh. Anal. Khim. 37, 222 (1982).

[9] U. Olsher, J. Am. Chem. Soc. 104, 4006 (1982).

[10] M. Yamauchi, A. Jyo, and N. Ishibashi, Anal. Chim. Acta 136, 399 (1982).

[11] K.M. Aalmo and J. Krane, Acta Chem. Scand. A36, 227 (1982).

[12] S. Kamata, M. Higo, T. Kamibeppu, and I. Tanaka, Chem. Letters 1982, 287.

[13] J. Gajowski, B. Rieckemann, and F. Umland, Fresenius Z. Anal. Chem. 309, 343 (1981).

[14] F. Vögtle, E. Weber, and U. Elben, Kontakte (Merck), 1980 (2), 36.

[15] W. Bussmann, J.-M. Lehn, U. Oesch, P. Plumeré, and W. Simon, Helv. Chim. Acta 64, 657 (1981).

[16] A.P. Thoma, A. Viviani-Nauer, K.H. Schellenberg, D. Bedeković, E. Pretsch, V. Prelog, and W. Simon, Helv. Chim. Acta 62, 2303 (1979).

[17] D.J. Cram and J.M. Cram, Science 183, 803 (1974).

[18] J.-M. Lehn, Accounts Chem. Res. 11, 49 (1978).

[19] F. Vögtle, H. Puff, E. Friedrichs, and W.M. Müller, J. Chem. Soc. Chem. Comm. 1982, 1398.

[20] A. Shanzer, C.E. Felder, and S. Lifson, Biopolymers 22, 409 (1983).

[21] Wissenschaftliche Tabellen Geigy, Vol. 2, Ciba-Geigy AG, Basel, 1979.

[22] W. Simon, D. Ammann, M. Oehme, and W.E. Morf, Ann. New York Acad. Sci. 307, 52 (1978).

[23] W.E. Morf, The Principles of Ion-Selective Electrodes and of Membrane Transport, Akadémiai Kiadó, Budapest, 1981; Elsevier, Amsterdam, 1981.

[24] U. Oesch and W. Simon, Anal. Chem. 52, 692 (1980).

[25] U. Oesch, O. Dinten, D. Ammann, and W. Simon, in Recent Advances in the Theory and Applications of Ion Selective Electrodes in Physiology and Medicine, M. Kessler, D.K. Harrison, and J. Höper (Eds.), Springer Verlag, Berlin, 1984, in press.

[26] U. Oesch, O. Dinten, and W. Simon, in preparation.

[27] P. Anker, D. Ammann, and W. Simon, Mikrochim. Acta I, 1983, 237.

[28] R.P. Buck, J.H. Boles, R.D. Porter, and J.A. Margolis, Anal. Chem. 46, 255 (1974).

[29] W. Simon, D. Ammann, P. Anker, U. Oesch, and D.M. Band, Ann. New York Acad. Sci. 428, 279 (1984).

[30] E. Metzger, D. Ammann, E. Pretsch, and W. Simon, Chimia, in preparation.

[31] K. Kimura, S. Kitazawa, and T. Shono, Chem. Letters 1984, 639.

[32] P. Anker, H.-B. Jenny, U. Wuthier, R. Asper, D. Ammann, and W. Simon, Clin. Chem. 29, 1508 (1983).

[33] D. Ammann, P. Anker, E. Metzger, U. Oesch, and W. Simon, in Recent Advances in the Theory and Applications of Ion Selective Electrodes in Physiology and Medicine, M. Kessler, D.K. Harrison, and J. Höper (Eds.), Springer Verlag, Berlin, 1984, in press.

[34] T. Shono, M. Okahara, I. Ikeda, and K. Kimura, J. Electroanal. Chem. 132, 99 (1982).

[35] K. Cammann, Working with Ion-Selective Electrodes, Springer Verlag, Berlin, Heidelberg, New York, 1979.

[36] P. Anker, H.-B. Jenny, U. Wuthier, R. Asper, D. Ammann, and W. Simon, Clin. Chem. 29, 1447 (1983).

[37] K. Kimura, T. Maeda, H. Tamura, and T. Shono, J. Electroanal. Chem. 95, 91 (1979).

[38] K. Kimura, T. Tamura, and T. Shono, J. Chem. Soc. Chem. Comm. 1983, 492.

[39] E. Lindner and E. Pungor, private communication.

[40] F. Behm, K. Brunfeldt, J. Halstrøm, D. Ammann, and W. Simon, Helv. Chim. Acta, in press.

[41] K. Kimura, K. Kumami, S. Kitazawa, and T. Shono, J. Chem. Soc. Chem. Comm. 1984, 442.

[42] J. Růžička, E.H. Hansen, and J.C. Tjell, Anal. Chim. Acta 67, 155 (1973).

[43] C. Fuchs, Ionenselektive Elektroden in der Medizin, Thieme Verlag, Stuttgart, 1976.

[44] K. Hartman, S. Luterotti, H.F. Osswald, M. Oehme, P.C. Meier, D. Ammann, and W. Simon, Mikrochim. Acta 1978 II, 235.

[45] M.W. Wise, United States Patent 3 723 281, 1973.

[46] J.A. Greenberg and M.E. Meyerhoff, Anal. Chim. Acta 141, 57 (1982).

[47] R.L. Coon, N.C.J. Lai, and J.P. Kampine, J. Appl. Physiol. 40, 625 (1976).

[48] R.J.J. Funck, W.E. Morf, P. Schulthess, D. Ammann, and W. Simon, Anal. Chem. 54, 423 (1982).

[49] W.E. Morf, I.A. Mostert, and W. Simon, Anal. Chem., submitted for publication.

Table 1. Physiological ranges of ions in blood and urine

	blood		urine	
	95% normal concentration range [1] [$mmol \cdot L^{-1}$]	emf range [a] [mV]	daily excretion [21] [$mmol \cdot d^{-1}$]	emf range [b] [mV]
H^+	$4.3 \cdot 10^{-5}$ - $5.6 \cdot 10^{-5}$	6.8	$6.3 \cdot 10^{-6}$ - $3.2 \cdot 10^{-2}$	219
Li^+	< 0.01			
Na^+	135 - 150	2.4	80 - 560	48.9
K^+	3.5 - 5.0	9.1	40 - 100	23.2
Mg^{2+}	0.45 - 0.8	7.4	2.5 - 8.3 [c]	15.4
Ca^{2+}	1.0 - 1.2	2.4	2.0 - 8.4 [c]	18.4
OH^-	$2.6 \cdot 10^{-4}$ - $4.1 \cdot 10^{-4}$	11.7		
Cl^-	95 - 110		80 - 270	31.3
Br^-	0.009 - 0.17	75.3	0.037 - 0.107	27.3
HCO_3^-	21.3 - 26.5	5.6		
SCN^-	0.007 - 0.017	22.7	< 0.1	
$H_2PO_4^-/HPO_4^{2-}$	0.3 - 1.0	d	1.6 - 38.4	d
SO_4^{2-}	0.3 - 1.0	15.4	2.5 - 45	37.1
CO_3^{2-}	< 0.1			

[a] calculated according to the corresponding activity range

[b] calculated on the basis of a daily excretion volume of 1 L, although it may vary between 0.5 L and 2 L [21]

[c] excretion of total calcium and magnesium, respectively

[d] the ranges of the individual species depend on the pH

Table 2. H^+-selective electrodes

		required characteristics [a]		characteristics of available electrodes [b]	
		for blood	for urine	Lig 1 (1.0%) DOS (65.6%) KTpClPB (0.6%) PVC (32.8%) [27]	glass-membrane [28]
selectivity					
$\log K_{HJ}^{Pot}$	$J=Na^+$	< -8.5		-10.4	< -13
	$J=K^+$	< -7.0		- 9.8	< -13
	$J=Ca^{2+}$	< -7.7		< -11.1	
stability					
drift [$mV\,h^{-1}$]					
standard deviation [mV]		< 0.35		0.05 [c]	0.06 [c]
reproducibility [mV]					
life time					
$\log P_{TLC}$ [a]	carrier	> 8.4	> 2.3	11.6	-
	plasticizer	> 12.8	> 4.1	10.1	-

[a] for more detailed explanations see text

[b] membrane compositions given in w/w percentage ; DOS: dioctyl sebacate; KTpClPB: potassium tetrakis(p-chlorophenyl)borate; PVC: poly(vinyl chloride)

[c] in aqueous buffer solution

Table 3. Li^+-selective electrodes

		required characteristics [a]		characteristics of available electrodes [b]	
		for blood	for urine	Lig 2 (1.3%) KTpClPB (0.4%) o-NPOE (65.3%) PVC (33.0%) [30]	Lig 3 (1.0%) KTpClPB (0.7%) o-NPOE (70.0%) PVC (28.3%) [31]
selectivity					
$\log K^{Pot}_{LiJ}$	$J=H^+$	< 2.1		1.0	-3.4
	$J=Na^+$	< -4.3		-2.3	-2.2
	$J=K^+$	< -2.8		-2.6	-2.0
	$J=Mg^{2+}$	< -3.5		-4.0	-4.7
	$J=Ca^{2+}$	< -3.6		-2.7	-4.4
stability					
drift [$mV\,h^{-1}$]					
standard deviation [mV]		< 1.9			
reproducibility [mV]					
life time					
$\log P_{TLC}$ [a]	carrier	> 8.4	> 2.3	7.2	
	plasticizer	> 12.8	> 4.1	5.9	5.9

[a] for more detailed explanation see text

[b] membrane composition is given in w/w percentage ; KTpClPB: potassium tetrakis(p-chlorophenyl)borate; o-NPOE: o-nitrophenyl octyl ether; PVC: poly(vinyl chloride)

Table 4. Na^+-selective electrodes

		required characteristics [a]		characteristics of available electrodes [b]			
		for blood	for urine	Lig 5 (1.1%) BBPA (66.1%) PVC (32.8%) [32]	Lig 4 (1.2%) ETH 469 (66.3%) PVC (32.5%) [33]	Lig 6 (6.5%) o-NPOE (66.7%) PVC (26.8%) [34]	glass-membrane NAS_{11-18} [35]
selectivity							
$\log K^{Pot}_{NaJ}$	$J=H^+$	< 4.4	< 1.4	0.7	0.1	c	3.0
	$J=Li^+$	< 0.3		-1.7		-3.0	-3.0
	$J=K^+$	< -0.6	< -2.1	-0.4	-1.5	-2.0	-3.0
	$J=Mg^{2+}$	< -1.2	< -0.6	-3.3	-3.2	< -3.7	
	$J=Ca^{2+}$	< -1.3	< -0.6	-3.1	-1.8	< -3.7	
stability [d]							
drift [$mV\,h^{-1}$]					0.03		
standard deviation [mV]		< 0.12			0.12		
reproducibility [mV]					0.23		
life time							
$\log P_{TLC}$ [a]	carrier	> 8.4	> 2.3	4.6	7.8	5.6	-
	plasticizer	> 12.8	> 4.1	9.3	10.8	5.9	-

[a] for more details see text

[b] membrane compositions given in w/w percentage; BBPA: bis(1-butylpentyl)adipate; ETH 469: 1,10-bis(4'-(5"-nonyloxy-carbonyl)-butyryloxy)-decane; for further abbreviations see footnotes in Table 3

[c] no interference in solutions of $1\text{-}10^{-6}$ M NaCl in the pH range 3-9

[d] in aqueous solutions

Table 5. K^+-selective electrodes

		required characteristics [a]		characteristics of available electrodes [b]				
		for blood	for urine	Lig 7 (2.5%) SR (97.5%) [36]	Lig 7 (1.3%) ETH 469 (68.3%) PVC (30.4%) [33]	Lig 9 (2.8%) DBP (69.4%) PVC (27.8%) [37]	Lig 8 (3.8%) o-NPOE (64.2%) PVC (32.0%) [38]	Lig 10 (5%) o-NPOE (60%) PVC (35%) [5]
selectivity								
$\log K^{Pot}_{KJ}$	$J=H^+$	< 2.8	< -0.04	-4.4	-3.4			
	$J=Li^+$	< -1.3		-4.3				
	$J=Na^+$	< -3.6	< -3.1	-4.0	-4.1	-3.2	-3.7	-3.0
	$J=Mg^{2+}$	< -2.8	< -0.9	-4.3	-5.7			
	$J=Ca^{2+}$	< -2.9	< -0.9	-4.2	-5.2			
stability [c]								
drift [$mV\,h^{-1}$]					0.01			
standard deviation [mV]		< 0.46			0.03			
reproducibility [mV]					0.16			
life time								
$\log P_{TLC}$ [a]	carrier	> 8.4	> 2.3	8.6	8.6	2.4		2.5
	plasticizer	> 12.8	> 4.1	-	10.8	4.6	5.9	5.9

[a] for more details see text

[b] membrane composition given in w/w percentage; SR: silicone rubber (85% siloprene K1000, 15% crosslinking agent KA 1); DBP: dibutyl phthalate; for further abbreviations see footnotes in Tables 3 and 4

[c] in aqueous solutions

Table 6. Mg^{2+}-selective electrodes

		required characteristics [a]		characteristics of available electrodes [b]
		for blood	for urine	Lig 11 (1.6%) KTpClPB (0.3%) o-NPOE (63.9%) PVC (34.2%) [40]
selectivity				
$\log K_{MgJ}^{Pot}$	$J=H^+$	< 8.9		0.5
	$J=Li^+$	< -2.3		-2.6
	$J=Na^+$	< -3.9		-2.3
	$J=K^+$	< -0.9		-1.0
	$J=Ca^{2+}$	< -2.4		-0.1
stability				
drift [$mV\,h^{-1}$]				
standard deviation [mV]		< 0.37		
reproducibility [mV]				
life time				
$\log P_{TLC}$ [a]	carrier	> 8.4	> 2.3	3.6
	plasticizer	> 12.8	> 4.1	5.9

[a] for more details see text

[b] membrane composition given in w/w percentage; for abbreviations see footnotes in Table

Table 7. Ca^{2+}-selective electrodes

	required characteristics [a]		characteristics of available electrodes [b]		
	for blood	for urine	Lig 12 (3.3%) KTpClPB (2.1%) ETH 469 (63.7%) PVC (30.9%) [33]	Lig 13 (3.0%) KTpClPB (1.0%) o-NPOE (66.0%) PVC (30.0%) [41]	Lig 14 (6.5%) DOPP (64.9%) PVC (28.6%) [42]
selectivity					
$\log K^{Pot}_{CaJ}$ $J=H^+$	< 9.3		-2.9		4.2
$J=Li^+$	< -1.9			-4.0	-4.2
$J=Na^+$	< -3.6		-3.5	-5.0	-5.2
$J=K^+$	< -0.6		-3.7	-4.5	-5.7
$J=Mg^{2+}$	< -1.9		-4.7	-5.0	-3.6
stability [c]					
drift [mV h^{-1}]			0.01		
standard deviation [mV]	< 0.12		0.03		
reproducibility [mV]			0.13		
life time					
$\log P_{TLC}$ [a] carrier	> 8.4	> 2.3	7.5		
plasticizer	> 12.8	> 4.1	10.8	5.9	

[a] for more details see text

[b] membrane compositions given in w/w percentage; DOPP: dioctyl phenyl phosphonate; for further abbreviations see footnotes in Tables 3 and 4

[c] in aqueous solutions

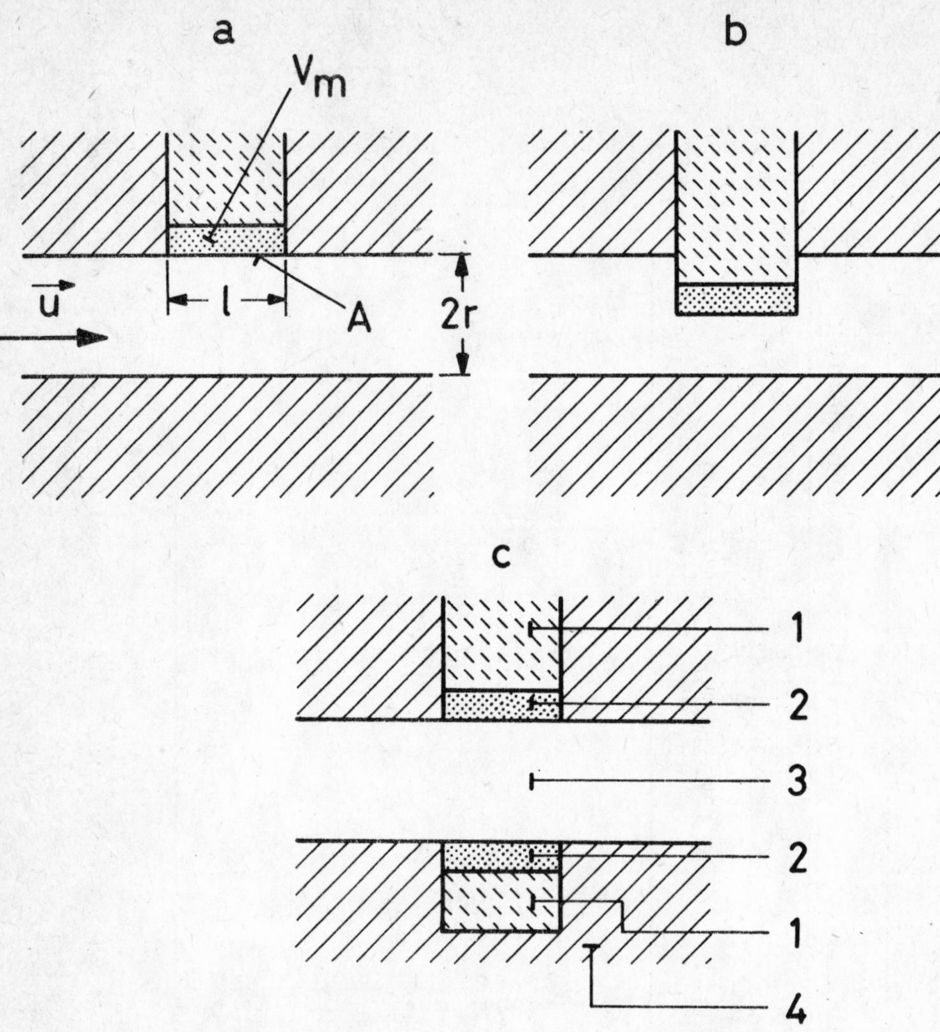

Fig. 1. Typical arrangements for solvent polymeric membrane electrodes in commercial clinical flow analyzers.

a: membrane is moulded into a small opening on the side of the sample channel (e.g. AVL 980, Boehringer ISE 2020, NOVA 6),

b: conventional electrode is inserted into the channel through an opening (e.g. Radiometer KNA1 and ICA1),

c: membrane of a tubular shape as a part of the channel wall (e.g. Radelkis OP-266).

1: internal reference solution, 2: ion-selective membrane, 3: sample channel, 4: channel housing.

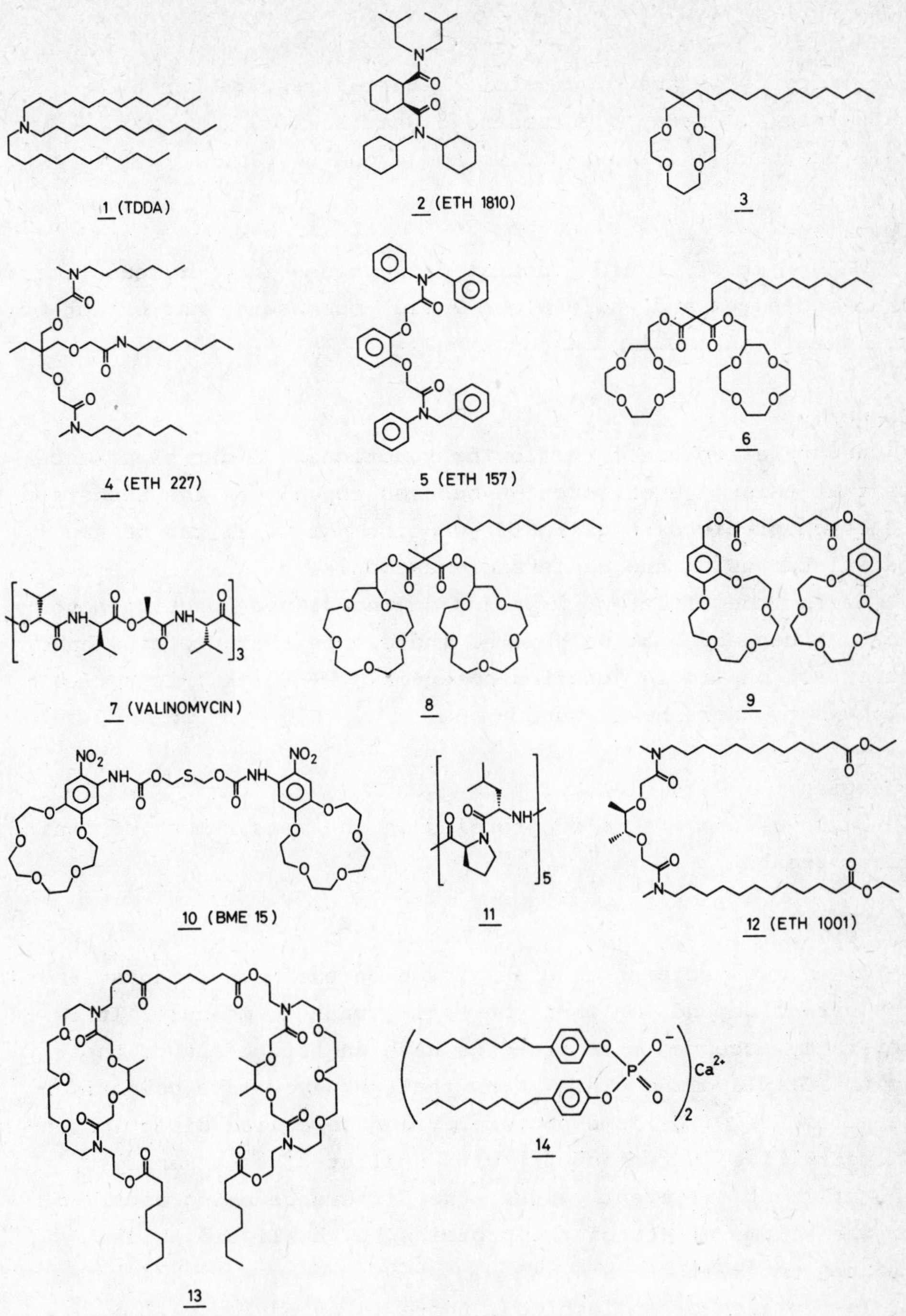

Fig. 2. Constitutions of ion-selective components discussed.

QUESTIONS AND COMMENTS

Participants of the discussion: L.Bartalits, R.A.Durst, G.Horvai, J.Janata, H.Marsoner, Z.Noszticzius, E.Pungor, K.Reichenbach, W.Simon, J.D.R.Thomas and C.C.Young

Question:

What sort of liquid junction did you use in your measurements in blood serum and whole blood ? Did you observe any effect of the sample on the liquid junction?

Answer:

We prefer to use free-flowing junctions and our experience is that this type of junction behaves roughly as the Henderson equation predicts, the liquid junction potential can be nicely calculated using the Henderson equation.

As for the effect of proteins on the liquid junction potential, I can say that we have no indication that proteins have an effect on liquid junction potentials. I think this agrees with what others have found.

Question:

What was the effect of proteins on the measurement of ionic constituents ?

Answer:

Of course proteins have an effect on the measurements, and the effect depends on what you really want to measure. It is generally accepted that proteins have an effect due to their volume displacement. This means that when you make comparison measurements with flame photometry and so called direct potentiometry, i.e. potentiometry with undiluted serum samples, you will get different values, the difference being ascribed to the volume effect of the proteins in the liquid which amounts to 6-7%.

In the case of whole blood, there is an effect called suspension effect. This means that when you measure ionic activity in whole blood, the value you get will differ slightly from what you measure after removing the blood cells by centrifuging. This effect is rather small, it amounts to a hundredth of a pH unit, if you measure pH.

Question:

Have you observed any interference by metabolites ?

Answer:

The interference by metabolites and chemicals is a very difficult issue. We cannot make any predictions, we can, however, make an educated guess. But in every case you have to check it. I draw your attention to the fact that Clinical Chemistry has a volume each year, which is devoted exclusively to interferences by drugs in the determination of parameters which are of clinical relevance.

Question:

What was the effect of the blood constituents on alkali ion determination in blood samples?

Answer:

We have done extensive measurements of sodium in blood serum and whole blood. Sodium has been determined both by flame photometry and direct potentiometry with undiluted samples. With direct potentiometry we found values 5-6% higher than by flame photometry, which is due to the volume displacement of proteins and lipids. We corrected for this effect for 30 samples, and obtained values which were 1% lower than the flame photometric results. I have no explanation for this.

However, some people believe that these low values are due to the fact that sodium is complexed by some components in blood serum, such as bicarbonate, phosphate or proteins.

Comment:

When we decided to use ion-selective instruments, we investigated a number of designs of reference electrodes and found that properly designed open-type liquid junctions, not only free-flowing junctions but also static open liquid junctions with KCl at least 1.5 mol/l do not show significant suspension effect in whole blood measurements. We checked this in the following way: We used fresh whole blood samples, split the sample and made blood serum. Then we measured almost simultaneously the whole blood and blood serum. With a large number of samples we found that with this type of reference electrode there is no significant suspension effect or difference between the measurements in whole blood and blood serum. The difference was 0.5-1.0 mV, which is about equivalent to the 0.01 pH unit mentioned earlier.

Comment:

We have done sodium and potassium measurements with ISEs in whole blood and blood serum, and I agree completely with what has been said about the importance of the design of the liquid junction.

As far as the suspension effect is concerned, I may add that we also studied the difference between whole blood and plasma. We started by measuring in plasma and then resuspended the blood cells, and measured again. We did not observe any significant difference in sodium and potassium results.

The other problem was the importance of the plasma protein volume. This may be studied by making measurements on diluted serum samples.

There may arise problems due to the standard we are using, if it does not match the physiological sample. If we use aqueous standards, the residual liquid junction is not the same.

Comment:

I agree totally. But I can give an example from a different field, colloid chemistry. I have made a lot of measurements on inorganic colloids. And the potential differences you measure are tremendous. And this is true in fields where inorganic colloids may be involved.

Comment:

In soil samples, for example, the suspension effect is tremendous. And there are different views on the explanation of this phenomenon. Some people say that it is essentially a liquid junction potential, others claim that it is a phase boundary potential.

Comment:

I made measurements on some oxides, HgO, MnO_2 and Al_2O_3, and the potential variation on sedimentation was rather small with HgO and much bigger with MnO_2. With Al_2O_3, it was found to depend on the way of preparation.

Question:

I would like to ask a question about the effect of macroglobulins, which may form a complex with Ca, for example. And also, they may influence the electrode surface, thus causing a change in potential. Have you observed this?

Answer:

There are certainly problems with poorly designed ion-selective electrodes. There are some proteins which interfere with some ion-selective electrodes quite dramatically. Heparin is one of these compounds and it causes problems with some chloride selective membranes. If they are in a heparin solution long enough, they become perfect electrodes for heparin. The reason is that the heparin enters the membrane phase. But I can say that with electrodes for Ca^{2+}, K^+, Na^+ and H^+ we have no evidence that there is such a problem involved.

Comment:

We found exactly the same effect by exposing the ion-selective field effect transistor to protein solution. There was an offset and a change in the slope if we recalibrated the electrode after three hours. We attributed this to the very strong binding of the albumin to the electrode.
One most also consider the effect of fatty acids present in

biological samples on the electrodes. They act as detergents and may wash the active components out of the electrode membrane.

We have found that if we coat the electrode with albumin by a brief exposure of about 30 sec, give about 30 sec dip in glutaraldehyde to crosslink the albumin to the surface of the electrode, it perfectly stabilized. Then we took the electrode out of the aqueous solution, put it back into serum many times, without observing any further change.

Comment on the sterilization of the electrodes:

It is possible to sterilize Ca and Na neutral carrier based membranes with ethylene oxide without any adverse effect as long as the membrane remains completely dry. And that is one advantage of ISFETS that you do not need an internal reference solution and therefore the whole structure can be dry.

Question:

You gave an explanation of the leaching effects for the pvc membranes. At the end you showed a slide on the use of the potassium membrane in urine which was based on silicone rubber. What was the purpose of that ?

Answer to the comments:

I fully agree with the statement on the effect of proteins on the membrane surface layer. There are certain ion-selective electrodes which can be contaminated or covered by a surface layer of proteins. I do not know of this effect for neutral carrier based membranes, but I know that this effect exists with classical ion exchanger membranes. E.g. a chloride--selective electrode is really coated by a protein layer. You may peel off the layer, do an amino acid analysis and you will find that the composition of this is roughly the composition of human amino acids. I am glad to learn that you are very successful in sterilizing with ethylene oxide. We have also tried to do this, but found the electrode structure to change slightly - the pvc tubing was deformed because the ethylene oxide treatment was carried out at 60°C.

Answer to the question:

Neutral carrier based ion-selective electrodes often suffer from an anion interference, which is basically due to the capability of the membrane phase to extract electrolytes. If you reduce this extraction property of the membrane, you will reduce the anion interference. Silicone rubber is an excellent medium to reduce this extraction. That is why we use silicone rubber membranes for measurements in urine. We do not know exactly which anion interferences in urine, but we assume that it is an anion.

Comment:

On the earlier comment concerning the cross-linking of the albumin and the effect of surface active agents. I am not sure that this experiment really proves the fact that you are preventing the washing out by the fatty acids of the ionophors. It could still be due to the fact that these charged surfactants are penetrating into the hydrophobic surface of the membrane and therefore changing the phase boundary potential. And the cross-linking of albumin may actually just prevent this kind of penetration.

Comment:

I think that this is simply a denaturation of proteins, which, coming close to a surface with a rather high electrolyte concentration, such as a classical ion exchanger, are simply denaturated, forming a surface layer which can be peeled off.

Comment:

On the other hand, you may have bactericides and fungicides in the solution, that can poison a liquid membrane electrode. And these certainly do not undergo a denaturation at the membrane surface. It has to be a certain exchange process or solubility process. And I still do not know what causes this poisoning effect.

Comment on the activity of ions in clinical solutions:

I think this problem will be solved by the development of appropriate standards. We know that what we are measuring in terms of the pH in blood is not the activity of hydrogen ions because there is definitely a residual liquid junction potential associated with it, but this has not been a problem because everybody has been on the same pH scale. So we can achieve the same situation with the other electrolytes, Na, K and Ca. The problem will be resolved by everyone being on the same activity scale, whether it is thermodynamically accurate or not. And I think our problem right now is that atomic absorption and flame photometry have been the techniques used by clinical chemists and we are trying to come up with a scale that matches the total concentration scales that are measured by these techniques. We do not want to have completely new numbers for the electrolytes, and it is a question of introducing a factor into the measurements with the electrodes to give us values that will be comparable to the results flame photometric and atomic absorption methods.

This approach has the additional advantage that physiological parameters of the patient will be more closely related with these ISE measurements, which produce activity than with the total amounts yielded by atomic absorption or flame photometric methods. I hope in a short time when we come up with appropriate standards these problems will be solved.

Question:

How can one check whether a calcium ion-selective electrode works properly in blood serum ? As long as no standards are available for ionized Ca, one can only check one electrode against another.

Answer:

You are right. And if you read our papers, you see that we never give any reference to a correct value. When we shall have any standard, we shall be able to refer to that. And even if it turns out that the standard is wrong, we can do corrections later.

What we do is: we calibrate our ion-selective electrodes with aqueous solutions having roughly the composition of normal extracellular blood electrolyte - the inorganic electrolytes in blood serum. After calibration we make measurements in blood serum and whole blood and we assume that these measurements can be interpreted using the Nernst, Nicolsky, Henderson etc. equations. We made studies by measuring in NBS $CaCl_2$ solutions with Ca-ISE, then transferred the electrode into an extracellular electrolyte. We computed what the activity was supposed to be using the equations, and the match of the emf we got was just about perfect.

Question:

If someone buys or builds an instrument for Ca measurements in blood, how can he check that he obtains results in agreement with your results, without sending samples to you?

Answer:

He cannot check.

Comment:

People are working in NBS and in other institutes on the ionized Ca problem, wanting to come up with a reference material for ionized Ca. And once this is done, the problem is solved. I think we are only one year from that.

Comment:

The problem is now that some instrument manufacturers calibrate their instruments by making extracellular electrolyte compositions, and they specify a concentration of ionic Ca, for example. And then any change in the emf is interpreted as a change in the concentration of Ca. This is wrong, because if there is a change in ionic strength, this also changes the emf, but the concentration remains unchanged, it is the activity which changes. This maybe is not a major error, but an error which has to be dealt with.

instruments, which when you contact them with aqueous electrolytes, and then with serum, and then again with the aqueous electrolyte, have about 2 mV change in E^o, which is just about the physiological range. And if Ladenson used this type of electrode, then of course he had problems. But now, as I tried to tell you in my lecture, we have electrodes which we can contact with aqueous solutions, then with blood serum, and with aqueous solution again, and we measure roughly the same value.

Question:

For Ca, specifically, is there any membrane composition that would avoid this problem ?

Answer:

You will find it in the written version of my paper.

Question:

to Dr.Marsoner about pH measurements in whole blood and serum. You mentioned a difference of 0.01 pH unit or 0.5-1.0 mV. For sodium, this is about 1/3 of the normal range. Would you comment on that ?

Answer:

I said that it was usually not larger than 0.5 mV.

Comment:

That would mean that if you took a blood sample from a patient, and measured sodium in the whole blood, and then, after centrifuging, in the serum, you would report to the doctor two values for the same patient which differ by 1/3 of the normal range. Normally you do not make both measurements in the same laboratory, but maybe the central laboratory makes the serum measurement and the doctor in the intensive care unit may do the measurement on whole blood.

Question:

Did you use pH buffer in the solution you used for calibration? Which buffer?

Answer:

Yes, we used TRIS

Comment:

I think that people tend to use other buffers, too. Some comment related to that may be helpful.

Comment:

This is a question which has to be decided upon.

Comment:

At the present time the NBS is working on a Ca activity standard. The direction has not been finalized in terms of what is going to be there. Presently we are leaning towards TRIS as a buffer and I found that at levels we need to buffer in the physiological range is not high enough to cause liquid junction problems that you often get with higher levels of TRIS. But we are also considering the possibility of using some of the other types of buffer, which do not cause any liquid junction problem. But I think we are still very early in the stage of developing this, and I think the 1 year mentioned earlier is too optimistic.

Question:

Ladenson wrote a paper in Clinical Chemistry about the disagreement between the calculated values of sodium, calculated on the basis of the protein volume and the liquid volume, and the values measured directly. I think he found more than 1% difference, it was about 4%. He made also assumptions about the possible reasons. Would you please comment on this ?

Answer:

I think there has been some progress in membrane technology. I know about Ca electrodes that have been used in clinical

Comment:

I think the difference is mainly due to the liquid junction potential. With the new sodium and potassium analyzers which are now on the market, you do not see any difference between the sodium result in serum and whole blood.

Question:

You have used a three-electrode system in your measurements, the third electrode being Pt. Would you explain why you used it?

Answer:

It need not be Pt, you can use any contactor, it is simply an electronic trick. You measure the potential between the ISE and the common electrode, and simultaneously between the reference electrode and the common electrode, and a differential amplifier will give you the potential difference between the ISE and the reference electrode. And this serves to reduce certain electronic noises.

Comment:

We have already discussed a number of questions concerning standards, that we were planning to discuss on Friday, during the special session devoted to standardization.

KEYNOTE LECTURES

4th Symposium on Ion-Selective Electrodes
Mátrafüred, 1984

CHEMICALLY MODIFIED ELECTRODE SENSORS

R.A. DURST

Organic Analytical Research Division
Center for Analytical Chemistry
National Bureau of Standards,
Washington, DC 20234 USA

ABSTRACT

Electroanalytical sensors based on amperometric measurements at chemically modified electrodes are in the early stages of development. The modes of modification can take many forms, but the most common approach at the present time is the immobilization of electrocatalyts in polymer films which are applied to bare metal and carbon electrodes.

This review gives a brief summary of the types of chemically modified electrodes, their fabrication, and some examples of their uses. The incorporation of biochemical systems should greatly extend the usefulness of these devices for analytical purposes.

INTRODUCTION

In recent years, considerable effort has gone into the development of a new class of electrochemical devices called chemically modified electrodes. While conventional electrodes are typified by generally nonspecific electrochemical behavior, i.e., they serve primarily as sites for heterogeneous electron transfer, the redox (reduction-oxidation) characteristics of chemically modified electrodes may be tailored to enhance desired redox processes over others. Thus, the chemical modification of an electrode surface can lead to a wide variety of effects including the retardation or acceleration of electrochemical reaction rates, protection of electrodes, electro-optical phenomena, and enhancement of electroanalytical specificity and sensitivity. As a result of the importance of these effects, a relatively new field of research has developed in which the electrochemical behavior of the attached substances is being studied as well as the influence of these modified surfaces on the catalysis and inhibition of a variety of electrochemical processes. The functionalization of an electrode surface may ultimately benefit a diverse array of processes such as electroorganic synthesis, electrocatalysis, semiconductor stabilization, photosensitization, photoelectrochemical energy conversion, electrochromism, and, the topic of this paper, selective chemical analysis.

To a large extent, the discovery and application of adsorption phenomena for the modification of electrode surfaces has been an empirical process with few highly systematic or fundamental studies being employed until recent years. Of more recent origin is the approach whereby modifiers are selected on the basis of known and desired properties and deliberately immobilized on an electrode surface to convert the properties from those of the bare electrode to those of the immobilized substance.

SURFACE IMMOBILIZATION TECHNIQUES

There are three principal approaches used for the immobilization of electroactive substances onto surfaces: chemisorption, covalent bonding, and film deposition.

In chemisorption, the electrochemically reactive material is strongly (and to a large extent irreversibly) adsorbed onto the electrode surface. Lane and Hubbard [1] were among the first to use this approach when they chemisorbed quinone-bearing olefins on platinum electrodes and demonstrated a pronounced effect of the absorbed molecules on electrochemical reactions at the metal surface.

Covalent attachment schemes have been developed for both mono-and multi-molecular layers of electroactive sites on semiconductor, metal oxide, and carbon electrodes. Since its introduction by Murray and co-workers [2], covalent immobilization by organosilanes has become the most widely used technique for preparing modified electrodes. Other covalent linking agents have been used including cyanuric chloride by Kuwana and co-workers [3], thionyl chloride [4], and acetyl chloride [5]. While it is sometimes possible to attach the electroactive substance directly to the electrode surface, covalent linking agents are much more commonly used because they allow the use of a greater variety of terminal functional groups than could be produced on the electrode surface alone. Covalent attachment is also usually more tolerant of exposure to different types of solvents.

Film deposition refers to the preparation of polymer (organic, organometallic, and metal coordination) films which contain the equivalent of many monomolecular layers of electroactive sites. As many as 10^5 monolayer-equivalents may be present [6]. The polymer film is held on the electrode surface by a combination of chemisorptive and solubility effects. Since the polymer film bonding is rather non-specific, this approach can be used to modify almost any type of electrode material. Polymer films have been applied to electrode surfaces by dip and spin coating, bonding via covalent linking agents, electrochemical precipitation and polymerization, adsorption from solution, and plasma discharge polymerization [7]. There are several reasons for the appeal of polymer modification: immobilization is technically easier than working with monolayers; the films are more stable; and because of the multiple layers of redox sites, the electrochemical responses are larger. Questions remain, however, as to how the electrochemical reactions of multimolecular layers of electroactive sites in a polymer matrix occur, e.g., mass transport and electron transfer processes by which the multilayers exchange electrons with the electrode and with reactive molecules in the contacting solution [6].

Electrocatalysis at a modified electrode is usually an electron transfer reaction, mediated by an immobilized redox couple, between the electrode and some solution substrate which proceeds at a lower overpotential than would otherwise occur at the bare electrode. This type of mediated electrocatalysis process can be represented by the scheme:

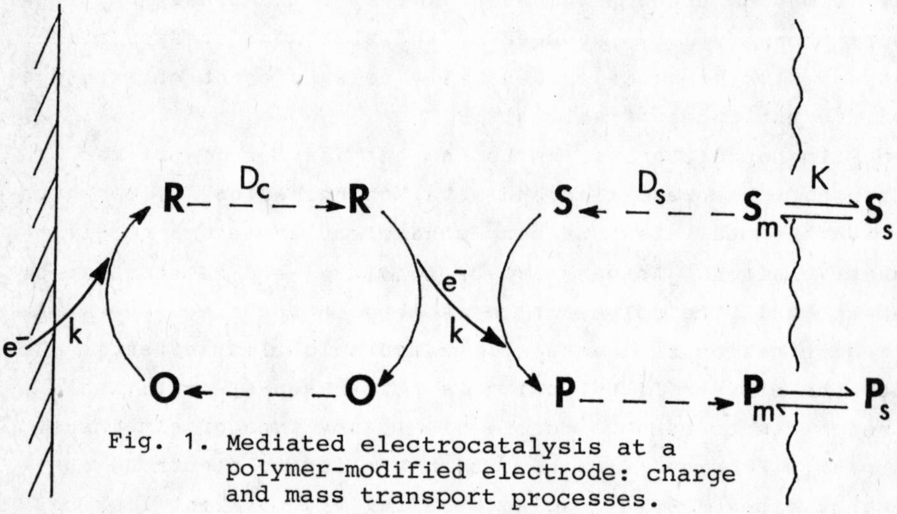

Fig. 1. Mediated electrocatalysis at a polymer-modified electrode: charge and mass transport processes.

In this scheme, the substrate, S, which is irreversibly reduced (or electroinactive) at the bare electrode is transported across the polymer film-solution interface (partition coefficient, K) and diffuses into the polymer film (diffusion coefficient, D_s). The electrocatalyst or mediator, R/O, undergoes rapid heterogeneous electron transfer and charge is transported through the polymer film at a rate given by the diffusion coefficient, D_c. In the bulk of the film, the substrate and mediator undergo a homogeneous electron transfer regenerating the oxidized form of the mediator and forming the reaction product, P. If the rate of electron transfer between the electrode and mediator is faster than the rate of the mediator-substrate reaction, then the substrate will undergo electrolysis at a potential near the surface formal potential of the mediator couple. The catalytic redox potential also depends on the rate constant for the mediator-substrate reaction. This type of electrocatalytic scheme has been the focus of numerous modified electrode studies.

This approach can be further extended to photoelectrochemical reactions at modified semiconductor electrodes. In such cases the immobilized substance(s) may serve several functions: mediation of the redox process, photosensitization of the semiconductor, and photocorrosion protection.

TECHNIQUES FOR MODIFIED ELECTRODE STUDY

In order to characterize the properties of molecules and polymer films attached to an electrode surface, a wide variety of methods have been used to measure the electroactivity, chemical reactivity, and surface structure of the electrode-immobilized materials. These methods have been primarily electrochemical and spectral as indicated in Table 1. Suffice it to say that a multidisciplinary approach is needed to adequately characterize chemically modified electrodes, combining electrochemical methods with surface analysis techniques and a variety of other chemical and physical approaches.

POLYMER-FILM ELECTRODES

Chemical modification of electrode surfaces by polymer films offers the advantages of inherent chemical and physical stability, incorporation of large numbers of electroactive sites, and relatively facile electron transport across the film. Since the polymer films usually contain the equivalent of one to more than 10,000 monolayers of electroactive sites, the resulting electrochemical responses are generally larger and thus more easily observed than those of immobilized monomolecular layers. Also, the concentration of sites in the film can be as high as 5 mol/L and may influence the reactivity of the sites because their solvent and ionic environments differ considerably from dilute homogeneous solutions [6].

Table 1. Techniques for Modified Electrode Study

Electrochemical Methods

- cyclic voltammetry
- differential pulse voltammetry
- chronamperometry and chronocoulometry
- chronopotentiometry
- coulostatics
- AC voltammetry
- rotated disk and ring-disk voltammetry

Spectral Methods

- x-ray photoelectron and Auger electron spectroscopy
- transmission spectroscopy at optically transparent electrodes
- reflectance spectroscopy
- photoacoustic and photothermal spectroscopy
- Raman spectroscopy
- inelastic electron tunneling spectroscopy
- secondary ion mass spectrometry
- second harmonic generation
- ellipsometry

Other Methods

- film thickness and roughness
- scanning electron microscopy
- elemental analysis
- contact angles
- radiolabels

In most polymer films, the electroactivity of the redox centers depends on the ionic conductivity of the film. This is usually achieved either by the penetration of supporting electrolyte ions through pores in the film or by the presence in the film of numerous fixed charge sites plus mobile counterions. In many cases, solvent permeation into the polymer facilitates ionic penetration and mobility. If the polymer film possesses electronic, rather than ionic, conductivity, then the electron-transfer reactions will most likely occur at the polymer/solution interface and the advantage of a three-dimensional reaction zone will be reduced.

Most of the redox centers in a polymer film cannot rapidly come into direct contact with the electrode surface. The widely accepted mechanism proposed for electron transport is one in which the electroactive sites become oxidized or reduced by a succession of electron-transfer self-exchange reactions between neighboring redox sites [8]. However, control of the overall rate is a more complex problem. To maintain electroneutrality within the film, a flow of counterions and associated solvent is necessary during electron transport. There is also motion of the polymer chains and the attached redox centers which provides an additional diffusive process for transport. The rate-determining step in the electron site-site hopping is still in question and is likely to be different in different materials.

ANALYTICAL APPLICATIONS

The relative fragility and preparative difficulty associated with monolayer-modified electrode surfaces hampered significant analytical progress for some time, and it was not until polymer-film electrodes were developed that the utility of modified electrodes in analysis could be demonstrated.

One approach utilized the polymer-modified surface as a preconcentrating surface in which the analyte or some reaction product is collected and concentrated by reactive groups attached to the electrode. The preconcentrated analyte is then measured electrochemically. Ideally, the collection process will be selective for the analyte species of interest. If not, the analyte must be electroanalytically discriminated from other collected species. The capacity of the polymer film on the modified electrode should be sufficient to prevent saturation by the analyte, and the electrochemical response during measurement should provide good sensitivity to the collected analyte [6].

Electrostatic binding [9] may provide another very useful approach to preconcentration analysis. Enhancement of the redox ion concentration in the ion-exchange polymer volume should permit very sensitive analysis when combined with an appropriate electroanalytical method [10,11]. However, the sensitivity of the ion-exchange equilibrium to the sample solution electrolyte composition and concentration and the necessity of having a multiply charged analyte ion may limit the usefulness of the electrostatic binding approach.

Many biological redox systems undergo very slow heterogeneous electron transfer at electrodes and consequently exhibit quasi-reversible and irreversible electrochemical behavior. One approach to circumvent this problem is to add to the solution an electroactive species, called a mediator, which acts as an electron shuttle to provide redox coupling between the electrode and the redox center in the biological compound [12]. In principle, the enhancement of the electrochemical behavior of the biocomponent can be accomplished by attachment of the mediator directly to the electrode surface. This approach has been demonstrated, for example, by Wrighton and co-workers for the electrocatalysis of cytochrome c at platinum electrodes modified with various organosilane polymers [13,14]. Schemes using immobilized-mediator electrodes will probably be extended to a variety of previously intractable biological redox reactions, both for analytical purposes and for fundamental studies of the biological couples.

The future holds considerable promise for the attachment of various types of biocomponents to provide new approaches for both amperometric and potentiometric biosensors. Systems based on enzyme/substrate and immunological reactions are the most promising based on their high specificity and possibilities for chemical amplification of the signal. One example of this approach (illustrated in figure 2) is the direct attachment of vesicles or liposomes (closed bilayer structures formed from synthetic surfactants or phospholipids) containing high concentrations of electroactive marker compounds. The membrane is sensitized by the incorporation, in this example, of an antibody, Y. The antigen, A, reacts with the antibody forming a complex, AY, which activates complement, C. The activated complement, $\bar{C}$, lyses the liposome by a series of enzymatic reactions thereby releasing the markers which undergo redox reaction at the electrode. This scheme could be further

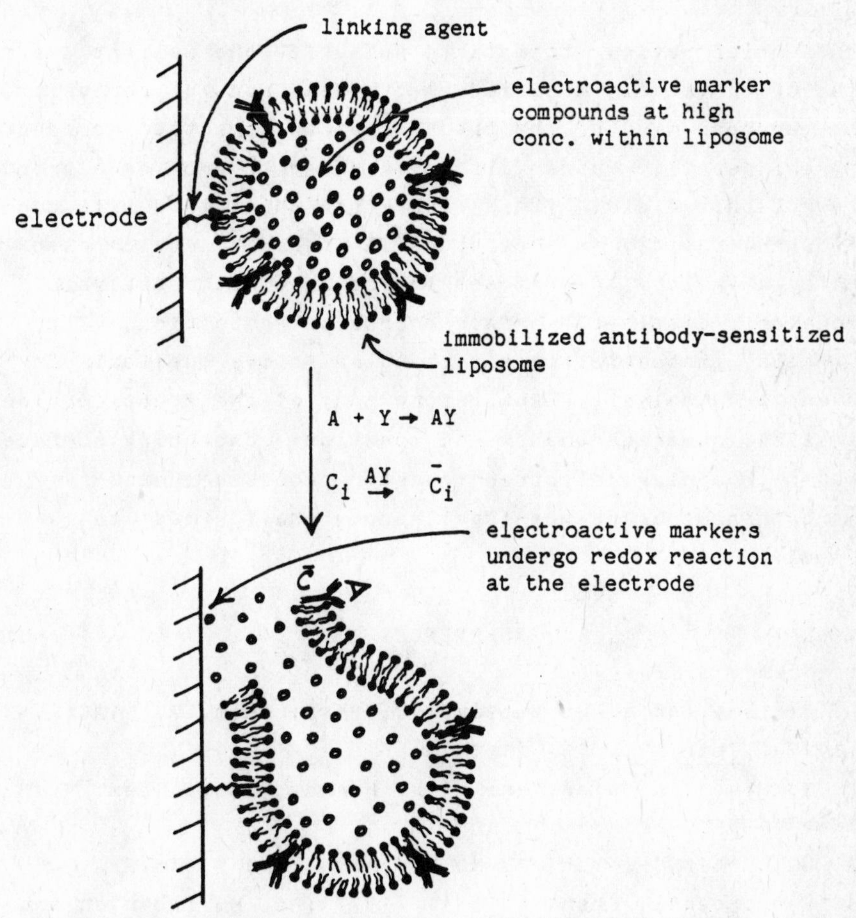

Fig. 2. Immune lysis of a sensitized liposome immobilized on an electrode surface.

extended in sensitivity by incorporating in the vesicles enzymes which, when released, enzymatically produce even larger quantities of electroactive products by reaction with some substrate in the external solution. The possibilities of these approaches, considering the large numbers of potentially useful biochemical reactions, are enormous. It is expected that significant advances will occur in the field of biosensor development in the near future, especially when newer biotechnological approaches are adapted to sensor development.

CONCLUSION

This brief review attempts to summarize the salient features of chemically modified electrodes, and, of necessity, does not address many of the theoretical and practical concepts in any real detail. Rather, it is intended to provide a broad overview of this field. It is clear that this field will continue to grow rapidly in the future in order to provide electrodes for a variety of purposes including electrocatalysis, electrochromic displays, surface corrosion protection, electrosynthesis, photosensitization, and selective chemical concentration and analysis. But before many of these applications are realized, numerous unanswered questions concerning surface orientation, bonding, electron-transfer processes, mass-transport phenomena and non-ideal redox behavior must be addressed.

REFERENCES

1. R. F. Lane and A. J. Hubbard, J. Phys. Chem. 77, 1401 (1973); ibid. 77, 144 (1973).
2. P. R. Moses, L. Wier, and R. W. Murray, Anal. Chem. 47, 1882 (1975).
3. A. W. C. Lin, P. Yeh, A. M. Yacynych, and T. Kuwana, J. Electroanal. Chem. 84, 411 (1977); A. M. Yacynych and T. Kuwana, Anal. Chem. 50, 640 (1978); M. J. Dautartas, J. F. Evans, and T. Kuwana, ibid. 51, 104 (1979).

4. B. J. Watkins, J. R. Behling, E. Kariv, and L. L. Miller, J. Am. Chem. Soc. 97, 3549 (1975).
5. R. D. Rocklin and R. W. Murray, J. Electroanal. Chem. 100, 271 (1979).
6. R. W. Murray, Chemically Modified Electrodes, in Electroanalytical Chemistry, A. J. Bard, ed., Vol. 13, Marcel Dekker, NY, 1983.
7. R. W. Murray, Acc. Chem. Res. 13, 135 (1980).
8. F. B. Kaufman, A. H. Schroeder, E. M. Engler, S. R. Kramer, and J. Q. Chambers, J. Am. Chem. Soc. 102, 483 (1980).
9. N. Oyama and F. C. Anson, J. Electrochem. Soc. 127, 247 (1980).
10. M. N. Szentirmay and C. R. Martin, Anal. Chem. 56, 1898.
11. G. Nagy, G. Gerhardt, A. Oke, R. N. Adams, R. B. Moore, M. N. Szentirmay, and C. R. Martin, J. Electroanal. Chem. (in press.)
12. M. L. Fultz and R. A. Durst, Anal. Chim. Acta 140, 1 (1982).
13. N. S. Lewis and M. S. Wrighton, Science 211, 944 (1981).
14. S. Chao, J. L. Robbins, and M. S. Wrighton, J. Am. Chem. Soc. 105, 181 (1983).

QUESTIONS AND COMMENTS

Participants of the discussion: R.P.Buck, R.A.Durst, M.Jänchen, J.D.R.Thomas and L.Tomcsányi

Question:

I can imagine that chemically modified electrodes can be very well used in practice. However, I wonder what the situation is if someone wants to measure charge transfer coefficients or other parameters of electrochemical processes. Organic molecules usually produce a layer on a carbon electrode, and this may be eliminated by using chemically modified electrodes. But if we modify the electrode, we have another charge transfer in the layer and another chemical reaction also. How can we evaluate the result ?

Answer:

This is one of the areas that is receiving a lot of attention right now. People in different laboratories are trying to understand the kinetics of charge transport processes and what happens under varying conditions of permeability of the substrate into the modified electrode, and the transport of the charge from the electrode to the surface of the electrode. Theories are being developed to describe the phenomena. A series of models have been devised to describe many of the possibilities that can occur. It turned out that the fastest process occuring is the charge transport from the electrode to the outer surface, and substrate transport actually does not occur, everything occurs at the surface of the polymer film. The curious thing is that it is better to have a thick or relatively thick multilayer polymer film on the electrode than to have a monolayer or a few multilayers. This is difficult to understand because if the reaction is all occuring on the outer surface of the membrane, why a thick layer should be better than a monolayer. This has not been explained yet. But this field is relatively new, and a lot of work needs to be done in devising these models. This was a very good question, I wish we had better answers.

Question:

It is very difficult to prepare an electrode with a completely clean surface. Iron electrodes have oxide, carbon electrodes carbon oxide at the surface. Should these be considered as chemically modified electrodes ? You mentioned that you considered electrodes on the surface of which specific adsorption had taken place, also chemically modified electrodes. This definition, I think, is not exact.

Answer:

I used to begin my talks by pointing out that there is no such thing as a bare metal electrode, that everything we use in electrochemistry, in one way or another, is a modified electrode. Usually there would be an oxide film or adsorption of some material in the solution. But what is meant by the present definition of chemically modified electrode is where you are putting something onto the electrode rather than just having it happen by some unknown mechanism in the system you are studying. You are applying a polymer film onto the electrode surface or incorporating in some way something that will change the electrochemical behaviour of your naked electrode.

Question:

It would then be an artificial modification of the electrode with a special aim.

Answer:

Yes. Initially, I think, a lot of people made accidental discoveries of materials that would have certain types of electroactivity. Now we are beginning to do it in a more scientific and systematic way. We know electrocatalytic reactions that occur in solution. And it turns out that if you take the same electrocatalytic materials and attach them to the electrode surface, it does not change their electrocatalytic behaviour. It is the same with an enzyme, if you immobilize an enzyme, usually it still behaves in the same way.

Question relating to enzyme electrodes:

Would you consider a system where you bind a reductase through a polymerization mediator onto the surface of an electrode, also a modified electrode ?

Answer:

I think, any time you do attach these to an electrode, it would be considered chemical modification. What people are working more toward now is actually putting oxy-reductase type enzymes onto the electrode where the electron transfer occurs right through the enzyme and electrode. The enzyme acts as a cofactor in the enzyme reaction, rather than having some chemical cofactor. And this is the ideal situation where you do not need other extraneous molecule in the solution. But I think even in the case where you go through other processes, where you are looking at products, these seem to be considered chemically modified electrodes.

Comment:

I think it is a very bad precedent you said that interposed chemical reactions on ion-selective electrodes are already called sensitized ISEs. But there the product is always ionic and you sense an ion. I think people working on chemically modified electrodes are always dealing with electron transfer reactions. Some of them are working on ion-exchange membranes, but they never put simple ions on ion-exchange membranes, they always put redox ions there, because the emphasis is on catalyzing electron transfer and not on catalyzing production of ions. I still think IUPAC will keep sensitized chemical reaction electrodes separate, as a separate idea from the chemically modified electrodes.

Comment:

I was given the job in the Electroanalytical Commission of IUPAC to come up with a document on chemically modified electrodes, and take your recommendation into consideration. But I should add that things are confused in other cases, too. We have sensors that are considered ISEs which are actually amperometric devices, all the oxygen electrode-based systems. I do not see how this is going to be resolved.

4th Symposium on Ion-Selective Electrodes
Mátrafüred, 1984

CHEMICALLY SENSITIVE FIELD EFFECT TRANSISTORS: THEIR APPLICATION IN ANALYTICAL AND PHYSICAL CHEMISTRY

JIŘÍ JANATA

Department of Bioengineering, University of Utah, Salt Lake City, Utah 84112 USA

INTRODUCTION

With ever increasing capacity to process information one has to ask a question: "Where is the new data going to come from?" There are two possible sources: modern mathematical and computational techniques will allow us to extract more information from the existing sensors (1). The second source will be new types of sensors such as solid state chemical devices of which chemically sensitive field effect transistors (CHEMFET) represent a small fraction (2).

There are two discrete steps in sensing: recognition and amplification. The first one provides the necessary selectivity but lacks the power; in other words, it is a soft signal source which must be coupled to an amplifier in order to obtain the information about the chemical composition of the sample. In conventional potentiometric arrangement this amplification is usually accomplished by means of a high input impedance electrometer (a pH meter). In CHEMFET the solid state part of the device is coupled directly to the chemically sensitive part by the electric field in the region of the transistor called a gate. Thus, in designing a new CHEMFET, one has to introduce the chemistry into the gate proper.

There are certain advantages which are inherent in solid state chemical sensors. They are invariably related to the mode of their preparation i.e., solid state fabrication (3). This technology has been developed and perfected over a period of the last thirty years to the point of truly incredible manufacturing precision and accuracy. The key position in this process belongs to photolithography which allows structural definition on micron scale. With electron beam and X-ray lithography, the structural features on the order of a few hundred angstroms are possible. Such dimensions do not seem to be necessary for preparation of even more sophisticated solid state sensors at present. However, the "routine" micron structures allow construction of microsensors which can be arranged in multichannel arrays suitable for chemometrics data processing (4) with overall power requirements of less than one milliwatt and the overall weight a few grams. Such parameters are important, for example, in biomedical and space applications. These applications fall broadly into the category of analytical measurements.

The direct coupling of chemistry and electric fields in the gate opens new classes of sensors. Thus, changes in electron work function can be analytically utilized (5-7) and the work function of various materials can be directly measured. Investigation of dielectric behavior of polymers (dielectric loss spectrometry) has been performed (8) and used in materials research. Direct measurement of charge at the solution/solid interface can be done (9), however, so far without an analytical application. Corrosion of various materials, including insulators, and measurement of exchange current densities can be studied by analysis of the electrical fluctuations induced by these processes (10-12).

There are, however, limitations and outstanding problems which are again related to the solid state fabrication technology. Integrated circuits based on silicon and related materials have been designed for a specific task: information processing. They can be, and always are, rigorously protected from the environment in order to insure their long lifetime and optimum performance. On the other hand, solid state chemical sensors based on the same family of silicon materials must be exposed, by definition, to the environment which they monitor. The single most adverse species is water which is almost totally incompatible with the proper functioning of silicon-based electronics. Ions which represent mobile charges also pose serious problems although less than the water itself. Mainly for this reason the solid state sensors have not yet found a commercial application and their use and development has been limited to few research laboratories.

The discussion of the issues which have been touched upon in this introduction is beyond the scope of this lecture and some selectivity will have to be applied. The remainder of the lecture will be divided into two sections: CHEMFET Operation in which selected examples of the device physics and chemistry will be discussed and Device Preparation, in which our attempt at solving the packaging problem will be outlined.

CHEMFET OPERATION

Ion Sensitive Field Effect Transistor (ISFET)

Historically, the first CHEMFET described in the literature was pH sensitive FET (13,14) which used the gate insulator as a pH sensitive layer. Later, (15) other ion selective membranes have been added as shown in Figure 1. The relationship of this sensor to ion selective electrode (ISE) or coated wire electrode (CWE) is simple: they are all the same and the only difference is the thickness (or length) of the internal conductor. This argument has been explored in detail elsewhere (16) and is reiterated briefly in Figure 2. A potentiometric sensor can be represented by an equivalent electrical circuit in which each boundary (interface) between adjacent layers is represented as an impedance (a parallel combination of charge transfer resistance and a double layer capacitance). Traditionally, a good ISE or a good CWE would be designed in such a way that the charge transfer resistance at each interface be as low as possible (i.e. Nernstian or non-polarized). In such a case the existence of the double layer capacitance is unimportant (for practical purposes) because the whole equivalent circuit reduces to a voltage divider. On the other hand, if any of the interfaces have a high charge transfer resistance (polarized or Non-Nernstian), then the value of the corresponding capacitance becomes significant. If this capacitance is furthermore valuable, the whole sensor exhibits unstable behavior. In an ISFET where capacitances are small and constant (by design) the issue of the magnitude of the charge transfer resistance is almost unimportant.

Let me briefly discuss the issue of potentiometric immunochemical sensors. The schematic diagram of this type of sensor is shown in Figure 3 and its equivalent electrical circuit is shown in Figure 4. The basic principle of this device is simple: the excess charge of the solution/membrane interface is dominated by the antigen-antibody reaction. This charge is measured directly with the FET in the capacitive arrangement shown in Figures 4. Unfortunately, in all materials which we have tested, the membrane/solution interface exhibited a non ideal behavior in which the double-layer charge on capacitor C_{dl} was shorted out by one or more charge transfer resistances R_{ct} (Figure 4) corresponding to the the movement of small inorganic ions across this interface. Although the idea of IMFET seems to be correct, in principle, its implementation may not be possible because of the lack of the electrochemically suitable membrane material. A detail analysis of this problem can be found in the original paper (17).

Enzymatically Coupled Transistor (ENFET)

The only true solid state electrochemical biosensor, seems to be enzymatically coupled transistors (ENFET), at present (18-20). The enzymatic diagram of this sensor is shown in Figure 5. It is composed of two parts: the potentiometric solid state sensor and an enzyme containing layer in which an enzymatically coupled reaction of the type

$$S + E \underset{k_{-1}}{\overset{k_1}{\rightleftarrows}} SE \xrightarrow{k_2} P + E$$

takes place. If there is an ion I related to either the substrate S or the product P, for example:

$$SI \text{ (or PI)} \rightleftarrows S \text{ (or P)} + I$$

then the ISFET can measure the surface concentration of this ion and produce an electrical signal which is related to the concentration of substrate S. Since both hydrodynamic (equilibrium) and kinetic (non-equilibrium) processes are involved in the generation of the response. It is meaningless to assign any specific value to the slope. The numerical analysis of the steady-state and transient response of the glucose ENFET has shown that the slope, detection limit and the dynamic range are governed by the concentration of the buffer, thickness of the enzyme containing layer, transport properties of the all involved species and by the ion response of the transistor itself (21). This analysis has been verified experimentally and extended to a single substrate, pH-independent kinetics of penicillin ENFET (22). It can be seen from this model that the detection limit for potentiometric enzyme sensors does not extend below 10^{-5}M for typical values of diffusion coefficients and for molecular weight of the substrate in the range 100–300. The example of the glucose ENFET response curve is shown in Figure 6 and the transient behavior is illustrated in Figure 7.

There are several advantages in using a field effect transistor as an enzymatic sensor. It is possible to subtract the common interferences by differential measurement (19). The amount of enzyme used is minute ($< 10^{-4}$ I.U.) which makes the use of even most expensive enzymes practical. The small area of the enzyme layer allows immobilization without a retaining membrane which shortens the time response. In the fabrication of the ENFET, the geometry of the sensor can be tightly controlled which enables us to predict its behavior using a mathematical model. With hundreds of enzymes available, this type of sensor is likely to be developed into a multisensor in the near future.

Work Function FET

The general response equation of a field effect transistor is

$$I_D = K\, V_D(V_G - V_T - \frac{V_D}{2}) \quad \text{for} \quad V_D < V_G - V_T$$

and

$$I_D = \frac{K}{2}(V_G - V_T)^2 \quad \text{for} \quad V_D > V_G - V_T$$

The drain current I_D is thus dependent on the applied gate voltage V_G (Figure 1), drain voltage V_D, and so called threshold or turn-on voltage V_T. The constant K depends on the geometry and materials used to build the transistor. The distribution of electrons between the two plates of the gate capacitor (Figure 8) depends on V_G and on the threshold voltage V_T which can be expanded

$$V = \Delta\phi_{LS} - Q_i/C_i + 2\ \phi_F - Q_B/C_i$$

where ϕ_F is the Fermi level of the semiconductor, Q_i and Q_B are the insulator and depletion charge respectively and C_i is the overall gate capacitance. The work function difference $\Delta\phi_{LS}$ between the chemical layer L and the semiconductor S is

$$\Delta\phi_{LS} = (\mu_e^L - \mu^S) - F(\Sigma_i\ n_i\ \cos\theta_i\ \eta_i + \chi_{SS})$$

The terms μ_e^L, μ_e^S refer to the chemical potential of electron in the layer L and semiconductor S. The difference of the work function $\Delta\phi_{LS}$ relates to the gate structure shown in Figure 8. The chemical species can enter into the gate proper and either adsorb on the surface of the chemical layer, in which case the dipolar potential of the surface χ_{SS} changes as the dipoles η_i adsorb and assume some preferred geometrical orientation ($\cos\theta_i$). If the chemical species interacts with the bulk of the layer L the electron chemical potential of that layer μ_e^L changes and the $\Delta\phi_{LS}$ term is again affected, the underlying chemical change in this device is thus related to the work function. The first device of this type has been described and extensively studied by Lundstrom and his coworkers (5). Lundstrom's transistor is based on palladium gate which selectively absorbs hydrogen. A more general structure has been made recently by two other groups (6,7), so called suspended gate field effect transistor (SGFET). The selectivity of this device for different species depends on the interaction of the species of interest with layer L. In our preliminary study we have shown (23) that electrochemically deposited layers can be made selectively reactive for various gaseous species (Figure 9) and to polar constituents in dielectric fluids (6).

CHEMFET PREPARATION

From the solid state device physics point of view, CHEMFETs are very simple devices to fabricate up to the chip level. The application of the chemically sensitive layer and overall packaging are the two most irreproducible and labor consuming steps. Yet, without reliable packaging it is pointless to do any measurements with these devices. For research purposes, and in order to demonstrate the feasibility of sensing concepts, it is possible to package the CHEMFETs by hand. It is also mandatory to test the encapsulation integrity of the whole package. The most unambiguous test is to measure the leakage current through the fully encapsulated device in the range of $\pm$ 2 V after several hour immersion in an electrolyte solution. If a step increase of current occurs in this range, the device is suspect.

In order to improve the yield of the packaging operation, we have developed and tested (24) the procedure which is outlined in Figure 10. It is quite clear that other, possibly better approaches can be found.

The adhesion of the membrane to the transistor surface is another problem which is responsible for device failure, particularly in biomedical, in vivo, applications in which the device is subjected to a mechanical stress. A successful solution to this problem has been found by constructing, over the gate area, a suspended polyimide bridge which anchors the membrane to the surface of the chip (25). The schematic diagram of this structure is shown in Figure 10. With these two improvements average lifetimes of membrane-based ISFETs in excess of 60 days have been obtained.

CONCLUSIONS

From the historical perspective the field of chemically sensitive solid state devices has been pioneered by electrical engineers and by solid state physicists. Later, electro-analytical chemists and bioengineers contributed to the development of this area to its present state. What solid state chemical sensors need most now is the attention of the materials scientists. A decision has to be made whether to continue with the use of silicon based substructures and protect them from the adverse effects of water and electrolytes or whether it would be worthwhile to search for different (than silicon) materials for fabrication of the basic circuits. The first alternative seems to be more feasible at present, but a concerted effort in design of protective materials must be done expertly.

ACKNOWLEDGEMENTS

The results of the work, which originated from the University of Utah, have been obtained by my graduate students and postdoctoral associates whose names appear in the list of publications. It is my pleasure to acknowledge long term collaboration with Professor R.J. Huber. Over the years this work has been supported by grants and contracts from the following institutions: National Institutes of Health, National Science Foundation, Office of Naval Research and Critikon Inc.

REFERENCES

(1) J.W. Frazer, D.J. Balaban, H.R. Brand, G.A. Robinson and S.M. Lanning, Anal. Chem. 55(1983), 855.

(2) J. Janata, in "Solid State Chemical Sensors", J. Janata and R.J. Huber Eds., Acad. Press, in print.

(3) R.J. Huber, in "Solid State Chemical Sensors", J. Janata and R.J. Huber Eds., Acad. Press, in print.

(4) M.F. Delaney, Anal. Chem. 56(1984) 261R.

(5) I. Lundstrom and C. Svensson, in "Solid State Chemical Sensors", J. Janata and R.J. Huber, Eds., Acad. Press, in print.

(6) G.F. Blackburn, M. Levy and J. Janata, Appl. Phys. Lett., 43,(1983) 700.

(7) M. Stenberg and B.I. Dahlenback, Sensors and Actuators, 4,(1983) 273.

(8) N.F. Sheppard, D.R. Day, H.L. Lee and S.D. Senturia, Sensors and Actuators, 2,(1982) 263.

(9) R.M. Cohen, and J. Janata, J. Electroanal. Chem. 151,(1983) 33.

(10) A. Haemmerli, J. Janata and J.J. Brophy, J. Electrochem. Soc. 151,(1983) 2306.

(11) Zheng Kang Li, J.M. Reijn and J. Janata, J. Electrochem. Soc. (1984), in print.

(12) Zheng Kang Li, J.M. Reijn and J. Janata, J. Electrochem. Soc. (1984), in print.

(13) P. Bergveld, IEEE Trans. BME-19, (1970) 70.

(14) T. Matsuo, M. Esashi and K. Inuma, Digest of Joint Meeting of Tohoku Sections of IEEE (1971).

(15) S.D. Moss, J. Janata and C.C. Johnson, Anal. Chem. 47,(1975) 2238.

(16) J. Janata, Sensors and Actuators, 4,(1983) 255.

(17) G.F. Blackburn and J. Janata, Ann. N.Y. Acad. Sci. 428,(1984) 286.

(18) B. Danielsson, I. Lundstrom, K. Mosbach and L. Stilbert, Anal. Lett. Pt. B 12,(1979) 1189.

(19) S. Caras and J. Janata, Anal. Chem. 52,(1980) 1935.

(20) Y. Hanazato and S. S. Shono, Proc. Chem. Sens., Fukuoka, Sept. 19-22 (1983), Elsevier, N.Y., 513.

(21) S. Caras and J. Janata, Anal. Chem. (1984) submitted.

(22) S. Caras and J. Janata, Anal. Chem.. (1984) submitted.

(23) M. Levy, J. Cassidy, S.B. Pons and J. Janata, 11th FACSS Meeting, Philadelphia (1984).

(24) N.J. Ho, J. Krotochvil, G.F. Blackburn and J. Janata, Sensors and Actuators 4(1983) 413.

(25) G.F. Blackburn and J. Janata, J. Electrochem. Soc. 129(1982) 2580.

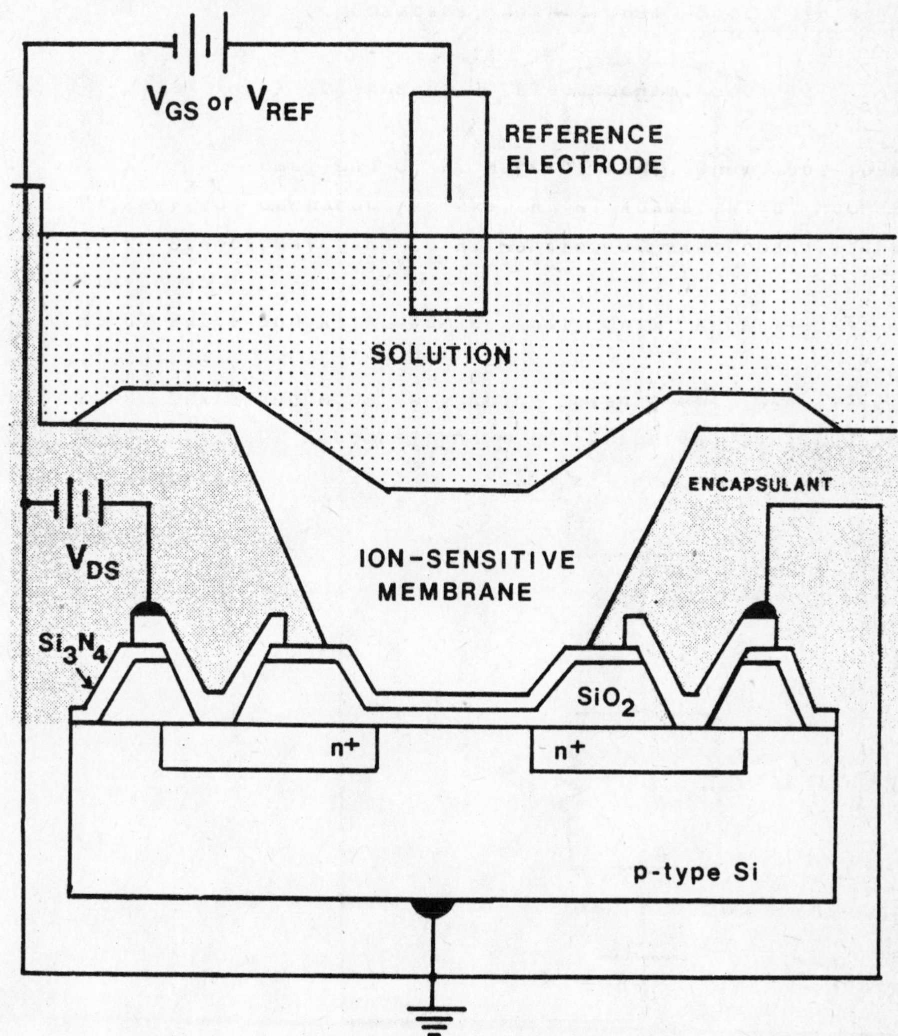

Fig. 1. Schematic diagram of ISFET.

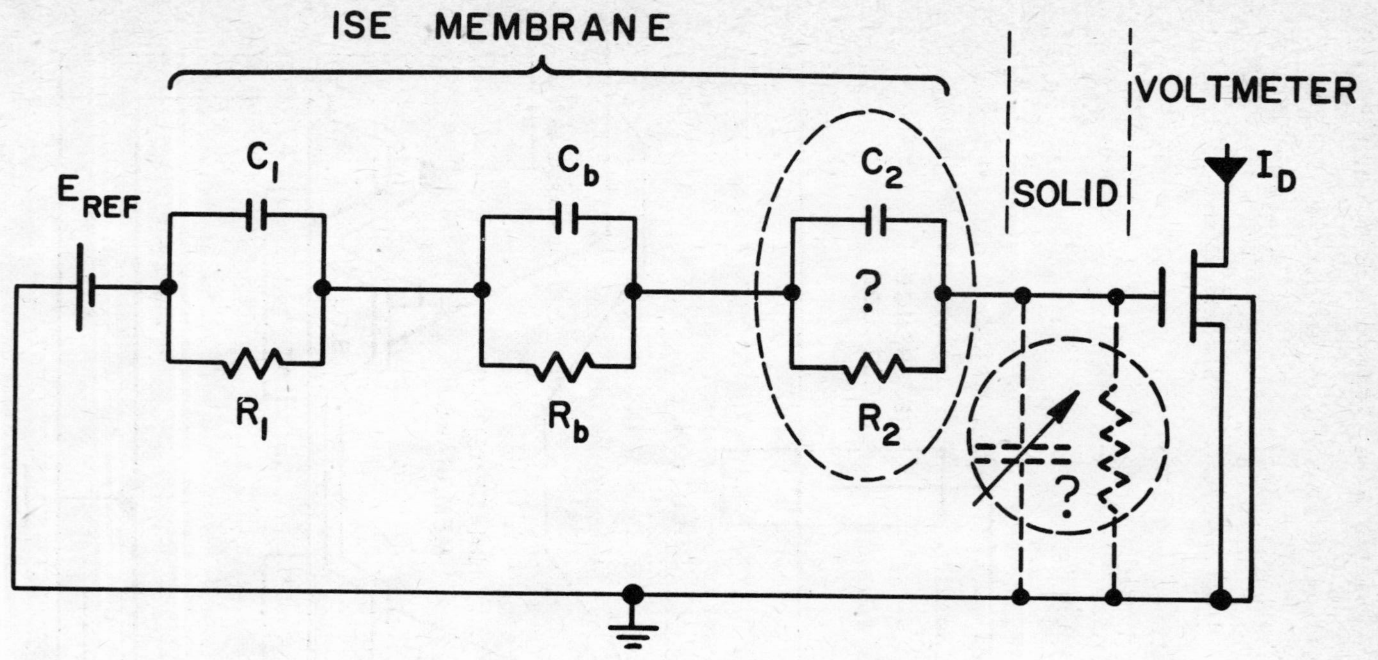

Fig. 2. Equivalent electrical circuit of a non-symmetrical potentiometric ion-selective device.

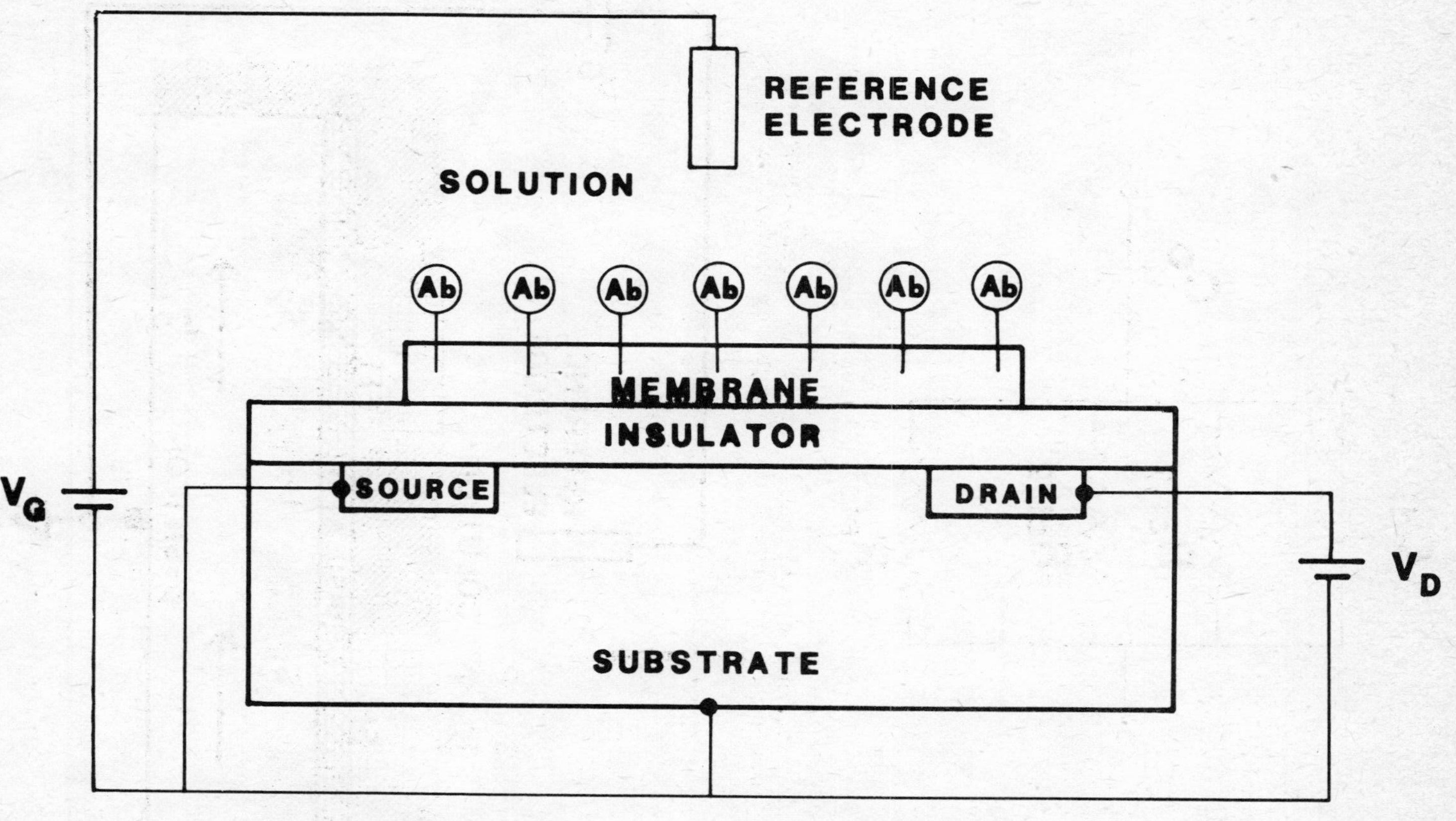

Fig. 3. Schematic diagram of immunochemically sensitive field effect transistor.

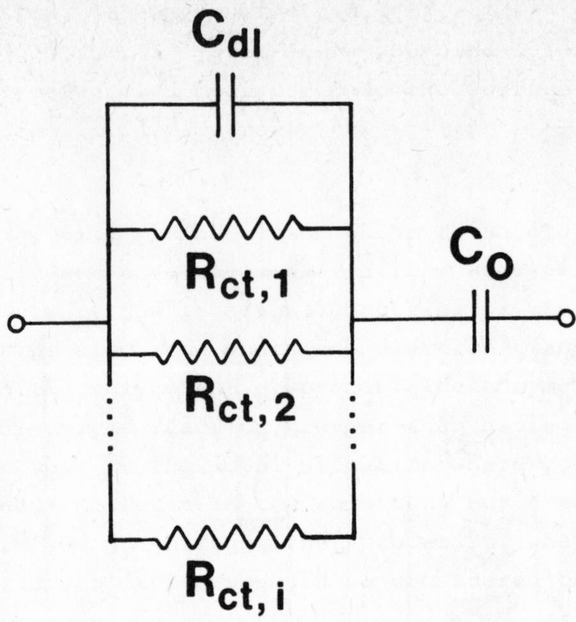

Fig. 4. Equivalent electrical circuit of a real IMFET. The immunochemical reaction charges up capacitor C_{dl}.

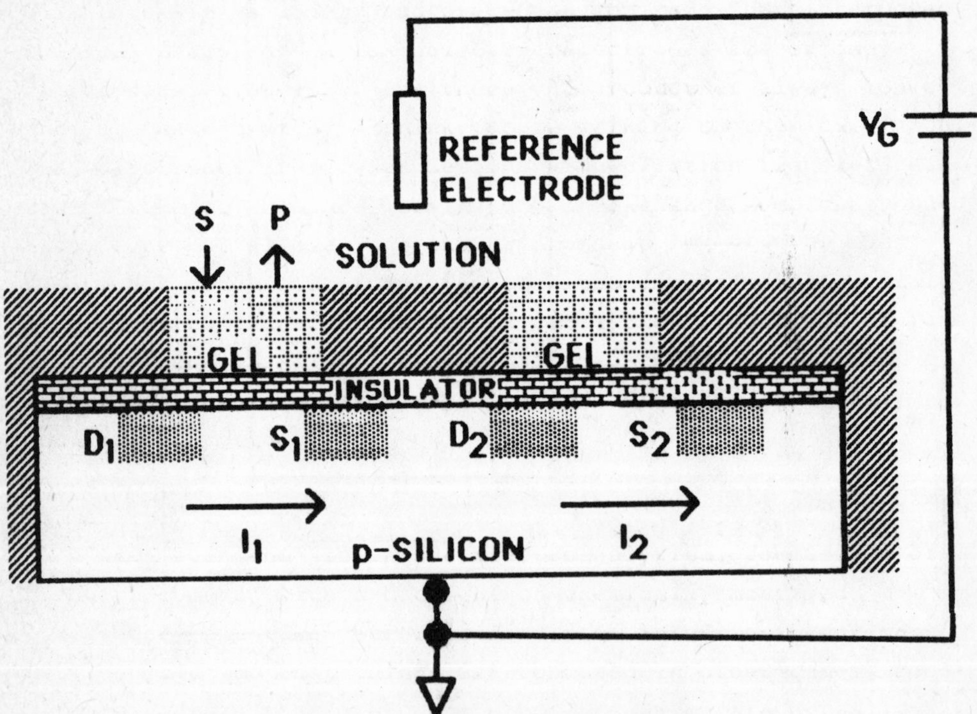

Fig. 5. Schematic diagram of ENFET.

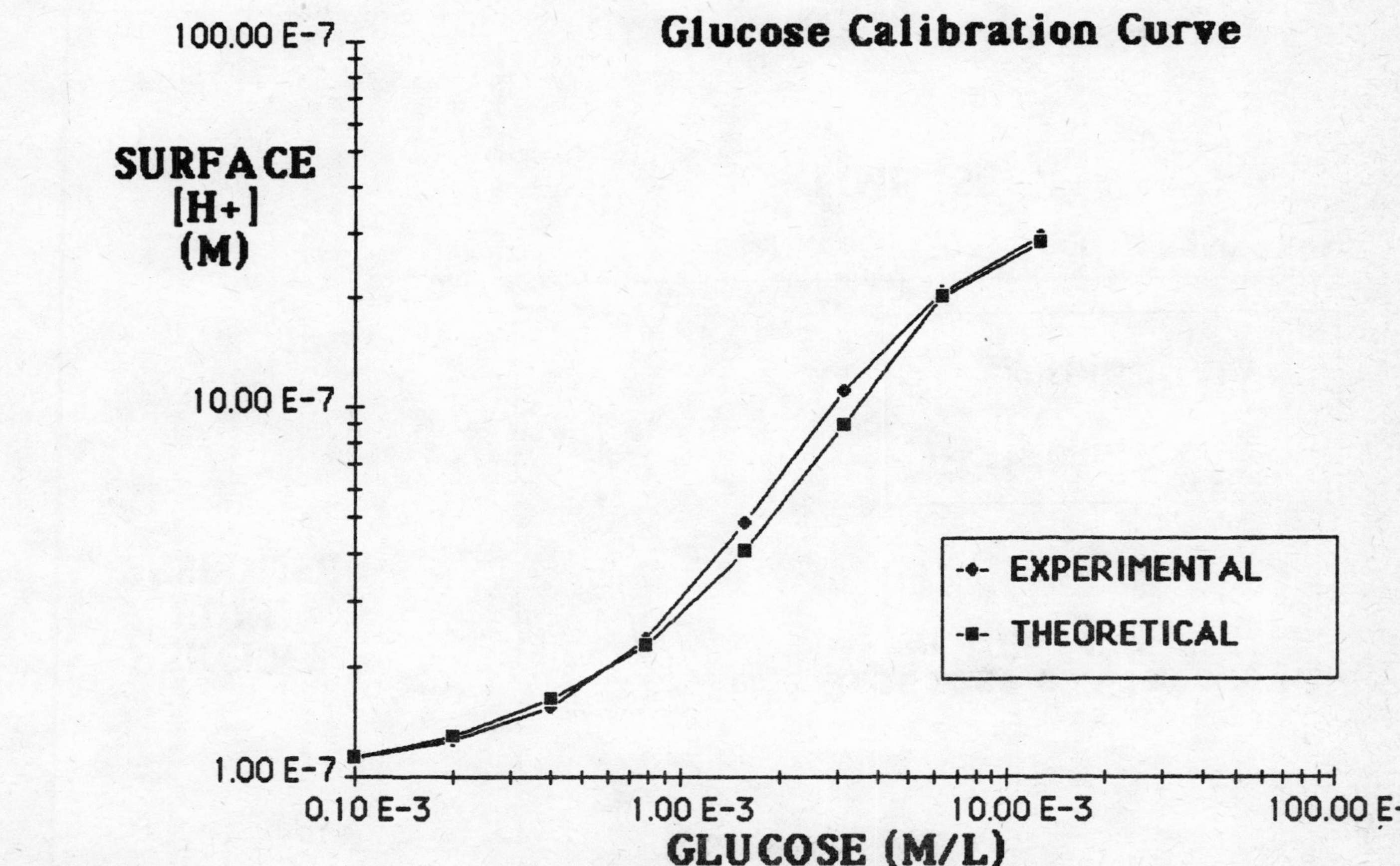

Fig. 6. Calibration curve of glucose ENFET.

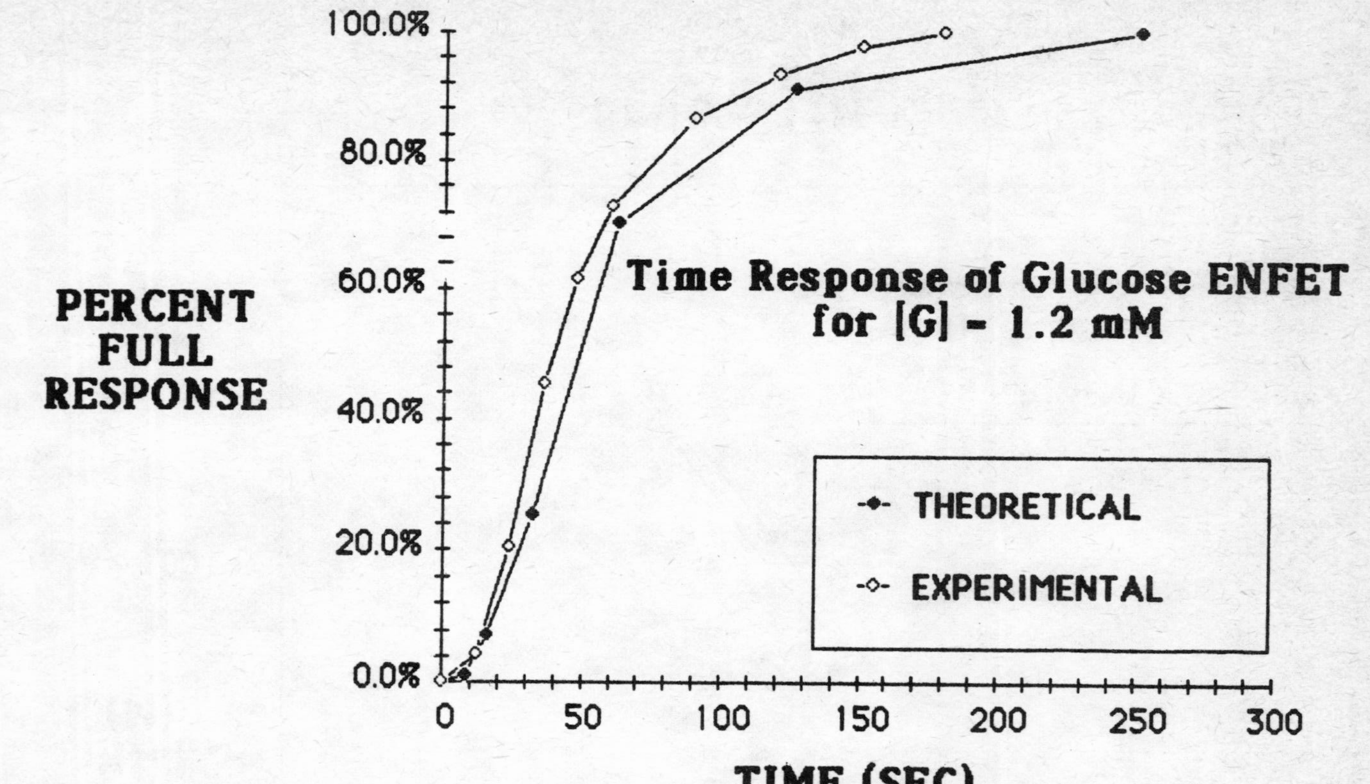

Fig. 7. Transient behavior of glucose ENFET.

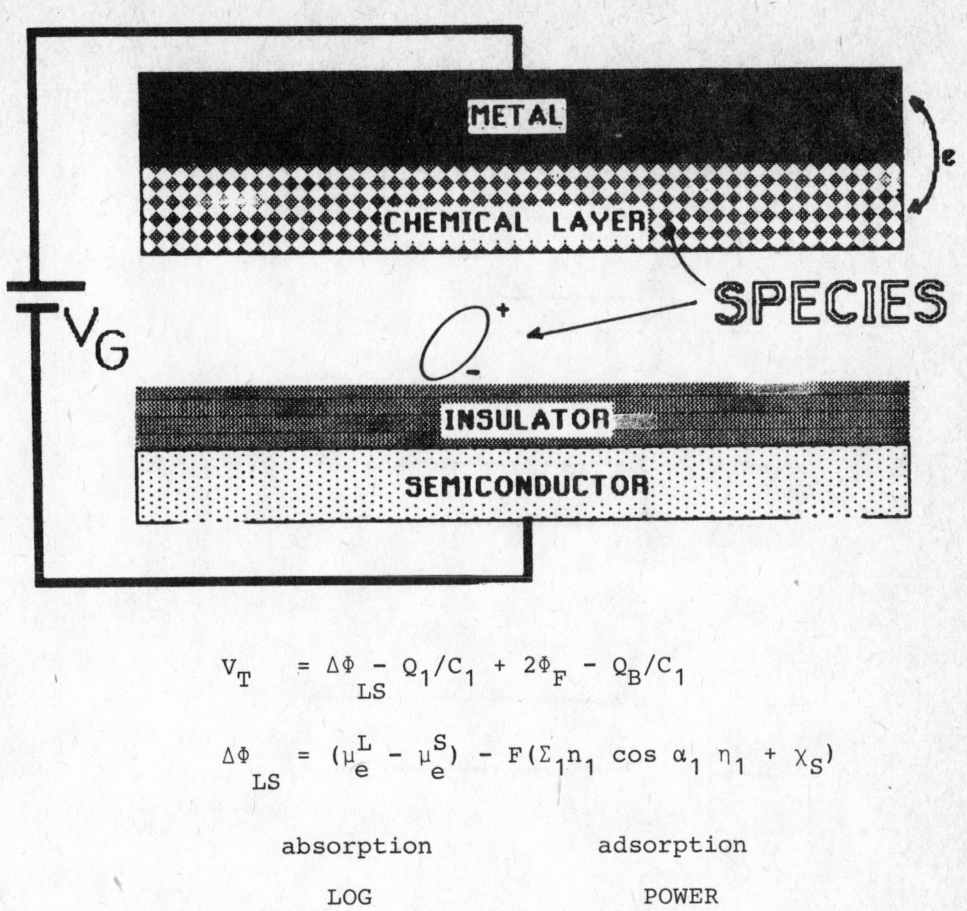

$$V_T = \Delta\Phi_{LS} - Q_1/C_1 + 2\Phi_F - Q_B/C_1$$

$$\Delta\Phi_{LS} = (\mu_e^L - \mu_e^S) - F(\Sigma_1 n_1 \cos\alpha_1 \eta_1 + \chi_S)$$

absorption adsorption

LOG POWER

Fig. 8. Schematic diagram of suspended gate field effect transistor.

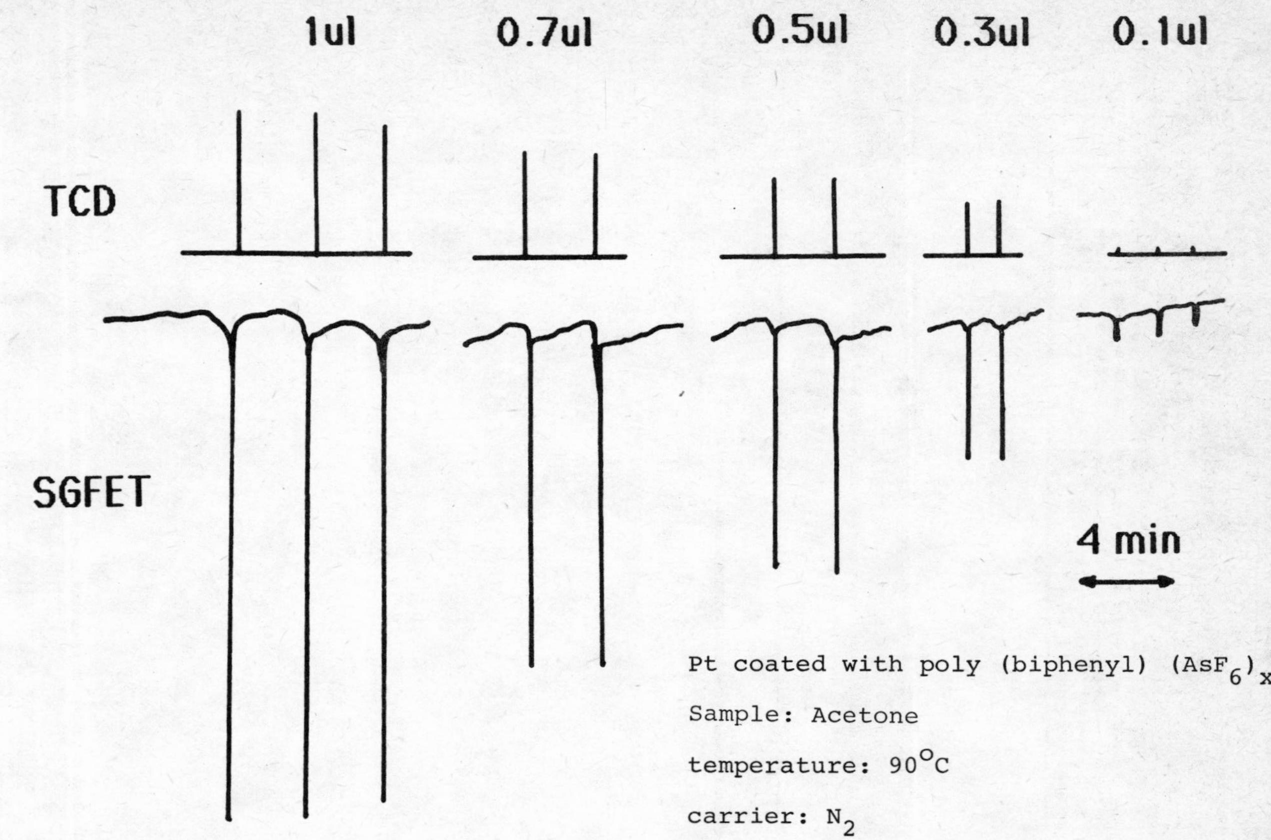

Fig. 9. Response of SGFET to acetone.

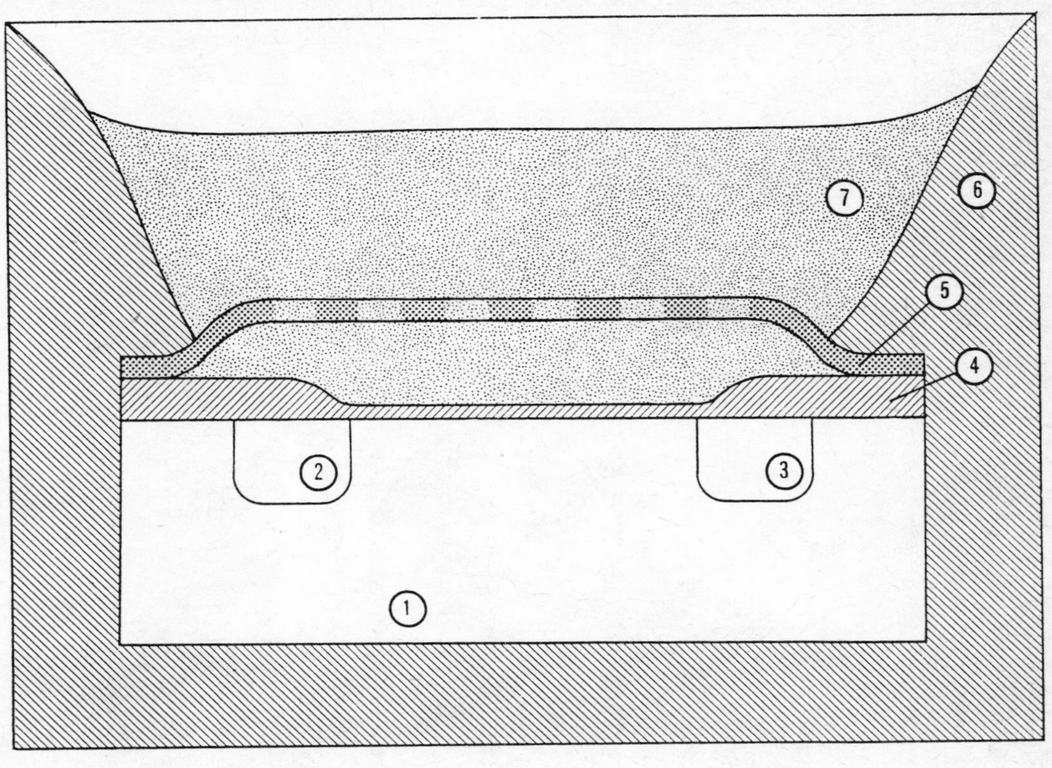

Fig. 10. Schematic diagram of suspended polyimide mesh field effect transistor. 1 - substrate, 2 - drain, 3 - source, 4 - insulator, 5 - suspended polyimide mesh, 6 - encapsulation, 7 - membrane.

QUESTIONS AND COMMENTS

Participants of the discussion: P.Becher, R.P.Buck, G.Horvai, J.Janata, G.Nagy, E.Pungor, W.Simon and J.D.R.Thomas

Question:

In the literature on ISFETs gate leakage currents of the order of 10^{-12} A are mentioned. On the other hand, if you combine a standard electrode with a very good high impedance electrometer, you may have curents in the order of 10^{-14} A, two orders of magnitude less than with the ISFETs. Could you give us a suggestion how these currents might flow and what their nature is? Is it connected with a mass transport, an ion-transport ?

A further question is connected with the impedance of silicon dioxide, silicon nitride and different oxides used in ISFETs. They have an impedance of 10^{16} Ω , and it is usually not defined how this arises, is it electronic or ionic conductivity. Could you comment on that ?

Answer:

As far as I know, nobody has established that the 10^{-12} A leakage current is flowing through the gate, it is always a current through the whole system. But les us suppose that you have a current of 10^{-12} A, and your electrode has an impedance of $10^{8}\Omega$, then the error will be 100 μV, by Ohm's law. So it is really inconsequential as long as it does not change very much. Your reference electrode is going to cause you more problem.

Comment:

In the first chapter of H.Freiser's book there is a section which points out that one should always use a membrane whose resistance is much less than the input impedance of the amplifier you are using to measure it. And after a while people stop writing that sentence, it turns out not to be important. Our first measurement on a FET device was to measure the impedance of the silicon dioxide before we ever put it into a solution. It was so high that we could not measure it. And still, we could measure a voltage across the silicon oxide layer, and called it a Volta potential.

Question:

I think the great interest in ISFETs is partly due to their micro-characteristics. What are the prospects of completing the micro-cell by a micro-reference electrode?

Answer:

I do not believe in a micro-reference electrode. The reference electrode, by definition, is any ISE which is put in the environment of constant chemical composition. You still have to make a connection to your sample through a liquid junction. It is just a question of how rapidly the outer space, i.e. the sample will exchange with the inner space of the reference electrode, which determines how long your electrode is going to be a reference electrode. One can read in the literature of microelectrodes fabricated from silicon chips, but I think they are not reference electrodes. They are, for five minutes but not for half an hour.

Question:

In the last but one slide you have shown a hole, and in the last one the hole filled with a membrane material. Does this membrane work and what is the lifetime of this membrane ?

Answer:

I have worries here. When we filled the hole with the corning ion exchanger for potassium, it had no response to potassium, but it measured chloride. However, if we fill the hole with a pvc membrane containing the ion exchanger, we do get a good potassium response. These are only preliminary results and we do not have any explanation yet.

Question:

1. Do you know of silicon nitride being used in a conventional membrane type electrode for pH measurement ?

2. Could you comment on the advantages of ISFETs over small size electrodes, or small electrodes with built-in small size electronics, such as those described by K.Nagy in his paper.

Answer:

1. You cannot make a glass-type electrode out of silicon nitride. The reason is of principle, because it is an extremely good insulator. Imagine the situation you would have: you would have a pH-sensitive surface which would then be separated by a finite thickness of silicon nitride which is a perfect insulator. And then you connect it to what ? If you connect it to a metal, then you have a capacitor, and at the other end the input into the electrometer is also a capacitor, so you have two capacitors in series, and you do not know what the distribution of charge on those two capacitors is unless you know what the capacitance values are.

The reason why a glass electrode works is that it has a mode of electronic conductivity through the membrane, through the dry portion of the glass electrode. With silicon nitride, being an insulator, you cannot transmit the information from the outer boundary of the electrode into the inside. And the only way it works in ISFETs is that you do not need the internal conductor. You are coupling the electric field directly to the layer. And with aluminium oxide, and tantalum oxide and other oxides, you are not really worried about the electronic or ionic conductivity of those materials. Whatever exchanges ions, you can couple it through the electric field into the transistor itself without having the prerequisite for the conductor which is normally found in ISEs.

2. As for the advantage of ISFETs over ISEs: at the present time I think that they do have advantage only if you can make them into multichannel sensors. Using ISEs, you really have to ask yourself how many good membranes you have. But with these devices, if you can make them into a multichannel system, using chemometrics and extracting the information even from a far non-Nernstian response, then you can get a tremendous amount of information from that, using statistical

methods. And you are doing it with a probe which has the same size as the one which would provide only one channel of information. This is important in medicine, where you normally do not have enough space to sample with many different electrodes and in space, because it takes 100 $ to launch 1 g of material into space. In those applications you certainly need very small weight, very small volume.

Comment:

My opinion is that even if you use chemometrics, pattern recognition, feature selection and so on, if you give bad data to the computer, the results cannot be any better. That is, without good sensors you cannot get really good results.

Answer:

I think that what is important is that those sensors, although they do not have to be characterized, have to be stable. This is the only requirement we have on those sensors.

Comment:

I can give you at least 6 or 7 membranes which are extremely stable. Make these work on ISFETs, and I will be quite happy.

Comment:

On the size-problem mentioned. The micropipette-type electrodes are much smaller than ISFETs, so the former still seem to be better in physiological studies. The other comment is on the multichannel system. With the micropipette type electrode, I really prefer to use just one sensor at a time, as two electrodes are going to cause more problem with noise pick-up than double.

Answer:

Some years ago we built a microelectrode which was coupled to the FET chip at a distance of less than 1 mm. This device had absolutely no noise pick-up, it could be used without the Faraday cage, without the operator being around. We make the electrode and our physiology department is using it, but I do not pretend that the general public will ever use that kind of device.

4th Symposium on Ion-Selective Electrodes
Mátrafüred, 1984

BIS(CROWN ETHER) BASED ION-SELECTIVE ELECTRODES

K. KIMURA and T. SHONO

Department of Applied Chemistry
Faculty of Engineering
Osaka University, Suita, Osaka 565, Japan

ABSTRACT

Various bis(crown ether) derivatives have been applied to neutral carrier type ion-selective electrodes of potassium, sodium, cesium, thallium(I), and calcium ions. The electrodes exhibited excellent ion selectivities, in which the specific cation-complexing properties of the bis(crown ether) derivatives were reflected. Generally, the bis(crown ether) based ion-selective electrodes are of great practical value. Lithium-selective electrodes based on lipophilic crown-4 derivatives are also described.

INTRODUCTION

Macrobicyclic polyethers containing crown ether moieties at the end of a short aliphatic chain, what they call bis(crown ether)s, possess attractive cation-complexing properties[1-5]. Bis(crown ether)s tend to complex particular metal ions strongly, a cation being sandwiched intramolecularly by two adjacent crown ether rings. High ion selectivity, therefore, has been realized in some bis(crown ether) derivatives as compared to the corresponding monocyclic analogs. This induced us to design novel bis(crown ether) derivatives and to apply to neutral carrier type ion-selective electrodes.

Bis(crown ether)s of 15-crown-5

A number of monocyclic crown ethers have been utilized as neutral carriers of polymeric membrane K^+-selective electrodes [6-11]. Most of the K^+-selective electrodes exhibited k_{KNa} (selectivity coefficient for K^+ over Na^+) values of about 10^{-2}, which are far from comparable to those for valinomycin-based K^+-selective electrodes, $\sim 10^{-4}$. Better k_{KNa} values were obtained in the electrodes based on dibenzo-30-crown-10 derivatives[8,10]. It also seems that high crown ether concentrations in electrode membranes improve K^+ selectivity over Na^+ in the ion-selective electrodes based on naphtho-15-crown-5[11].

We have synthesized bis(benzo-15-crown-5) derivatives as neutral carriers of PVC membrane K^+-selective electrodes[12-14], taking advantage of their outstanding selectivities for K^+. The electrode properties of the K^+-selective electrodes based on bis(crown ether)s 1 are summarized in Table 1, together with those of the corresponding monocyclic analog and valinomycin for comparison. Generally, the bis(crown ether) based electrodes are superior to those of the monocyclic benzo-15-crown-5 derivative in the K^+ selectivity. Especially, the K^+ electrode based on bis(crown ether) 1(m=5) possesses an excellent k_{KNa} value of 3×10^{-4} when *o*-nitrophenyl octyl ether (NPOE) is employed as the membrane solvent. The K^+ selectivity over Na^+ of the bis(crown ether) based electrode is comparable to that for the valinomycin-based electrode obtained here. Moreover, in these bis(crown ether) electrodes, Rb^+ does not interfere so seriously as the valinomycin electrode, in which the k_{KRb} value is not below unity. The length of the chain connecting two crown ether moieties in the bis(crown ether)s may change their cation-complexing abilities[1], which in turn affects more or less the K^+ selectivity of the bis-(crown ether) based electrodes. In K^+ electrodes based on bis(benzo-15-crown-5) derivatives incorporating cyclohexane ring 3[14], the 1,2-*cis* derivative functioned as an excellent K^+ neutral carrier of the PVC membrane electrodes in a similar way to 1, while the 1,2-*trans* derivative did not. This suggests that configuration of bis(crown ether)s also influences their ion-selectivities and, therefore, the electrode selectivities.

Bis(crown ether) 7 was also tested for its usefulness as a neutral carrier of K^+-selective electrode, but the electrode did not work well because of its poor solubility in membrane solvents[12]. Polymers carrying crown ether moiety at the side chain, poly(crown ether)s, manifest similar cation-complexing properties and selectivities to the corresponding bis(crown ether)s. Attempts were made to use some poly(crown ether)s as neutral carriers of K^+-selective electrodes, but the poly(crown ether) based ion-selective electrodes have proved rather impractical[12]. Recently, other workers also synthesized several bis(benzo-15-crown-5) derivatives to apply to K^+-selective electrodes and reported high K^+ selectivities over Na^+ of the electrodes analogous to those of our bis(crown ether) based electrodes[15-18].

Since bis(benzo-15-crown-5) derivatives 1 possess great affinity for Tl^+ as well as K^+, they can also be used as neutral carriers of Tl^+-selective electrodes[19]. The bis(crown ether) based electrodes showed near-Nernstian response in wide Tl^+ activity ranges. The electrodes naturally suffered considerably from K^+ interference, the k_{TlK} value being 3 x 10^{-1} in the Tl^+ electrode based on 1(m=5). However, the preference of Tl^+ over alkaline-earth and heavy metal ions such as Ag^+, Pb^{2+}, and Cd^{2+} is very high in the bis(crown ether) based electrodes unlike Tl^+-selective electrodes based on thallium salts[20].

Bis(15-crown-5) derivatives bearing a lipophilic chain 11 and 12 may be more excellent as neutral carriers of K^+-selective electrodes than bis(benzo-15-crown-5) derivatives 1[21]. The ion selectivity for the coated-wire type K^+-selective electrode based on 11 is equivalent or even higher than for the K^+ electrode based on 1(m=5). For instance, the k_{KNa} value for the K^+ electrode of 11 is 2 x 10^{-4}. Some comparisons between the K^+-selective electrodes based on bis(crown ether) derivatives 1(m=5) and 11 were made in the electrode property besides the ion selectivity. In both of the K^+-selective electrodes, the response time was within 10 seconds as estimated by an incremental method, which is acceptable for their practical use. A drastic difference between the K^+-selective electrodes based on 1(m=5) and 11 was observed in the electrode durability. Changes in maximal slope of calibration plots were followed in the bis(crown ether) based electrodes (Fig. 1). There was no appreciable change of the slope up to 200 times of measurement in the K^+ electrode of 11, whereas in that of 1(m=5) the calibration slope decreased significantly after 60 times of measurement. This deterioration of the electrode based on 1(m=5) is presumably accounted for dissolving out of the neutral carrier from the membrane phase to sample solutions. The K^+-selective electrode based on 11 is, therefore, superior in the electrode lifetime to that of 1(m=5) owing to the high lipophilicity. Another marked difference between two of the bis(crown ether) based electrodes is in effect of sample stirring on the e.m.f. readings (Fig. 2). It should be noted that the intermittent stirring of sample solutions caused serious drift of the e.m.f. value in the K^+ electrode of bis-(benzo-15-crown-5) derivative 1(m=5). In contrast the stirring scarcely affected the e.m.f. readings in the electrode of bis-(15-crown-5) derivative 11. Thus, the lipophilic bis(15-crown-5) derivative, 11, generally surpasses the bis(benzo-15-crown-5) derivative, 1(m=5), as neutral carriers for K^+-selective electrodes, being a promising alternative to valinomycin.

Bis(crown ether)s of 18-crown-6

Bis(benzo-18-crown-6) derivatives 2 form stable sandwich-type complexes with Cs^+, being Cs^+-selective neutral carriers for PVC membrane electrodes[22] (Table 2). The Cs^+-selective electrodes showed linear Nernstian response in wide Cs^+ activity ranges. The Cs^+ selectivities over K^+ in the bis-(crown ether) based electrodes, although not so remarkable, are higher than in the electrode based on the corresponding monocyclic analog. Bis(benzo-18-crown-6) derived from maleic acid, *cis*-5, was expected to be highly Cs^+-selective on account of the pre-organized configuration suitable for the formation of the sandwich-type Cs^+ complexes[3]. Actually the *cis*-bis-(crown ether), when incorporated into electrode membranes, is more selective for Cs^+ than the corresponding *trans* isomer[23].

Also, the Cs^+ electrode based on *cis*-5 exceeds those of 2 slightly in the Cs^+ selectivity over K^+.

Lipophilic bis(18-crown-6) derivative 13 was also attempted to use as a neutral carrier of Cs^+-selective electrode[21]. The Cs^+ electrode of 13 was equivalent to those of 2 in the electrode selectivity, but the calibration plots in the former electrode possesses a maximal slope of only 51 mV $decade^{-1}$. In bis(crown ether) 13, two crown ether moieties are little too close to each other for the formation of stable sandwich-type Cs^+ complexes. This may be one of the reason for the poor sensitivity of the 13-based Cs^+-selective electrode.

Bis(crown ether)s of 12-crown-4

It was anticipated from molecular model examinations that bis(benzo-12-crown-4) derivatives such as 6 might be highly Na^+-selective. This was, however, not true due to the poor cation-complexing ability of benzo-12-crown-4 moiety. In contrast bis(12-crown-4) derivatives 14(m=1 or 2), which do not contain any benzene ring in the crown ring, exhibited rather high Na^+ selectivity[24]. Hence, the properties of the Na^+-selective electrodes based on bis(crown ether)s 9, 10, and 14(m=1, 2, or 3) have been elucidated [25]. The selectivity coefficients for Na^+ over K^+ in the Na^+ electrodes (Table 3) indicate that the Na^+ selectivity of the bis(12-crown-4) derivatives are markedly dependent on the length of the chain connecting two crown ether moieties. The electrodes of bis-(12-crown-4) derivatives 14(m=1 or 2) are much more selective for Na^+ than that of the glutarate derivative 14(m=3), which resembles even the electrode based on the monocyclic analog, 12-crown-4, in the k_{NaK} value. The difference in Na^+ selectivity between the bis(12-crown-4) derivatives appears to be related to the ease of cooperation of two adjacent crown ether rings on forming the sandwich-type Na^+ complexes. More striking are excellent electrode properties of the Na^+-selective electrode based on α,α -disubstituted malonate derivative of bis(12-crown-4) 10. One of the advantages for 10 over 14(m=1 or 2) is the enhanced Na^+ selectivity by the back strain attributable to the geminal substituents at the *a* position of the malonate. Another advantage is the high lipophilicity caused by the incorporation of a long aliphatic chain, which in turn brought about the prominent electrode properties of the 10 -based Na^+ electrode, *i.e.*, stable e.m.f readings, short response time, and long electrode lifetime. Thus, the lipophilic bis(12-crown-4) derivative, 10, has proved to be a promising Na^+ neutral carrier for the ion-selective electrodes.

Attempts have been made to use practically the Na^+- and K^+-selective electrodes based on bis(12-crown-4) derivative 10 and bis(15-crown-5) derivatives 1(m=5) or 11, respectively[26, 27]. Concentrations of Na^+ and K^+ in human blood serum and urine samples were determined successfully by using the coated-wire type ion-selective electrodes of the bis(crown ether) derivatives. The correlation of the results of Na^+ and K^+ determinations in human urine between the ion-selective method and flame photometry represent that both of the results are in good agreement (Fig. 3). This indicates that the bis(crown ether) based ion-selective electrodes are of great practical use.

Bicyclic polyether-amide derivatives

Since cyclic polyether-amide derivative complex preferentially alkaline-earth metal ions, especially Ca^{2+}, they have been applied to Ca^{2+}-selective electrodes[28-30]. The Ca^{2+}-selective electrodes exhibited rather high Ca^{2+} selectivity over Mg^{2+}, the k_{CaMg} values ranging from 10^{-3} to 10^{-4}. We recently reported that Ca^{2+} selectivity was enhanced in bicylic polyether-amide derivatives such as 15 and 16, as compared to the corresponding monocyclic analogs 17 and 18[31,32]. The selectivity coefficients for Ca^{2+} over Mg^{2+} and Na^+ in the Ca^{2+}-selective electrodes based on 15 and 16 are lower by about one order of magnitude than in those based on 17 and 18 (Fig. 4). It is considered to be partly because the cyclic polyether-amide are most likely to complex Ca^{2+} with 2:1 stoichiometry of the ring and cation[28]. The Ca^{2+} preference over $H^+(H_3O^+)$ in the Ca^{2+}-selective electrodes of the bicyclic polyether-amide derivatives is quite high ($k_{CaH} = \sim 1 \times 10^{-3}$ as determined by the fixed interference method), although there is little enhancement on the k_{CaH} value by changing the neutral carrier from the monocyclic derivatives to the bicyclic ones. The k_{CaH} values for the Ca^{2+} electrodes are significantly lower than for those based on organic phosphates[33,34]. By using the Ca^{2+}-selective electrode based on bicyclic polyether-amide 16, total Ca^{2+} in blood serum were determined successfully.

Lipophilic crown-4 derivatives as lithium neutral carriers

Li^+-selective electrodes should be very useful for determination of Li^+ activity in biological systems, *e.g.*, Li^+ monitoring during therapy of maniacal psychosis. Crown ethers are great candidates as neutral carriers for Li^+-selective electrodes. Several crown ether derivatives have been so far attempted for their usefulness as neutral carriers of Li^+-selective electrodes [35-37], but only poor Li^+ selectivities have been attained in the Li^+ electrodes. The k_{LiNa} values, which are not lower than 10^{-1}, are insufficient for the practical use. Nevertheless, crown ethers containing 4 oxygen atoms, what they call crown-4 derivatives, may be still expected to show Li^+ affinity.

We have designed lipophilic crown-4 derivatives with 13- through 16- member rings, in order to obtain highly Li^+-selective neutral carriers[38,39]. The selectivity coefficients for the Li^+-selective electrodes based on 19 through 22(R=H or CH_3) are summarized in Fig. 5. In general, the 14- and 15-member ring macrocycles 20 and 21 are much more selective for Li^+ than the 13- and 16-member ring ones 19 and 22. Molecular model examinations for the crown-4 derivatives revealed that Li^+ fits into the 15-crown-4 ring and more favorably into the 14-crown-4 one. The Li^+ selectivities of the crown-4 derivatives were reflected in the selectivity coefficients for the Li^+-selective electrodes based on the crown ethers. It is worth noting that dodecyl-methyl-14-crown-4 20($R=CH_3$) is the most excellent of all in the Li^+ selectivity of the electrode. Especially, the Li^+ selectivity over Na^+ in the 20($R=CH_3$)-based electrode is exceedingly high, the k_{LiNa} values being 7×10^{-3}

for the PVC-NPOE membrane and 4 x 10^{-3} for the PVC- *o*-nitrophenyl phenyl ether(NPPE) one. H^+ interference is significantly small in the crown ether based Li^+-selective electrodes as compared to those based on the previous neutral carriers[35,37, 40]. The Li^+-selective electrode based on the lipophilic crown-4 derivative, 20($R=CH_3$), showed linear Nernstian response in a wide activity range of 1 x 10^{-5} - 1 M Li^+. The combined use of a small quantity of trioctylphosphine oxide(TOPO)[41] and the crown ether in the PVC membranes enhanced the Li^+ selectivity over Na^+ more or less, although the TOPO addition to the membrane caused marked increase in the k_{LiH} value of the electrode.

A practical use of the Li^+-selective PVC membrane electrode based on 20($R=CH_3$) was tested by Li^+ determination in artificial blood serum samples containing various concentrations of Li^+. Correlation of the actual values for the Li^+ concentration in the samples and the results obtained potentiometrically is given in Fig. 6. Despite of a high background of Na^+(145 mM) in the samples, the found values of the Li^+ concentration are in good agreement with the actual ones down to 1 mM. Thus, the high ion selectivity of the Li^+-selective electrode based on the lipophilic crown-4 derivative, 20($R=CH_3$), has been substantiated by the successful Li^+ determination in the artificial blood serum samples.

REFERENCES

1 M.Bourgoin, K.H.Wong, J.Y.Hui, and J.Smid, J.Am.Chem.Soc., 97/1975/3462.

2 K.Kimura, T. Maeda, and T.Shono, Talanta, 26/1979/945.

3 K.Kimura, T.Tsuchida, T.Maeda, and T.Shono, Talanta, 27/1980/801.

4 K.H.Wong and H.L.Ng, J.Coord.Chem., 11/1981/49.

5 M.J.Calverley and J.Dale, J.Chem.Soc., Chem.Commun., 1981/684.

6 G.A.Rechnitz and E.Eyal, Anal.Chem., 44/1972/370.

7 O.Ryba, E.Knižáková, and J.Petránek, Coll.Czech.Chem.Commun., 38/1973/497.

8 J.Petránek and O.Ryba, Anal.Chim.Acta, 72/1974/375.
9 M.Mascini and F.Pallozzi, Anal.Chim.Acta, 73/1974/375.
10 Sh.K.Norov, A.K.Tashmukhamedova, and N.Zh.Saifullina, Zh.Anal.Khim., 37/1982/222.
11 M.Yamauchi, A.Jyo, and N.Ishibashi, Anal.Chim.Acta, 136/1982/399.
12 K.Kimura, T.Maeda, H.Tamura, and T.Shono, J.Electroanal. Chem.Interfacial Electrochem., 95/1979/91.
13 H.Tamura, K.Kimura, and T.Shono, Bull.Chem.Soc.Jpn., 53/1980/547.
14 K.Kimura, A.Ishikawa, H.Tamura, and T.Shono, Bull.Chem. Soc.Jpn., 56/1983/1859.
15 K.W.Fung and K.H.Wong, J.Electroanal.Chem.Interfacial Electrochem., 111/1980/359.
16 B.Ágai, I.Bitter, É.Csongor, and L.Tőke, Acta Chim. Acad. Sci.Hung., 110/1982/29.
17 T.Ikeda, A. Abe, K. Kikukawa, and T. Matsuda, Chem.Lett., 1983/369.
18 D.Huang, J.Zhang, C.Zhu, D.Wang, H.Hu, T.Fu, H.Ou, Z.Shen and Z.Yu ,Huaxue Xuebao, 42/1984/101.
19 H.Tamura, K. Kimura, and T.Shono, J.Electroanal.Chem. Interfacial Electrochem., 115/1980/115.
20 W.Szczepaniak and K.Ren, Anal.Chim.Acta, 82/1976/37.
21 K.Kimura, A.Ishikawa, H.Tamura, and T.Shono, J.Chem.Soc., Perkin Trans II, 1984/447.
22 K.Kimura, H.Tamura, and T.Shono, J.Electroanal.Chem. Interfacial Electrochem., 105/1979/335.
23 H.Tamura, K.Kimura, and T.Shono, Nippon Kagaku Kaishi, 1980/1648.
24 T.Maeda, M. Ouchi, K.Kimura, and T.Shono, Chem.Lett., 1981/1573.
25 T.Shono, M.Okahara, I.Ikeda, K.Kimura, and H.Tamura, J.Electroanal.Chem.Interfacial Electrochem., 132/1982/99.
26 H.Tamura, K.Kimura, and T.Shono, Anal.Chem., 54/1982/1224.
27 H.Tamura, K.Kumami, K.Kimura, and T.Shono, Mikrochim.Acta, 1983 II/287.
28 J.Petránek and O.Ryba, Anal.Chim.Acta, 128/1981/129.
29 J.Petránek and O.Ryba, Coll.Czech.Chem.Commun., 48/1983/1944.

30 I.J.Borowitz, J.D.Readio, and V.-S. Li, Tetrahedron, 40/1984/1009.
31 K.Kimura, K.Kumami, S.Kitazawa, and T.Shono, J.chem.Soc., Chem.Commun., 1984/442.
32 K.Kimura, K.Kumami, S. Kitazawa, and T. Shono, Anal.Chem., in press.
33 J.Růžička, E.H.Hansen, and J.Chr.Tjell, Anal.Chim.Acta, 67/1973/155.
34 A.Craggs, G.J.Moody, and J.D.R.Thomas, Analyst, 104/1979/412.
35 K.M.Aalmo and J.Krane, Acta Chem.Scand., A36/1982/227.
36 U.Olsher, J.Am.Chem.Soc., 104/1982/4006.
37 V.P.Y.Gadzekpo and G.D.Christian, Anal.Lett., 16/1983/1371.
38 K.Kimura, S.Kitazawa, and T.Shono, Chem.Lett., 1984/639.
39 S.Kitazawa, K.Kimura, H.Yano, and T.Shono. J.Am.Chem.Soc., in press.
40 A.F.Zhukov, D.Erne, D.Ammann, M.Güggi, E.Pretsch and W.Simon, Anal.Chim.Acta, 131/1981/117.
41 M.Katahira, T.Imatoh, and N.Ishibashi, Anal.Chim.Acta, in press.

Table 1. Properties of K^+-selective PVC membrane electrodes based on bis(crown ether)s 1

Neutral Carrier	Plasticizer	Maximal Slope $mV \cdot decade^{-1}$	Range pK	k_{KM}^{Pot}			
				Na^+	Rb^+	Cs^+	NH_4^+
1(m=1)	NPOE [a]	56	4.5 - 1	4×10^{-4}	1×10^{-1}	2×10^{-2}	1×10^{-2}
1(m=3)	NPOE	58	5 - 1	5×10^{-4}	1×10^{-1}	2×10^{-2}	1×10^{-2}
1(m=5)	NPOE	58	5 - 1	3×10^{-4}	2×10^{-1}	1×10^{-2}	1×10^{-2}
1(m=7)	NPOE	57	5 - 1	4×10^{-4}	1×10^{-1}	1×10^{-2}	1×10^{-2}
Benzo-15-crown-5 derivative [c]	NPOE	58	5 - 1	1×10^{-3}	7×10^{-1}	1×10^{-1}	5×10^{-2}
Valinomycin	NPOE	59	4.5 - 1	2×10^{-4}	1.4	3×10^{-1}	1×10^{-1}
1(m=1)	DPP [b]	59	5 - 1	5×10^{-4}	2×10^{-1}	2×10^{-2}	8×10^{-3}
1(m=3)	DPP	55	5 - 1	2×10^{-3}	2×10^{-1}	2×10^{-2}	1×10^{-2}
1(m=5)	DPP	56	5 - 1	7×10^{-4}	2×10^{-1}	1×10^{-2}	9×10^{-3}
1(m=7)	DPP	58	5 - 1	9×10^{-4}	2×10^{-1}	1×10^{-2}	1×10^{-2}
Benzo-15-crown-5 derivative [c]	DPP	58	5 - 1	4×10^{-3}	9×10^{-1}	3×10^{-1}	8×10^{-2}
Valinomycin	DPP	59	5 - 1	1×10^{-4}	2.1	5×10^{-1}	2×10^{-2}

a) *o*-nitrophenyl octyl ether, b) dipentyl phthalate, c) hexanoyloxymethylbenzo-15-crown-5.

Table 2. Properties of Cs^+-selective PVC membrane electrodes based on bis(crown ether)s 2 and 5

Neutral Carrier	Plasticizer	Maximal Slope $mV \cdot decade^{-1}$	Range pCs	k^{Pot}_{CsM}			
				Na^+	NH_4^+	K^+	Rb^+
2(m=3)	NPOE	59	6 - 2	9×10^{-4}	1×10^{-2}	8×10^{-2}	8×10^{-2}
2(m=5)	NPOE	59	6 - 2	9×10^{-4}	9×10^{-3}	8×10^{-2}	8×10^{-2}
2(m=7)	NPOE	59	6 - 1.5	1×10^{-3}	9×10^{-3}	8×10^{-2}	8×10^{-2}
5(*cis*)	NPOE	57	5.5- 2	4×10^{-4}	2×10^{-2}	4×10^{-2}	6×10^{-2}
5(*trans*)	NPOE	45	5 - 2.5	5×10^{-3}	2×10^{-1}	8×10^{-1}	5×10^{-1}
Benzo-18-crown-6 derivative a)	NPOE	59	5.5- 1.5	4×10^{-3}	2×10^{-2}	5×10^{-1}	1×10^{-1}
2(m=3)	DPP	59	5.5- 1.5	8×10^{-3}	2×10^{-2}	2×10^{-1}	2×10^{-1}
2(m=5)	DPP	59	5.5- 1.5	8×10^{-3}	2×10^{-2}	2×10^{-1}	2×10^{-1}
2(m=7)	DPP	59	5.5- 1.5	6×10^{-3}	1×10^{-2}	2×10^{-1}	2×10^{-1}
5(*cis*)	DPP	57	5 - 2	7×10^{-3}	3×10^{-2}	9×10^{-2}	1×10^{-1}
5(*trans*)	DPP	51	5 - 1.5	2×10^{-1}	3×10^{-1}	5	1
Benzo-18-crown-6 derivative a)	DPP	59	5.5- 1.5	5×10^{-2}	6×10^{-2}	2	3×10^{-1}

a) hexanoyloxymethylbenzo-18-crown-6.

Table 3. Properties of Na^+-selective PVC membrane[a] electrodes based on bis(crown ether)s 9, 10, and 14.

Neutral Carrier	Maximal Slope mV·decade^{-1}	Range pNa	$k^{Pot}_{Na\,M}$ Li^+	K^+	Rb^+	Cs^+	NH_4^+
9	53	4.5- 1	1×10^{-3}	2×10^{-2}	2×10^{-2}	2×10^{-2}	2×10^{-2}
10 [b]	55	4 - 1	1×10^{-3}	9×10^{-3}	4×10^{-3}	1×10^{-2}	1×10^{-3}
14 (m=1)	50	4 - 0	---	4×10^{-2}	4×10^{-2}	4×10^{-2}	2×10^{-2}
14 (m=3)	55	4 - 1	---	6×10^{-2}	2×10^{-2}	4×10^{-2}	3×10^{-2}
14 (m=3)	55	4 - 1	---	8×10^{-1}	7×10^{-1}	7×10^{-1}	3×10^{-1}
12-crown-4	52	4 - 1	---	6×10^{-1}	2	1	8×10^{-1}

a) NPOE was employed as the membrane solvent.

b) The selectivity coefficients for Na^+ over alkaline-earth metal ions were below 2×10^{-4}.

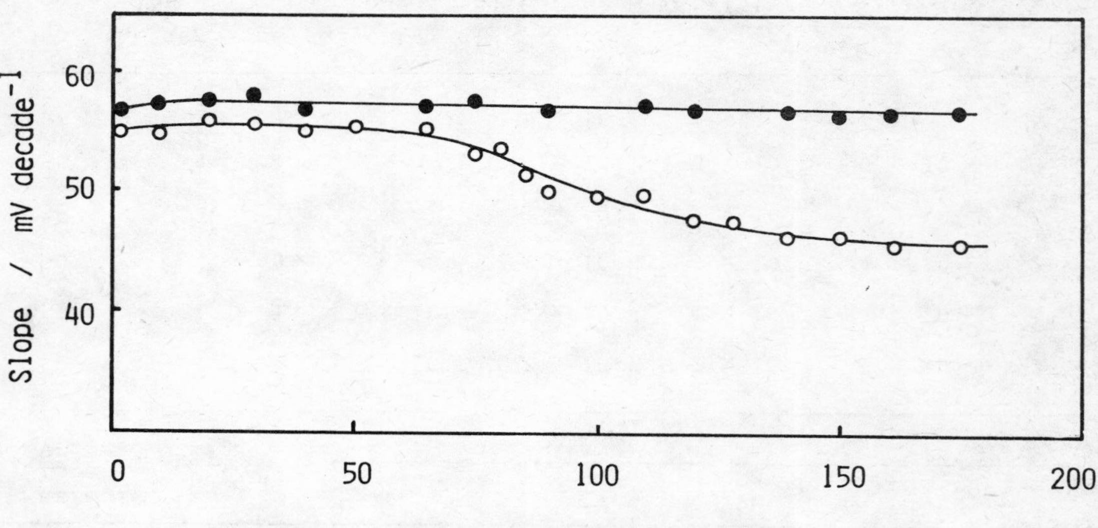

Fig. 1 Change in maximal slope of calibration plots with measurement number in K^+-selective electrodes based on bis(crown ether)s 1(m=5) (○) ans 11 (●).

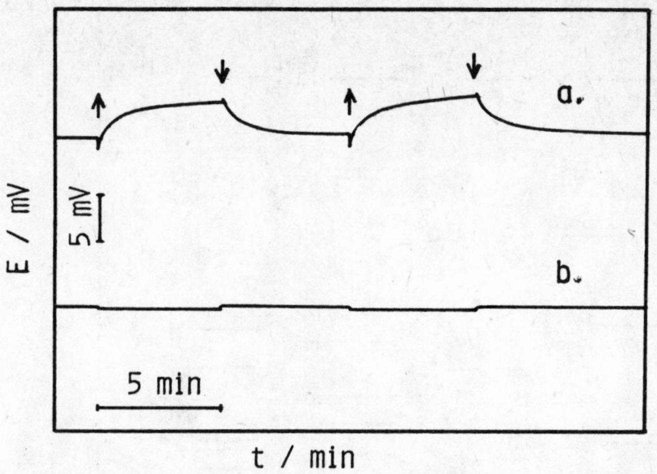

Fig. 2 Effect of intermittent stirring of sample solution upon e.m.f. reading by K^+-selective electrodes of bis(crown ether)s. sample: a mixture of 1×10^{-3} M KCl and 7.5×10^{-3} M NaCl. "↑" and "↓" stand for turning "on" and "off" of magnetic stirrer. a. 1(m=5)-based electrode, b. 11-based electrode.

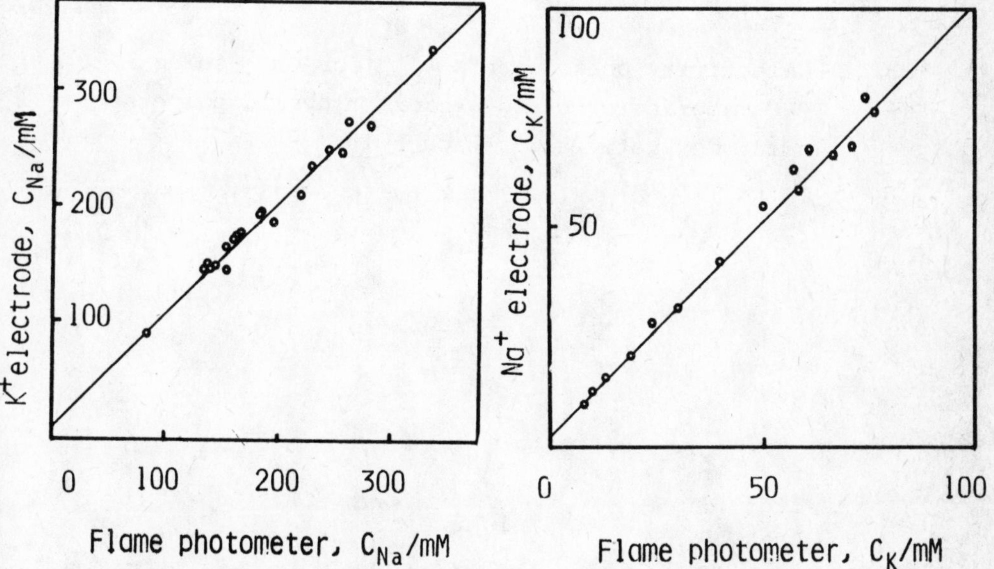

Fig. 3. Correlation of results of Na^+ and K^+ determinations in human urine samples obtained by flame photometry and potentiometry (by using coated-wire type Na^+- and K^+-selective electrodes based on 10 and 11).

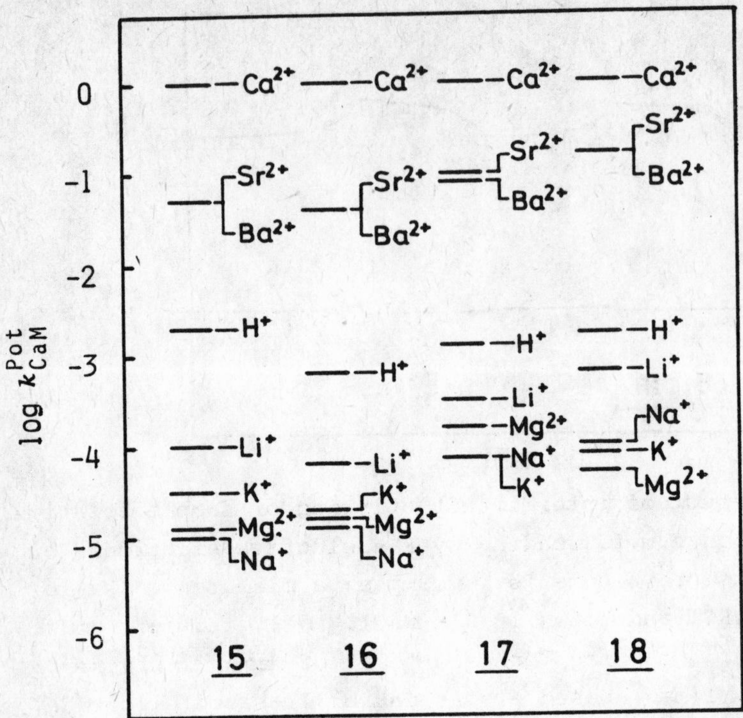

Fig. 4 Selectivity coefficients k_{CaM} for Ca^{2+}-selective PVC membrane electrodes based on cyclic polyether-amide derivatives 15 through 18.

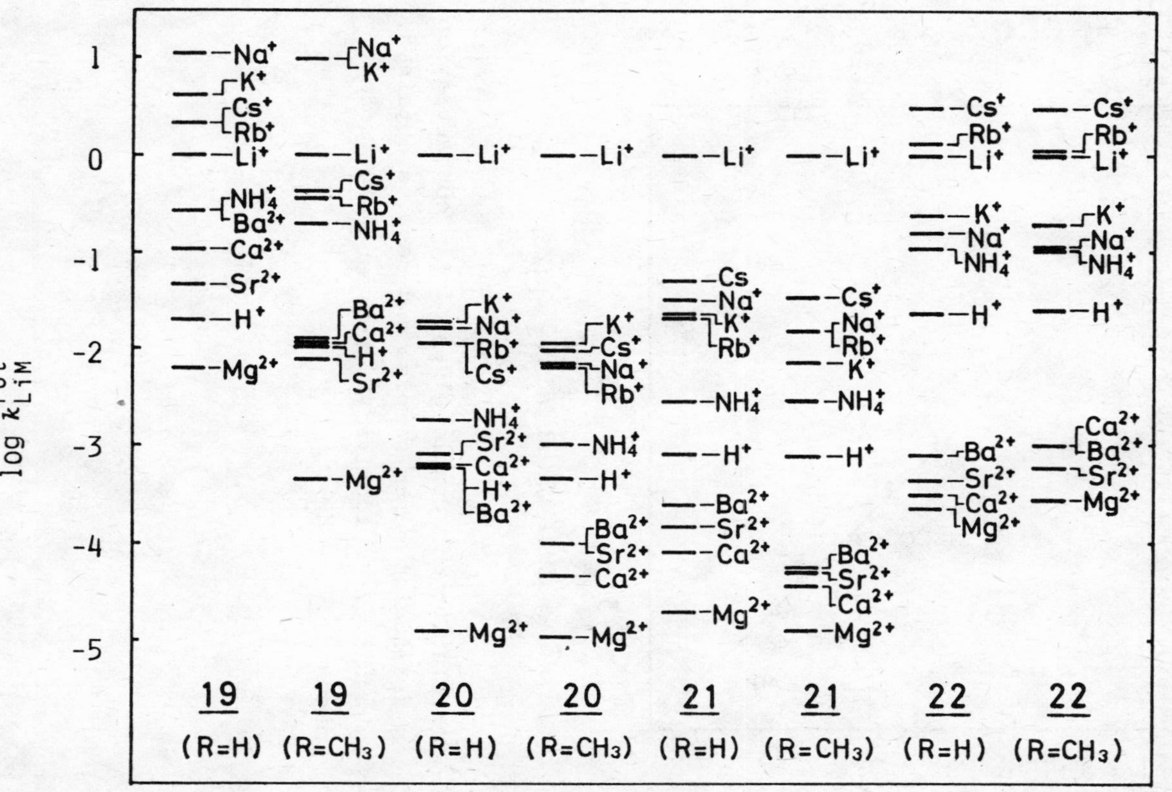

Fig. 5 Selectivity coefficients k_{LiM} for Li^+-selective PVC membrane electrodes based on lipophilic crown-4 derivatives 19 through 22.

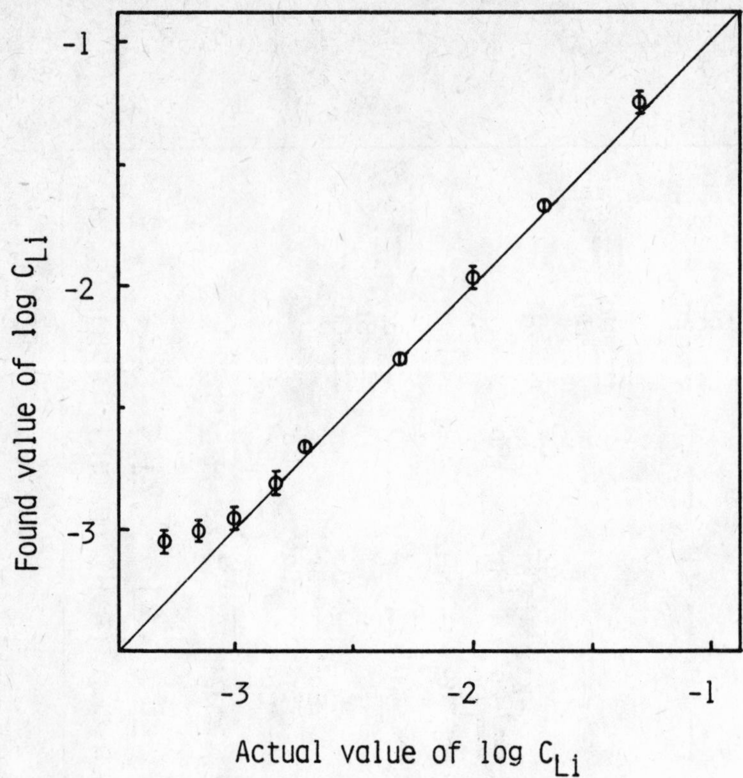

Fig. 6 Correlation of actual values of Li^+ concentration in artificial serum samples and results obtained by using Li^+-selective PVC membrane electrodes based on <u>20</u>(R=CH_3).

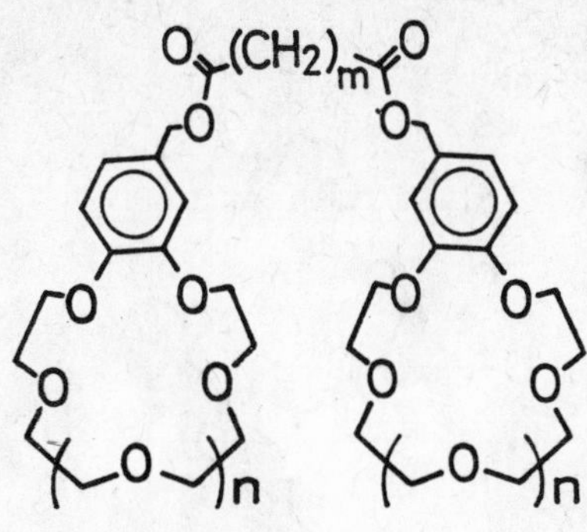

1(n=1;m=1,3,5,or 7)
2(n=2;m=3,5,or 7)

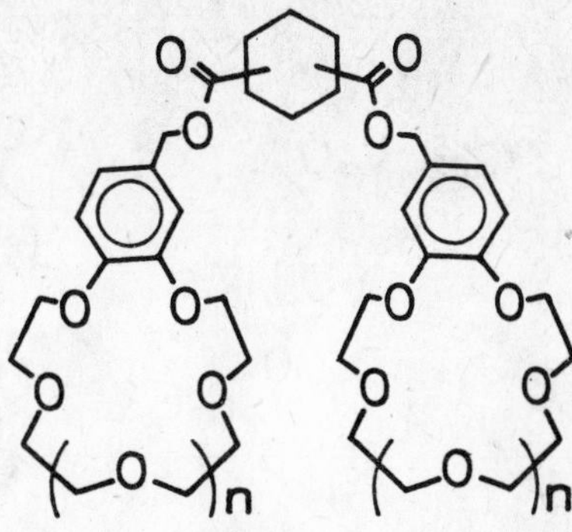

3(1,2-cis,1,2-trans, or 1,4-cis)

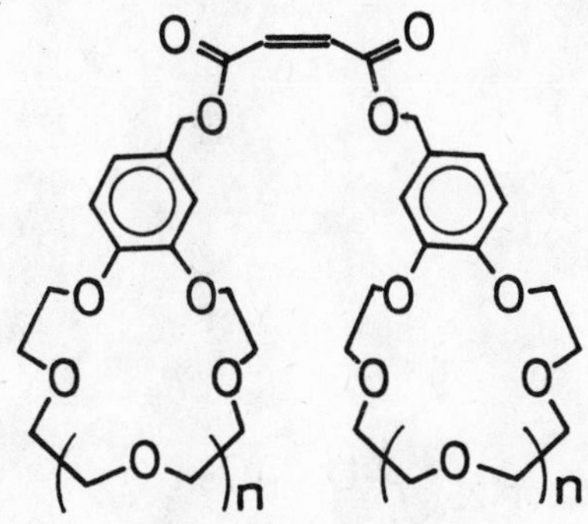

4(n=1;cis or trans)
5(n=2;cis or trans)

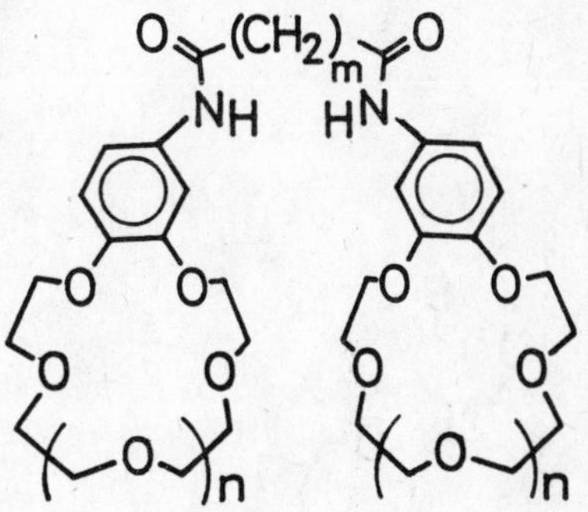

6(n=0;m=3)
7(n=1;m=3)
8(n=2;m=3)

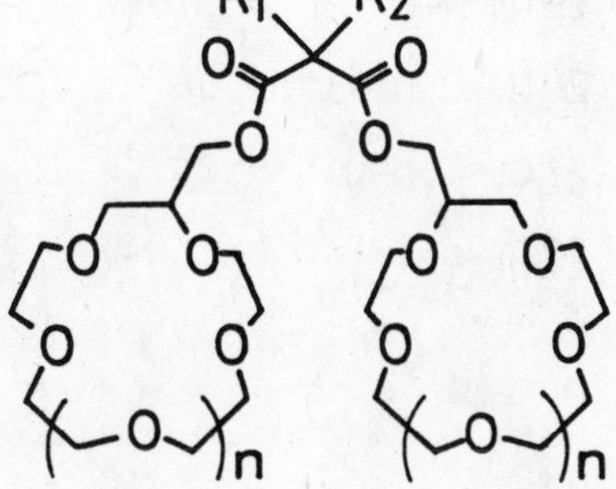

9(n=0; $R_1=R_2=C_2H_5$)
10(n=0; $R_1=CH_3$, $R_2=C_{12}H_{25}$)
11(n=1; $R_1=CH_3$, $R_2=C_{12}H_{25}$)
12(n=1; $R_1=H$, $R_2=C_{12}H_{25}$)
13(n=2; $R_1=CH_3$, $R_2=C_{12}H_{25}$)

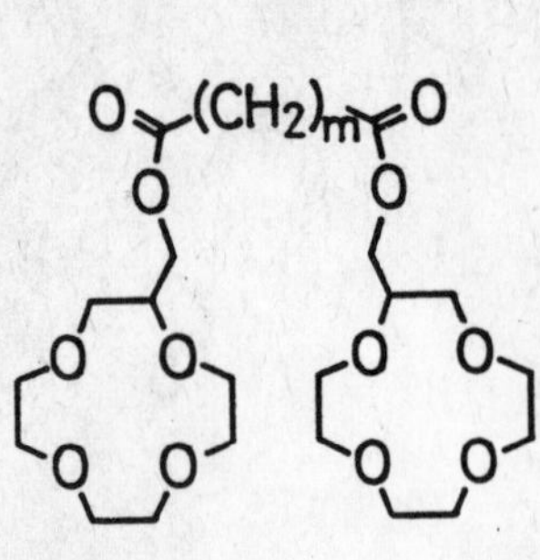

14(m=1,2,or 3)

15(R=$CO_2C_7H_{15}$)

16(R=$CO_2C_{11}H_{23}$)

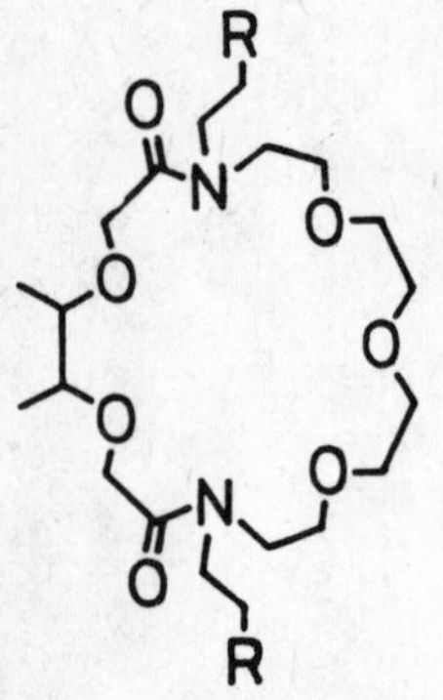

17(R=$CO_2C_7H_{15}$)

18(R=$CO_2C_{11}H_{23}$)

19(m=n=0; R=H or CH_3)

20(m=0, n=1; R=H or CH_3)

21(m=1, n=0; R=H or CH_3)

22(m=n=1; R=H or CH_3)

QUESTIONS AND COMMENTS

Participants of the discussion: R.P.Buck, K.Cammann, K.Kimura, E.Lindner, W.Simon, J.D.R.Thomas and A.Zhukov

Question:

You mentioned that you have done some work with crown ether systems that were incorporated into polymer chains. You also mentioned that the response was very much slower. That would presumably be due to the inability of the crown ether part of the sensing molecule to move about in the membrane. Is that so, and would you comment on that?

Answer:

Probably it is very hard for the polymer to move in the membrane, because it has such a big molecule, and that is why the electrode response is so slow.

Comment:

This really means that, taking organic sensors in general, there is little to gain in immobilizing the sensor into the polymer matrix.

Answer:

I have never tried to get a film of the poly-crown itself, I only tried a mixture of pvc and polycrown ether. But I think that the polycrown ether membrane will not work as an ISE.

Question:

In your earlier publications you always spoke about benzo-crown ethers, which had disadvantages, concerning lifetime etc. Later you turned to these lipophilic compounds but neither of them contains the benzene ring. This means that you changed two parameters simultaneously: you increased the chain length and omitted the benzene ring. Why did not you prepare the benzo-crown part with this lipophilic chain ?

Answer:

By choosing the structure we used our purpose was to increase the lipophilicity, and by omitting the benzene ring we wanted to achieve better selectivity due to steric effects.

Remark:

I have some concern about converting liquid membranes into solid membranes. I can say two examples: in 1969 valinomycin was mixed with solid paraffin and the electrode prepared did not respond to potassium. And this led some people to saying: it has to have mobility. And at the same time there was a so called solid calcium exchanger by taking the calcium salt of dioctyl hexaphosphonic acid and mixing it with collodium, and it did work. It was said that the reason it worked was that it had a trace of plasticizer in it and the ions had mobility.

Comment:

There may be a kind of difference between the various systems. In the crown ether system you may have to have the whole of the crown ether complex to migrate, whereas with some other systems it may be sufficient to have the ions just transferring from one side to the other. We should discuss each system individually.

Question:

You showed an example of bis-crown compounds where the two mono-crown units could have cis- or trans conformation. In the cis form, they can form a sandwich, whereas in the trans form they cannot. And the selectivity of the trans form is poorer than that of the cis form. What strikes me is that the selectivity of the trans form is poorer than that of the mono-crown compound. Have you an explanation for this ?

Comment:

If we assume that potassium ions form 1:2 complexes with the mono-crown ether, whereas the trans bis-crown cannot, we can explain the poorer selectivity for the trans-bis compound.

Comment:

However, I do not see the reason why the trans-bis compound cannot form a 1:2 complex.

Comment:

But there must surely be some effect, electrical effect imposed by the second ring in the double trans system.

Comment:

It is possible that two molecules with the trans configuration can join and complex one metal ion, but it might be kinetically hindered and the selectivity is poorer.

Question:

Can you say anything about the stability constants of the complexes of these crowns with metal ions ?

Answer:

No, I have never tried to measure them.

4th Symposium on Ion-Selective Electrodes
Mátrafüred, 1984

THEORETICAL TREATMENT OF THE DYNAMIC BEHAVIOUR OF PRECIPITATE BASED ION-SELECTIVE MEMBRANE ELECTRODES IN THE PRESENCE OF INTERFERING IONS

E. LINDNER, M. GRATZL and E. PUNGOR

Institute for General and Analytical Chemistry, Technical University of Budapest, Gellért tér 4, 1111 Hungary

ABSTRACT

Precipitate-based ion-selective electrodes respond to sudden changes in the interfering ion activity with non-monotonous, overshoot-type transient signals, when a certain amount of primary ion is also present in the solution and when the respective selectivity factor is much less than one. A simplified and a more detailed quantitative description of these signals are presented in terms of diffusion processes in the adherent solution layer, and of adsorption/desorption equilibria on the electrode membrane surface. The validity of this description is proved by excellent fittings to experimental signals in cases of: /i/ increasing and decreasing interfering ion activity steps; /ii/ subsequent changes in interfering ions activity; /iii/ interfering ion activity steps at different primary ion activity level. The selectivity factor, diffusion layer thickness values and the amount of ions adsorbed or desorbed providing good fitting were in agreement with the experimental values determined by other methods, within the respective experimental and calculation errors.

INTRODUCTION

During the past few years a considerable number of papers have been devoted to the time dependent selectivity behaviour and transient response of ion-selective electrodes in the presence of primary and interfering ions /1-10/. The current theories were summarized in an earlier publication of Lindner et al./1/ where it was demonstrated for the first time that also precipitate based ion-selective electrodes respond to sudden changes in interfering ion concentration by potential overshoot type transient functions. By interpreting the non-monotonous transient functions a new model was set up based on the assumption that following the sudden increase in activity of interfering ions, interfering ions are adsorbed and in parallel to that primary ions are desorbed from the electrode surface and consequently the ionic activities in the adhering solution layer - which determine the instantaneous value of the electrode potential -, temporarily differ from those in the bulk of sample. Similarly when the activity of interfering ions suddenly decreases interfering ions are desorbed and simultaneously primary ions are adsorbed on the electrode surface resulting again in a concentration change of the boundary layer. The difference in primary ion concentration thus created by the interfering ion activity step initiates diffusion processes towards equalization. In our model it is assumed that the sorption and ion exchange processes are much faster than the diffusion processes arising after the change in interfering ion activity. Hence, a dynamic description is needed for the diffusion only. The processes supposed to take place consecutively or simultaneously at the electrode solution interface are listed on a schematic representation showing all characteristic features of the transient signals recorded in the "two ion range" /Fig.1./.

The aim of the present paper to support our earlier qualitative model /1/ by independent analytical techniques and by the mathematical description of the curves. The quantitative description discussed in the present paper, however, is limited to the first two sections of the transient signals /Fig.1, phases A and B/ because in the slowest section of the curves /Fig.1. phase C/ most likely a chemical change of the membrane surface occurs and the rate determining role of the diffusion cannot be assumed anymore. Neither do we want to deal here with cases where the interfering ions form, with the cations or anions of the membrane, precipitates of lower solubility than that of the membrane material itself. Such cases /i/ were discussed in detail by Morf /2/ and others /3-5/, /ii/ do not result in non-monotonous signals but rather the opposite happens i.e. the electrode potential varies according to a monotonous asymptotic function; /iii/ irreversible transformation of the membrane material occurs analoguous to the processes determining the third phase of transient signals.

EXPERIMENTAL

In this work the experimental results of our earlier publication /1/ were used to control the efficiency and characteristic features of the mathematical description, i.e. the newly derived potential-time functions were fitted to the earlier experimental curves. For the calculations a HP-85 type desk-top computer was used.

RESULTS AND DISCUSSION

The potential response of precipitate based ion-selective electrodes in the presence of interfering ions can be given by the Nicolsky equation:

$$E = E_i^o - \frac{RT}{z_i F} \ln /a_i' + K_{ij} a_j'/ \qquad /1/$$

where E is the electrode potential, E_i^o is a reference potential term, independent from the sample solution activities, K_{ij} is the theoretical selectivity factor, which can be calculated from the solubility products of the respective precipitates, z_i, R, T, F have their usual electrochemical meanings and a_i' ; a_j' are the ion activities of primary and interfering ions respectively in the boundary solution layer being in direct contact with the membrane surface.

The theoretical selectivity factor, under the experimental conditions discussed here $/K_{ij} \ll 1/$ is practically the same with its measured value $/K_{ij}^{pot}/$, in contrast to the $K_{ij} \gg 1$ case which was discussed in detail by several authors /2-4/.

To describe the non-monotonous potential-time functions the time dependent surface activities $(a_i/x=o,t/;\ a_j/x=o,t/)$ should be inserted into the Nicolsky equation:

$$E_{/t/} = E_i^o - \frac{RT}{z_iF} \ln [c_i^o + c_i/x=o,t/ + K_{ij} \cdot c_j/x=o,t/] \qquad /2/$$

where x is the distance from the electrode surface, c_i^o is the bulk concentration level of the primary ions and $c_i/x=o,t/$ is the increase in the primary ion concentration with respect to c_i^o in the vicinity of the electrode surface, $c_j/x=o,t/$ is the concentration of the interfering ions at the membrane surface /x=o/ following the activity change at $x = \delta$.
However to describe the surface activities as function of time, the differential equations describing the counterflowing diffusion processes of primary and interfering ions should be solved simultaneously. This task is made more difficult by the fact that the boundary condition of one of the differential equations depends on the solution of the other. Hence several assumptions were made in interest of mathematical simplifications.

As first approximation an instantaneous desorption of primary ions was supposed and the drained amount of interfering ions /chemisorbed by the electrode surface/ was neglected,moreover $c_j/x=o,t/$ was replaced by the $c_j/x=o,t=\infty/$ term in Eq.2. These simplifications could be supported by the experimental conditions. According to these assumptions the concentration of primary ions with fast convergence in the small time range $/\sqrt{Dt} \lll \delta/$ can bw given:

$$c_i/x,t/ = \frac{M_i}{\sqrt{\pi D't}} \left\{ \exp\left(-\frac{x^2}{4\,D't}\right) + \sum_{k=1}^{\infty} /-1/^k \left[\exp\left(-\frac{/2k\delta + x/^2}{4\,D't}\right) + \right.\right.$$
$$\left.\left. + \exp\left(-\frac{(2\,k\delta - x)^2}{4\,D't}\right)\right]\right\} \qquad /3/$$

where x is the distance from the electrode surface, x=o at the electrode surface and $x=\delta$ at the other boundary of the diffusion layer as it is shown on Fig.2, M_i is the quantity of primary ions forced to desorb from the electrode surface, and D' in the mean diffusion coefficient.

Next after inserting Eq./3/ into Eq/2/the potential-time function thus obtained was fitted to experimental transient curves. E^o and S were determined by prior calibration, c_i^o and $c_j/x=o,t=\infty/$ were given by the experimental conditions, while D' was taken to be 1.86×10^{-5} cm^2/s. During fitting K_{ij}, M_i and δ were varied and the best solution was searched for. The results /Figs.3,4,5/ seemed to justify the assumptions.

In a more comprehensive approach the surface concentration change of interfering ions $(c_j/x=o,t/)$ was taken quantitatively into consideration /Eq.4//11,12/, furtheron simultaneously two diffusion processes /of finite rate and reverse direction/ were considered /Fig.2/, by determining the quantity of primary ions desorbed from the electrode surface as function of the time dependent activity of the interfering ions.

$$c_j/x=o,t/ = c_j/x,t=\infty/ \left[1- \frac{4}{\pi} \sum_{k=o}^{\infty} \frac{/-1/^k}{2k+1} \exp \left(- \frac{D' /2k+1/^2 \pi^2 t}{16 \delta^2} \right) \right] \qquad /4/$$

This more detailed problem involving finite diffusion rates for both components can be given as the mathematical convolution of the solution of the separate diffusion problems /13/; one of them corresponds to the instantaneous desorption of primary ions and their diffusion into the bulk of the sample solution /Eq.3/ while the other to the diffusion of interfering ions towards the electrode surface /Eq.4/:

$$c'_i /x=o,t/ = \int_o^t c'_i /t-\tau/ \, . \, f /\tau/ \, d\tau \qquad /5/$$

where c'_i is given by Eq.3 for unit amount of desorbed ions $/M_i=1/$ and

$$f/\tau/ = \left(\frac{\partial M_i}{\partial t} \right)_{t=} = - K \frac{\partial c_j /x=o,t/}{\partial t} \qquad /6/$$

where K is the slope of the adsorption isotherm and $c_j/x=o,t/$ is given by Eq.4.

Thus Eq.3 and the derivative of Eq. 4 have to be inserted into Eq.5 and combined with Eq.4 and Eq. 2 to obtain the time dependence of the electrode potential:

$$E/t/ = E_i^o - s\log\left\{ c_i^o + \frac{K\sqrt{\pi D'}\; c_j/x=o,t=\infty/}{4\delta^2\sqrt{t}} \cdot \right.$$

$$\cdot \int_o^t \left[/1+2\sum_{k=1}^{\infty} /-1/^k \exp\left(-\frac{k^2\delta^2}{D'/t-\tau/}\right)\right] \cdot \qquad /7/$$

$$\left[\sum_{k=o}^{\infty} /-1/^k\; /2k+1/ \exp\left(-\frac{D'/2k+1/^2\pi^2 t}{16\delta^2}\right)\right] +$$

$$\left. + K_{ij}\; c_j\; /x=o,t=\infty/ \left[1-\frac{4}{\pi}\sum_{k=o}^{\infty} \frac{/-1/^k}{2k+1} \exp\left(-\frac{D'/2k+1/^2\pi^2}{16\delta^2}t\right)\right]\right\}$$

The fitting of Eq./7/ to the experimental data /Fig.6/ support the earlier qualitative model /1/ and the assumptions used in the derivation of the more simplified approach. This model clears up the concentration dependence of the transient signals /Fig.3/, can explain the effect of the direction of activity change /Fig. 3 and 4/ and appropriate to describe transient signals following activity increase or decrease in several subsequent steps /Fig. 5/. The K_{ij}, M_i and δ values determined in the course of fitting to various experimental curves are fairly similar to the values determined by other methods /14,15/.

Attempt has been made to find also an experimental proof for our model assumptions. For this the amount of iodide ions adsorbed at the surface of AgI pellets from 10^{-5} mol/l labelled iodide solution as well as the desorbed part of this quantity by placing the pellets into solutions containing different bromide ion concentration /16/ were determined. The experimental data thus obtained and those measured by AAS technique on CuS based copper/II/ electrodes /17/ supported also the validity of the model assumptions.

ACKNOWLEDGEMENT

The authors are indebted to K.Tóth for her valuable comments and contribution.

REFERENCES

1 E.Lindner, K.Tóth, E.Pungor, Anal.Chem., 54, /1982/ 202
2 W.E.Morf, Anal.Chem., 55, /1983/ 1165
3 A.Hulanicki, A.Lewenstam, Anal.Chem. 53, /1981/ 1401
4 A.Hulanicki, A.Lewenstam, Talanta, 24, /1977/ 171
5 R.K.Rhodes, R.P.Buck, Anal.Chim.Acta, 113, /1980/ 67
6 J.Bagg, R.Vinen, Anal.Chem., 44, /1972/ 1773
7 F.A.Schultz, R.E.Reinsfelder, Anal.Chim.Acta, 65, /1973/ 425
8 D.E.Mathis, F.S.Stover, R.P.Buck, J.Membr.Sci., 4, /1979/ 395
9 K.Cammann, "Das Arbeiten mit ionenselektiven Elektroden" Springer Verlag, Berlin, Heidelberg, New York, 1977
10 W.E.Morf, Anal.Letters, 10, /1977/ 87
11 W.E.Morf, E.Lindner, W.Simon, Anal.Chem., 47, /1975/ 1596
12 R.P.Buck "Ion-Selective Electrodes in Analytical Chemistry" H.Freiser ed. Plenum, New York, 1978, Chapter 1
13 M.Gratzl, E.Lindner, E.Pungor, send for publication to Anal.Chem.
14 A.Hulanicki, A.Lewenstam, M.Maj-Zurawska, Anal.Chim.Acta, 107, /1979/ 121
15 E.G.Harsányi, K.Tóth, L.Pólos, E.Pungor, Anal.Chem., 54, /1982/ 1094
16 E.Lindner, A.Farkas, K.Tóth, E.Pungor, in preparation
17 E.G.Harsányi, K.Tóth, E.Pungor, Paper presented on the Fourth Scientific Section on Ion-Selective Electrodes, Mátrafüred 8-12 October, 1984

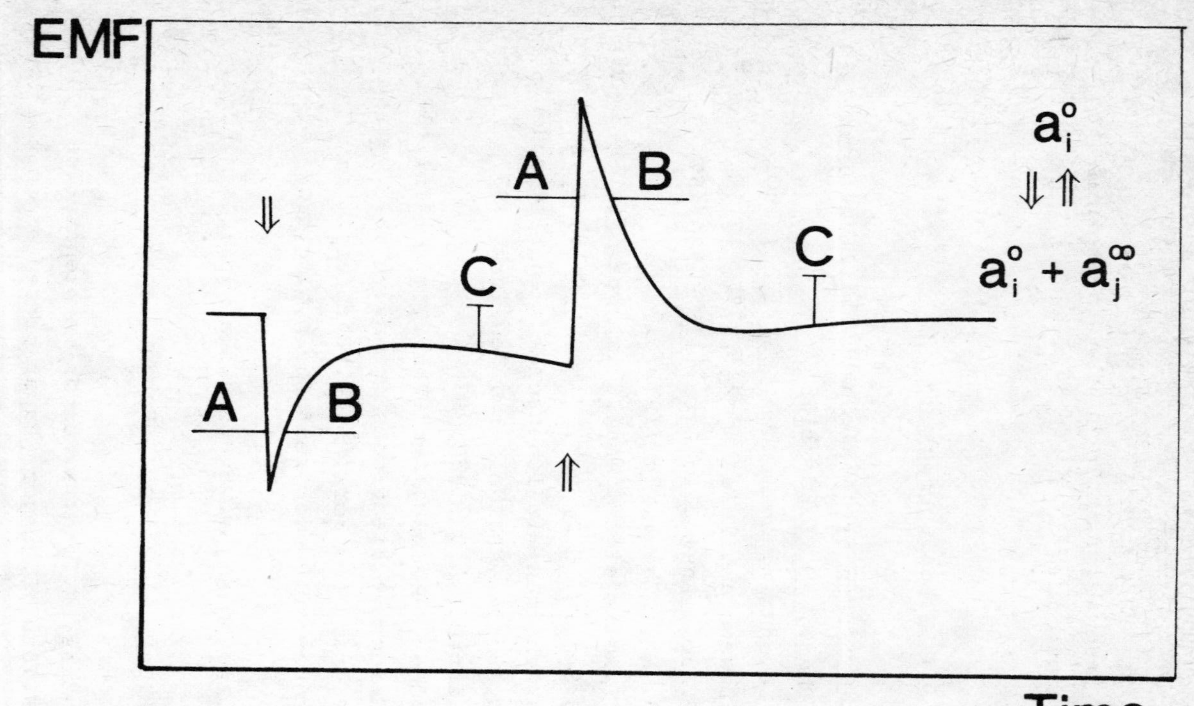

Fig. 1. Typical non-monotonous transient signals following interfering ion activity increase or decrease
A: the overshoot phase /ms range/,
B: the relaxation phase /s range/,
C: slow surface transformation phase /min range/

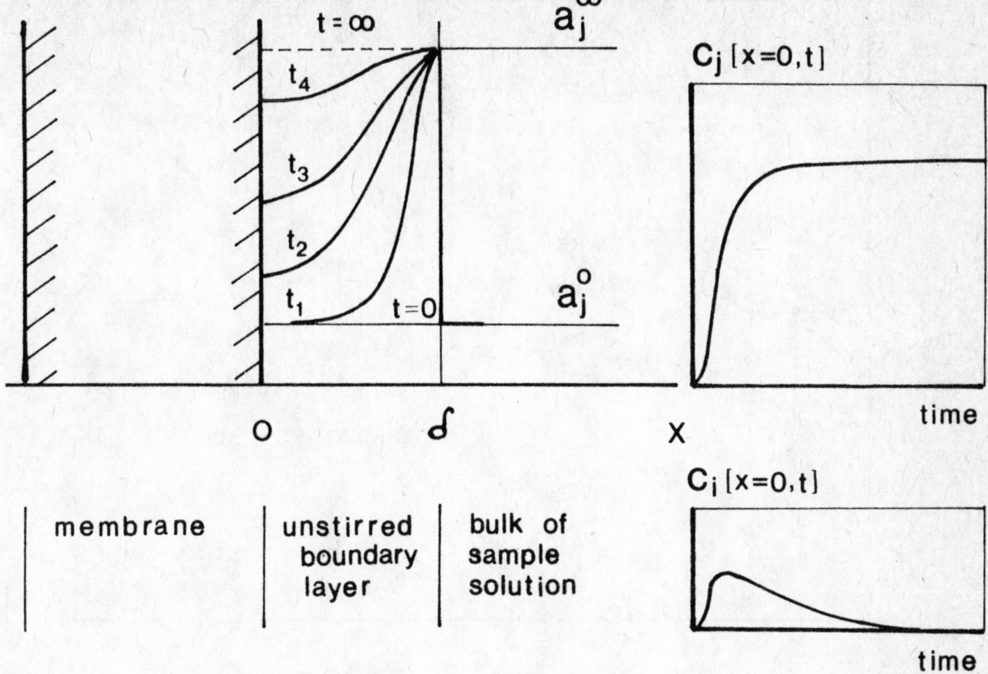

Fig. 2. Schematic diagrams displaying the assumptions used for the mathematical description

x: length coordinate perpendicular to the membrane surface /x=o at the surface, x= δ at the other boundary of the diffusion layer/

t: time /t=o at the stepwise change in the interfering ion activity/

For other symbols see the text

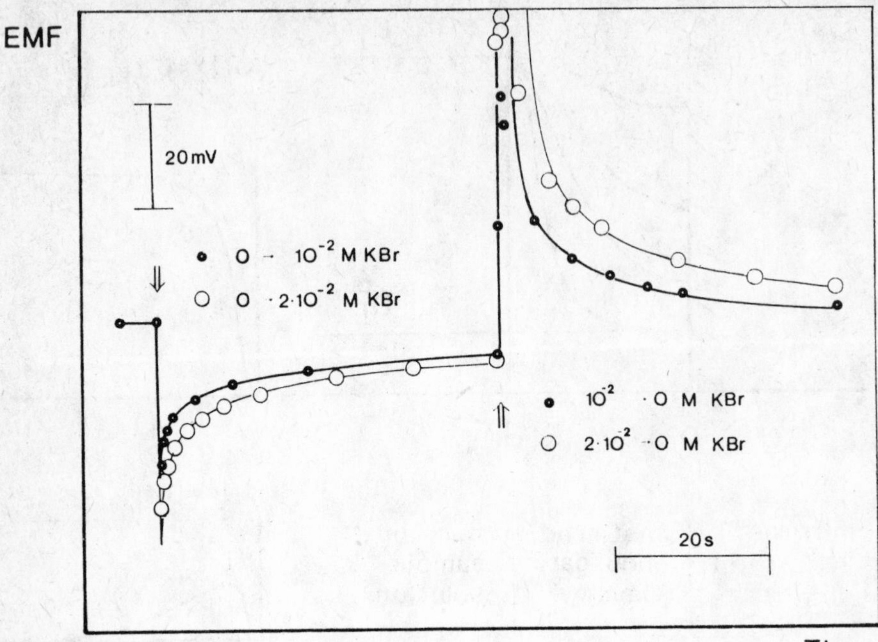

Fig. 3. Time dependent response of a silver iodide precipitate based electrode to solutions containing 10^{-5} M KI and different bromide concentrations: /solid lines/ theoretical curves calculated from eq. 6 and 7 using the following fitting parameters: $\delta = 2{,}9.10^{-2}$ cm, $K_{ij} = 2{,}5.10^{-4}$, $M_i = 9.3.10^{-8}$ mmol/cm^2 /for a bromide activity step: $0 \Leftrightarrow 10^{-2}$ M KBr/ and $\delta = 4{,}95.10^{-2}$ /$4{,}1.10^{-2}$/ cm, $K_{ij} = 4.10^{-4}$, $M_i = 1{,}8.10^{-7}$ /$1{,}4.10^{-7}$/ mmol/cm^2 /for a bromide activity step: $0 \Rightarrow 2.10^{-2} \Rightarrow$ 0 M KBr respectively/; /points/ experimental values at constant ionic strength /10^{-1} M KNO_3/

o activity step $0 \Leftrightarrow 10^{-2}$ M KBr

● activity step $0 \Leftrightarrow 2.10^{-2}$ M KBr

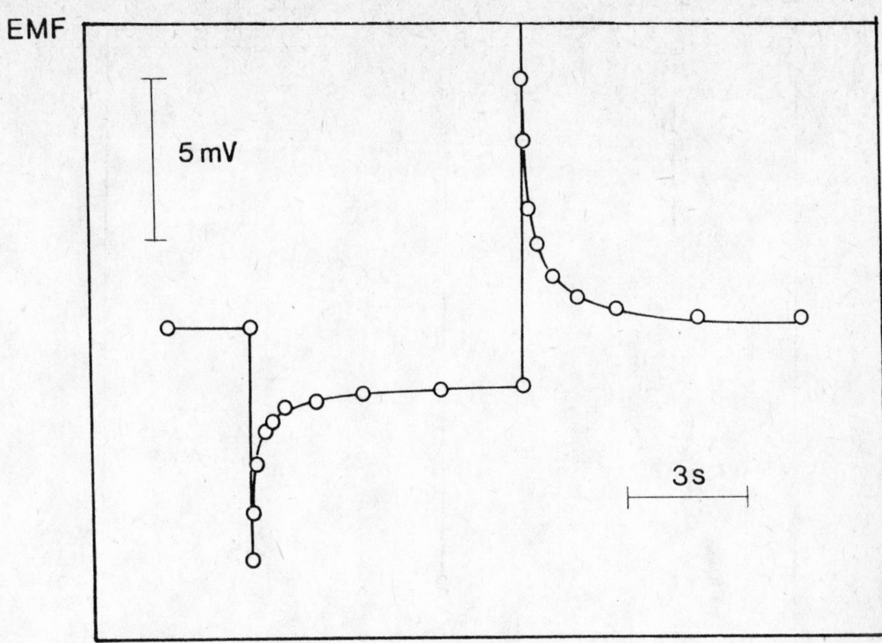

Fig. 4. Transient signals measured /O/ and calculated /solid line/ with Eqs. 6 and 7 at 10^{-4} M KI activity level and for a bromide activity step: 0 $\Leftrightarrow$ 10^{-1} M KBr. Fitted parameter values: $\delta = 8.10^{-3}$ cm, $K_{ij} = 7{,}9 \cdot 10^{-5}$, $M_i = 2{,}5.10^{-8}$ /$4{,}5.10^{-8}$/ mmol/cm^2. The fitted δ and K_{ij} values are identical for activity increase and decrease but slightly different for M_i /value in parentheses/.

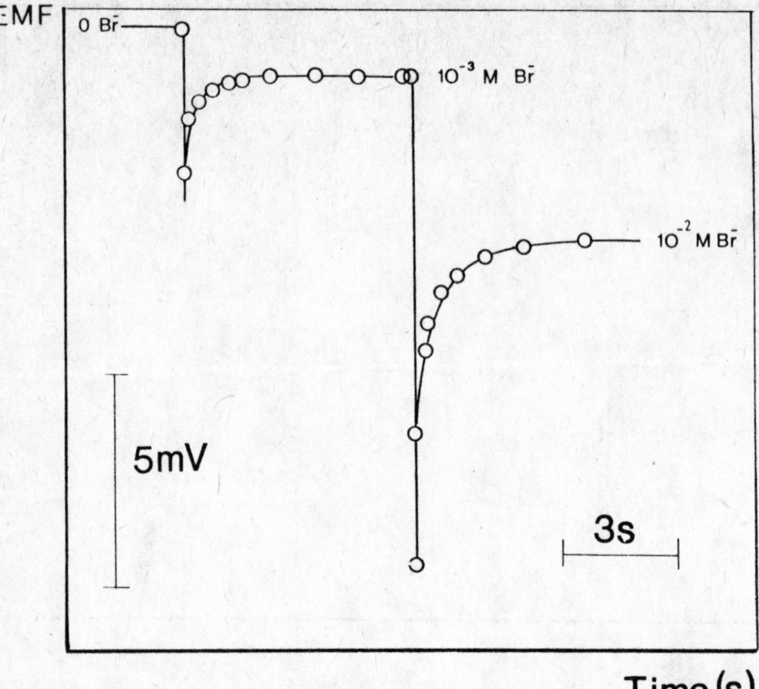

Fig. 5. Transient signals recorded with a precipitate based iodide electrode at 10^{-5} M KI level following subsequent changes in the interfering ion activity: /Solid line/ theoretical curve calculated with Eqs. 6 and 7; /points/ experimental values measured at constant activity steps: $0 \Rightarrow 10^{-3}$ M KBr; 10^{-3} M KBr $\Rightarrow 10^{-2}$ M KBr

Fitted parameters: $\delta = 3{,}2.10^{-3}$ cm, $K_{ij} = 5{,}2.10^{-4}$, $M_i = 2{,}5.10^{-9}$ mmol/cm^2 for $C^{\infty}_{Br} = 10^{-3}$ M KBr and $\delta = 3{,}2.10^{-3}$ cm, $K_{ij} = 2{,}5.10^{-4}$, $M_i = 1{,}8.10^{-8}$ mmol/cm^2 for $C^{\infty}_{Br} = 10^{-2}$ M KBr

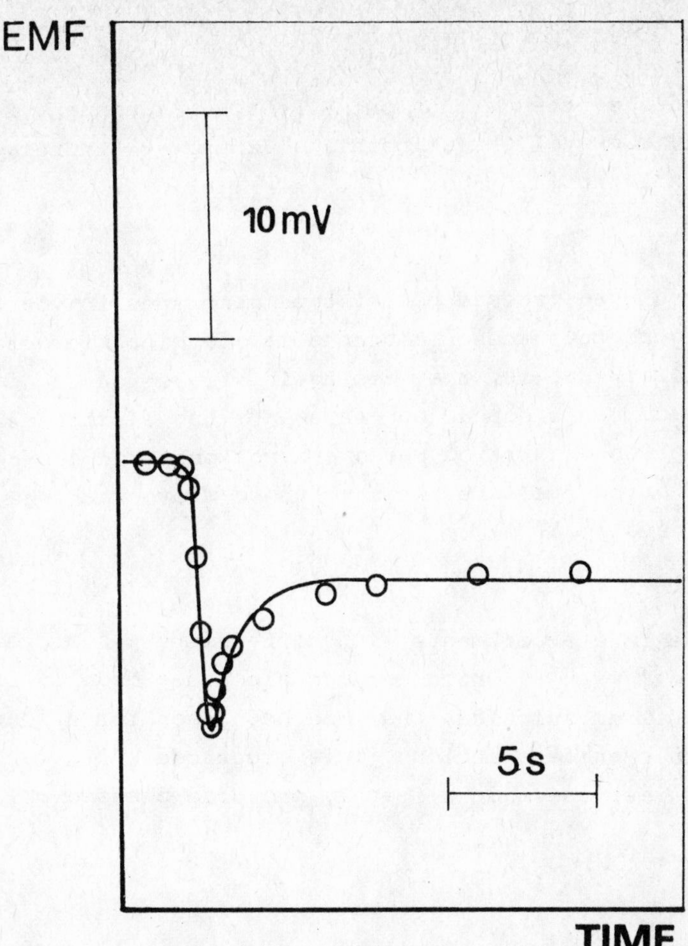

Fig. 6. Transient signal recorded at 10^{-5} M KI activity level and constant ionic strength /10^{-1} M KNO_3/: /solid line/ theoretical curve calculated according to the more detailed model /eq.17/. Experimental and fitting parameters: $C_I^o = 10^{-5}$ M KI; $C_{Br}^o = 0$ M KBr, $C_{Br}^{\infty} = 10^{-2}$ M KBr, $\delta = 1{,}1.10^{-3}$ cm, $K_{ij} = 2{,}05.10^{-4}$, $K = 1.85.10^{-6}$ cm; /points/ experimental values.

QUESTIONS AND COMMENTS

Participants of the discussion: R.P.Buck, K.Cammann, E.Gráf-Harsányi, M.Gratzl, N.Ishibashi, J.Janata, E.Lindner, Z.Noszticzius and E.Pungor

Question:

Have you used pressed pellet type membranes in your studies? We have found that these membranes may develop microcracks after a time, that is, they are ageing. Have you found the transient curve to depend on the age of the electrodes? Do the adsorption-desorption phenomena you observed depend on the state of the pellets, whether these microcracks have or have not developed?

Answer:

We made experiments with different types of pellets, first of all with an Orion cyanide electrode made of silver iodide + silver sulphide, that had been used for a long period of time in cyanide solutions. This electrode had a much slower transient response than a freshly prepared pressed-pellet membrane.

Comment:

A paper was already published on the so called corrosion electrodes and a similar effect was found.

Question:

You say that the transient potential change comes from the desorption and diffusion of the primary ion into the bulk of the solution. How about the diffusion into the membrane surface?

Answer:

These transient signals were found to consist of three sections. The first one running upwards is in the millisecond range, the second one in the second range. We discussed only the first two sections in the paper. It seems to me that in this short time no chemical change could take place, it could only occur in the third section, if we exposed the electrode to a solution of the interfering ion of high concentration.

Question:

Was it checked that no change occured on the membrane surface ?

Answer:

We studied freshly prepared silver iodide membranes and to check that there is really no change of the surface we prepared membranes by pressing a small amount of bromide onto the electrode surface. We did the same experiment with those and the transient signals obtained were completely different from those found with freshly prepared silver iodide membranes. From this we concluded that within the time of the first two sections no significant chemical change could take place at the electrode surface.

Question:

What justifies the assumption of the adsorption-desorption model ?

Answer:

The direction of the potential change could be explained by assuming adsorption or desorption. A potential change in the direction corresponding to more concentrated solution indicated a desorption, whereas the opposite process an adsorption. In addition to this, our assumptions have also been supported by measurements with atomic absorption spectroscopy and radioanalytical methods.

Question:

Do you know how many adsorbed layers you have? You can calculate it, since you know the amount adsorbed and the surface area of the electrode.

Answer:

The amount calculated by us is much less than the saturation level.

Comment:

Studying adsorption and desorption of ions at the surface of ISEs I have found that the maximum amount of adsorbed ions corresponds to a monolayer. When we find bound amounts higher than that corresponding to a monolayer, a transformation of the membrane surface is expected.

4th Symposium on Ion-Selective Electrodes
Mátrafüred, 1984

NEW DEVELOPMENTS AND CONSEQUENCES OF THE THEORY OF ION-SELECTIVE MEMBRANES

W.E. MORF, P. OGGENFUSS and W. SIMON

Department of Organic Chemistry
Swiss Federal Institute of Technology
CH-8092 Zürich, Switzerland

ABSTRACT

A unified approach to the theory of ion-selective liquid membrane electrodes is presented. The potential response and selectivity behavior towards monovalent cations and anions is described for membranes containing different ion-exchange sites and neutral carriers. Results are discussed for novel sensors responsive to nitrite, chloride, and enantiomeric organic ammonium ions.

INTRODUCTION

Ion-selective liquid membrane electrodes are among the most widely used potentiometric sensors [1-3]. In conventional ion-exchanger-based systems the organic membrane phase permanently contains highly lipophilic sites R^- (e.g. tetraphenylborate ions, in cation exchangers) or Q^+ (e.g. quaternary ammonium ions, in anion exchangers) which lead to a selective extraction of counterions M^{z_m} and X^{z_x}, respectively, and to a Donnan exclusion of coions. The selectivity behavior of typical neutral carrier membrane electrodes, on the other hand, is largely determined by the presence of membrane components L (carrier antibiotics or synthetic ionophores) which form charged complexes of the type $ML_n^{z_m}$ with certain cations. Various combined membrane systems were also introduced which incorporate both cation-selective neutral carriers and negatively charged sites [2-4]. Very recently, even anion-selective carriers L^* forming complexes $XL_n^{*z_x}$ were discovered [5,6].

A respectable number of theoretical treatments was devoted to the different types of ion-selective liquid membrane electrodes (for a review, see [3,7]). All these treatments are restricted to more or less idealized systems, however. Therefore, the present work offers a unified approach to the theory of liquid membrane potentials, and an intercomparison of the performance of different types of analytically relevant sensors. For simplicity, only membrane systems with two classes of ions, namely monovalent cations and monovalent anions, will be discussed.

THEORETICAL TREATMENT OF LIQUID MEMBRANE POTENTIALS

A general formal description of the membrane potential built up across liquid membranes was obtained by extension of Planck's theory of liquid junctions [3,8]. Accordingly, the membrane potential E_M is given as a function of the activities a of all permeating cations (index m) and anions (index x) present in the boundary zones of the two aqueous solutions (denoted by (') and (")), of the distribution coefficients k of the ions involved, and of the integral transference numbers τ of cations and anions ($\tau_m + \tau_x = 1$) [3,8]:

$$E_M = \tau_m \frac{RT}{F} \ln \frac{\sum_m k_m a'_m}{\sum_m k_m a''_m} - \tau_x \frac{RT}{F} \ln \frac{\sum_x k_x a'_x}{\sum_x k_x a''_x} \quad (1)$$

The potential-determining ions that may occur in liquid membranes are

$$m : M^+, ML_n^+, Q^+$$

$$x : X^-, XL_n^{*-}, R^-$$

Hence, the general result for the membrane potential assumes the form:

$$E_M = \tau_m \frac{RT}{F} \ln \frac{\sum_M k_M a'_M + \sum_M \sum_L \sum_n k_{ML_n} a'_{ML_n} + \sum_Q k_Q a'_Q}{\sum_M k_M a''_M + \sum_M \sum_L \sum_n k_{ML_n} a''_{ML_n} + \sum_Q k_Q a''_Q}$$

$$- \tau_x \frac{RT}{F} \ln \frac{\sum_X k_X a'_X + \sum_X \sum_{L^*} \sum_n k_{XL^*_n} a'_{XL^*_n} + \sum_R k_R a'_R}{\sum_X k_X a''_X + \sum_X \sum_{L^*} \sum_n k_{XL^*_n} a''_{XL^*_n} + \sum_R k_R a''_R}$$

respectively:

$$E_M = \tau_m \frac{RT}{F} \ln \frac{\sum_M K'_M a'_M + \sum_Q k_Q a'_Q}{\sum_M K''_M a''_M + \sum_Q k_Q a''_Q} - \tau_x \frac{RT}{F} \ln \frac{\sum_X K'_X a'_X + \sum_R k_R a'_R}{\sum_X K''_X a''_X + \sum_R k_R a''_R} \quad (2)$$

where the over-all distribution coefficients K_M and K_X account for the formation of cationic and anionic carrier complexes within the membrane (complex stability constants β_{ML_n} and $\beta_{XL^*_n}$; concentrations c_L and c_{L^*} of free carriers at the two membrane boundaries (0) and (d)):

$$K'_M = (1 + \sum_L \sum_n \beta_{ML_n} [c_L(0)]^n) k_M \quad (3a)$$

$$K''_M = (1 + \sum_L \sum_n \beta_{ML_n} [c_L(d)]^n) k_M \quad (3b)$$

$$K'_X = (1 + \sum_{L^*} \sum_n \beta_{XL^*_n} [c_{L^*}(0)]^n) k_X \quad (4a)$$

$$K''_X = (1 + \sum_{L^*} \sum_n \beta_{XL^*_n} [c_{L^*}(d)]^n) k_X \quad (4b)$$

These parameters evidently characterize the fundamental distribution equilibria:

free ions (aqueous) $\rightleftharpoons$ ionic species (membrane)

Equations (2) - (4) constitute the basis for specific theoretical descriptions of cation-selective and anion-selective membrane electrodes.

In many earlier treatments, discussions were restricted to liquid membranes with dissociated ion-exchange sites, R^- or Q^+, which were assumed to induce ideal permselectivity for cations or anions. The result for this simple case formally corresponds to Eq. (2) with τ_m=1 or τ_x=1 [3,7-9]:

$$E_M = \frac{RT}{F} \ln \frac{\sum_M K'_M a'_M}{\sum_M K''_M a''_M} \quad \text{or} \quad -\frac{RT}{F} \ln \frac{\sum_X K'_X a'_X}{\sum_X K''_X a''_X} \quad (5)$$

Ion-selective liquid membrane electrodes of this type would generally exhibit an EMF response according to the Nicolsky equation [10]:

$$EMF = E^0 + \frac{RT}{zF} \ln [a'_I + \sum_J K^{Pot}_{IJ} a'_J] \quad (6)$$

where I and J are primary and interfering ions of the same charge z, respectively, a'_I and a'_J are the ion activities in the sample film contacting the membrane, and E^o is the standard potential of the cell. From Eqs. (3)-(6) the potentiometric selectivities K_{IJ}^{Pot} are identified with the ratios of ionic distribution parameters:

$$K_{IJ}^{Pot} = \frac{K'_J}{K'_I} \cong \frac{(\sum_L \sum_n \beta_{JL_n} c_L^n)}{(\sum_L \sum_n \beta_{IL_n} c_L^n)} \cdot \frac{k_J}{k_I} \text{, for neutral carrier membranes} \quad (7)$$

$$K_{IJ}^{Pot} = \frac{k_J}{k_I} \text{, for pure ion-exchanger membranes} \quad (8)$$

There is ample evidence, however, that association between ions cannot be neglected for the liquid membranes applied in ion sensors, especially when relatively nonpolar solvents or plasticizers are used [2,3,11,12]. For example, Figure 1 documents that association constants in dibutyl sebacate are typically of the order of $10^4\ M^{-1}$. We should therefore focus on associated membrane systems that predominantly contain electrically neutral ion pairs MR (concentration c_{MR}, association constant K_{MR}) or QX (c_{QX}, K_{QX}) and complexes ML_nR (c_{ML_nR}, K_{ML_nR}) or QXL^*_n ($c_{QXL^*_n}$, $K_{QXL^*_n}$). A general description of such membranes can be derived from Eq. (2) using the relationships for the total concentration of ion-exchange sites, e.g., in the case of cation-selective systems:

$$c_R^{tot} \cong \sum_M c_{MR} + \sum_M \sum_L \sum_n c_{ML_nR} = \sum_M K_M(R)\, a_M \cdot k_R\, a_R \quad (9)$$

$$K_M(R) = (K_{MR} + \sum_L \sum_n K_{ML_nR}\, \beta_{ML_n}\, c_L^n)\, k_M \quad (10)$$

where the parameter defined in Eq. (10) is characteristic of the following distribution process:

free ions (aqueous) $\rightleftharpoons$ associated species (membrane)

Substitution for the terms $k_R a_R$ in Eq. (2) then leads to the final result:

$$E_M = \tau_m \frac{RT}{F} \ln \frac{\sum_M K'_M a'_M}{\sum_M K''_M a''_M} - \tau_x \frac{RT}{F} \ln \frac{\sum_R \frac{c_R^{tot}}{\sum_M K'_M(R)\, a'_M} + \sum_X k_X a'_X}{\sum_R \frac{c_R^{tot}}{\sum_M K''_M(R)\, a''_M} + \sum_X k_X a''_X} \quad (11)$$

This description encompasses cation-selective neutral carrier membranes, pure liquid ion-exchanger membranes, combined systems, homogeneous- and heterogeneous-site glass membranes (liquid-state approach), and even lipid bilayer membranes in the "equilibrium domain". The relation applies to ideally permselective membranes as well as to systems with failure of coion exclusion (anion interference). The former results for membranes with dissociated ion-exchangers can also be derived from Eq. (11) (e.g., for τ_X=0). An analogous relationship can be written for anion-selective liquid membranes. Hence, in the usual case of a permselective sensor membrane with only one sort of ionic sites, the EMF response to primary ions I and interfering ions J of the same charge z becomes:

$$\mathrm{EMF} = E^o + (1-\tau)\frac{RT}{zF} \ln [a'_I + \sum_J K_{IJ}^{(1)} a'_J] + \tau \frac{RT}{zF} \ln [a'_I + \sum_J K_{IJ}^{(2)} a'_J] \quad (12)$$

where the transference number τ is a measure of the relative mobility of the charged sites in the membrane, and the two selectivity coefficients $K_{IJ}^{(1)}$ and $K_{IJ}^{(2)}$ are given by the ratios of the basic distribution parameters (see Eqs. (3), (4), and (10)):

$$K_{IJ}^{(1)} \cong \frac{(\sum_L \sum_n \beta_{JL_n} c_L^n)\, k_J}{(\sum_L \sum_n \beta_{IL_n} c_L^n)\, k_I} \quad \text{or} \quad \frac{(\sum_{L*} \sum_n \beta_{JL*_n} c_{L*}^n)\, k_J}{(\sum_{L*} \sum_n \beta_{IL*_n} c_{L*}^n)\, k_I} \quad (13)$$

$$K_{IJ}^{(2)} \cong \underbrace{\frac{(\sum_L \sum_n K_{JL_nR}\, \beta_{JL_n} c_L^n)\, k_J}{(\sum_L \sum_n K_{IL_nR}\, \beta_{IL_n} c_L^n)\, k_I}}_{\text{cation-selective carrier membranes}} \quad \text{or} \quad \underbrace{\frac{(\sum_{L*} \sum_n K_{QJL*_n}\, \beta_{JL*_n} c_{L*}^n)\, k_J}{(\sum_{L*} \sum_n K_{QIL*_n}\, \beta_{IL*_n} c_{L*}^n)\, k_I}}_{\text{anion-selective carrier membranes}} \quad (14)$$

respectively:

$$K_{IJ}^{(1)} = \frac{k_J}{k_I} \tag{15}$$

$$K_{IJ}^{(2)} = \underbrace{\frac{K_{JR}\, k_J}{K_{IR}\, k_I}}_{\substack{\text{cation-ex-}\\\text{changer}\\\text{membranes}}} \quad \text{or} \quad \underbrace{\frac{K_{QJ}\, k_J}{K_{QI}\, k_I}}_{\substack{\text{anion-ex-}\\\text{changer}\\\text{membranes}}} \tag{16}$$

The results obtained for pure ion-exchanger-based membrane electrodes conform to the expressions derived from earlier treatments [3,7-9]. In contrast, no comparable description of neutral carrier membrane electrodes has been presented so far.

DISCUSSION OF RESULTS

Since the beginning of the era of sensor technology, numerous attempts were made to utilize the ion selectivity of electrically charged components in membrane electrodes (liquid ion-exchanger type). As is shown above, however, the ion binding affinity of such ligands, as characterized by the association constants in Eq. (16), cannot be fully exploited for electrode applications. The reason is that the basic requirement of an extremely high site mobility (i.e. $\tau \cong 1$) is hardly fulfilled in reality, except for glass electrodes where charge transport by a vacancy mechanism seems to be operative [3,13]. On the other hand, ion-exchange sites will exert no selectivity-enhancing effect when they are

a) predominantly dissociated,

b) poorly mobile or even chemically bound (Eq. (12) with $\tau \cong 0$), or

c) poorly selective (Eq. (12) with $K_{IJ}^{(1)} \cong K_{IJ}^{(2)}$).

In all three cases the EMF response is given by the Nicolsky equation (6), and the selectivity exhibited among different counterions simply reflects the extraction properties of the membrane solvent (see Eq. (8)) and can be correlated with the lipophilicity of the ions or with their hydration energies [3,14]:

$$\log K_{IJ}^{Pot} \cong \text{const}\cdot(\Delta G_H^o(J) - \Delta G_H^o(I)) \quad (17)$$

Such behavior is found for virtually all anion-selective liquid membrane electrodes reported so far (for a review, see [15]). As a rule, these sensors show the same selectivity sequence (Hofmeister series, see Figure 2 left)

$$ClO_4^- > SCN^- > I^- > NO_3^- > Br^- > Cl^- > HCO_3^- \sim OAc^- \sim SO_4^{2-} \sim HPO_4^{2-}$$

which corresponds to the order of increasingly negative hydration energies $\Delta G_H^o(J)$. Only few anion sensors with unusual selectivities are known. For example, a nitrite electrode was introduced very recently which makes use of the highly selective and sufficiently stable complex formation between the primary ion and a cationic Co(III)-complex (a derivative of vitamin B_{12}, see Figure 3) [16,17]. Due to this specific interaction, a preference of NO_2^- over NO_3^- and Cl^- by a factor of $\sim 10^3$ was achieved (Figure 3). Even more promising is the design of neutral carriers for anions which, according to Eqs. (12) - (14), permits anion sensors with widely different selectivities to be realized. This is underscored by results obtained on membrane electrodes with neutral tin-organic ligands [5,6] (see Figure 2) which may be very attractive for chloride determinations.

A large number of cation-selective sensors based on neutral carriers has been introduced [2,3]. It is well known that the selectivity of these devices mainly reflects the relative stabilities of the cation/carrier complexes formed, but the present treatment reveals that ion pair formation in the membrane may also have some influence (see Eq. (14)). In fact, the observed ion selectivities usually correlate with extraction data which include the effects of ionic association. Close correlations between selectivity coefficients and complex stability constants are obtained only for special carrier systems. For example, this is the case for membranes containing enantiomer-selective chiral ligands [18-21]. Since it is reasonable to assume that the complexes formed between the enantiomeric ligand L (or its counterpart $\bar{L}$) and the two enantiomeric ions I and J have the same stoichiometry and the same dimensions, one can establish the following relationship for the potentiometric selectivity of corresponding membrane electrodes with L or $\bar{L}$:

$$K_{IJ}^{Pot}(L) = \frac{\beta_{JL_n}}{\beta_{IL_n}} = \frac{1}{K_{IJ}^{Pot}(\bar{L})} \tag{18}$$

Although the highest enantiomer selectivity realized in sensors is only 4.2 resp. 0.24 [18] (L,$\bar{L}$: enantiomeric crown ethers; I,J: enantiomers of phenylglycine methyl ester · H^+), analytical measurements of biogenic ammonium ions became accessible by the introduction of membrane electrode assemblies of the type shown in Figure 4 [19-21] which yield the response:

$$EMF = \frac{RT}{zF} \ln \frac{a_I' + K_{IJ}^{Pot}(L)\, a_J'}{a_J' + K_{IJ}^{Pot}(L)\, a_I'} \tag{19}$$

Figure 5 illustrates that the EMF of such measuring cells is nearly linearly related to the enantiomeric excess ee of ions in solution, which agrees with theoretical expectations [20,21]:

$$ee = \frac{c_I - c_J}{c_I + c_J} \cdot 100\% \cong \frac{EMF}{EMF_{max}} \cdot 100\% \tag{20}$$

It was shown that potentiometric determinations of the enantiomeric excess can be performed even in samples containing heavy interferents [21].

In practice, it is often observed that the apparent potentiometric selectivities are time- and concentration-dependent even in cases where theory strictly predicts an EMF response behavior of the Nicolsky type. The reason is that the theoretical selectivity coefficient K_{IJ}^{Pot} is defined in Eq. (6) in terms of boundary activities and corresponds to the equilibrium constant of the interfacial exchange reaction,

I (sensor) + J (aq, interface) $\rightleftharpoons$ J (sensor) + I (aq, interface),

whereas the assessment of the practical selectivity K_{IJ}^{Pot}(app.) is based on the given bulk sample activities. Evidently, another process may become selectivity-determining before the final equilibrium between the sensor and the bulk of sample solution is established, namely the interdiffusion process

$$J\ (aq,bulk) + I\ (aq,interface) \rightleftharpoons I\ (aq,bulk) + J\ (aq,interface)$$

Accordingly, the initially observable selectivity can be identified with the permeability ratio of the two species between sample and sensing electrode surface (e.g., the ratio of diffusion coefficients in the aqueous boundary film [3,22-24]). Such time- and concentration-dependent changes in the apparent selectivities were found for liquid and solid-state membrane electrodes [3,22-24] as well as for gas sensors [25] (see Figures 6 and 7).

REFERENCES

[1] M.A. Arnold and M.E. Meyerhoff, Anal. Chem. 56, 20R (1984).

[2] D. Ammann, W.E. Morf, P. Anker, P.C. Meier, E. Pretsch, and W. Simon, Ion Selective Electrode Reviews 5, 3 (1983).

[3] W.E. Morf, The Principles of Ion-Selective Electrodes and of Membrane Transport, Studies in Analytical Chemistry, Vol. 2, Akadémiai Kiadó, Budapest, 1981; Elsevier, Amsterdam, 1981.

[4] P.C. Meier, W.E. Morf, M. Läubli, and W. Simon, Anal. Chim. Acta 156, 1 (1984).

[5] U. Wuthier, H.V. Pham, R. Zünd, D. Welti, R.J.J. Funck, A. Bezegh, D. Ammann, E. Pretsch, and W. Simon, Anal. Chem. 56, 535 (1984).

[6] U. Wuthier, H.V. Pham, D. Ammann, and W. Simon, in preparation.

[7] R.P. Buck, Crit. Rev. Anal. Chem. 5, 323 (1975).

[8] W.E. Morf, Anal. Chem. 49, 810 (1977).

[9] J. Sandblom, G. Eisenman, and J.L. Walker, Jr., J. Phys. Chem. 71, 3862 (1967).

[10] B.P. Nicolsky, Zh. Fiz. Khim. 10, 495 (1937).

[11] R. Büchi, E. Pretsch, W.E. Morf, and W. Simon, Helv. Chim. Acta 59, 2407 (1976).

[12] P. Oggenfuss, Dissertation ETH No. 7619, Zürich, 1984.

[13] O.K. Stephanova and M.M. Shults, Vestn. Leningrad. Univ., No. 4, 1972, 80.

[14] W.E. Morf, E. Pretsch, U. Wuthier, H.V. Pham, R. Zünd, R.J.J. Funck, K. Hartman, K. Sugahara, D. Ammann, and W. Simon, in Recent Advances in the Theory and Application of Ion Selective Electrodes in Physiology and Medicine, M. Kessler, D.K. Harrison, and J. Höper, Eds., Springer-Verlag, Berlin, Heidelberg, New York, 1984, in press.

[15] D. Wegmann, H. Weiss, D. Ammann, W.E. Morf, E. Pretsch, K. Sugahara, and W. Simon, Mikrochimica Acta [Wien] 1984, in press.

[16] P. Schulthess, D. Ammann, W. Simon, Ch. Caderas, R. Stepánek, and B. Kräutler, Helv. Chim. Acta 67, 1026 (1984).

[17] P. Schulthess and W. Simon, Anal. Chem., in preparation.

[18] A.P. Thoma, A. Viviani-Nauer, K.H. Schellenberg, D. Bedeković, E. Pretsch, V. Prelog, and W. Simon, Helv. Chim. Acta 62, 2303 (1979).

[19] W. Bussmann and W. Simon, Helv. Chim. Acta 64, 2101 (1981).

[20] W.E. Morf, W. Bussmann, and W. Simon, Helv. Chim. Acta 67 (1984), in press.

[21] W. Bussmann, W.E. Morf, J.-P. Vigneron, J.-M. Lehn, and W. Simon, Helv. Chim. Acta 67 (1984), in press.

[22] A. Hulanicki and A. Lewenstam, Anal. Chem. 53, 1401 (1981).

[23] W.E. Morf, in Ion-Selective Electrodes, E. Pungor and I. Buzás, Eds., Akadémiai Kiadó, Budapest, 1981, p. 267.

[24] W.E. Morf, Anal. Chem. 55, 1165 (1983).

[25] W.E. Morf, I.A. Mostert, and W. Simon, Anal. Chem., submitted for publication.

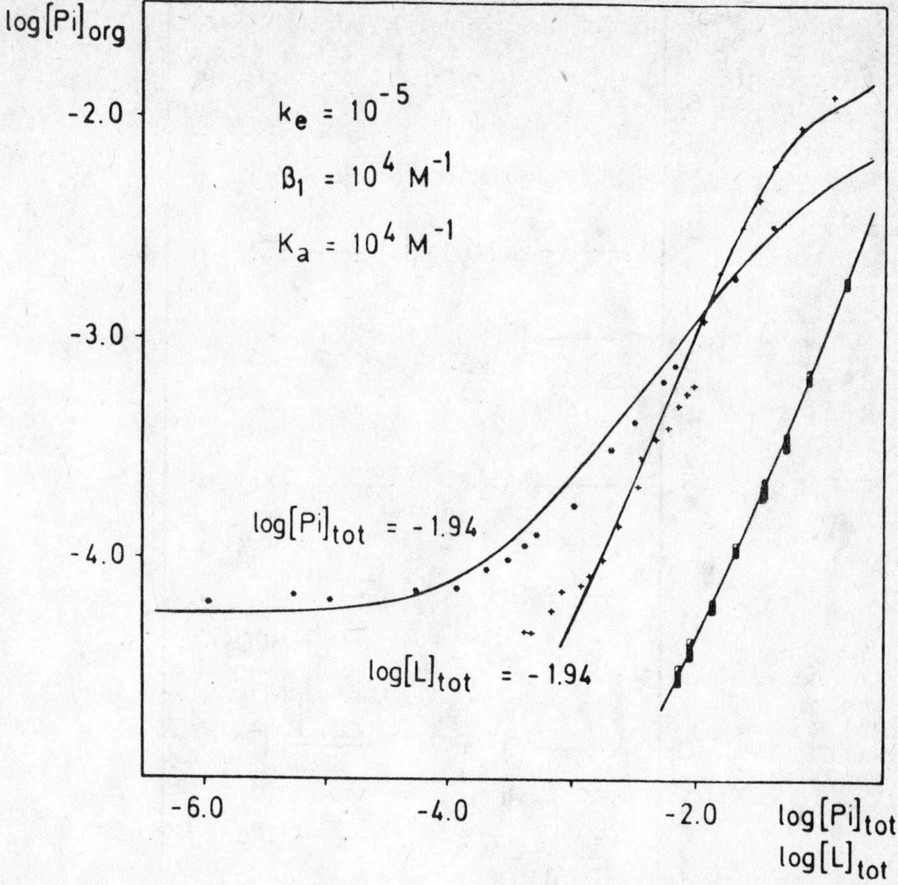

Fig. 1. Extraction of lithium picrate into the plasticizer dibutyl sebacate in the absence (lower curve) and in the presence (upper curves) of the lithium-selective carrier ETH 149 [12]. The measured concentration of extracted picrate in the organic solution, $[Pi]_{org}$, is given as a function of the initial salt concentration in the aqueous phase, $[Pi]_{tot}$, or as a function of the total carrier concentration, $[L]_{tot}$. The theoretical curves were calculated using a salt extraction constant of 10^{-5}, a stability constant of 1:1 cation/carrier complexes in dibutyl sebacate of $10^{4}\ M^{-1}$, and a formation constant of ion pairs and electrically neutral associates of $10^{4}\ M^{-1}$.

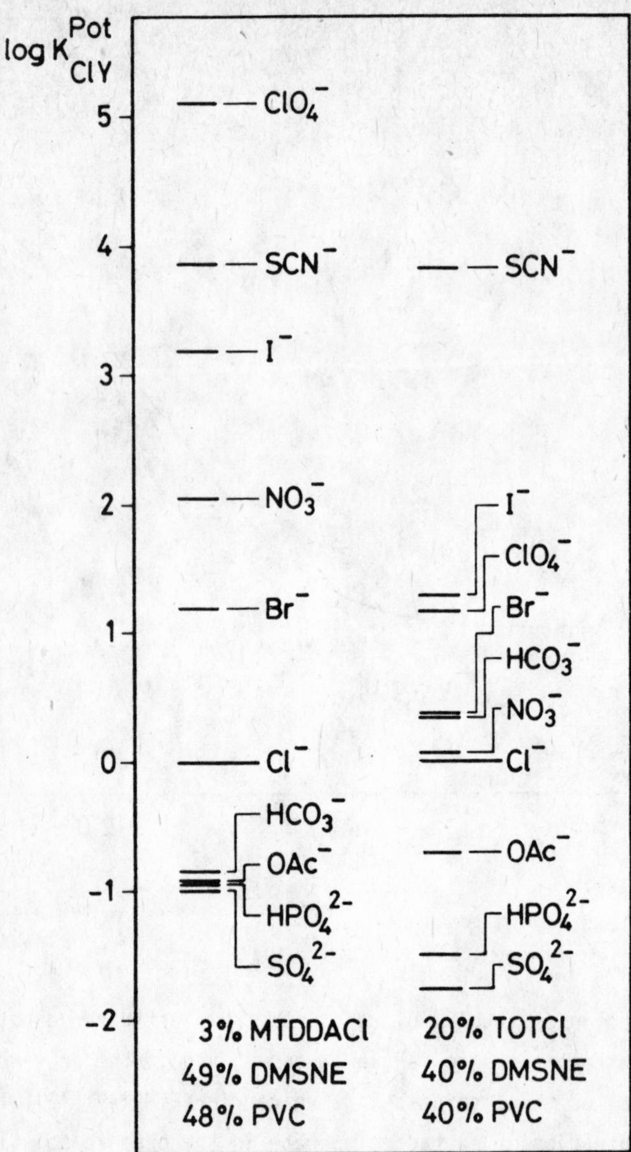

Fig. 2. Anion selectivities of a liquid membrane electrode based on a conventional anion-exchanger (methyl-tridodecylammonium chloride, left) in comparison with selectivities achieved for a sensor based on anion-selective neutral carriers (trioctyltin chloride, right) [5,6].

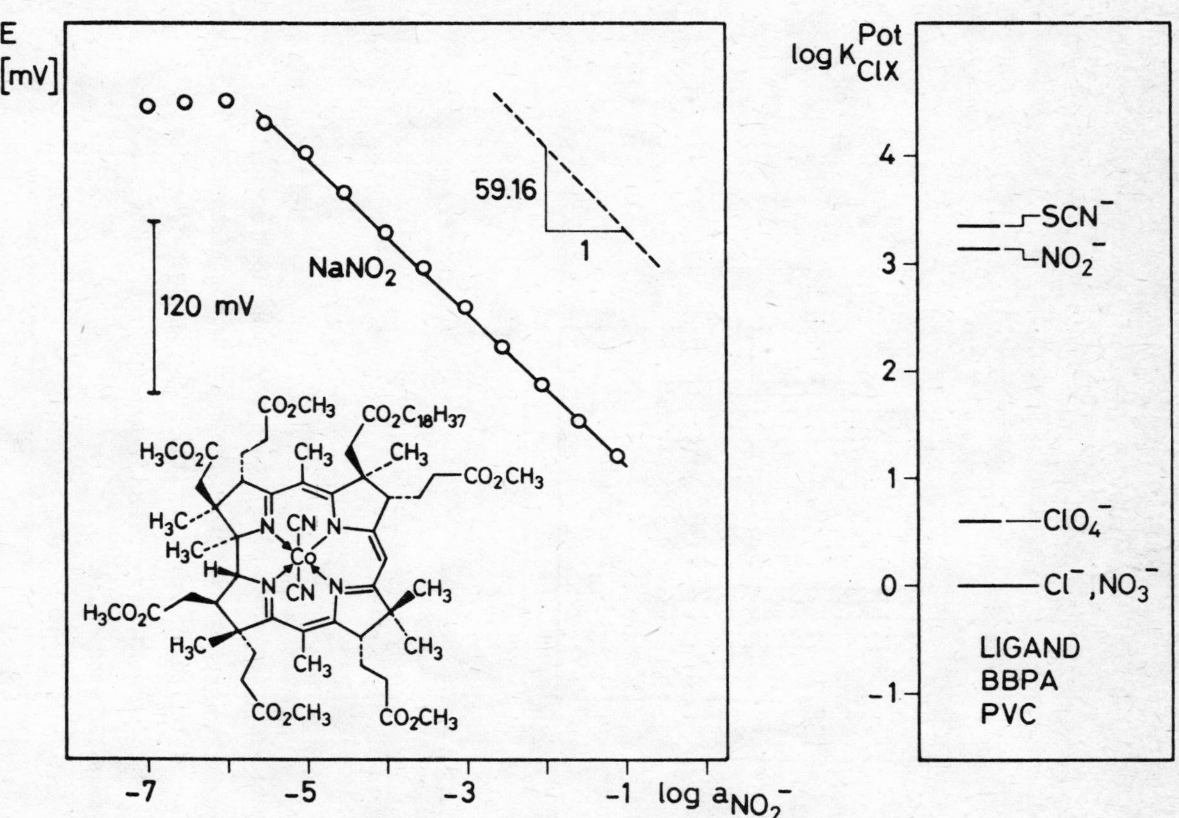

Fig. 3. EMF response and anion selectivities of a nitrite sensor based on a Co(III)-complex, a derivative of vitamin B_{12} [16,17].

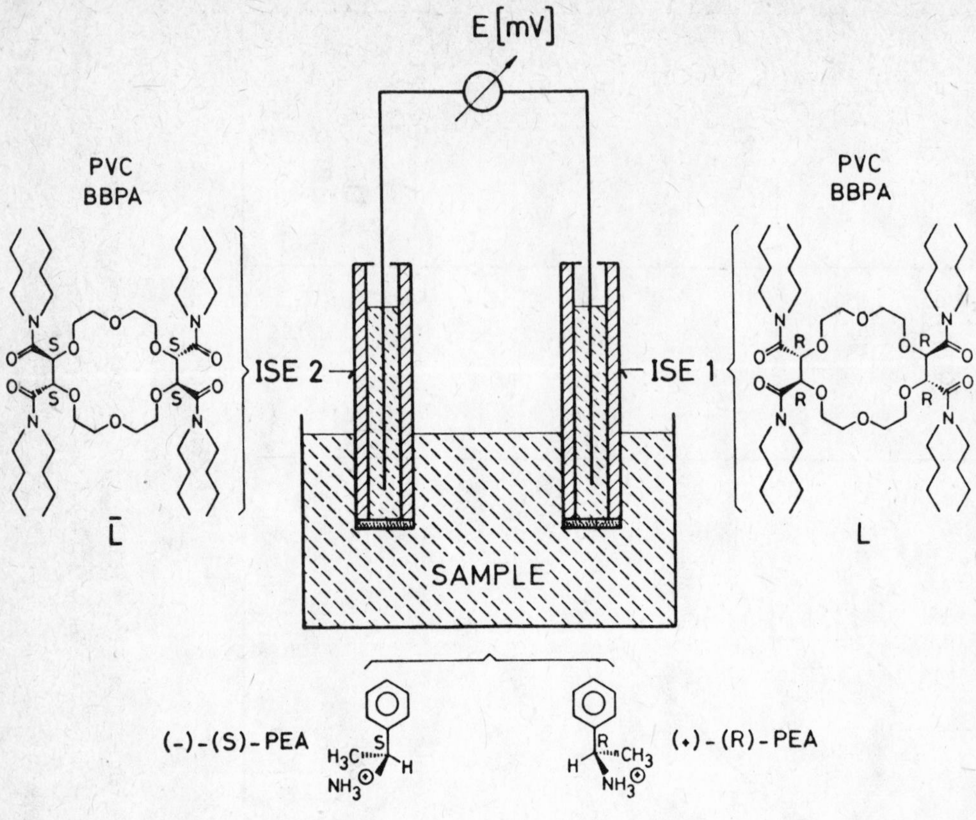

Fig. 4. Schematic representation of a liquid membrane electrode cell designed for the measurement of the enantiomeric excess (a measure of the enantiomeric purity) of 1-phenylethylammonium ions [21]. The enantiomer selectivity exhibited by the two electrodes is 2.7 resp. 0.37.

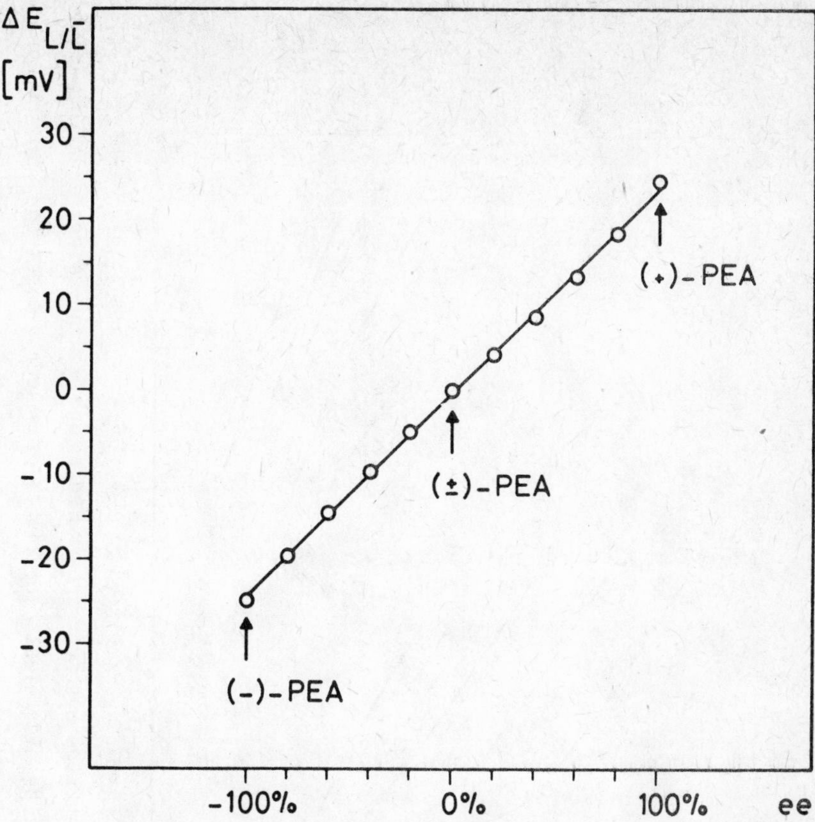

Fig. 5. Dependence of the EMF of an enantiomer-selective electrode cell (Fig. 4) on the enantiomeric excess ee of 1-phenylethylammonium ions in the sample solution [20].

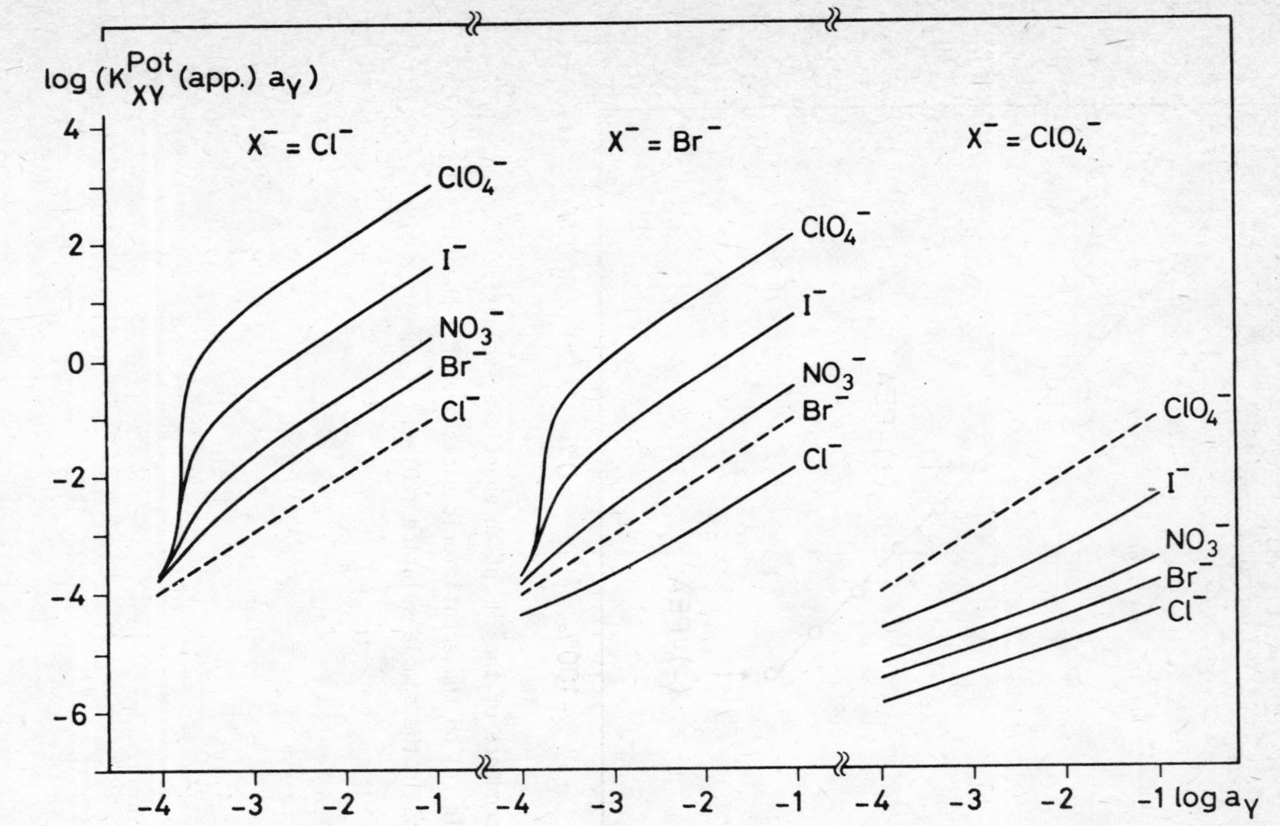

Fig. 6. Apparent selectivity behavior of liquid membrane electrodes based on anion-exchangers (quaternary ammonium salts). Left: chloride electrode; middle: bromide electrode; right: perchlorate electrode. The shown functions are related to the calibration plots for primary ions (dashed curves) and interfering ions (solid curves). For details see [3,15].

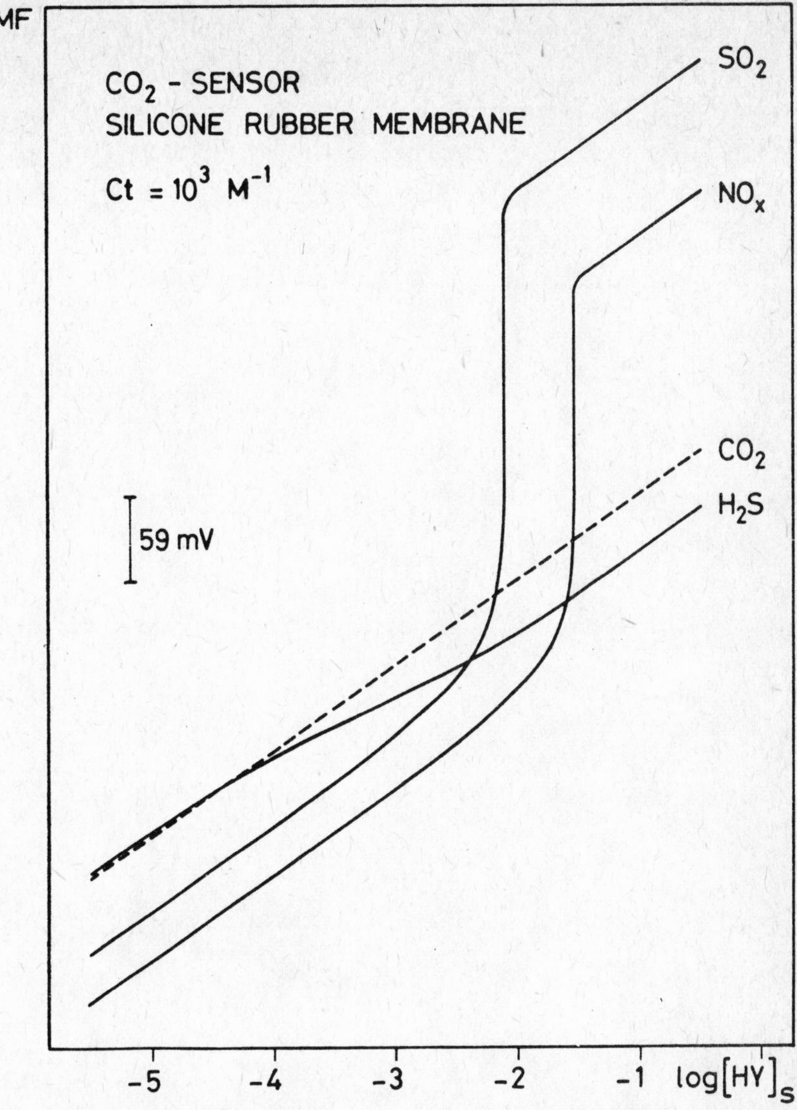

Fig. 7. Calculated EMF response to different gases of a CO_2 sensor based on a silicone rubber gas-permeable membrane. The concentration-dependence of the apparent selectivities is illustrated. For details see [25].

4th Symposium on Ion-Selective Electrodes
Mátrafüred, 1984

CONTINUOUS ANALYSIS WITH ION-SELECTIVE ELECTRODES: FLOW-INJECTION ANALYSIS AND MONITORING OF ENZYME-BASED REACTIONS

J.D.R. THOMAS

Applied Chemistry Department, Redwood Building
UWIST, PO Box 13, Cardiff CF1 3XF, Wales, UK

ABSTRACT

The behaviour of PVC matrix membrane calcium ISEs based on calcium bis di[4-1,1,3,3-tetramethylbutyl)phenyl]phosphate sensor with trioctyl phosphate solvent mediator is described for the flow-injection analysis of calcium in natural and potable waters and of calcium in soap powder solutions. The analyses were carried out with a tubular flow-through calcium ISE.

For sulphide ISEs, various FIA profile parameters were studied, including carrier stream composition, flow rate and sample volume parameters. It was deemed appropriate to have a carrier stream composition of 5 x 10^{-6} M sodium sulphide in SAOB, and results for the flow-through mode of EDT Research EES sulphide ISE was superior to those of an Orion 94-16A electrode used in the cascade mode. The Orion 94-16A sulphide ISE is appropriate for the continuous monitoring of growth and culture media parameters for Desulfovibrio species of sulphate-reducing bacteria.

The use of the iodide ISE, used in conjunction with PVC matrix membranes of glucose oxidase on its own, or in admixture with peroxidase, or with glucose oxidase chemically immobilized with glutaraldehyde on bovine albumin, is illustrated for the monitoring of glucose in solution.

Finally, potentiometric sensors are appropriate for monitoring microcomputerised titrations. An interesting facility with such an assembly is the ability to calibrate electrodes in terms of e.m.f. versus log(concentration) plots by means of a menu option available in the microcomputerised system.

INTRODUCTION

Ion-selective electrodes (ISEs) are versatile devices for the continuous monitoring of a wide range of processes and reactions, such as, industrial processes and samples, titrations, bacterial activity, and enzyme reactions. For these purposes, the ISEs, coupled to suitable reference electrodes, can be set up as direct monitors in association with potentiometric recording equipment, or they can be set up in a flow-injection mode for the rapid analysis of a succession of samples. This paper describes various aspects of work at UWIST on flow-injection analysis (FIA) and continuous analysis with ISEs. These relate to uses of calcium and sulphide ISEs for monitoring the primary ions in both the direct and FIA mode, uses of the iodide electrode for monitoring enzyme reactions and some aspects of ISEs in microcomputerised titrations.

FLOW INJECTION ANALYSIS

FIA was carried out by having the electrode in a "cascade flow" mode [1] with the units of carrier solution reservoir, peristaltic pump (Ismatec Model MP 13GJ-4), injection unit and ISE fitted for cascade flow of carrier/sample solution being arranged in series with interconnecting Solva-Tube brand of tubing. For the cell assembly set in the overflow vessel, the saturated calomel reference electrode was set to have the carrier solution flow on to the ceramic junction by means of a loop from the main carrier solution stream [2].

A similar sequence was used for FIA with flow-through or tubular membrane electrodes, except that the reference electrode, fitted with a polyethene cup, was set in a full by-pass loop of carrier solution.

Applications of FIA to Analysis of Calcium

Tubular "flow-through" PVC matrix membrane calcium ISEs based on calcium bis di[4-(1,1,3,3-tetramethylbutyl)phenyl]-phosphate sensors with either dioctyl phenylphosphonate (electrode A) or trioctyl phosphate (electrode B) solvent mediator have been used for the analysis of calcium in water and in soap powder solutions. The FIA carrier solution used in each case consisted of borax (10^{-3} M) and sodium hydroxide (10^{-3}M) in sodium chloride (0.14M) at a pH of 9.4. 50 mm^3 samples were injected through a stainless steel HPLC valve type of injection head and the carrier stream flow rate was 6 cm^3 min^{-1}. The injection head was just 5 cm away from the sensing electrode in order to control dispersion.

The alternative trioctyl phosphate solvent mediator (electrode B) was superior to dioctyl phenylphosphonate (electrode A) with regard to resistance from the deleterious effects of anionic surfactants [3] as can be seen from the data of Table I. Thus, electrode B is better able to achieve the responses in the presence of 10^{-4} M sodium dodecyl sulphate (SDS) anionic surfactant (79.7 to 99.8%) than were obtained for the samples even when no surfactant was present.

Electrodes A and B, used in the FIA mode for a series of water samples (Table II) gave similar results for calcium ion-activities, which were between 1 and 10% less than the total calcium (free and complexed) obtained by atomic absorption spectroscopy.

Table III shows that the calcium ISE based on electrode B can be used for calcium ion activity measurements in soap powder solutions, containing up to 3 g dm^{-3}. However, the FIA tubular membrane version of electrode B is not sufficiently sensitive for the low calcium ion levels that occur in solutions containing more than 3 g dm^{-3} of soap powder (not shown in Table III), nor for solutions containing less than ≈3 x 10^{-3} M total calcium for 3 g dm^{-3} soap powder solution. Conventional electrodes of membrane B composition are able to cope with these 'extreme' solutions [3].

FIA Studies with Sulphide ISEs

Practical difficulties in the application of ISEs to sulphide and thiol measurements relate to oxidation and frequently to volatility of the sample components [4,5]. Nevertheless, FIA systems can be used with the sulphide ISE as monitor, as has previously been shown for the monitoring of a process stream and for ethane thiol [6,7]. Measurements with the sulphide ISE are best carried out on samples and standards in a background of Sulphide (or Standard) Anti-Oxidant Buffer (SAOB) [8]. It is important that the SAOB, normally added in volume ratios equal to those of sample, is either freshly prepared or its quality maintained by storage under nitrogen, when its strength can be maintained for 4 weeks or more [9].

Normally, SAOB consists of sodium hydroxide (2M) with ascorbic acid (20 g dm^{-3}), but there are other compositions, e.g., a "double-strength" SAOB consists of sodium hydroxide (4M) with ascorbic acid (40 g dm^{-3}) [10].

Carrier Stream Studies - Studies have been made with sulphide ISEs of various FIA parameters [10]. thus, carrier stream parameters were studied for sample volumes of 40 mm^3 at a carrier flow rate of 2.23 cm^3 min^{-1}. Neat normal SAOB and small background levels of sulphide in SAOB were compared for their efficacy as carrier streams for obtaining good sulphide ISE calibrations.

Poor calibrations resulted from having either SAOB or 10^{-6} M sodium sulphide in SAOB as carrier streams for an Orion 94-16A sulphide ISE used in the "cascade flow" mode. On the other hand, a carrier stream of 5 x 10^{-6} M sodium sulphide for the "cascade flow" Orion electrode gave good calibrations for sulphide standards with a linear calibration range between 10^{-2} M and 5 x 10^{-5} M sulphide.

Similar calibration for an EDT Research EES flow-through sulphide ISE showed the carrier stream of 5 x 10^{-6} M sodium sulphide in normal SAOB to be effective and the linear calibration range extended to $<10^{-5}$ M sulphide.

Attempts to run FIA with a carrier stream of 10^{-3} M sodium sulphide in normal SAOB were aimed at analysing samples with greater or less than 10^{-3} M sulphide from the drop or increase in e.m.f., as appropriate [10]. The "cascade mode" Orion ISE gave no response for samples containing less than 10^{-3} M sulphide, accountable by slow electrode kinetics and/or sulphide loss by oxidation. For the EDT Research EES flow-through ISE there was a response for sulphide standards between 10^{-5} M and 10^{-1} M [sulphide], but with low response slope.

Flow Rate and Sample Volume Studies - These were carried out [10] by running a potentiometric chart recorder at a fast rate (12 cm^3 min^{-1}) in order to resolve the components of the profile of FIA for peaks of height, h, for samples into t_R

(time required from baseline to maximum e.m.f. change), w_p (peak width, i.e., time of maximum e.m.f. change) and t_A (time required for e.m.f. change to fall from maximum to baseline zero). For this purpose, the "flow rate" study was conducted on 40 mm^3 samples for flow rates of 0.76, 1.24 and 2.23 $cm^3\ min^{-1}$. The carrier stream was 10^{-4}M sodium sulphide in SAOB. The "sample volume" study was made for the same carrier stream at a flow rate of 2.23 $cm^3\ min^{-1}$ for sample volumes of 20, 30, 40, 60 and 80 mm^3.

A number of trends emerged from the two studies, namely:

(i) Faster response times (t_R) for 10^{-3}M sulphide standards than for 10^{-1}M standards because of the greater millivolt scan in the latter case. It should be noted that the electrode kinetics will be fast at these sulphide ranges, so that t_R is controlled predominantly by the recorder time constant.

(ii) Faster recovery time from sample response to carrier stream baseline (t_A) for 10^{-3}M versus 10^{-1}M suphide standards, for similar reasons to (i). However, t_A is considerably greater than t_R because of the longer response time in passing from concentrated to dilute solutions vis à vis the reverse direction. Also, t_A data are more erratic than t_R data.

(iii) The trend for peak plateaux (w_p) to be narrower for 10^{-1}M sulphide than for 10^{-3}M standards arises from the greater t_R for 10^{-1}M solutions, thus reducing the width for a given sample size regardless of flow rate.

(iv) In general, larger sample sizes give differences in peak height, $\underline{h}$, nearer to the 58.2 mV (20° C) expected for differences between 10^{-3} and 10^{-1}M standards. The EDT Research EES flow-through electrode is better in this respect, and also gives higher values of $\underline{h}$. Nevertheless, it is stressed that FIA calibrations be for multiple sample injection in order to obtain good average data.

(v) The t_R data (generally of the order of 2 to 4 s for 2.23 cm^3 min^{-1} flow rate) are relatively short, but exceed the 1.1 s required for a 40 mm^3 plug of sample to pass over the electrode surface (2.2 s for a 80 mm^3 sample). The t_R data reflect some of the effects of dispersion (to increase sample plug size), electrode kinetics, and instrumental time constants.

(vi) The w_p data reflect the short time that samples are in contact with the electrode surface. They also reflect electrode kinetics and memory, instrumental constants and wash-out times (especially for the flow-through mode). Both the w_p data and those of t_R are greater for the slow flow rates of 1.24 and 0.76 cm^3 min^{-1} than for 2.23 cm^3 min^{-1} because of the greater tendency towards diffusion created by the longer residence times of samples between the point of injection and electrodes.

CONTINUOUS MONITORING WITH SULPHIDE ISEs

Orion 94-16A sulphide ISEs have been successfully used for monitoring the growth and nutrient parameters of *Desulfovibrio* species of sulphate-reducing bacteria [11,12]. Such monitoring has been carried out in indirect and direct modes for *D.sulfuricans*, *D.gigas* and *D.vulgaris*. Sulphide determined by both modes matched amounts expected from the nutrient sulphate present in the culture media (Tables V and VI). It was possible to determine pauses in growth and features of nutrient starvation from the e.m.f. *versus* time data [11].

All three species of bacteria thrived on the metabolic intermediates of sulphite, thiosulphate, metabisulphite and dithionite [11]. The species, represented in these studies by *D.vulgaris*, also grow in certain organic sulphur species, e.g., cysteine, cystine and glutathione as alternative sulphur sources to sulphate [12]. However, they will not grow with methionine as the sulphur source, and this is attributed to the relative stability of C-S-C bonds which are not adjacent to the amino group [12].

Sodium tetraborate(III) (2% m/v) and 2,4-dinitrophenol (2% m/v) are confirmed as effective bactericides for D.vulgaris which is the most robust of the three species (D.sulfuricans, D.gigas and D.vulgaris) studies [11,12].

MONITORING WITH ENZYME ELECTRODES

Enzyme electrodes normally consist of a membrane containing an appropriate enzyme used in conjunction with suitable potentiometric or voltammetric sensors of a reactant or product. Among the studies at UWIST are those involving oxidases, illustrated here by glucose oxidase. Thus, glucose oxidase has been used in an enzyme electrode where an iodide ISE was the sensor [13]. The basis of such an enzyme electrode for sensing glucose is represented by the following equations:

$$\text{Glucose} + O_2 \xrightarrow{\text{Glucose oxidase}} \text{Gluconic acid} + H_2O_2 \qquad (1)$$

$$H_2O_2 + 2I^- + 2H^+ \xrightarrow[\text{or peroxidase}]{\text{Mo(VI) catalyst}} I_2 + 2H_2O \qquad (2)$$

For these purposes, glucose oxidase (E.C. 1.1.3.4) (50 mg) was immobilized in PVC (20 mg) and dioctyl phenylphosphonate (40 mg) when molybdenum(VI) was the catalyst for reaction (2). This membrane was fitted to an Orion Model 94-53 iodide ISE for monitoring iodide consumption by hydrogen peroxide reaction (2) which was related to glucose concentration from hydrogen peroxide production according to reaction (1).

An alternative PVC matrix enzyme membrane had the above composition supplemented by peroxidase (E.C. 1.11.1.7) (0.18 g). Both electrodes were compared with those for membranes of glucose oxidase chemically immobilized with glutaraldehyde on bovine albumin, and the data are summarized in Table VII. Chemically immobilized membranes may be prevented from cracking by incorporating a small proportion of high molecular mass polyethoxylate as humectant?[14].

Entrapment of glucose oxidase in a PVC matrix offers a simple method of constructing an enzymatic membrane that is convenient to handle in association with an iodide electrode

[13]. The co-immobilization of peroxidase with glucose oxidase in PVC yields better membranes of improved lifetimes [13]. The membranes with glucose oxidase chemically immobilized with glutaraldehyde have superior lifetimes and have more favourable conditioning and washtime features, but suffer from lower slopes than the PVC membrane systems [13]. Lower washtimes can be obtained with thinner membranes which also generally speed up response of enzyme electrodes; both features make for improved scope in FIA and continuous monitoring.

MICROCOMPUTERIZED TITRATIONS

The availability of personal microcomputers permits the opportunity for controlling titrations and general continuous monitoring with data processing at a low cost [15]. Furthermore, the versatility of the computer allows any type of potentiometric determination, including a Gran's plot.

A titration unit, controlled by a ZX microcomputer has been assembled [15] through a parallel bus interface system; this can also calibrate ISEs with respect to ion concentration. This system has been illustrated for a precipitation titration of silver nitrate with a silver wire indicator electrode and glass reference electrode [15].

The titration program was constructed in sections and the user chooses from a 'menu' of available options, based on "titration options" and "results options" [15].

the titration option is the main one and proceeds without further attention. It calls on the burette control and ADC input subroutines. When titrating, the system adds a quantity of titrant, waits a preset time and then starts taking readings. When these are stable within a set limit, the e.m.f., together with the titrant volume and derivative values, are stored in an array [15].

As the end-point approaches, the addition volume is reduced in response to the increasing slope of the titration curve. The volume of each aliquot is determined as follows

$$\text{New VOL} = \frac{\text{previous VOL x target e.m.f. change}}{\text{previous e.m.f. change}}$$

When the titration is complete, the endpoints are calculated from the second derivative values. The main menu is then re-presented.

Display and print-out of titration results constitute a second menu option and titration curve plotting is a third option. A further option is the drawing of electrode calibration curves [e.m.f. vs log(concentration)]. This is based on the knowledge of titrand concentration at each point, so that a plot of log(concentration) versus electrode e.m.f. will result in a conventional calibration curve [15]. This is illustrated in Fig. 1 for a silver ion calibration of a silver wire electrode used with a glass electrode as reference. However, the method is generally applicable to potentiometric indicating electrodes, including ISEs, and is less tedious than the conventional static procedure based on serially diluted standard solutions [15].

CONCLUSION

The approaches outlined here confirm and further illustrate the versatility of various types of ion-selective electrodes for flow-injection analysis and monitoring in various circumstances. Also demonstrated is the scope for electrochemical sensors to be an integral part of microcomputerised systems.

REFERENCES

1. E.H.Hansen, A.K.Ghose and J.Růžička, Analyst, 102, 795 (1977).
2. E.H.Hansen and J.Růžička, RSC International Symposium on Electroanalysis in Clinical, Environmental and Pharmaceutical Chemistry, 13-16 April 1981, UWIST, Cardiff, Paper 31.
3. A.J.Frend, G.J.Moody, J.D.R.Thomas and B.J.Birch, Analyst, 108, 1357 (1983).
4. D.J.Crombie, G.J.Moody and J.D.R.Thomas, Anal.Chim.Acta, 80, 1 (1975).
5. E.J.Duffield, G.J.Moody and J.D.R.Thomas, Anal.Proc., 17, 533 (1980).

6. E.J.Duffield, G.J.Moody and J.D.R.Thomas, unpublished data.
7. J.D.R.Thomas, in E.Pungor (Editor) "Modern Trends in Analytical Chemistry. Part A. Electrochemical Detection in Flow Analysis", Akadémiai Kiadó, Budapest (1984) pp 141.
8. Orion Research Inc., Applications Bulletin No. 12, 1969.
9. M.G.Glaister, G.J.Moody and J.D.R.Thomas, Anal.Chim.Acta., in the press.
10. M.G.Glaister, G.J.Moody and J.D.R.Thomas, to be published.
11. I.K.Al-Hitti, G.J.Moody and J.D.R.Thomas, Analyst, 108, 43 (1983).
12. I.K.Al-Hitti, G.J.Moody and J.D.R.Thomas, Analyst, 108, 1209 (1983).
13. I.K.Al-Hitti, G.J.Moody and J.D.R.Thomas, Analyst, 109, 1205 (1984).
14. K.K.Louie and J.D.R.Thomas, unpublished data.
15. S.W.Bateson, G.J.Moody and J.D.R.Thomas, J.Automatic Chem., 5, 174 (1983).

TABLE I EFFECT OF EXPOSING TUBULAR MEMBRANE ELECTRODES TO 10^{-4} M SDS DURING FIA (Data from Ref.3)

Sample injected	ΔE (mean of 6 determinations)/mV			
	Membrane A		Membrane B	
	Before	After*	Before	After*
10^{-2} M Ca^{2+}	57.3	46.9(81.8)	52.5	52.4(99.8)
10^{-3} M Ca^{2+}	28.2	24.0(85.1)	26.0	24.9(95.8)
Tap water	29.0	24.9(85.9)	26.1	25.4(97.3)
Tap water + 10^{-4} M SDS		16.7(59.6)+		20.8(79.7)+

* Data in parentheses are percentages of ΔE observed before exposure to 10^{-4} M SDS.

+ Data in parentheses are percentages of ΔE referred to tap water without background 10^{-4} M SDS.

TABLE II CALCIUM LEVELS FOUND IN WATER SAMPLES BY FIA (TUBULAR ELECTRODE) COMPARED WITH ATOMIC ABSORPTION SPECTROSCOPY (AAS) (Data from Ref.3)

Samples were taken on 6 separate days in each case

Water sample (Cardiff area)	Calcium content (s.d. for n=6), ppm		
	FIA (tubular electrode) Membrane A*	FIA (tubular electrode) Membrane B*	AAS
Tap water 1	33.5 (0.4)	32.0 (0.4)	34.5 (0.8)
Tap water 2	29.6 (0.5)	29.2 (0.3)	32.0 (0.7)
Tap water 3	36.7 (0.6)	37.1 (0.4)	37.5 (1.0)
River water	35.9 (0.4)	36.1 (0.6)	37.8 (0.8)
Lake water	31.4 (0.3)	31.6 (0.6)	32.7 (1.0)

* Calcium content expressed as $a_{Ca^{2+}}$ for electrodes with membranes A and B

TABLE III COMPARISON OF CALCIUM ION ACTIVITIES FOUND BY TUBULAR ELECTRODE BY FIA AND BY THE STATIC METHOD FOR MEMBRANE B ELECTRODES (Data from Ref.3)

Calcium added /mM	Concentration of model soap powder/g dm^{-3}	Experimental $a_{Ca^{2+}}$(s.d for n=6)/M	
		Static mode	FIA
1	1	$1.7(0.5) \times 10^{-5}$	$6.3(0.5) \times 10^{-5}$
2	1	$2.4(1.4) \times 10^{-4}$	$3.4(0.3) \times 10^{-4}$
3	1	$6.4(2.7) \times 10^{-4}$	$6.3(0.7) \times 10^{-4}$
4	1	$1.2(0.4) \times 10^{-3}$	$1.4(0.1) \times 10^{-3}$
5	1	$1.6(0.3) \times 10^{-3}$	$2.2(0.2) \times 10^{-3}$
4	3	$1.0(0.4) \times 10^{-4}$	$1.6(0.3) \times 10^{-4}$
5	3	$2.1(0.6) \times 10^{-4}$	$2.5(0.1) \times 10^{-4}$

TABLE IV FIA RESPONSE TIMES FOR TUBULAR ELECTRODES WITH MEMBRANES OF COMPOSITIONS A AND B (Data from Ref.3)

$[Ca^{2+}]$ injected/M	Response times/s	
	A	B
10^{-4}	4.9 (±0.3)	5.1 (+0.4)
10^{-3}	3.7 (±0.1)	3.6 (±0.2)
10^{-2}	3.0 (±0.5)	3.1 (±0.4)
10^{-1}	3.0 (±0.3)	3.2 (±0.4)

TABLE V SULPHIDE PRODUCTION BY DESULFOVIBRIO SPECIES FROM SODIUM SULPHATE (Indirect Monitoring Mode) (Data from Ref.11)

Desulfovibrio species	Na_2SO_4 /g	S_i in medium /g	S^{2-} in Monitoring flask: Sulphide /g	Gravimetric /g	Age of inoculum /d
D.desulfuricans	0.500	0.113	0.108	0.105	43
"	0.564	0.127	0.128	0.131	11
"	none	-	4×10^{-5}	none	22
D.gigas	0.500	0.113	0.106	0.108	24
"	0.564	0.127	0.136	0.131	28
"	none	-	3×10^{-5}	none	161
D.vulgaris	0.500	0.113	0.112	-	10
"	0.564	0.127	0.128	0.118	4
"	none	-	0.006	none	49

TABLE VI SULPHIDE PRODUCTION BY DESULFOVIBRIO SPECIES FROM VARIOUS INORGANIC SULPHUR (S_i) SOURCES (Direct Monitoring Mode)

Sulphur source	S_i in medium /g	S^{2-} by sulphide ISE/g	Desulfovibrio species	Age of inoculum /d
0.564g Na_2SO_4	0.127	0.123	D.vulgaris	93
No inorganic sulphur	-	0.0001	"	49
0.564g Na_2SO_4	0.127	0.134	D.desulfuricans	11
No inorganic sulphur	-	0.002	"	32
0.564g Na_2SO_4	0.127	0.136	D.gigas	28
0.500g $Na_2S_2O_3 \cdot 5H_2O$	0.129	0.122	D.desulfuricans	121
0.500g Na_2SO_3	0.127	0.123	D.vulgaris	254
0.377g $Na_2S_2O_5$	0.127	0.122	"	265

TABLE VII CHARACTERISTICS OF GLUCOSE OXIDASE/IODIDE ENZYME ELECTRODES FOR GLUCOSE (Data from Ref. 13)

Property	Electrode type		
	Glucose oxidase in PVC	Glucose oxidase + peroxidase in PVC	Glutaraldehyde + Glucose oxidase on bovine albumin
Range for glucose/M	$10^{-3}-10^{-2}$	$\approx 10^{-4}-\approx 10^{-2}$	$10^{-3}-10^{-2}$
Mean slope of electrodes made from different membranes/ mV decade^{-1}	72(sd=3.2) (n = 6)	74(sd=6.0) (n = 5)	43(sd=1.1) (n = 7)
Response times/min	2-12	2-10	8-10
Conditioning time for fresh enzyme membranes/h	≈12	≈24	<2
Wash times between samples/h	1	0.5	0.2
Lifetimes/d	3	7	>14

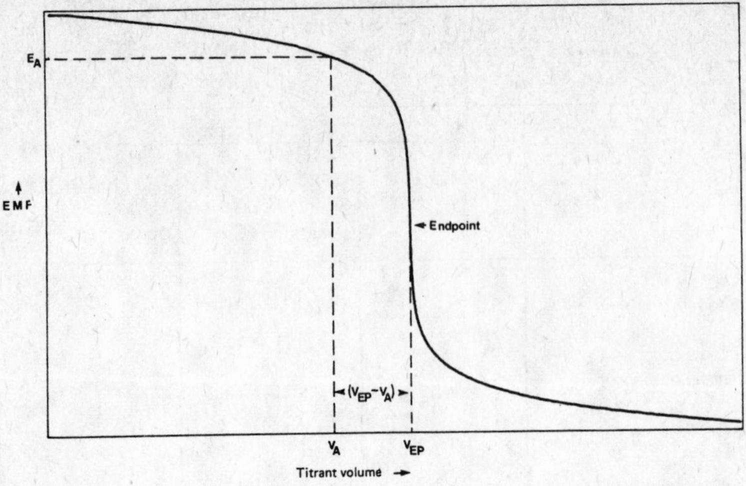

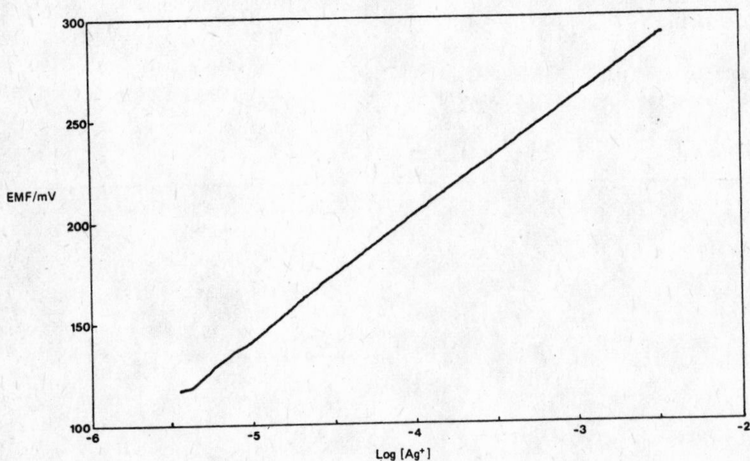

Figure 1. Calibration of a silver wire electrode with respect to silver ion concentration (bottom). Calibration calculations and calibration processed by microcomputer from microcomputer-controlled titration (top) of silver nitrate with potassium bromide (Data from Ref. 15).

QUESTIONS

Participants of the discussion: W.Eckert, G.Horvai, M.Jänchen, G.Nagy, E.Pungor, W.Simon and J.D.R.Thomas

Question:

Did you observe any effect of the pH on the hydrogen sulphide reading taken in the suspension of bacteria ?

Answer:

In the direct monitoring mode for sulphate reducing bacteria, we followed the pH of the system as well as the emf of the sulphide ion-selective electrode against the reference electrode, because we wanted the two separate informations. And of course the H_2S present would influence the pH but it is a fact.

Question:

What is the speciation of Ca at pH 9.4?

Answer:

The ion-selective electrode measures Ca^{2+} at pH 9.4. There may be other species present in the solution, but we believe that the Ca ISE does not respond to those.

Question:

You presented a method you used to immobilize enzyme on pvc. Do you intend to have the enzyme within the membrane, or to have the enzyme molecules sticking out of the membrane ? Why did you choose this method of immobilization ?

Answer:

It was curiosity on our part on whether pvc would be a functional immobilization system in this case. And it is more of a physical immobilization method rather than a heterogeneous immobilization system.

Question:

Do you have then an enzyme trapped in pvc, or is it a micro encapsulated enzyme, or do you have the enzyme only on the surface ?

Answer:

It is a physically immobilized system. These days we have moved over to chemically immobilized systems. But I can add a further comment in relation to the pvc system. It does help to preserve the membrane, to prevent cracking, which often occurs with enzyme membranes. Addition of a small amount of polyethoxylate also helps prevent membrane cracking.

Question:

You have shown a nice enzyme system, and at the end you detected iodide. Studying the same reaction we have found that the consumption of H_2O_2 by iodide is not too specific, there are many substances that may interfere. Did you notice that ?

Answer:

We were dealing with pure systems. It is possible that with real samples containing glucose oxidase there could be interfering materials. But again, as I mentioned, we believe that the amperometric detector systems have moie promise as a future prospect.

Question:

Did you observe any sorption of the bacteria on the electrode ?

Answer:

It is possible that this could happen in the direct monitoring mode. But we calculated the sulphide produced from the emf responses in relation to pH and everything, and again the total sulphide produced, as determined by the ion--selective electrode matches that determined by chemical, gravimetric and other means.

Question:

Did you use tetrahydrofurane for dissolving the glucose oxidase ?

Answer:

Yes. You may find the details in the written version of my paper.

Question:

Did you observe any denaturation ?

Answer:

It is interesting, but the answer is no. We have to realize that enzymes are really rather more robust than we think. But this is not to say that all enzymes are equally robust.

4th Symposium on Ion-Selective Electrodes
Mátrafüred, 1984

BIS (CROWN ETHER)S FOR POTASSIUM ION-SELECTIVE ELECTRODES

K. TÓTH, E. LINDNER, E. PUNGOR, B. ÁGAI*, I. BITTER* and L. TŐKE*

Institute for General and Analytical Chemistry
Technical University, Budapest, 1111 Hungary

*Department of Organic Chemical Technology
Technical University, Budapest, 1111 Hungary

INTRODUCTION

A wide range of neutral organic ligands have been proved to be excellent as ionophores for highly sensitive and selective potentiometric sensors /1/. The antibiotics valinomycin based potassium electrodes for example are by now the most widely used sensor for monitoring and high precision determination of potassium in different matrices.

Since the pioneering work on the synthesis of macrocyclic polyethers and the discovery that these molecules "crown ethers" were capable to form stable complexes with alkali- and alkaline-earth metal ions /2/ extended efforts were devoted towards the preparation and the use of synthetic ionophores for ion-selective electrode purposes /e.g. 3-8/.

Since benzo-15-crown-5 forms 1:2 complex with potassium, it was attractive to study how bis-benzo-15-crown-5 derivatives behave as ionophores for potassium selective electrodes /9-12/. Different bis-15-crown-5 derivatives gained application in potassium ion-selective electrodes are listed in Fig.1.

In our studies which dates back to the end of seventies it was aimed first to study the correlation between the chemical structure and the ion-selectivity of different bis-benzo-15-crown-5 derivatives with urethane linkages as ionophores in liquid ion-selective electrodes and then, to develop a new synthetic, bis-crown-ethers based ion-selective electrodes of relevant analytical importance. Accordingly the effect of the nature of the substituents in the benzo-rings and that of the chains connecting the two crown-ether moieties were studied.

EXPERIMENTAL

Preparation of PVC membrane electrodes

5 mg ligand /Fig.2./, 65 mg PVC powder /Aldrich 18956-1 high molecular weight/ and 120 mg orto-nitro-phenyl-octyl-ether /o-NPOE, Fluka 73732/ were dissolved c.a. 2 ml tetrahydrofuran /THF, Fluka 87368, puriss p.a./. This solution was poured into a glass ring / Ø = 28 mm/ fixed on a glass plate /13/ and let the THF evaporate under controlled conditions.

From the membranes Ø=7 mm rings were cut out and build into a Philips IS-561 electrode body. As internal filling 10^{-2} M KCl was used.

Electrochemical cell:

Ag,AgCl|1 M KCl‖0.1 M Li-acetate‖sample solution|ion selective electrode

Radelkis OP 8201

Measurements

For calibration of the electrodes potassium chloride solutions made by serial dilution were used. The selectivity coefficients were determined by the separate solution method using 10^{-1} M solutions.

All potential measurements were carried out at room temperature.

The measured potential data were corrected by the diffusion potential according to the Henderson formalism /14/, while the solution activities were calculated according to the extended Debye Hückel equation /15/.

RESULTS AND DISCUSSION

From the comparative studies carried out with bis-benzo--15-crown-derivatives with urethane linkages it became clear that the presence of NO_2-group on the benzo-rings has a pronounced effect on the analytical parameters of ion-selective electrodes based on bis-crown-derivatives with urethane linkages /Fig.3/. It means ion-selective electrodes based on bis-benzo--crown-ethers with NO_2 substituent on both benzo-rings proved only to be useful for practical analytical purposes. Moreover, it was observed that the properties of ion-selective electrodes based on urethane-linked bis-benzo-crown-derivatives without NO_2-substituents were far beyond than that of sensors based on ionophores synthetised by Shono's group /10/ or Wong et.al /9/.

Besides, from this study it became clear that if a nitro--substituted bis-benzo-15-crown-5 ligand with urethane linkage is employed as ionophore for liquid membrane electrodes than the nature of the chain connecting the two crown-ether moieties did not effect the analytically important parameters of such ionophores based electrodes /Fig.2/.

The experimental data i.e. the selectivity increasing effect of NO_2-group on the benzo-rings of an urethane linked bis--benzo-15-crown-5 derivatives, could be explained either by the decreased electron donating ability of the two O-atoms, close to the aromatic ring due to the electron withdrawing effect of NO_2-group, or by the conformation of the two crown-ether rings providing a special arrangement i.e. an ion-trap, which is stabilised inter moleculare H-boundings /Fig.4/. In the former case the two 15-crown-5 rings would behave as an 18--crown-6 from the points of view of coordination, forming 1:1 complexes with potassium ions.

The existance of cross H-boundings in the bis-nitro-benzo--15-crown-5 derivatives with urethane linkages has been supported by the experimental facts that the $N\text{-}CH_3$ substitution in the urethane linkage or the absence of the NO_2-group in the benzo-rings drastically decreased the analytical parameters of such ligand based ion-selective electrodes /Fig.5./. Moreover C_{13} relaxation time measurements provided evidences for the above assumption.

With the aim to differentiate between the two effects we have synthetised bis-benzo-15-crown-5 derivatives with esther' and ether linkages published in the literature /9,10/ as well as the nitro-derivatives of the corresponding ligands. The comparison of the electrochemical data of ion-selective electrodes based on, an such ligands /i.e. with or without a nitro-substituent/ showed no pronounced difference in the potentiometric selectivity / $-\log K_{K,M}$/ or slope values /s/ of the sensors. This also proved in an indirect way our second assumption, i.e. the role of the formation of ion-trap stabilised by cross H-boundings.

The most important parameters of a potassium ion-selective electrode based on ligand BME-15 with some other ligand based potassium electrodes have been compared /Fig.6/.

The calibration properties of a BME-15 ligand based ion--selective electrodes in blood serum electrolyte have been tested /7/.

The life time of the BME-15 ligand based electrode in aqueous 10^{-3}M KCl in continuous use exceeds 1 year.

CONCLUSION

A great number of bis-benzo-crown derivatives were prepared and studied as ionophores in liquid membrane electrodes in order to obtain information on the structure - ion-selectivity relationship among them the ligand BME-15 exhibited similar potassium ion-selectivity as the antibiotics valinomycin.

From the study it can be concluded the significant potassium ion-selectivity requires the following essential parts to be present in the molecule of the ligand used as ionophores:

1. Two crown ether units connected with a flexible chain;
2. $-NO_2$ group on the benzo-rings of the crown units;
3. NH-CO group near to the $-NO_2$ group.

REFERENCES

1. W.E.Morf and W.Simon, in Ion-Selective Electrodes in Analytical Chemistry, Ed. H.Freiser, Plenum Press, New York, London, 1978. Chapter 3.
2. J.M.Lehn, in Structure and Bonding, Springer, Vol.16. p.1. /1973/
3. I.M.Kolthoff, Anal.Chem. 51, 1R /1979/

4. C.P.Pedersen, J.Am.Chem.Soc. 89, 7017 /1967/
5. F.Vögtle, E.Weber and U.Elben, Kontakte /Merck/ /2/,36/1980/
6. G.A.Rechnitz and E.Eyal, Anal.Chem. 44, B70 /1972/
7. O.Ryba and J.Petranek, J.Electroanal.Chem. 44, 425 /1973/
8. M.Mascini, and F,Pallozzi, Anal.Chim.Acta 73, 375 /1974/
9. K.W.Fung, K.H.Wong, J.Electroanal.Chem. 115, 115 /1980/
10. K.Kimura, T.Maedo, H.Tamura, T.Shono, J.Electroanal.Chem. 95, 91 /1979/
11. B.Ágai, I.Bitter and L.Tőke, Acta Chim.Hung.Acad.Sci. 110 /1/, 29 /1982/
12. L.Tőke, B.Ágai, I.Bitter, E.Pungor, K.Tóth-Szepesvárynee, E.Lindner, M.Horváth, J.Havas /Magyar Tudományos Akadémia/ Crown ether compounds and ion-selective membrane electrodes containing the same
PCT Int.Appl. WO 8300, 149 /Cl. C07J5/18/, 20 Jan. 1983, Hu.Appl.81/1.999, 09 Júl. 1981; 34 pp.
13. G.J.Moody, R.B.Oke and J.D.R.Thomas, Analyst, 95, 910 /1970/
14. P.Henderson, Z.Physikal.chem.63, 325 /1908/
15. M.S.Mohan, R.G.Bates, Clin.Chem.21, 864 /1975/

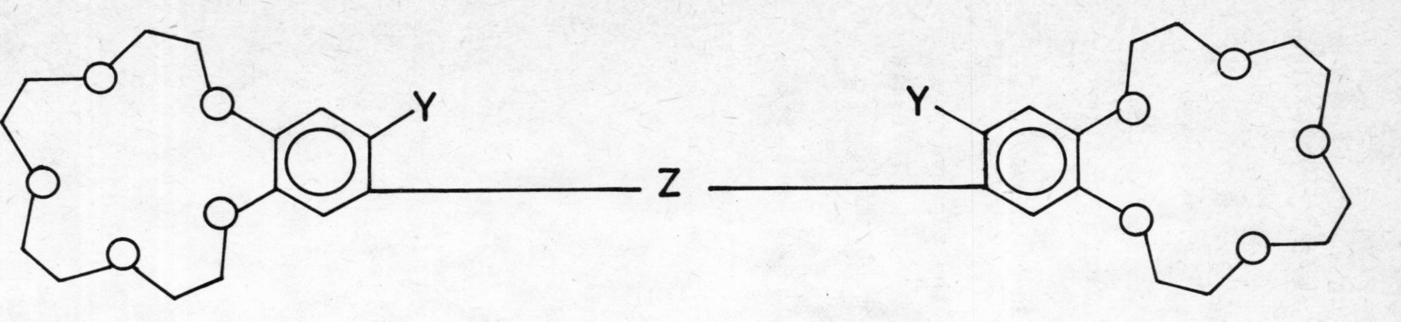

	Z
BME	$-NH-CO-O-X-O-CO-NH-$
KIMURA et al.	$-CH_2-O-CO-X-CO-O-CH_2-$
WONG et al.	$-CH_2-(O-X)_n-O-CH_2-$ n = 0, n = 1, n = 2
SMIDT et al.	$-CO-O-CH_2-CH_2-O-CH_2-CH_2-O-CO-$

Fig. 1 Bis-benzo-15-crown-5 derivatives as ionophores for potassium selective electrode

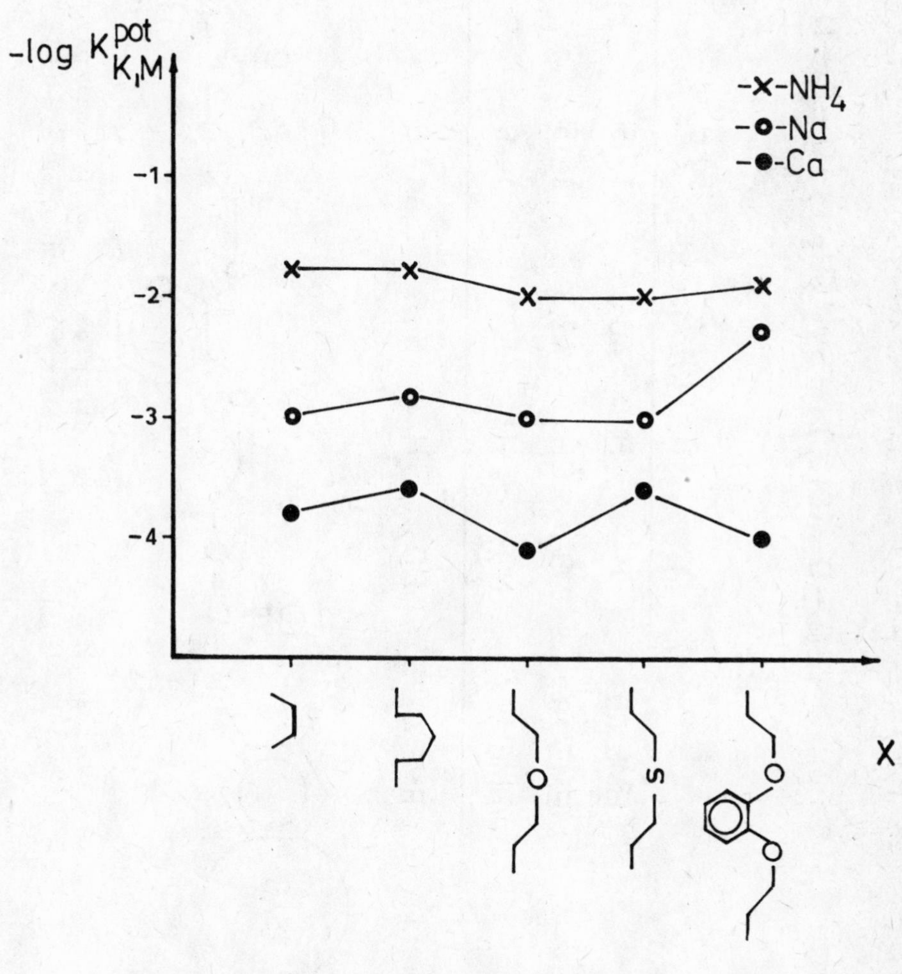

Fig. 2 Potentiometric selectivity of different bis-benzo-15--crown-5 derivatives with urethane linkages

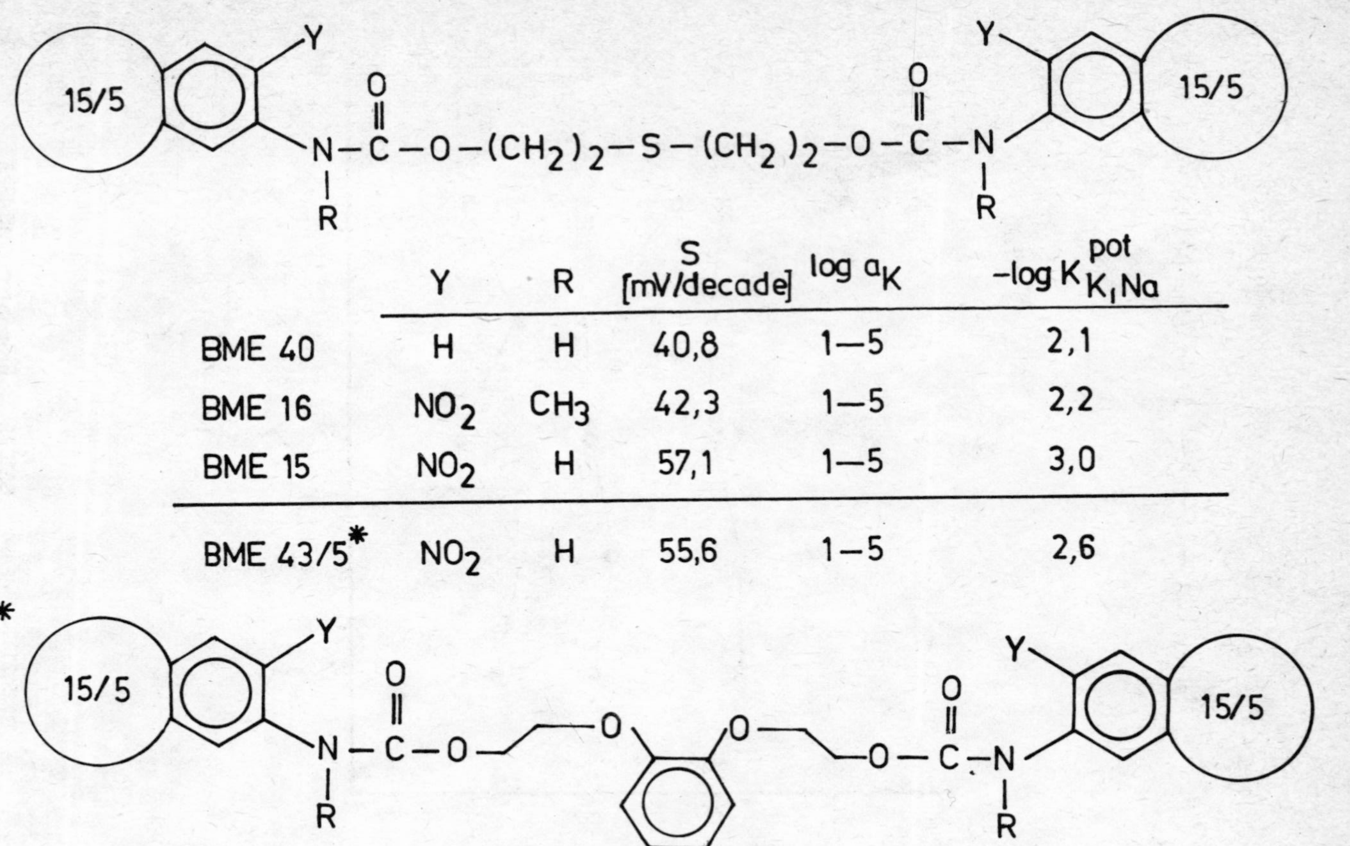

	Y	R	S [mV/decade]	log a_K	$-\log K^{pot}_{K,Na}$
BME 40	H	H	40,8	1–5	2,1
BME 16	NO_2	CH_3	42,3	1–5	2,2
BME 15	NO_2	H	57,1	1–5	3,0
BME 43/5*	NO_2	H	55,6	1–5	2,6

Fig. 3 Effect of $-NO_2$ group on the analytical parameters /slope, S and $-\log K^{Pot}_{K,M}$ / of some bis-benzo-15-crown-5 derivatives with urethane linkages

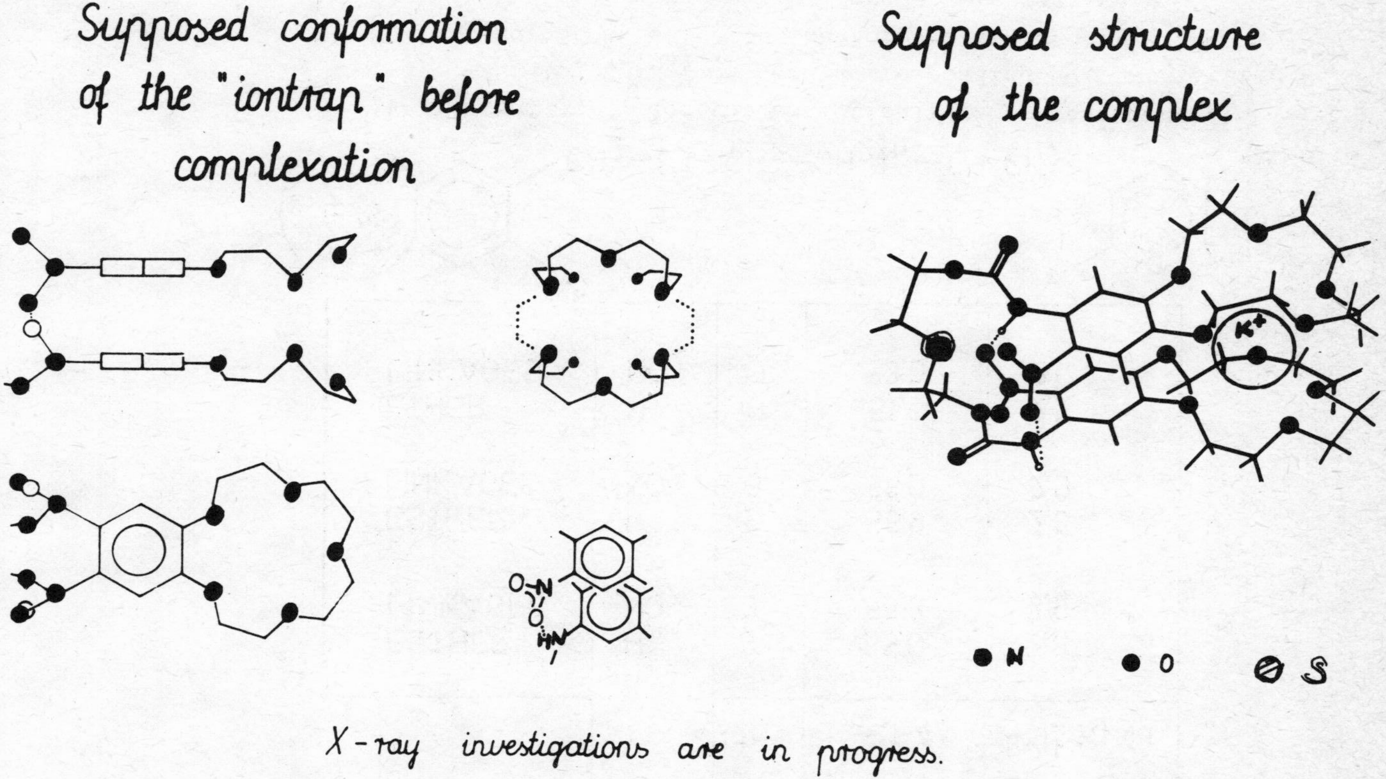

Supposed conformation of the "iontrap" before complexation

Supposed structure of the complex

O-N
O
HN

N O S

X-ray investigations are in progress.

Fig. 4 Supposed conformation of the ligand and the structure of the K-ligand complex

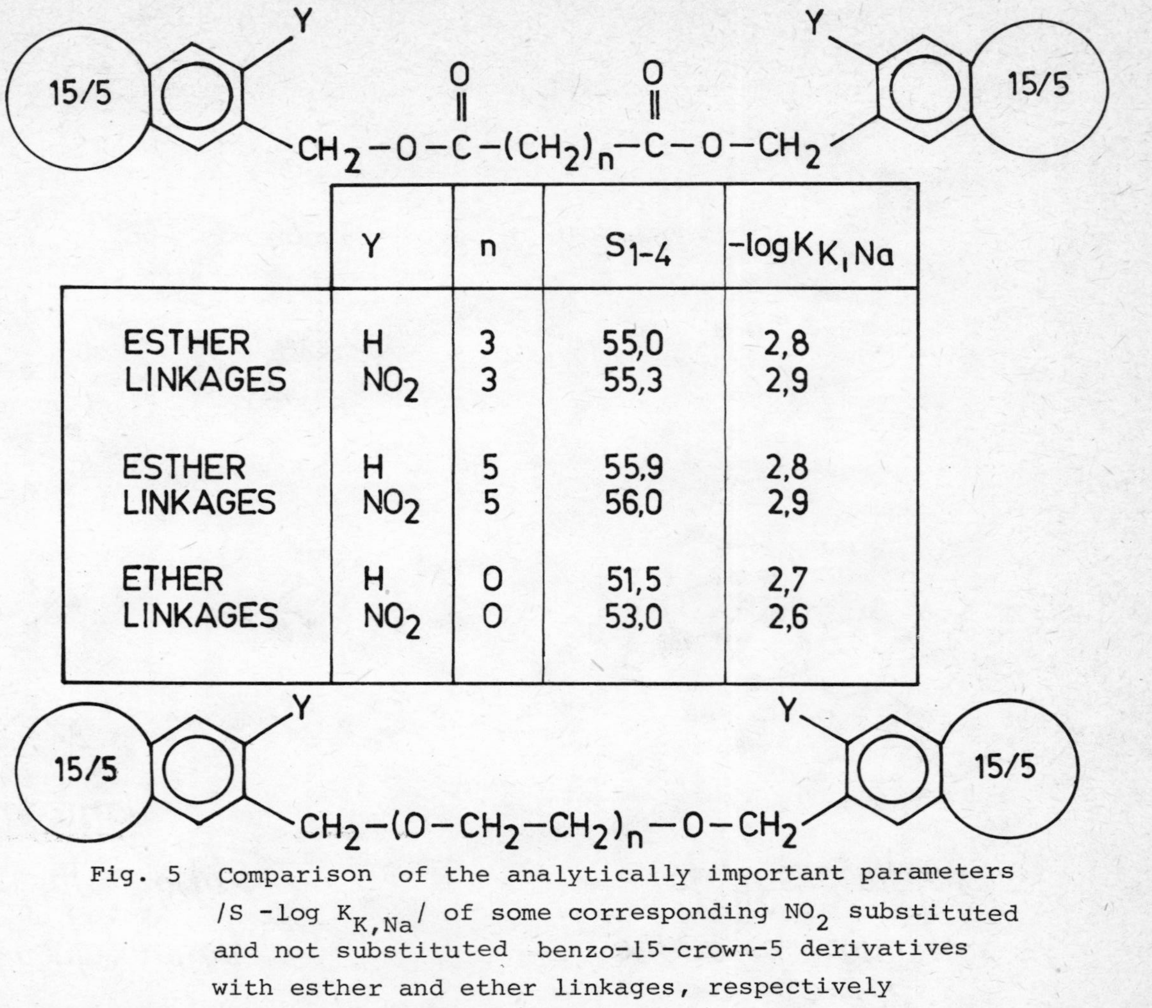

	Y	n	S_{1-4}	$-\log K_{K,Na}$
ESTHER LINKAGES	H	3	55,0	2,8
	NO_2	3	55,3	2,9
ESTHER LINKAGES	H	5	55,9	2,8
	NO_2	5	56,0	2,9
ETHER LINKAGES	H	0	51,5	2,7
	NO_2	0	53,0	2,6

Fig. 5 Comparison of the analytically important parameters /S $-\log K_{K,Na}$/ of some corresponding NO_2 substituted and not substituted benzo-15-crown-5 derivatives with esther and ether linkages, respectively

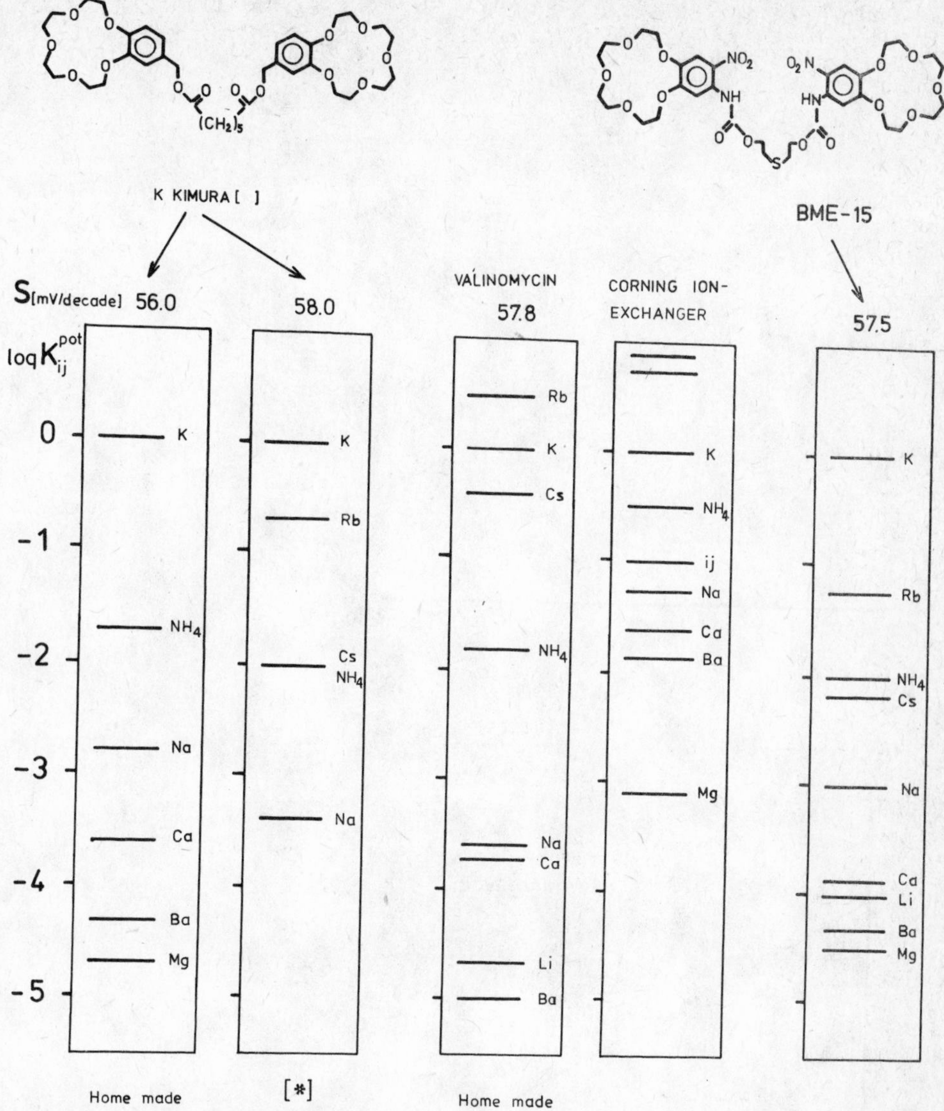

* KIMURA, K., MAEDA, T., TAMURA, H., SHONO, T., J.Electroanal.Chem., 95 (1979) 91-101.

Fig. 6 Comparison of the selectivity parameters of the BME-15 ligand based potassium sensor with that of other ligand based sensors

CALIBRATION IN BSE

BSE: Na^+ 140,0 mmol/l
Ca^{2+} 1,1 -"-
Mg^{2+} 0,6 -"-
Cl^- 119,4 -"-
HCO_3^- 24,0 -"-

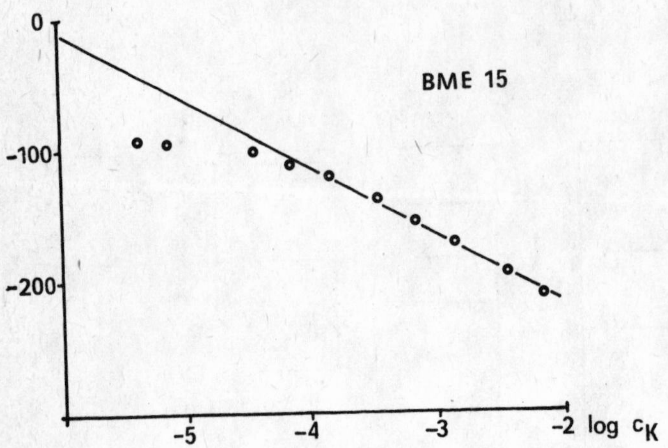

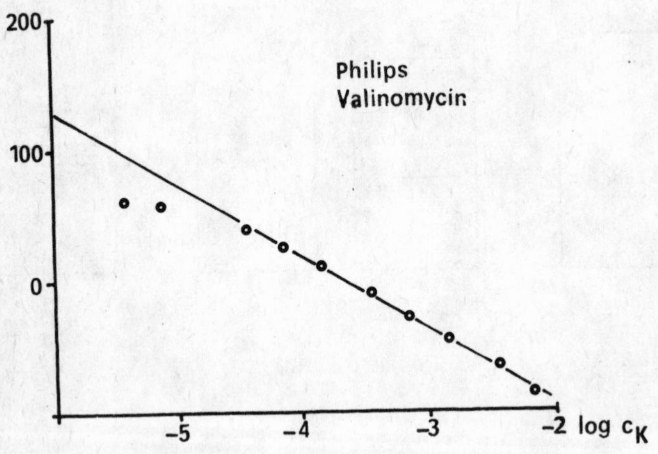

Fig. 7 Comparison of the calibration properties of a BME-15 ligand based potassium ion-selective sensor with that of a Philips Valinomycin based sensor

QUESTIONS AND COMMENTS

Participants of the discussion: K.Kimura, E.Lindner, W.Simon and K.Tóth

Question:

I was puzzled by the effect of the nitro group, I really do not understand this effect. Do you have any spectroscopic evidence that there is a direct interaction of the nitro group with the ortho-functional group in the molecule ?

Answer:

We made ^{13}C and proton NMR studies. In the proton NMR a band at 10,47 ppm indicated a strong hydrogen bridge in the NH-chelate. The bridge remains unchanged even in protic solvents.

Comment:

That is good evidence. I have asked this question because the effect of the nitro group was just opposite than one would expect from purely electronic effects on the basicity of the ether-oxygen atoms.

Answer:

It is true and that is why we could not get any improvement with other types of compounds. We have some secondary proof as well. With the N-methyl derivative, where the hydrogen bond cannot be formed, we got a very poor selectivity.

Question:

Did you experience any problem with the solubility of your crown ether ? Because I think that compounds with urethane structure tend to have low solubility.

Answer:

We had no problem with the solvents used, e.g. o-nitrophenyl octylether. Normally we had only 0.4-4 mg of the active material in the membrane, so the solubility was not really crucial.

Question:

Why do you use the sulphur atom as the linkage between the two units ?

Answer:

We had also tried oxygen and methylene linkages. As long as the NH and NO_2 groups were also present, they behaved practically in the same way in respect of selectivity and slope values.

4th Symposium on Ion-Selective Electrodes
Mátrafüred, 1984

ION-SELECTIVE FIELD-EFFECT TRANSISTORS: DIFFERENT TYPES AND PROBLEMS

Yu.G. VLASOV

Department of Chemistry, Leningrad University,
Leningrad 199164, USSR

CONTENT

1. INTRODUCTION

Nearly fifteen years ago P.Bergveld reported /1/ about the ion sensitivity of insulated-gate field-effect transistor (IGFET) without metal gate electrode immersed into aqueous solution. This paper initiated the development and investigation of the new type of ion-selective electrodes (ISE) called "ion-sensitive field-effect transistors" (ISFET). ISFET comprises electronic properties of MISFET and electrochemical properties of ion-selective electrode. Due to the achievements of microelectronics and integrated-circuit technology on one hand and potentiometry on the other the possibilities for the development of microsize chemical sensors for biomedical investigations and microanalysis have been created.

Along with a small size ISFET's, comparing them with common ISE,have some advantages and peculiarities: all-solid-state design; in-situ transformation in most sensors of high input to low output impedance; fast response time; possibility of multiprobe sensor fabrication on a small semiconductor chip.

All these specific features of ISFET's encouraged scientists in many countries and the number of scientific groups dealing with this problem is enlargening. At present such investigations are peformed in USA (J.Janata, R.P.Buck, J.N.Zemel), USSR (Yu.G.Vlasov), the Netherlands (P.Bergveld), Japan (T.Matsuo, T.Akiyama), Norway (K.Nagy, T.A.Fjeldy), England (A.Sibbald), Poland (W.Torbicz), GDR (W.Leimbrock). Since the first report ISFET's with different ion-sensitive membranes have been constructed to determine the pH of aqueous solutions and to measure the concentration of some cations (Na^+, K^+, NH_4^+, Ag^+, Ca^{2+}) and anions (F^-, Cl^-, Br^-, I^-, CN^-) (see Table 1). Various methods are used for the deposition and post-deposition treatment of ion-selective membrane thin films upon the gate insulator: CVD-technique /2/, doping of insulator films with ions from solution using anodic polarization /3/, ion implantation /4,5/, chemical deposition from solution /6/, vacuum evaporation /7,8/, various sputtering techniques /9,10/, LB technique /11/ etc. Several types of ISFET's have been proposed for chemical analysis and medical studies including in-

-vivo measurements. For example, ISFET's have been used to measure the concentration of some ions in blood /12,13/, serum /14/, dental plaque /15/. Application of ISFET's in flow-injection analysis system /16,17/, in rapid acid-base titrations /18/, for double layer study /19,20/ have been reported.

Further progress has been succeeded in the development of multisensors. At first, ISFET for simultaneous mesurement of Na^+ and H^+ ions had been proposed /21/ and recently Sibbald reported about triple- and quadruple-function ISFET's for the determination of K^+, H^+, Na^+ and Ca^{2+} ions in solutions /22, 23/. More studies of GasFET's /24/ and ChemFET's /25,26/ had been carried out and the new type of chemically sensitive semiconductor devices (CSSD) called ion-controlled diod appeared as reported by Zemel /27/. Theory, design and application of ISFET's had been reviewed by several authors /22,28-30/ and theoretical principles discussed in /31-33/.

Despite of the successes in the ISFET study mentioned above there remain some problems - technological and theoretical ones that are to be solved. Microelectronic technology can easily be used for the fabrication of ISFET's and ChemFET's but methods of ion-selective membrane deposition require further research. For some kinds of materials the exact ion sensitivity mechanism is still unknown and that hinders the creation of new materials for ion-selective membranes and improvement of known ones.

This paper is concerned with the classification system of electrodes used to measure the concentration of ions or molecules in solution and the place of ISFET's and ChemFET's in this system. Also the main trends in ISFET investigation and some problems are discussed.

2. THE PLACE OF ISFET IN THE CLASSIFICATION SYSTEM

Since 1976 when IUPAC recommendations on ISE classification had been proposed new types of ISE appeared and some additions and refinements were made /34,35/. Our discussion will be based on the classification system which we described in detail earlier /35/. Due to the latest achievements in the development of CSSD we must include some new elements in this system

(Fig.1). Three main levels are distinguished in this system. The first one determines the species to which electrode is sensitive (ions, molecules), the second characterises the membrane material (solid or liquid; homogeneous or heterogeneous) and the third level concretizes the different types of materials used as membrane (glass, chalcogenide glass, crystalline ionic conductors, insulator films, etc). According to the principals of the system it is sensible to put the CSSD (that include ISFET, ChemFET, ICD) on a sublevel just below the first main level to show that although these devices are a kind of ISE they have some specific features. What differs these two kinds of ISE in principle is (i) the existence of polarizible interfaces in case of ISFET, (ii) no charge transfer across the membrane and (iii) a possibility to use insulator films as ion-sensitive membrane materials. As it is seen from Fig.1 CSSD is connected with the elements of two other levels to show that all the membrane materials used in conventional ISE can also be applied in semiconductor microsensors.

3. ISFET AS A MULTIPHASE SYSTEM

3.1. From MISFET to ISFET

A usual method of ISFET and ChemFET fabrication is to combine a field-effect transistor with a membrane sensitive to some ion or molecule species. It is interesting to note that term "membrane" is now commonly applied to ISFET's although no ion penetration or transfer across the membrane is needed in that case. Field-effect transistor consists of semiconductor substrate (for e.g. p-Si) in which n-type zones of source and drain are formed by impurity diffusion and thin (~ 1000 Å) layers of insulator (SiO_2, Si_3N_4) are deposited. In MISFET structures (Fig.2a) metal electrode is evaporated on top of the insulator gate region. The potential V_G of appropriate sign being applied to the gate the channel between source and drain is formed due to the field effect and the drain current flowing through the channel will depend on the gate potential.

The ion-sensitive membrane may be depositied right upon the gate metal film /36/ or connected to the gate terminal of the MISFET with short metallic bond /37/ (Fig.2b). When an insulator film is playing the role of membrane it may be deposited just on the surface of semiconductor (Fig.2c). But the most common case is a situation when ion-selective membrane is deposited upon the gate insulator layer (Fig.2d). A solution/ /membrane interface potential difference (p.d.) φ_{SM} is an additive in a total p.d. applied to the gate but it is φ_{SM} that depend on concentration (activity) of potential determining ions in solution and in ideal case it must obey the Nernst equation:

$$\varphi_{SM} = \varphi_o \pm \frac{RT}{z_i F} \ln a_i \qquad (1)$$

where φ_o is a standard electrode potential, R is a gas constant, z_i is electrical charge of ion i in the solution, F is Faraday constant, T is thermodynamic temperature, and a_i is the activity of species i in the bulk of the solution.

Fig.3 shows ISFET in simplest electrochemical cell. Measuring changes of drain current , I_D, (V_G, V_D = const) or gate voltage, V_G, changes (I_D, V_D = const) one may find the variation of interface p.d., φ_{SM}, and, thus, according to (1) the activity of ion i in the solution.

3.2. ISFET is a multiphase system

Description of ISFET given above reveals that ISFET is actually a multiphase system. When the role of ion-selective membrane is played by insulator itself the system consist of three phases (electrolyte/insulator/semiconductor or EIS-system). If an ion-selective membrane is deposited over the insulator film then we are dealing with 4-phase system (electrolyte/membrane/insulator/semiconductor or EMIS-system). It is obvious that to understand the behaviour of ISFET as an ion-selective electrode one must study the ISFET as a system, that means investigation of interfacial and bulk properties of each phase (Fig.4).

Let us consider in brief each of the phases comprising ISFET and outline the class of problems connected with them.

Electrolyte solution. To know the activity coefficients one must take into account properties of the electrolyte species in the bulk of a solution and their interaction with a solvent. That are usually discussed in terms of the Debye-Huckel theory described in detail in /38/. Much more problems arise concerning the double layer, its structure, parameters, charge distribution and interaction of ions with solid phase.

Membrane. The electrochemical properties of the ion-selective membrane mainly depend on the physical and the chemical properties of the membrane material /33/. To elucidate the exact mechanism of ion sensitivity one must study in detail bulk properties of the membrane material such as chemical composition, crystalline structure, type of conductivity (electronic, ionic), the state of charge carriers. It is also essential to clarify the interfacial properties: the double layer structure in solid phase at the solution/membrane and membrane/insulator interfaces; the distribution of charge and potential; the chemical composition of interfaces. We also have to note that comparing electrochemical properties of ISFET's and conventional ISE we must take into consideration that some characteristics of thin film and bulky membrane may differ and that (unlike ISE) ISFET's have a blocked membrane/insulator interface.

Insulator. Insulator in ISFET structure may play a double role - as a passive gate insulator and as an active ion-sensitive membrane material. But certain restrictions exist concerning physical and electrical properties of the insulator film. Fig.5 shows the types of charges in the insulator/semiconductor structure /39/. They are as follows: 1) Q_{fs}, fixed charge is a positive stable charge in insulator, very close to the insulator/semiconductor interface related to the structure of the interface region; 2) surface states charge, Q_{ss}, results from the defects in the structure of the interface region; 3) Q_i, mobile impurity ions in the insulator, usually positive alkali ions or H^+. Drift of these ions under bias causes instabilities; 4) Q_t, electrons or holes trapped in the insulator. The distribution and the value of each type of charge alters the insulator/semiconductor structure. The process of silicon thermal oxidation is a well developed one and gives the most stable electrical characteristics of the insulator/semiconductor interface than any other method of insulator deposition. So it is a common practice to use a thin (50-500 Å) underlayer of SiO_2 before the deposition of some other insulator. If in ISFET ion-selective membrane with high ionic conductivity is used some "ion barrier" layer (for e.g. Si_3N_4) is to be deposited to prevent the ion penetration into silicon dioxide film in which alkali ions have rather high diffusion coefficients.

Semiconductor. The most suitable semiconductor material for ISFET is Si, but Ge is also used /40/. Nevertheless, semiconductor material stable at high temperatures and pressures for ion-control in some technological processes is required.

3.3. Methods of ISFET investigation.

Two main methods of investigation are used to study the influence of ion concentration of the solution on the parameters of the multiphase system-ISFET. The first way is to measure the dependence of ISFET electrical characteristics on the solution composition.

In ISFET structures a change in drain current, I_D, is assumed /28/ to be related to a change in solution/membrane potential difference:

$$I_D = \frac{\mu_n C_o W V_D}{L}(V_G - V_T^* + \Delta\varphi - E_{Ref} - \frac{V_D}{2}) \quad (1)$$

for operation in the non-saturation region and

$$I_D = \frac{\mu_n C_o W}{2L}(V_G - V_T^* + \Delta\varphi - E_{Ref})^2 \quad (2)$$

for operation in the saturation region where μ_n is the electron mobility in the ISFET channel, W is the wigth and L is the length of the channel; C_o is the capacitance per unit area of the gate insulator; V_G is the gate voltage applied through the reference electrode with E_{Ref} potential; V_D is the drain voltage; and V_T^* is the threshold voltage which is a function of several parameters: (i) values of charges presenting in the insulator and insulator/semiconductor interface; (ii) the insulator gate capacitance; (iii) the flat band voltage; and (iv) various potential drops at the interfaces /41/. If as a result of electrolyte solution/membrane interaction some of above parameters change, I_D will also change its value.

Thus, the only result this method gives is I_D vs. concentration dependence, I_D being the measure of ion concentration, when all other parameters are constant. So, this method gives poor information about the ion sensitivity mechanism of the device.

The second way of ISFET investigation is to study the capacitance-voltage (C-V) characteristics of the EIS or EMIS system (Fig.4) in solutions at various ion concentrations. This method is an informative one and can be applied not only for the ion concentration measurements but also for the investigation of surface state density, charge distribution in the insulator, charge value and polarity; the electrolyte/insulator interface potential difference. Method of high frequency and low frequency C-V measurements was used by different authors /42-47/ for investigation of the ISFET ion sensitivity mechanism and recently was applied for investigation of oxide/electrolyte double layer characteristics /20,32,48/.

4. THE C-V METHOD FOR ISFET INVESTIGATION

The C-V technique is nondestructive, accurate and fast method by which many properties of the insulator film, semiconductor layer, and interfaces in electrolyte/insulator/semiconductor system can be determined. It is widely used for investigation of MIS-structures and parameters control in IC-technological processes /39,49/ and is also applicable for the EIS-system investigation /42,46/.

4.1. Advantages of C-V technique

Now in brief the C-V method for ISFET investigation will be considered. Fig.6 shows as an example the equivalent circuit of the EIS-system. The common situation is that insulator capacitance (C_i) and space charge capacitance (C_{sc}) are small enough and give the main contribution to the EIS-system impedance in solutions with high ionic strength /46/.

Typical C-V curves for MIS-structure are shown in Fig.7. Curve (b) results from high frequency (HF) measurements when minority carriers in the inversion layer follow dc bias but not ac signal. Curve (a) is a low frequency (LF) or quasistatic C-V curve when minority carriers follows both ac signal and dc bias. The maximum capacitance value (C_i) being that of insulator itself, any changes in insulator thickness or dielectric constant caused by electrolyte solution/insulator interactions or some kind of treatment are easely registrated as changes in C_i value. On condition that electrolyte/insulator interactions are limited to surface ion-exchange processes, any variations in the potential-determining ion concentration will result in the C-V curve shift as it is shown in Fig.8. Calculating the theoretical value of flat band capacitance, C_{FB}, /39/ and defining the flat band potential value for each curve, V_{FB} vs. concentration can be plotted, V_{FB} being the measure of the device ion-sensitivity.

The C-V measurements combined with the sample etching can also give some information about charge distribution in the insulator bulk. It is rather difficult to obtain the absolute value of charge presenting in the insulator film in the EIS-system as far as electrochemical and physical energy scales are different but some attempts to solve this problem have been made /41/.

The shape of C-V curves can also give some information about the interface state properties of the EIS-system but to have more detailed data LF or quasistatic C-V measurements are to be used /43-45/.

Another advantage of the C-V method is a simplification of the sample fabrication and encapsulation processes. The application of C-V method will be demonstrated using the results of EIS-systems with SiO_2, Si_3N_4, $SiO_2(F^-)$ and ZrO_2 investigation.

4.2. SiO_2 films. pH-response

The physical and electrochemical properties of SiO_2 insulator films in Metal/Insulator/Semiconductor (MIS) and Electrolyte/Insulator/Semiconductor (EIS) systems have been studied and reviewed in detail /45,48,50/.

The problem of interest concerning electrolyte/SiO_2/Si system used in ISFET's is the mechanism of the pH sensitivity. The peculiarities of this system are: non-Nernstian response to hydrogen ions which depends on the deposition conditions; both interfaces (electrolyte/SiO_2 and SiO_2/Si) and the whole film volume could take part in the potential generating process. It has been proposed /51,52/ that SiO_2 ion sensitivity mechanism is based on some charged species diffusion process. Detailed high frequency (HF) and quasistatic C-V investigations /43,45,46/ revealed that pH changes do not alter the shape of HF C-V curve and result only in its parallel shift, SiO_2/Si surface state density being immutable. Thus these results allow one to conclude that changes in the electrolyte/ /SiO_2 interface potential difference with pH are responsible for the observed shift and V_{FB} vs. pH dependence. Fast response time, independent on the oxide film thickness, gives further support to ion-exchange mechanism at the electrolyte/ /SiO_2 interface.

The properties of the oxide/electrolyte interface are usually discussed in terms of the Site-Binding Model /42/ which states that hydrogen-ion sensitivity of the interface potential difference depends first of all on the concentration of surface

$$\equiv M - O^- \;;\; \equiv M - OH \qquad \text{and} \qquad \equiv M - OH_2^+$$

groups and the value of their dissociation constants. Vlasov et al /54/ showed that pH-response of SiO_2 depends on the type of preparation and treatment of silicon dioxide films (Fig.9). For example, Si oxidation in the presence of water vapor results in near-Nernstian behavior (55-60 mV/pH) of the obtained SiO_2 films connected with the increase in Si-OH group density. It is well known, that SiO_2 has a poor selectivity to Na^+. But we have found (Fig.10), that after P-treatment SiO_2-films become unsensitive to Na^+. This results are in agreement with the paper /60/.

4.3. Si_3N_4 films. pH-response

Silicon nitride films are widely used as ISFET pH-sensitive membrane material. The C-V technique was applied to investigate the electrochemical properties of the electrolyte/Si_3N_4/Si system /46,47,55/, a correlation between pH-sensitivity and silicon nitride hydrogen content being established. Fig.11 shows V_{FB} vs. pH dependences for samples with different concentration of hydrogen which mainly exist as N-H groups. The hydrogen content in silicon nitride being increased, stability and the slope of electrode respose also increases.

Silicon nitride is known as a very chemically inert material due to the $p\pi - d\pi$ interaction between silicon 3d orbitals and nitrogen lone pair located after hybridization on p orbitals /56/. In case of Si-NH groups such electron density redistribution does not occur. The lone pair is localized on nitrogen atom and thus can react with hydrogen ions in solution. So two types of functional groups exist on silicon nitride surface, the potential generating reactions being as follows /57/:

$$SiOH \rightleftharpoons SiO^- + H^+ \qquad (x)$$

$$SiOH + H^+ \rightleftharpoons SiOH_2^+ \qquad (xx)$$

$$SiNH + H^+ \rightleftharpoons SiNH_2^+ \qquad (xxx)$$

Thermal treatment in vacuum or in Ar-atmosphere of Si_3N_4 leads to the decrease of NH-groups concentration /56/ and to the decrease of pH-sensitivity especially in acidic solutions (Fig.11). Reactions (x) and (xx) may be assumed to play the main role for low NH-content Si_3N_4 films and pH-response becomes much like that of SiO_2 (Fig. 9). Reaction (xxx) is important at high hydrogen concentration and leads to better pH-response. The obtained results enabled to develop pH-ISFET /57/ with silicon nitride film containing approriate amount of hydrogen. The sensor had near-Nerstian slope (54 mV/pH) and fast response time.

4.4. SiO_2 films doped by F^-. pF response

The application of insulator (SiO_2, Si_3N_4) films as ISFET membranes sensitive to different ions (not only to hydrogen) is of great interest, adhesion problem being solved and technological process of membrane deposition being simplified. As we showed above (see 4.2 and 4.3) a correlation existing between the value of the pH-sensitivity and insulator film hydrogen content /54,55/ may result from ion-exchange processes at electrolyte/insulator interface. Thus, experimental results lead us to the assumption that doping of insulator films with some appropriate impurity might create some new ion-sensing layers.

In order to obtain F^--sensitive membranes the doping of SiO_2 layers by fluoride ions has been carried out by anodic polarization of SiO_2/Si structures in 0,01 M NaF solution, fluoride ions being injected in strong ($\sim 10^7$ V/cm) electric field at a room temperature /3/.

The C-V technique was used in that case for two reasons. On one hand, from HF C-V curves distribution of charge in SiO_2 films after anodic polarization have been obtained (Fig.12); on the other hand, the dependence of the flat band voltage, V_{FB}, on concentration of fluoride ions in solution (Fig.13) has been acquired. Moreover, SiO_2/Si interface surface state density has been controlled with the help of quasistatic C-V measurements.

The penetration of fluoride ions into the SiO_2 film under anodic polarization results in C-V curve shift (Fig.14). Fig.12 shows the profile of fluoride ion concentration in 120 nm thick SiO_2 film doped by F^-. The membrane being etched, different F^- surface concentration could be obtained according to the presented profile. From Fig.13 where the dependence of V_{FB} of different samples vs. fluoride ion concentration in the solution is presented it is seen that: (i) initial SiO_2 film is insensitive to the F^- concentration in the solution; (ii) after anodic polarization treatment SiO_2 film with fluorine profile shown in Fig.12 reveals F^--sensitivity of approximately 30 mV/pF; (iii) the etching of the surface layer ($\sim$ 20 nm) leads to maximum surface concentration of fluorine estimated as $\sim 10^{19}$ cm^{-3} and results in changes of the V_{FB} vs pF dependence. Now it has complex character and is composed of two sections with -40 and +50 mV/pF slopes. Two parallel mechanisms of ion-exchange and specific adsorption have been proposed /3/ but further experiments are required to explain these data.

4.5. Metal oxide films as ISFET membrane material. ZrO_2 pH response

Along with silicon nitride different metal oxide films have also been studied /9,10,83,58/ as hydrogen-ion-sensitive ISFET membranes. Results have been compared with those obtained in colloid chemistry investigations /42/ and have been discussed in terms of surface complexation model /42,58/. For electronically conducting oxides some reaction mechanisms have been proposed and discussed /59/. Mixed oxide films have been prepared and studied /2,9/ as Na^+ and K^+ ion-sensing layers.

As it was mentioned above, several methods of oxide films deposition exist. For Al_2O_3 and Ta_2O_5 gate ISFET's Abe et al /9/ used CVD technique and reported of ideal pH-sensitivity and excellent stability. Ta_2O_5 films made by sputtering the Ta_2O_5 target and studied by Akiyama et al /58/ also showed good results. So in this case ion-sensing characteristics are independent of the deposition method. On the other hand it is interesting to compare the results on ZrO_2 film deposited by two different methods. In /58/ ZrO_2 films have been obtained by plasma oxidation of the Zr film evaporated by electron beam, research being carried with the help of ISFET. The ZrO_2-gate ISFET had the pH sensitivity of approximately 50 mV/pH and had very slow pH-response (10 min to get 90 % response). In our research /10/ 110 Å and 930 Å thick films of ZrO_2 have been deposited by the electron-beam sputtering method. Electron diffraction and X-ray photoelectron spectroscopy (ESCA) were used to study the film crystalline structure and chemical composition. To study the influence of the solution composition on electrolyte/ZrO_2/SiO_2/Si system C-V method was used. Fig.6 shows the equivalent circuit of the system under consideration. pH changes shift the C-V curve along the gate voltage axis and the shift in flat band voltage, V_{FB}, can be regarded as a measure of the ion-sensitivity.

C-V curves of the structures with thin (110 Å) films of ZrO_2 showed no peculiarities, but those of the structures with 930 Å thick films of ZrO_2 were shifted along the voltage axis in the negative direction and distorted. Such changes can be attributed to the changes in the insulator fixed charge value and the increase of the surface states density at the insulator/semiconductor interface. In accordance with /10/ thick (930 Å) ZrO_2 films showed poor pH-response (long-term drift, nonlinear pH vs. V_{FB} function, slow response (15-20 min for 90 % response). Thin (110 Å) ZrO_2 films, on the contrary, showed ideal (58 ± 2 mV/pH) pH-response and (Fig.15) fast (few seconds) response time. Results of electron diffraction and ESCA studies reveal that both films have identical chemical composition but differ in crystalline structure. Thin films (110 Å) have been found to be amorphous while thick (930 Å) films were crystalline. The observed difference in thick and thin films electrochemical behaviour might be attributed in terms of Site-Binding Model to the differences in hydration ability of crystalline and amorphous ZrO_2 films.

5. ISFET FOR DIFFERENT ION DETERMINATION: pH, CATIONS, ANIONS

In this chapter ISFET-determined ions are divided in three parts: hydrogen, cations and anions. They are represented in Table 1 (parts 1,2 and 3). The Table is constructed in chronological way and shows different membranes or prepared-in-a--different-way membranes used in ISFET. All details concerned the ISFET fabrication and their parameters of operation can be taken from the literature cited in the Table.

6. ISFET PROBLEMS

ISFET development and application gave rise to some problems that are to be solved to provide mass production of these devices. The main difficulties are: sensor encapsulation, temperature sensitivity, development of a small size reference electrode, adhesion of thin films to the semiconductor or insulator surface. Medical application of ISFET put forward problems of thrombogenicity and adsorption of proteins on the surface of ion-selective membrane. Many of these problems are far from their solution but some interesting projects and investigations have been published.

6.1. Temperature sensitivity of ISFET

It should be noted that considering ISFET thermal sensitivity it must be taken into account that the whole system consists of reference electrode, electrolyte solution, and ISFET itself and that not only ISFET parameters are temperature dependent but activity coefficients of ions in the solution and the reference electrode potential also vary with temperature. ISFET temperature-dependent semiconductor characteristics have been described in detail in /60,77/. The overall temperature sensitivity of the whole system is about 1 mV/°C and yeilds an error of approximately 0.02 pH/°C. So for the precise measurements temperature compensation must be used. Several techniques have been proposed. Janata /76/ used differential-current pH-measurements provided with the help of two ISFETs fabricated on the same substrate one of which was pH-sensitive and the other played a role of reference electrode. Thermal characteristics of both transistors having been just the same, the temperature effect was compensated (Fig.16).

It has been shown /60,77/ that ISFET has an atermal point (Fig.17) i.e. a point where the drain current of the transistor is not thermally dependent. Choosing the operating conditions of the ISFET in the vicinity of this point we minimize the error caused by the temperature sensitivity of the semiconductor device but we could not eliminate it at all because of the ion activity and reference electrode potential thermal sensitivity.

Electronic system for ISFET temperature control and compensation has been proposed by Bergveld /77/.

6.2. Reference electrode

Theoretical considerations of the EIS system established the necessity of the reference electrode for ISFET measurements and this problem is now of historical interest only but the practical problem of the compact and convenient reference electrode development still remains. Naturally, there always exists a possibility of using a conventional reference electrode (calomel or Ag/AgCl), but to exploit the advantages of ISFET it is essential to have small-size reference electrode of the ISFET dimensions. Several approaches have been made. Moss et al /84/ used small Ag/AgCl reference electrode with liquid junction incorporated into the catheter with ISFET on its tip. Janata /76/ described an integrated on-chip reference field-effect transistor used in conjunction with pH-sensitive ISFET. The reference electrode FET is pH-sensitive ISFET with Si_3N_4 gate insulator and has a compartment filled with 1.3 10^{-4} ml of buffered agarose which is connected with the solution by liquid junction via glass capillary. The advantages of this construction are: pH-ISFET thermal and optical sensitivity compensation, small size and on-chip location of the reference electrode, and the problem lies in incompatibility of the fabrication procedure which requires a lot of carefulness and skill with conventional IC-technology.

T.Matsuo, who discovered in 1978 that ISFET with thin Parylen (poly-p-xylylene) gate film is insensitive to the pH changes, proposed /78/ to use this device as a solid-state reference electrode. These results have been reviewed /29/, however, they appeared to be erroneous and lately it was shown /79/ that parylen films exibit some pH-sensitivity that depends on the deposition and annealing conditions. Moreever, ISFET with parylen gate film appeared to be sensitive to macromolecule adsorption, particularly to albumin /85/. Another approach to the reference electrode problem is to use two ISFETs with high selectivity constants sensitive to two different ions in the solution. Maintaining the constant activity of one of the potential generating ions in the solution to be measured, the second ion concentration can be determined. This approach can find its application in some fields of laboratory chemical analysis but, surely, for complex system (e.g. biological) some universal reference electrode is required.

6.3. Adhesion

Adhesion plays a crucial role in ISFET membrane fabrication. Poor adhesion leads to the limited lifetime of a device submitted to prolonged immersion in aqueous solution. To increase the membrane adhesion suitable substrate cleaning procedures should be used and some special conditions should be choosen when ion-selective film is deposited by means of ion- or electron-beam sputtering, thermal evaporation or some other method. Surface chemical treatment can also be used in order to introduce some surface chemical groups that would link the substrate with the deposited film. Another way is to use some known highly adhesive material impregnated with some organic or inorganic ion-exchanger. This method has been realised by Zemel /81/ and Kawakani /86/. For pK-sensor membrane they used valinomycin-doped negative photoresist layers with good adhesive properties towards insulator (SiO_2, Si_3N_4) films. pK-sensitive membrane has been deposited from the mixture of photoresist, valinomycin and plasticizer after the photopolymerization under UV radiation. Thus, K^+-sensitive transducer - ion-controlled diode /81/ has been developed.

Another approach has been made by Blackburn and Janata /80/ who proposed the suspended mesh modification to the polymer-membrane ISFET structures. The polyimide mesh of 10 μm square holes was suspended 1 μm above the gate insulator and then 50 μm thick potassium selective membrane was cast over the gate area. The mesh increased membrane adhesion which led to the 20-40 fold increase in device effective lifetime.

6.4. Thrombogenicity

Advantages of ISFET small size can be realised in the development of catheter-tip electrodes for the medical purposes, particularly for in-vivo measurements. However, first studies have revealed a problem of silicon nitride high thrombogenicity, which prevents the usage of pH-ISFETs with Si_3N_4 gate for in-vivo measurements in blood.

Several approaches have been made to solve this problem. Janata et al /68/ used pH membrane cast from a chlorobenzene--dichloromethylene solution, pH-function being linear in the 4-10 pH range.

To improve the electrode and blood compatibility Shimada et al /82/ used pH-ISFET coated with hydrogel of comb-like copolymer of polymethyl-methacrylate and polyhydroxyethyl-methacrylate. They found that this polymer is the best in terms of its mechanical properties and nonthrombogenicity. The hydrogel layers of different thickness were obtained and it was found that 0.8 nm thick film provides time constants approximately 3 s for the 99 % change in output signal. The thick (2 nm) layers showed down time constant up to 60 s. The continious in-vivo measurements of pH carried out in arteries and muscles of animals showed long-term stability of the developed ISFET.

REFERENCES

1 P.Bergveld, IEEE Trans.Biomed.Eng., 17, (1970) 70.

2 T.Matsuo. M.Esashi, Sensors and Actuators, 1, (1981) 77.

3 Yu.G.Vlasov, Yu.A.Tarantov, V.P.Letavin et al, Zh.Prikl.Khim., 55, (1982) 1310.

4 Y.Sanada, I.Akiyama, E.Niki, Bunseki Kagaku, 30, (1981) 678; Fresenius Z.Anal.Chem., 312, (1982) 526.

5 M.Pham, W.Hoffmann, in "Ion-selective electrode,4", ed.E.Pungor, Akad.Kiadó, Budapest, 1985.

6 N.Sujkovskaya, Chemical Methods for thin films deposition, Khimiya, Leningrad, 1971.

7 R.Buck, D.Hackleman, Anal.Chem., 49, (1977) 2315.

8 Yu.G.Vlasov, D.Hackleman, R.Buck, Anal.Chem., 51, (1979) 1570.

9 H.Abe, M.Esashi, T.Matsuo, IEEE Trans.Electr.Dev., ED-26, (1979) 1939.

10 Yu.G.Vlasov, A.V.Bratov, Yu.A.Tarantov, Zh.Prikl.Khim., (1985) in press.

11 P.S.Vincett, G.G.Roberts, Thin Solid Films, 68, (1980) 135.

12 Y.Ohta, S.Shoji, M.Esashi et al, Sensors and Actuators, 2, (1982) 387.

13 K.Shimada, M.Yang, K.Shibatani et al, Med.and Biol.Eng.and Comput., 18, (1980) 741.

14 P.Cheng, W.Ko, D.Fung et al, in "Theory, Design and Biomedical Application of Solid State Chemical Sensors", ed.P. Cheng, CRC Press, Cleveland, 1978.

15.K.Igarashi, K.Kamiyama, T.Yamada, Archs.Oral Biol., 26, (1981) 203.

16 A.Ramsing, J.Janata, J.Ruzicka et al, Anal.Chim.Acta, 118, (1980) 45.

17 A.Haemmerli, J.Janata, H.M.Brown, Anal.Chim.Acta, 144, (1982) 115.

18 M.Bos, P.Bergveld, A.Van-Veen-Blaaw, Anal.Chim.Acta, 109, (1979) 145.

19 J.F.Schenck, J.Coll.and Interface Sci., 61, (1977) 569.

20 L.Boussi, N.F.DeRooij, P.Bergveld, Surf.Sci., 135, (1983) 479.

21 M.Esashi, T.Matsuo, IEEE Trans.Bio-Med.Eng., BME-25 (1978) 184.

22 A.Sibbold, IEE Proc., 130, (1983) 233.

23 A.Sibbold, P.Whalley, A.Covington, Anal.Chim.Acta, 159, (1984) 47.

24 J.Lundstrom, C.Nylander, A.Spetz, Electron.Lett., 19, (1983) 249.

25 A.Covington, T.Harbunson, A.Sibbold, Anal.Lett., 15, (1982) 1423.

26 M.Esashi, Kagaku Kogyo, 33, (1982) 481.
27 C.Wen, T.Chen, J.Zemel, IEEE Trans., ED-26 (1979) 1945.
28 J.Janata, R.Huber, Ion-Selective Electrode Rev., 1, (1979) 31.
29 Yu.G.Vlasov, Zh.Prikl.Khim., 52, (1979) 3.
30.P.Bergveld, N.De Rooij, Sensors and Actuators, 1, (1981) 5.
31 J.Lanks, Sensors and Actuators, 1, (1981) 261.
32 L.Bousse, N.F.De Rooij, P.Bergveld, IEEE Trans.Electr.Dev., ED-30 (1983) 1263.
33 R.Buck, Sensors and Actuators, 1, (1981) 197.
34 W.Morf, The principles of ion-selective electrodes and membrane transport, Akad.Kiadó, Budapest, 1981.
35 Yu.G.Vlasov, in "Ion-selective electrode,3", ed.E.Pungor, Akad.Kiadó, Budapest, 1981.
36 R.Smith, J.Janata, R.Huber, J.Electrochem.Soc., 127, (1980) 1599.
37 M.Afromowitz, S.S.Yee, J.of Bioengineering, 1, (1977) 55.
38 J.O.M.Bockris, A.K.N.Reddy, Modern Electrochemistry, Plenum/Rosetta Edition, New York, 1970.
39 S.M.Sze, Physics of Semiconductor Devices, Wiley, New York, 1969.
40 Yu.G.Vlasov, Yu.A.Tarantov, V.P.Letavin, Zh.Prikl.Khim., 58, (1985) in press.
41 L.Bousse, J.Chem.Phys., 76, (1982) 5128.
42 W.Siu, R.Cobbold, IEEE Trans.Electr.Dev., ED-26 (1979) 1805.
43 P.Barabash, R.Cobbold, IEEE Trans.Electr.Dev., ED-29 (1982) 102.
44 N.De Rooij, P.Bergveld, Thin Solid Films, 71, (1980) 327.
45 Yu.G.Vlasov, Yu.A.Tarantov, A.P.Barabash et al, Zh.Prikl. Khim., 53, (1980) 1980.
46 Yu.G.Vlasov, A.V.Bratov, V.P.Letavin, in "Ion-selective electrode,3", ed.E.Pungor, Akad.Kiadó, Budapest, 1980.
47 Yu.G.Vlasov, A.V.Bratov, Elektrokhim., 17, (1981) 601.
48 L.Bousse, P.Bergveld, J.Electroanal.Chem., 152, (1983) 25.
49 K.Zaininger, F.Heiman, Solid State Technol., (1970), May, 49; June, 46.
50 The physics of SiO_2 and its interfaces. Ed.S.Pantelides, Pergamon Press, New York, 1978.

51 A.G.Revesz, Thin Solid Films, 41, (1977) L43.
52 A.G.Revesz, J.Electrochem.Soc., 126, (1979) 122.
53 J.Davis, R.James, J.Leckie, J.Coll.and Interf.Sci., 63, (1978) 480.
54 Yu.G.Vlasov, Yu.A.Tarantov. V.P.Letavin et al, Zh.Prikl. Khim., 55, (1982) 459.
55 Yu.G.Vlasov, Yu.A.Tarantov, V.P.Letavin, Zh.Prikl.Khim., 53, (1980) 2345.
56 Silicon nitride in electronics, Ed.A.Rzhanov, Nayka, Novosibirsk, 1982.
57 Yu.G.Vlasov, A.V.Bratov, Elektrokhim., 20, (1984) 2033.
58 T.Akiyama, Y.Ujihika, Y.Okabe et al, IEEE Trans.Electron. Dev., ED-29, (1982) 1936.
59 A.Fog, R.Buck, Sensors and Actuators, 5, (1984) 137.
60 O.Leistiko, Physica Scripta, 18, (1978) 445.
61 M.Esashi, T.Matsuo, Proc.of the 6th Conf.on Solid State Dev., Tokyo, 1974.Suppl.to the J.of the Japan Soc.of Appl. Phys., 44, (1975) 339.
62 S.Moss, J.Smith, P.Comte et al, J.of Bioengineering, 1, (1976) 11.
63 P.Comte, J.Janata, Proc.30th Ann.Conf.on Eng.in Medicine and Biology, Los-Angeles, California, v.19, 298, abstr.41.3 (1977).
64 P.Cheung, W.Ko. D.Fung et al, ibid., abstr.41.4 (1977); Theory, design and biomedical application of solid state chemical sensors, ed.P.Cheng, CRC Press, Cleveland, Ohio, 1978.
65 M.Esashi, T.Matsuo, IEEE Trans.Biomed.Eng., BME-25 (1978) 184.
66 T.Akiyama, T.Sugano, E.Niki, Bunseki Kagaku, 29, (1980) 584.
67 A.R.Olszyna, W.Wlosinski, D.Sobczynske et al, Materialy Elektroniczne, 2, (1982) 28.
68 P.T.McBridge, J.Janata, P.A.Comte et al, Anal.Chim.Acta, 101, (1978) 239.
69 T.Matsuo, M.Esashi, K.Iinuma, Digest of Joint Meeting of Tohoku, Section I.E.E.J., 28, abstr.2A-9 (1971).
70 T.Matsuo, M.Esashi, Proc.30th Ann.Conf.on Eng.in Medicine and Biology, Los-Angeles, California, v.19, 297, abstr.41.2

(1977). See also /9/.

71 Y.Sanada, T.Akiyama, E.Niki, Bunseki Kagaku, 30, (1981) 678; Fresenius Z.Anal.Chem., 312, (1982) 526.

72.S.Moss, J.Janata, C.Johnson, Anal.Chem., 47, (1975) 2238.

73 J.Topich, D.Fung, S.Wong, Extended Abstracts of the Electrochemical Soc., Spring Meeting, Seattle, Washington, abstr.85, 206 (1978).

74 S.Yee, M.Afromowitz, P.Galen, ibid., abstr.109, 257 (1978).

75 H.Nakajima, M.Esashi, T.Matsuo, J.Electrochem.Soc., 129, (1982) 141.

76 P.Comte, J.Janata, Anal.Chim.Acta, 101, (1978) 247.

77 P.Bergveld, Med.and Biol.Eng., 17, (1979) 655.

78 T.Matsuo, M.Esashi, Extended Abstracts of the Electrochemical Soc., Spring Meeting, Seattle, Washington, abstr.83, 202 (1978).

79 M.Fujihira, M.Fukui, T.Osa, J.Electroanal.Chem., 106, (1980) 413.

80 G.Blackburn, J.Janata, J.Electrochem.Soc., 129, (1982) 2580.

81 C.C.Wen, I.Lanks, J.Zemel, Thin Solid Films, 70, (1980) 333.

82 K.Shimada, M.Yano, K.Shibatani, Med.and Biol.Eng.and Comp., 18, (1980) 741.

83 T.Akiyama, T.Sugano, E.Niki, Bunseki Kagaku, 29, (1980) 584.

84 S.Moss, J.Smith, P.Comte et al, J.Bioeng., 1, (1976) 11,

85 H.Nakajima, M.Esashi, T.Matsuo, Nip.Kag.Kai, 1980 (1980) 1499.

86 S.Kawakami, T.Akiyama, Y.Ujihira et al, private communication.

87 P.Bergveld, IEEE Trans.Bio-Med.Eng., 19, (1972) 342.

88 U.Quesh, S.Caras, J.Janata, Anal.Chem., 53, (1981) 1983.

Table 1. Ions determined by ISFET. Part 1. pH

Ion	Membrane	Authors	Ref.
H^+	SiO_2	Bergveld 1972	/87/
	Si_3N_4	Esashi, Matsuo 1974	/61/
	Si_3N_4	Moss, Smith, Comte 1976	/62/
	Si_3N_4	Comte, Janata 1977	/63/
	Si_3N_4	Cheung, Fung, Wong 1977	/64/
	Corning 0150	Afromowitz, Yee 1977	/37/
	Al_2O_3	Matsuo, Esashi 1978	/65/
	SiO_2 (psg)	Leistico 1978	/60/
	polymer	McBridge, Janata 1978	/68/
	SiO_2	Matsuo, Esashi 1979	/9/
	Ta_2O_5	Akiyama, Sugano, Niki 1980	/66/
	SiO_2	Vlasov, Tarantov, Letavin 1980	/45/
	$Si_xN_yH_z$ ($x \leq 3$, $y \leq 4$, $z > 0$)	Vlasov, Bratov, Letavin 1980	/46/
	ZrO_2	Akiyama, Ujihira, Niki 1982	/58/
	BN	Sobezynska, Torbicz 1982	/67/
	ZrO_2	Vlasov, Bratov, Tarantov 1984	/10/

Table 1. Ions determined by ISFET. Part 2. Cations

Ion	Membrane	Authors	Ref.
Na^+	SiO_2	Begveld 1970	/1/
	SiO_2	Matsuo, Esashi, Iinuma 1971	/69/
	Al_xSi_yO	Matsuo. Esashi 1977	/70/
	ion-exchanger in PVC	Quesch, Caras, Janata 1981	/88/
	Al_2O_3 implanted by Li^+, Si^+	Sanada, Akiyama, Niki 1981	/71/
K^+	valinimycin in PVC	Moss, Janata, Johnson 1975	/72/
	valinomycin in PVC (ion sputtering)	Topich, Fung, Wong 1978	/73/
	valinomycin in photoresist	Wen, Lanks, Zemel 1980	/81/
	valinomycin in photoresist	Kawakami, Akiyama 1982	/86/
NH_4^+	organic ion-exchanger in PVC	Quesch, Caras, Janata 1981	/88/
	Teflon	Nakajima, Esashi, Matsuo 1982	/75/
Ag^+	AgBr	Buck, Hackleman 1977	/7/
	AgCl - AgBr	Vlasov, Hackleman, Buck 1979	/8/
Ca^{2+}	inorganic ion-exchanger in PVC	Yee, Afromowitz 1978	/74/
	polymeric	McBridge, Janata, Comte 1978	/68/

Table 1. Ions determined by ISFET. Part 3. Anions

Ion	Membrane	Authors	Ref.
Br^-	AgBr	Buck, Hackleman 1977	/7/
Cl^-	AgCl - AgBr	Vlasov, Hackleman, Buck 1979	/8/
	AgCl - Ag_2S	Shiramizu, Janata, Moss 1979	/28/
I^-	AgI - Ag_2S	Shiramizu, Janata, Moss 1979	/28/
CN^-	AgI - Ag_2S	Shiramizu, Janata, Moss 1979	/28/
F^-	SiO_2, doped by F^-	Vlasov, Tarantov, Letavin 1982	/3/

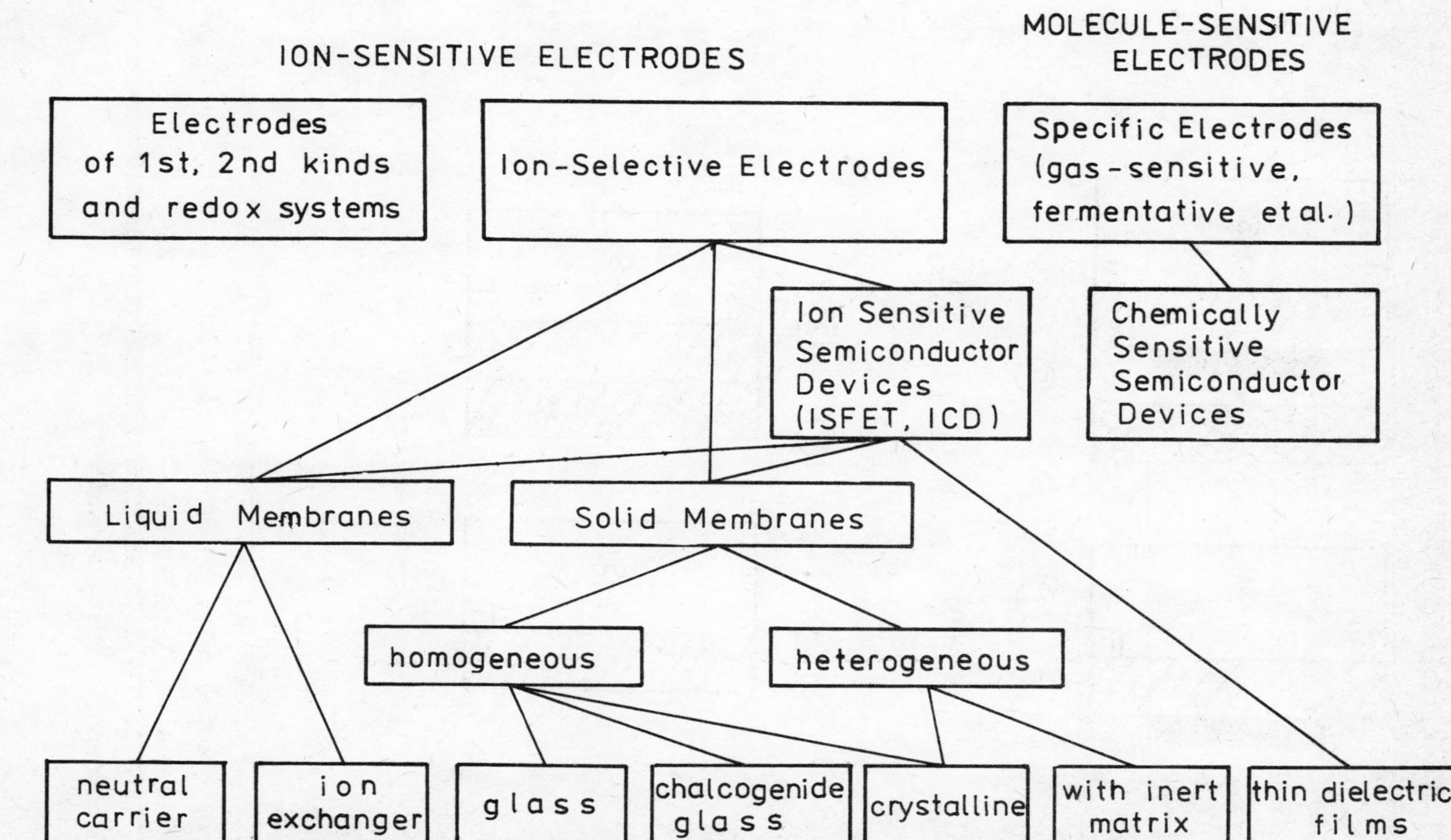

Fig. 1. Chemically sensitive semiconductor devices (CSSD) in the classification system of electrodes for determination of concentration of different species.

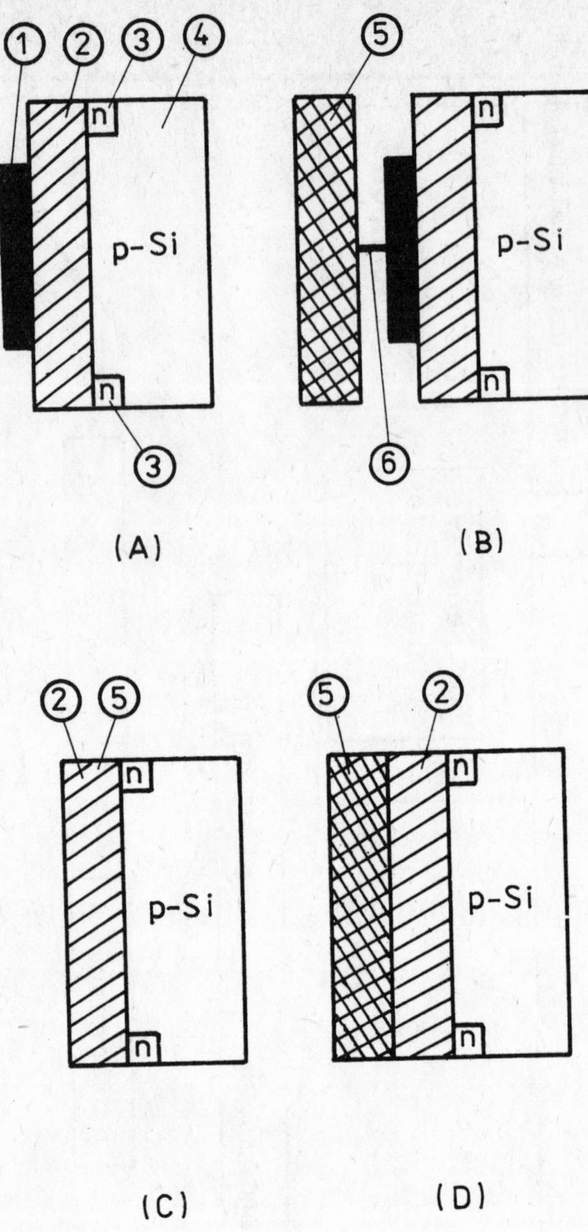

Fig.2. From MISFET (A) to ISFET (B, C, D).
1 - metal gate; 2 - dielectric (SiO_2, Si_3N_4); 3 - n-type regions (source and drain); 4 - p-type substrate (p-Si); 5 - ion-sensitive membrane; 6 - metal wire or conductive past contact.

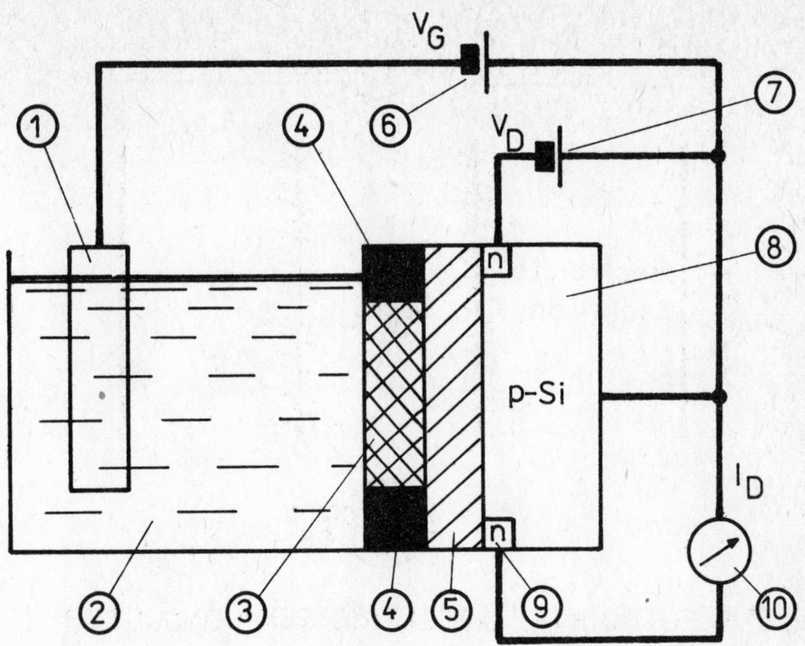

Fig.3. The simplest electrochemical cell for ISFET measurements.
1 - reference electrode; 2 - electrolyte solution; 3 - - membrane; 4 - encapsulator material; 5 - insulator; 6 - V_G power supply; 7 - V_D power supply; 8 - silicon substrate; 9 - source and drain n^+-zones; 10 - ammeter.

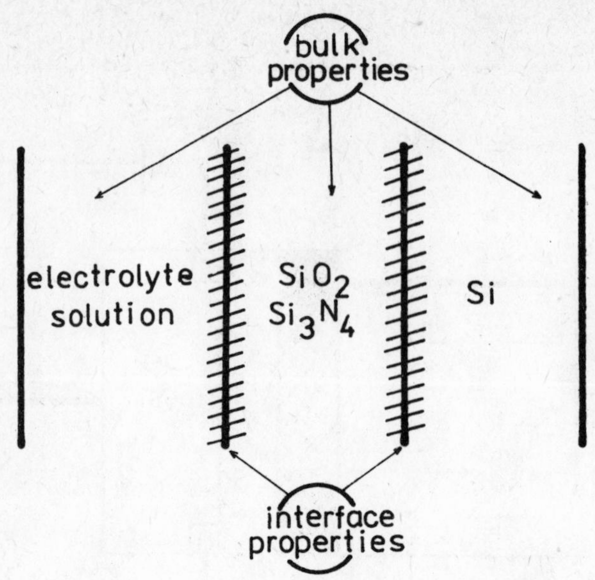

(a) ELECTROLYTE - DIELECTRIC-SEMICONDUCTOR (EDS)

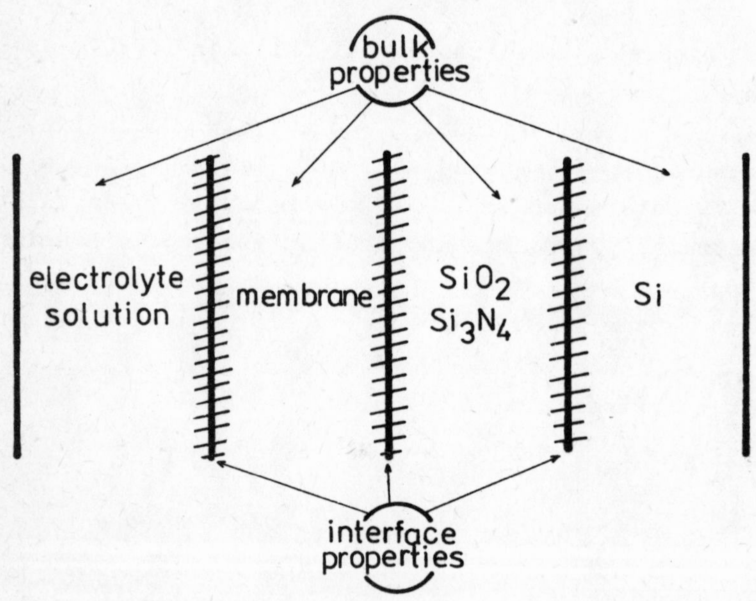

(b) ELECTROLYTE-MEMBRANE-DIELECTRIC-SEMICONDUCTOR (EMDS)

Fig.4. ISFET as a multiphase system: phases and interfaces.

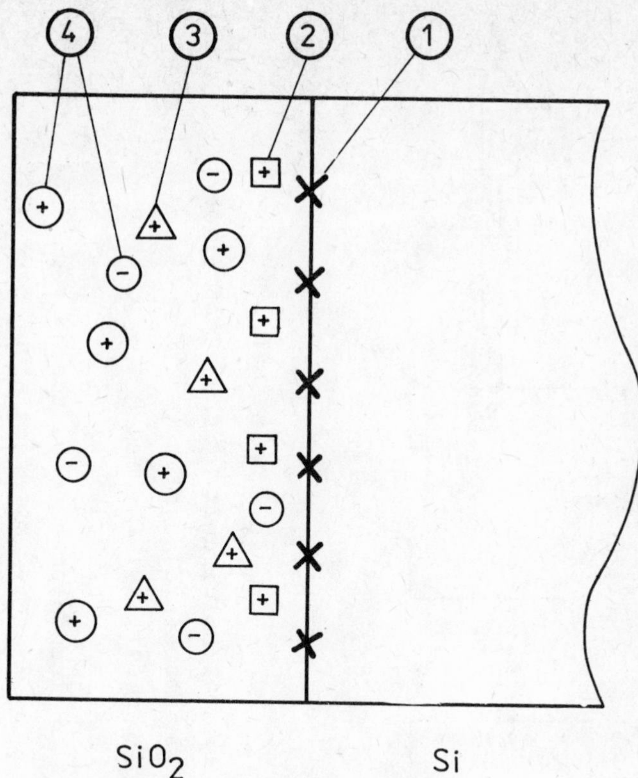

Fig.5. Charge distribution in DS-structure.
1 - surface state charge, Q_{ss}; 2 - fixed charge, Q_{fc}; 3 - mobile ions charge, Q_i; 4 - trapped electrones and holes charge, Q_t.

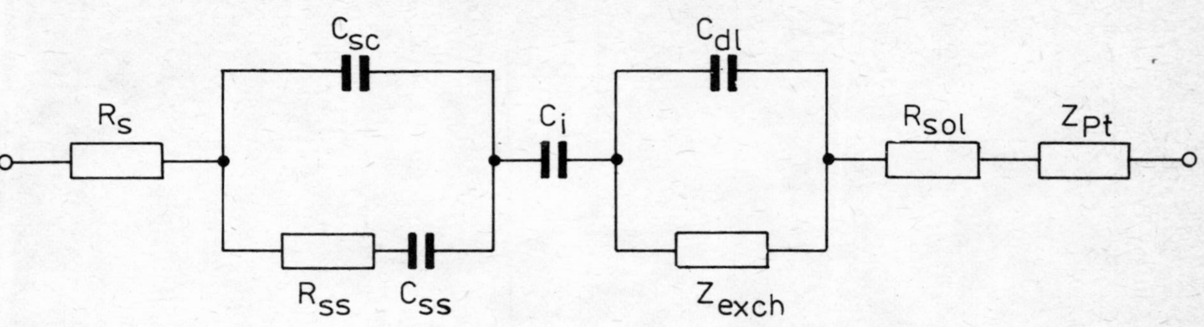

Fig.6. The C-V method. Equivalent circuit of the electrolyte/ /insulator/semiconductor (EIS) system (see Fig.4). R_s - silicon bulk resistivity, C_{ss} and R_{ss} - the surface states capacitance and resistivity, C_{sc} - space charge capacitance, C_i - insulator capacitance, C_{DL} - double layer capacitance, Z_{exch} - impedance associated with the ion exchange processes at the solution/insulator interface, R_{sol} - solution resistivity, Z_{Pt} - impedance of the platinum/solution interface.

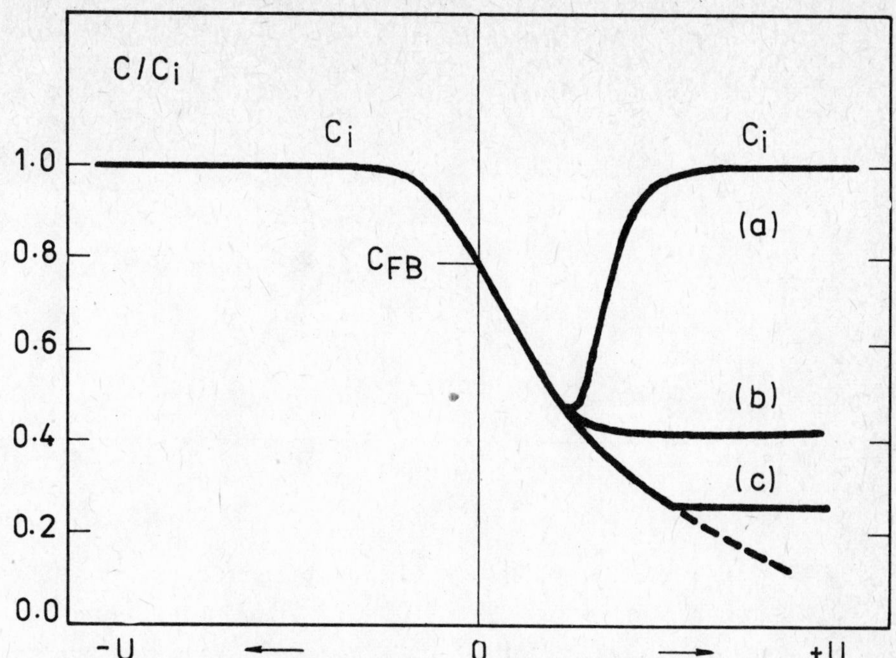

Fig.7. The C-V curves of ideal MIS-structure: a) - low frequency, b) - high frequency, c) - special case. C_i - insulator capacitance, C_{FB} - flat band capacitance.

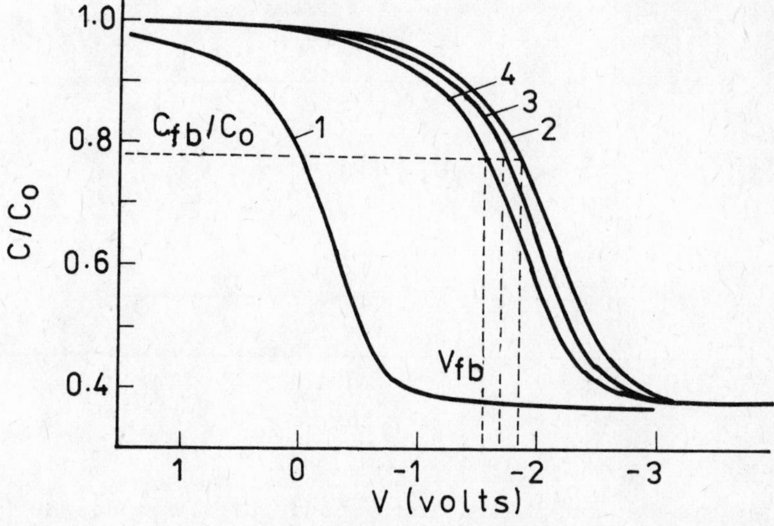

Fig.8. HF C-V curves for the electrolyte/SiO_2/Si system at different pH values (2-4) compared with the ideal (1). C_{FB} - flat band capacitance, C_o - insulator capacitance, V_{FB} - flat band voltage.

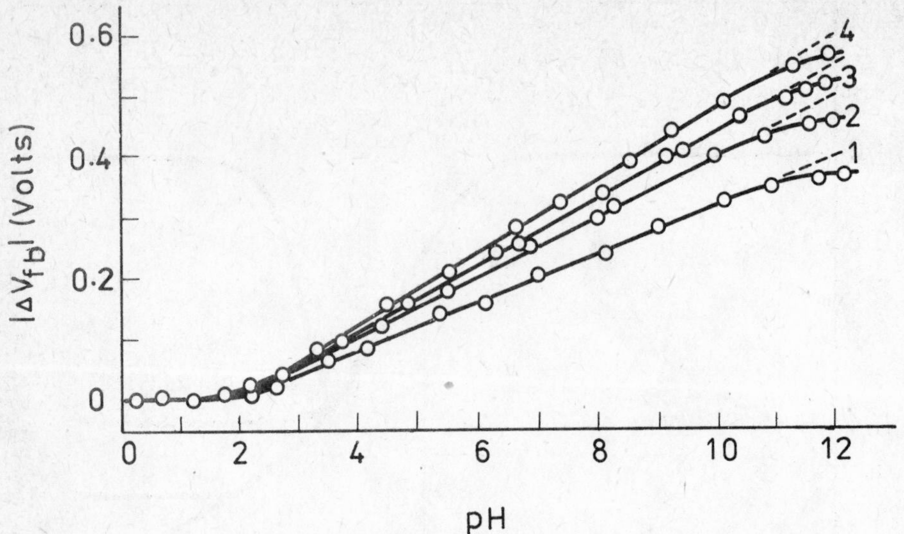

Fig.9. The pH-response of SiO_2 films obtained by thermal oxidation in dry oxigen at 1100°C (1) and in water vapors at 1050°C (2), at 950°C (3), at 900°C (4).

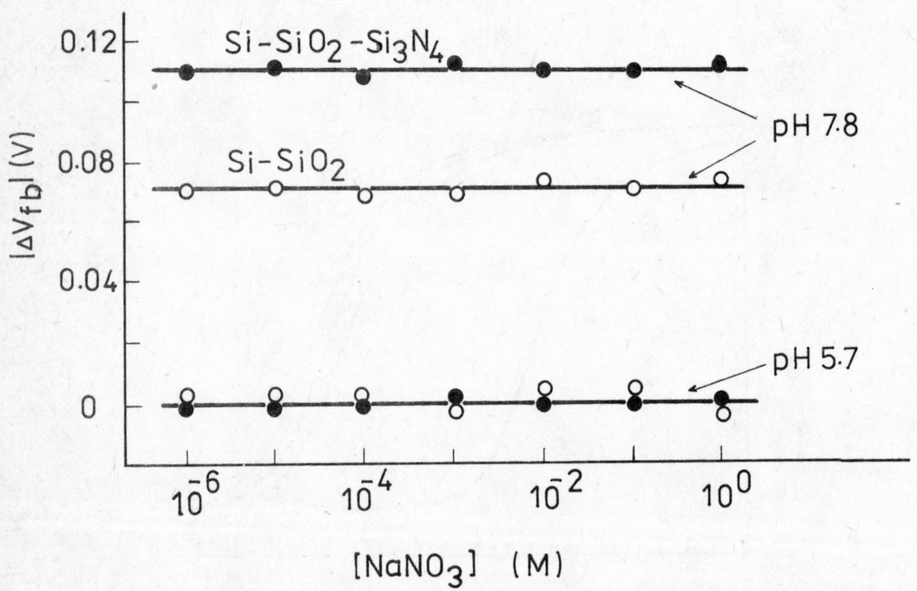

Fig.10. V_{FB} vs Na^+-concentration dependences for phosphorus silicate glass layer/SiO_2/Si (o) and Si_3N_4/SiO_2/Si (•) structures obtained in 2 N KNO_3 solution at different pH-values. 1 - pH 5,7; 2 - pH 7,8.

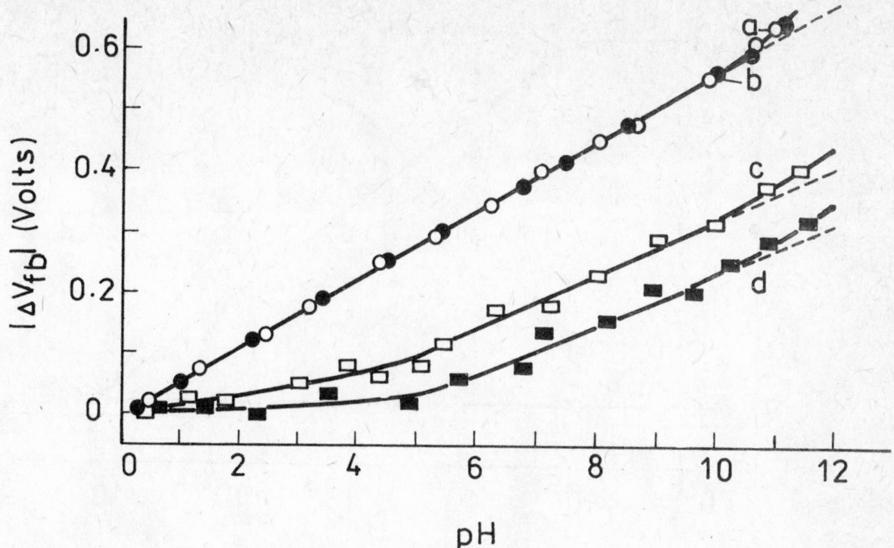

Fig.11. The pH-response of Si_3N_4 films with different content of NH-groups. a, b - 10^{22} cm^{-3} NH-groups; c, d - - 3 10^{21} cm^{-3} NH-groups.

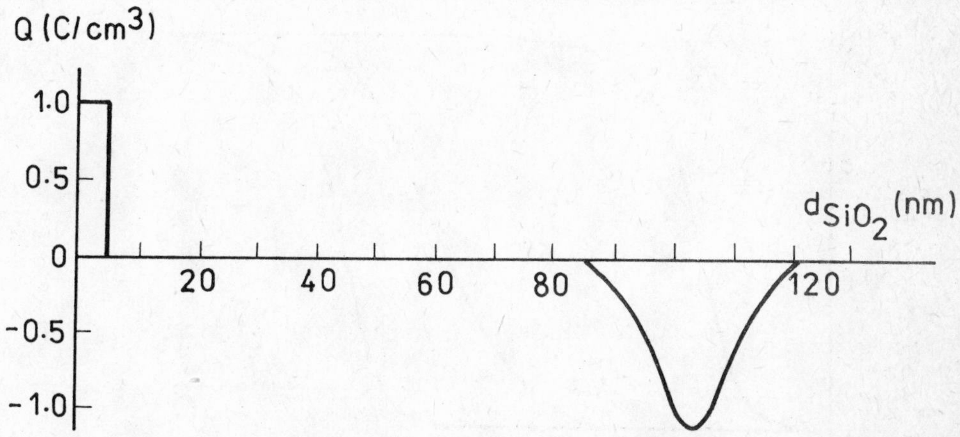

Fig.12. Charge distribution in 120 nm thick SiO_2 film after anodic polarization in 0.01 M NaF. The value of negative charge depends on F^- concentration in SiO_2 film. d_{SiO_2} - the distance from the semiconductor/insulator interface.

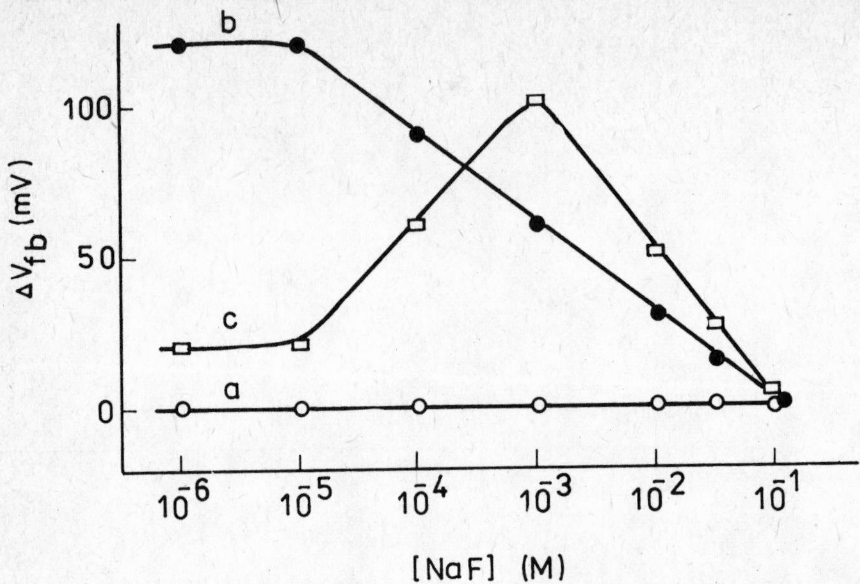

Fig.13. The V_{FB} vs F^- concentration dependence for the SiO_2--Si structure: a - before the polarization; b - after 20 min of polarization; c - after etching of 20 nm layer.

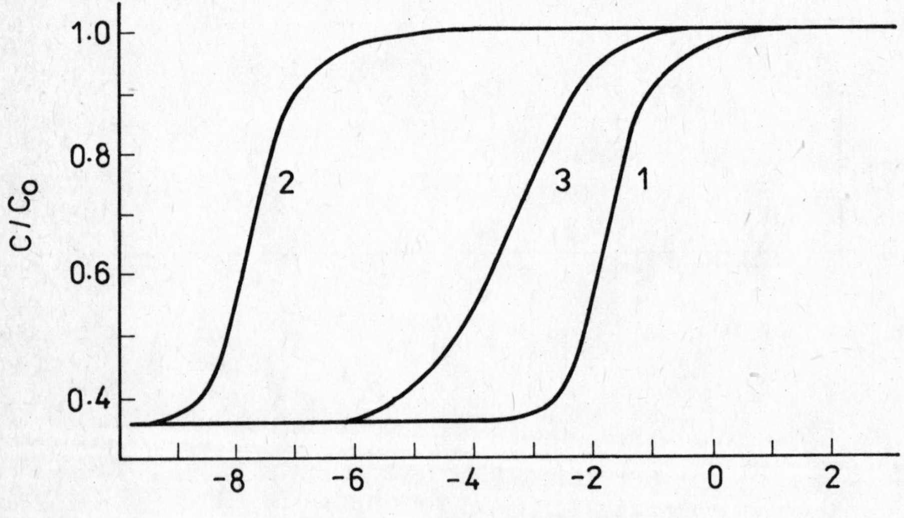

Fig.14. SiO_2/Si structure C-V curve shifts.
1 - initial curve; 2 - after the anodic polarization in 2 N $NaNO_3$; 3 - after the anodic polarization in 5 10^{-2} M NaF solution.

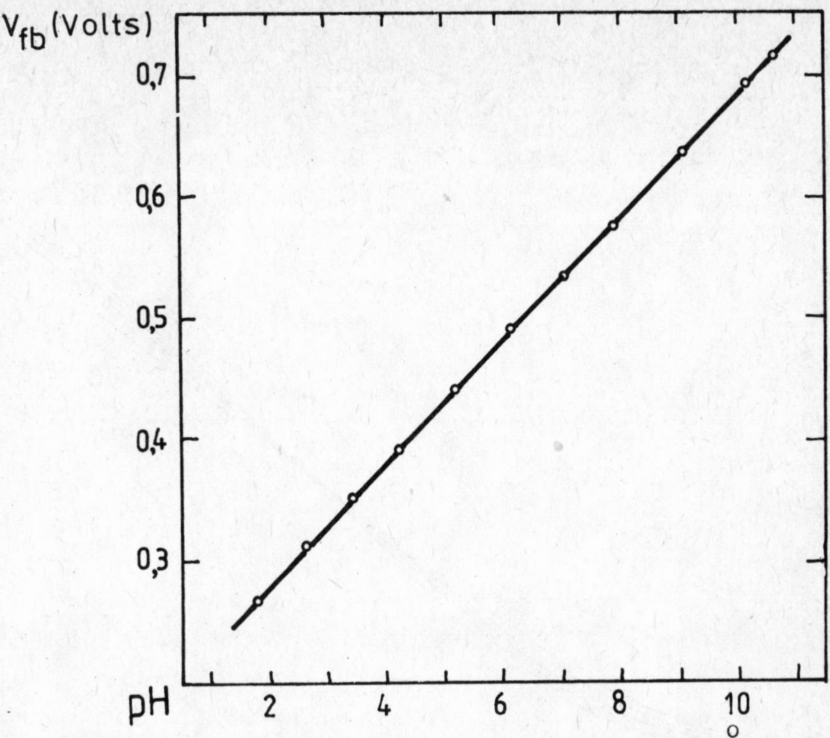

Fig.15. The pH-response of the ZrO_2 thin (110 Å) film.

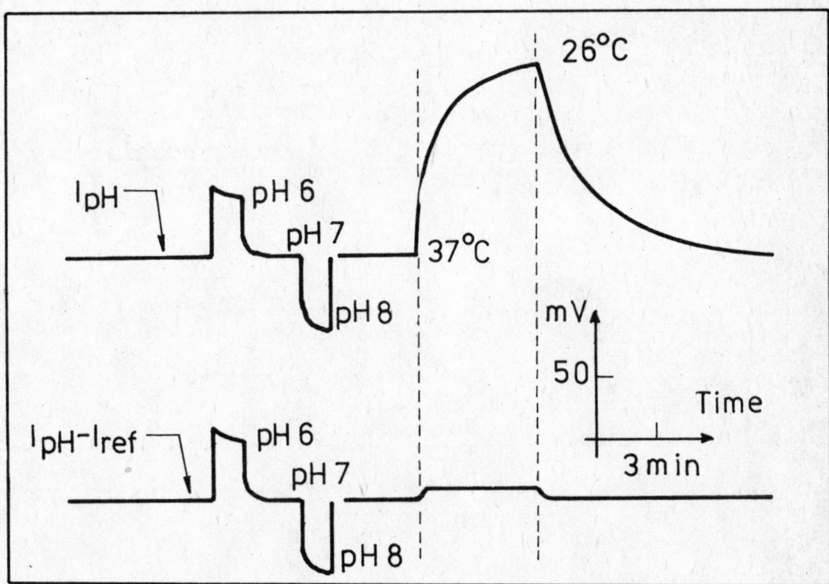

Fig.16. The pH and thermal response of pH-ISFET with conventional reference electrode (upper curve) and FET reference electrode /76/.

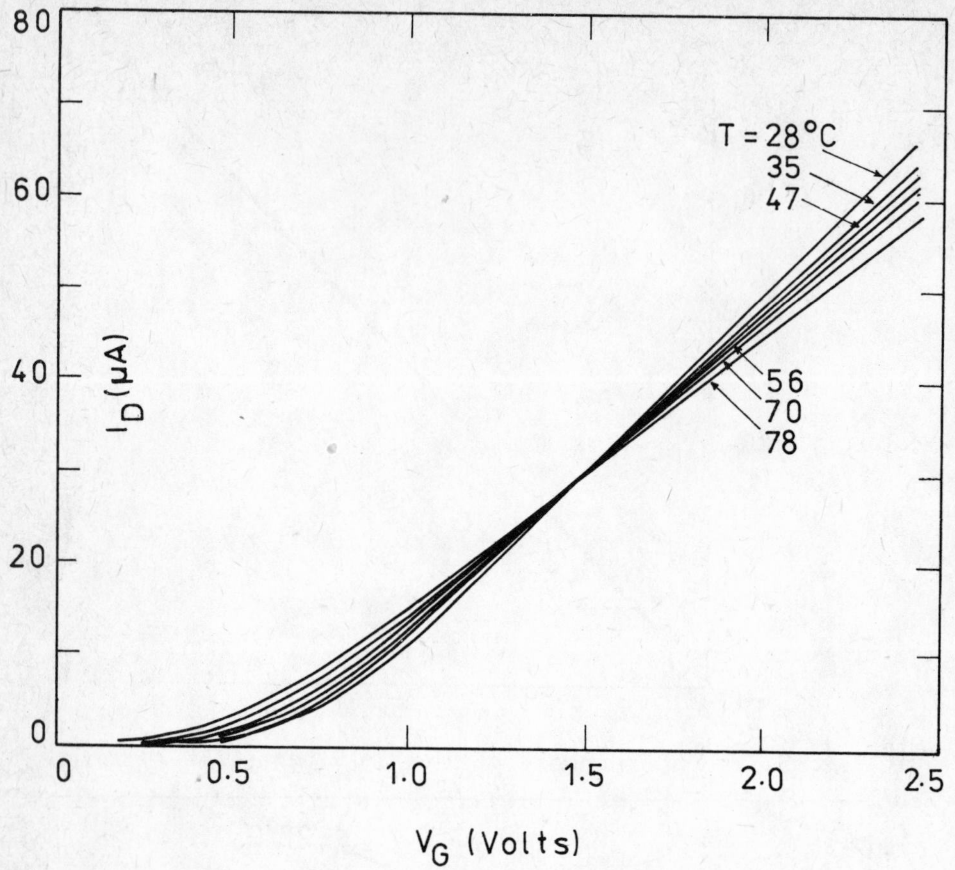

Fig.17. The ISFET drain current, I_D, thermal sensitivity /60/. V_G - gate voltage.

QUESTIONS

Participants of the discussion: K.Nagy, E.Pungor and Yu.G.Vlasov

Question:

You have described a fluoride electrode based on a silicon dioxide chip. Could you please give some details of the polarization technique you used ?

Answer:

The anodic polarization method is not very complicated but very useful for the purposes of changing the properties of these thin films, because you can receive a very strong electric field in the thin dielectric film. The field we used was approximately 10^7 V/cm. It is exactly what is needed to introduce fluoride into the thin film of the dielectric. By using a stronger field you may destroy the system. At a weaker field the fluoride will not enter the film. We used a solution approximately 0.1 mol/l in fluoride, with a background electrolyte to ensure good conductivity. By this technique we can dope the film.

Question:

What kind of coating do you get at the surface with this treatment ?

Answer:

I cannot tell exactly what happened there, but the behaviour of the sensor is different depending on the depth of fluoride penetration. If the fluoride is inside the film, the response is sub-Nernstian. The reason for this may be that we do not have enough centres on the surface to give a response. We have calculated the concentration of the fluoride inside the films and it was approximately 10^9 atoms/cm^3. But from experiments with silicon nitride we calculated a fluoride concentration of 10^{22} atoms/cm^3 for a sensor giving theoretical response. If you treat this silicon nitride under special conditions so that this concentration gradually diminishes, e.g. to 10^{21} atoms/cm^3, the response becomes poorer.

Question:

You have shown a calibration curve for fluoride which was of peak type.

Answer:

We tried to get maximum concentration of fluoride in the dielectric film, to have maximum sensitivity for the fluoride in solution. As for the peak-type calibration curve, in our opinion there are two different mechanisms for the fluoride response in different concentration ranges which depend on the concentration of fluoride in the dielectric film. This mechanism may be an ion exchange, a specific adsorption, but we need additional experiments to prove either of them.

DISCUSSION LECTURES

4th Symposium on Ion-Selective Electrodes
Mátrafüred, 1984

TRANSFERENCE NUMBER OF IONS IN GLASS ELECTRODES

Z. BOKSAY, J. HAVAS* and E. ROHONCZY-BOKSAY*

Department of General and Inorganic Chemistry
Eötvös Loránd University, Budapest, Hungary

*RADELKIS Electrochemical Instruments,
Budapest, Hungary

ABSTRACT

Silver, sodium and hydrogen ions were introduced by electrolysis into glass electrodes of various composition and ion sensitivity, and the shifting of concentration profile curve was determined. From this shifting the transference number of ion as a function of local concentration was calculated on the theoretical bases presented here. Usually the introduced guest ion exhibits higher mobility, even if it is larger than the host ion. The reversed electrolysis indicated that the transference number of ion in the range of concentration gradient may depend on the direction of electric field applied.

INTRODUCTION

A glass electrode membrane usually comprises a surface layer which contains at least two charge carriers with different mobilities. In this layer the concentration of the so called guest ions, replacing the host ions, decreases with the depth and tends to zero. Owing to the uneven distribution of different charge carriers, diffusion potential arises the contribution of which to the overall electrode potential may not be neglected except for special instances.

The transference number, Θ, is closely related to the ion mobilities, u, and concentrations, c,

$$\Theta_i = \frac{c_i u_i}{\sum_j c_j u_j} \qquad (1)$$

i refers to a particular ion and j is a running index.

The early hypotheses concerning the ion mobilities and transference number will be omitted from here since all of them have lost their importance. First Baucke (1-3) emphasized that the ion mobility is a matter of investigation. He introduced hydrogenion and different mixtures of hydrogen ion and lithium ion into lithium silicate based membranes and determined the distribution of ions in the surface layer. As the front of the hydrogen ion was relatively sharp in his experiments the mean velocity and mobility of ions could easily be calculated.

When the front is, however, diffuse another method is required. Such a method for determination of transference number has been elaborated (5) and will be presented here.

THEORETICAL

The Nernst-Planck equation should be applied to a particular ion

$$\phi = - D \frac{\partial c}{\partial x} - \frac{cu}{F} \frac{\partial \varphi}{\partial x} \qquad (2)$$

where ϕ is the flux of ion, D the diffusivity, c and u the concentration and the mobility, resp., of the ion, φ denotes the local potential and x is the coordinate perpendicular to the surface. Equation (1) can be rewritten concerning that

$$cu = \Theta \kappa \qquad (3)$$

$$\kappa \frac{\partial \varphi}{\partial x} = - i \qquad (4)$$

where Θ denotes the transference number, κ is the conductivity and i the current density. Inserting eq. 3 and eq. 4 into eq.

(2) gives

$$\phi = - D \frac{\partial c}{\partial x} + \frac{i}{F} \Theta. \qquad (5)$$

A derivation according to x at constant t yields on the left side $(\partial\phi/\partial x)_t$ which is equal to $-(\partial c/\partial t)_x$ according to Fick's pattern. Concerning also the basic relation between the partial derivatives;

$$-\left(\frac{\partial c}{\partial t}\right)_x = \left(\frac{\partial c}{\partial x}\right)_t \left(\frac{\partial x}{\partial t}\right)_c \qquad (6)$$

the equation 5 will have the form

$$\left(\frac{\partial x}{\partial t}\right)_c = - \frac{\partial}{\partial c} D \left(\frac{\partial c}{\partial x}\right)_t + \frac{i}{F} \left(\frac{\partial \Theta}{\partial c}\right)_t \qquad (7)$$

If the effect of the diffusion on the shifting of the profile curve is negligible compared to that of electric field we obtain

$$\left(\frac{\partial x}{\partial t}\right)_c = \frac{i}{F} \frac{d\Theta}{dc} \qquad (8)$$

An integration according to time from the beginning to the end of the electrolysis gives

$$\Delta x_c = \frac{q}{F} \frac{d\Theta}{dc} \qquad (9)$$

where Δx_c is the variation of x coordinate at constant c, and q the total electric charge passed through a unite area of the sample. Integration according c yields the final formula:

$$\Theta(c) = \Theta(c_o) + \frac{F}{q} \int_{c_o}^{c} \Delta x_c \, dc \qquad (10)$$

EXPERIMENTAL

The transference number measurements were performed at 40 ^{o}C on electrodes made of sodium containing McInnes-Dole glass, and RADELKIS commercial electrodes of two types. The latters are made of different multicomponent lithium silicate based glasses, used for pH and pNa determination.

In each experiment an electric charge of 200-500 mC passed the electrode wall of 0.02-0.05 cm thickness and of 3-5 cm^2 uncovered area. The current due to the voltage of 500-520 V was of 10 μA order of magnitude. Silver, sodium and hydrogen ions were introduced into the glass membranes investigated from AgI powder, Na_2CO_3 and HCl solution, resp. After the electrolysis the surface layer was removed gradually by HF solution, and the solution fractions obtained were analysed for the monovalent cations and silicon. The distribution of the guest and host ions was calculated from the analytical data.

Fig. 1. shows lithium ion concentration profile curves after introducing different amount of silver ions into the sodium ion sensitive glass electrode. It is evident from equation (9) that at a given c value Δx_c is proportional to q. So if we multiply Δx by q_n/q, where q_n is a chosen constant value, the data for individual samples with different q values determine a unique curve, as it is seen on Fig. 2. The curves refering to $q_n = 10^3$ C m^{-2} will be quoted as normalised profile curves, which are very convenient for various comparisons of results.

RESULTS

When sodium ions were replaced by silver ions during the electrolysis the latter proved to be faster. Fig. 3. referring to McInnes-Dole glass shows the transference number of sodium ion and the ratio of mobility of sodium ion to that of silver ion as a function of the relative concentration of sodium.

It is evident from Fig. 4. that the replacement of

lithium ion in RADELKIS pH electrodes and pNa electrodes by the largest silver ion is limited to an extent of about 70 % of the total lithium ion concentration. The silver ions entering the membrane at the phase boundary can occupy the left places of the moving silver ions only at the very surface. Consequently, the transference number of silver is equal to 1.0 there which place corresponds to $c_{Li}/c_o \approx 0.3$. With increasing lithium concentration the transference number of lithium increases from zero to the unity. Even in this interval the mobility of silver ion is higher than that of the lithium ion, Fig. 5.

The distribution of sodium ion in the RADELKIS pNa electrode is compared with that of silver ion in Fig. 6. As normalised profile curve for the sodium ion exhibits two inflection points while the curve for silver has no one. We may conclude that the bonding state and transport mechanism for the two types of ion are rather dissimilar.

To our great surprise, the composition dependence of the transference number of hydrogen ion is rather uniform for all glasses investigated i.e. we have not found any clear correlation between the transference number determined and the electrode function /Fig. 7 /. However, the state of surface layer during the fast electrolysis on the one hand and in quasi equilibrium with an aqueous solution on the other may not be identical. But if the outher part of the surface layer contains water molecules the transference number of hydrogen ion in pH electrode is approximately equal to one while in pNa electrodes is negligible.

The experimental data presented indicate that the mobility of guest ion is always higher compared to that of host ion irrespectively of the diameter and chemical properties of the ions involved. We have supposed that this statement in which the ions are either guest or host ion, cannot be a final rule and a supplementary distinction is needed. The transference number of ion may also depend on whether it is a preceeding ion or a following one. In all experiments discussed so far the guest ion was the following one as well. Now let us consider the effect of the change of role by inspecting Fig.8. Previously, silver ions were introduced into a McInnes-Dole glass sample and subsequently an inverse electric field was applied. The transference number of host ions moving outwards was calculated from the shifting of the profile curve and plotted against the local composition in Fig.9. A comparison with the result of direct electrolysis indicates a remarkable difference. Instead of giving a detailed explanation we would like to draw an indisputable conclusion. Since the transference number depends on the direction of electric field in the range of concentration gradient, so do the ion mobilities and the conductivity. With other words, typical scalar quantities may depend on the direction of current when a concentration gradient exists in a glass.

We do not intend to be involved in a detailed explanation in which the oriented environment of vacancies concerning the ion distribution plays an outstanding role. Instead, we finish this paper with the hope that the results presented offers new points of view for the better understanding of high resistance in glass electrodes (4,6) .

REFERENCES

1. F.G.K. Baucke, In: Mass Transport Phenomena in Ceramics, eds. A.R. Cooper and A.H. Heuer,/Plenum, New York, 1975/ p 337.
2. F.G.K. Baucke, In: The Phisics of Non-Cristalline Solids, ed. G.H. Frischat /Trans Techn. Publ., Aedermannsdorf, 1977/ p 503.
3. F.G.K. Baucke, J. of Non-Crist. Solids,40/1980/ 159-169.
4. R.P. Buck, J. Electroanal. Chem., 18/1968/ 363.
5. Z. Boksay, G. Bouquet, I.S. Ivanovskaya, Experimental Determination of Transference Number in Glass Electrode membrane /in Russian/, In: Stekloobraznoe Sostoyanie Nauke, Leningrad, 1983., p 108-111.
6. A. Wikby, J. Electroanal. Chem. 38/1972/ 429.

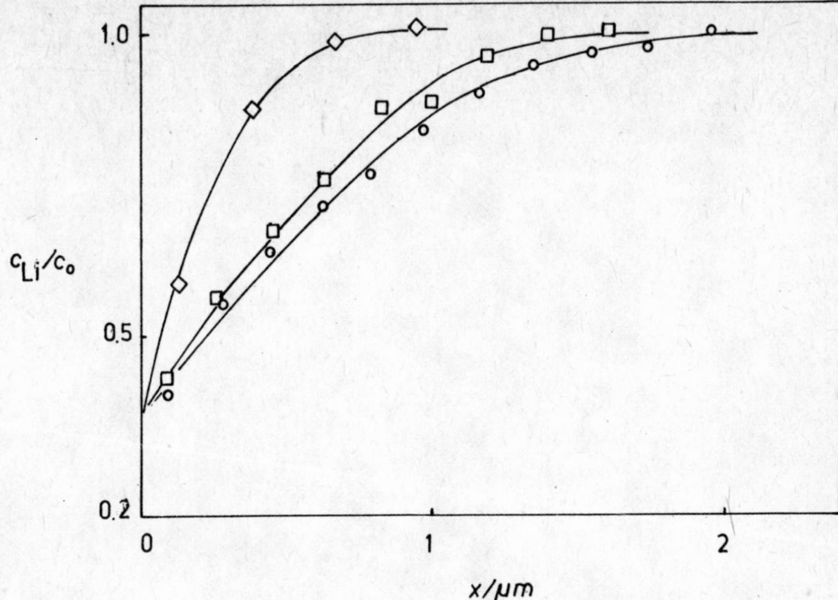

Fig. 1. The relative concentration of lithium ion plotted against the depth after introducing different amount of silver ion into RADELKIS pNa electrode.

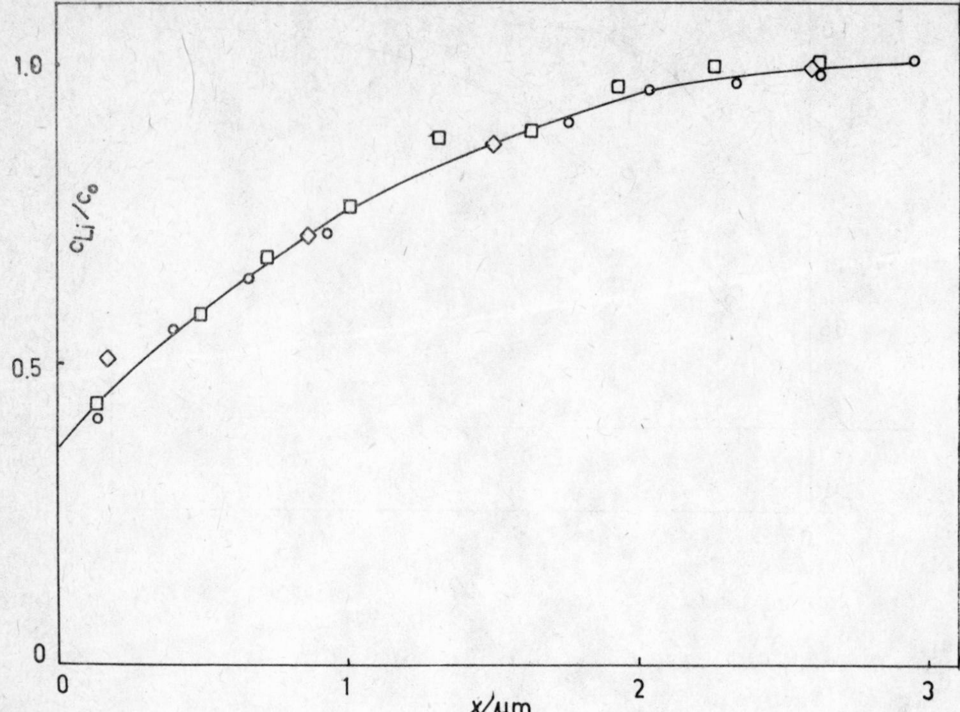

Fig. 2. Normalized profile curve for samples indicated in Fig. 1.

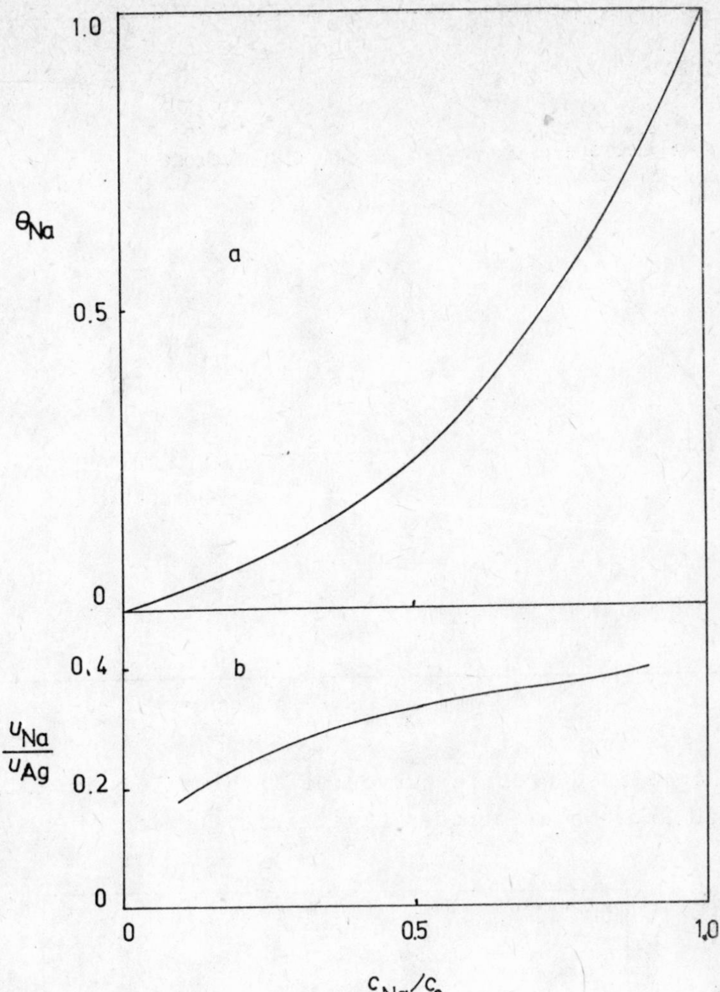

Fig. 3. Transference number of sodium ion, diagram a, and ratio of mobility of sodium ion to that of silver ion, diagram b, in McInnes-Dole glass /22% Na_2O, 6% CaO, 72% SiO_2 by weight/ plotted against the relative concentration of sodium ion.

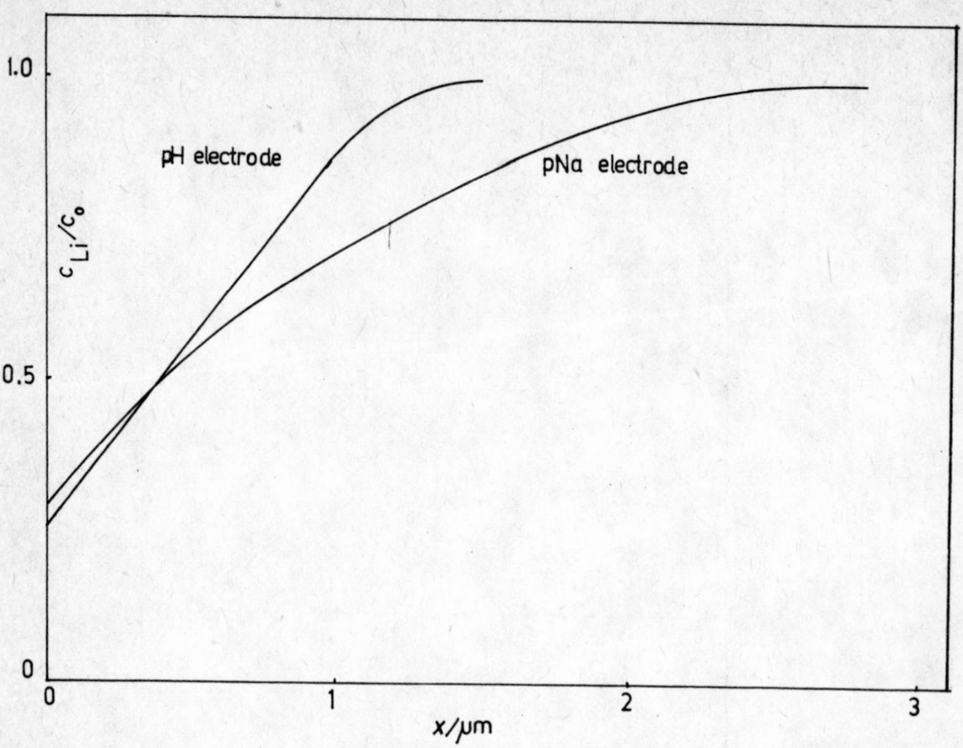

Fig. 4. Normalized profile curves of lithium ion for RADELKIS pH and pNa electrodes after introduction of silver ions.

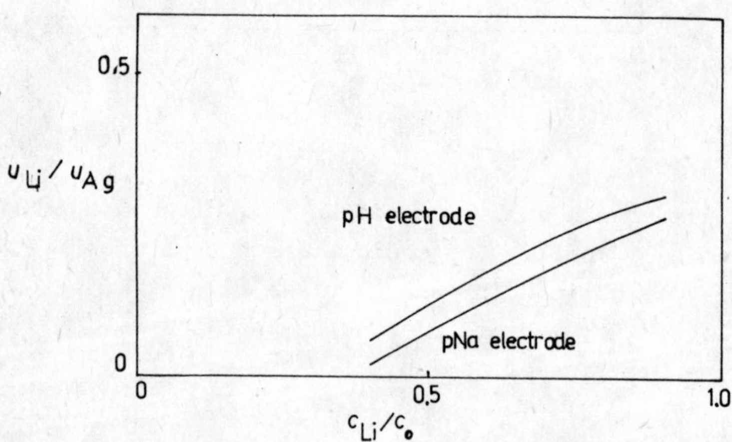

Fig. 5. The ratio of mobility of lithium ion to that of silver ion in RADELKIS pH and pNa electrodes.

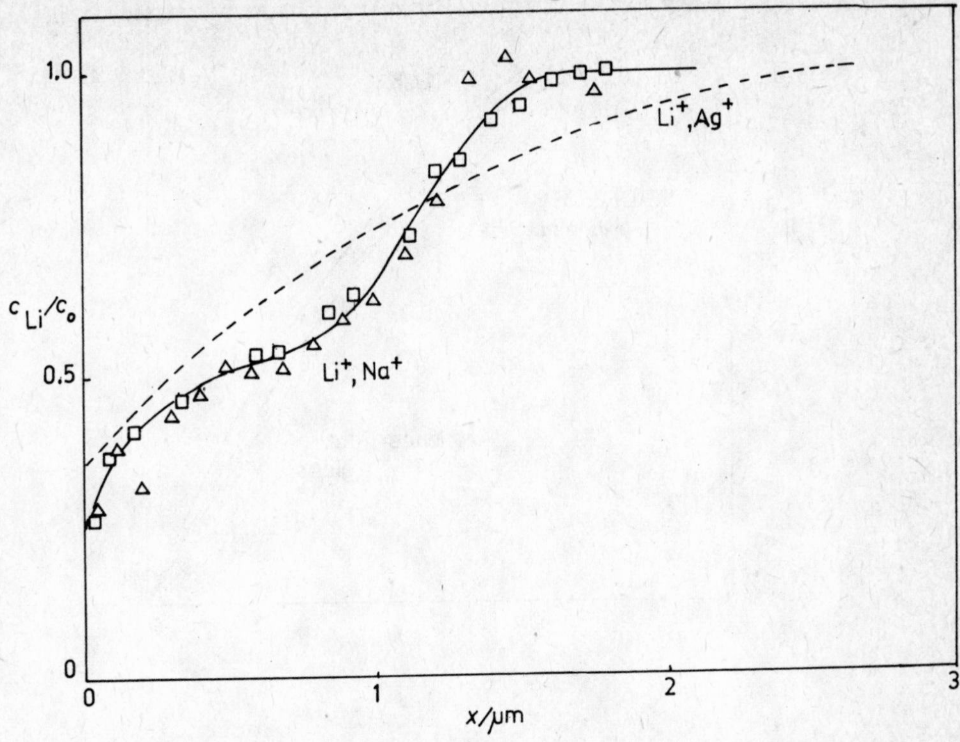

Fig. 6. Normalized profile curves of RADELKIS pNa electrode after introduction of sodium ion /solid line/ and silver ion /dashed line/, respectively.

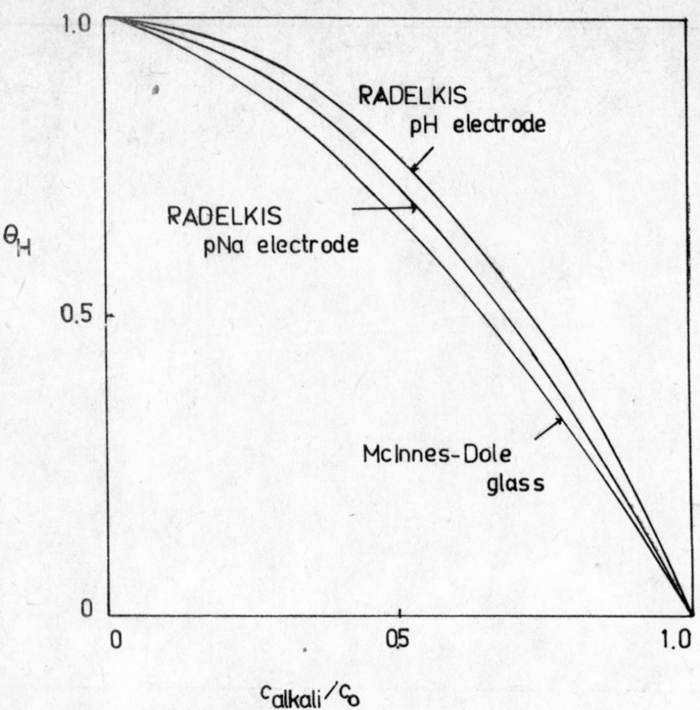

Fig. 7. The transference number of proton in samples indicated plotted against the alkaline ion relative concentration.

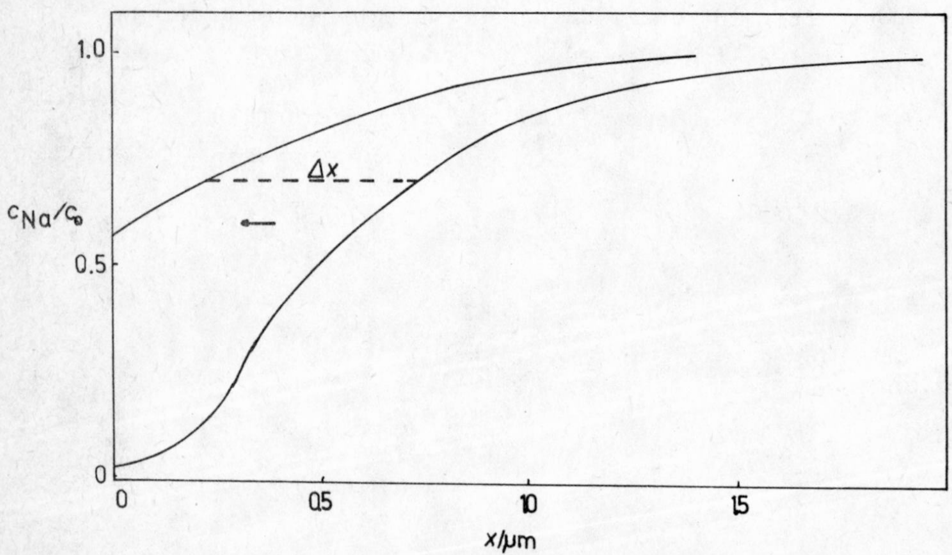

Fig. 8. The shifting of profile curve for McInnes-Dole glass on the effect of the inversion of current direction.

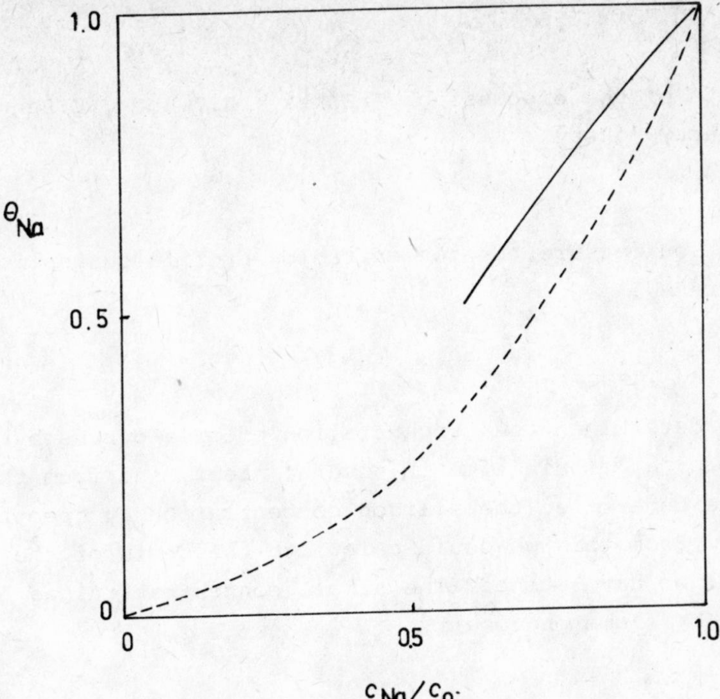

Fig. 9. Transference number of sodium ion plotted against the relative sodium ion concentration before and after the changing of electric field indicated by dashed and solid line, respectively.

QUESTIONS AND COMMENTS

Participants of the discussion: Z.Boksay, R.P.Buck, E.Pungor and E.Rohonczy-Boksay

Question:

How did you measure the concentration profile curves in the glass membrane ?

Answer:

We have determined the concentration profile curves by etching the glass for a certain time in several steps, and from the solution we determined the silicon concentration by spectrophotometry. From this we could calculate the depth of the small shell we have etched. The alkali concentration has been determined by flame photometry.

Question:

Had it not been better to use the SIMS technique for this purpose, to sputter the surface and measure the components by MS?

Answer:

The method we used is very well suited to the determination of the ion distribution in the given depth range.

Question:

I did not understand the equation

$$\Delta x_c = \frac{q}{F} \frac{d\theta}{dc}$$

What are the dimensions?

Answer:

Δx_c is the variation of the profile curve during electrolysis [m]

q is the electric charge passed through unit surface area, cm^{-2}

F   is the Faraday constant C mol^{-1}

Θ has no dimension

c   is the concentration mol m^{-3}

Comment:

Δx_c is an arbitary value for the comparison of different experiments. We call the curves to which this value was applied normalized curves. It does not depend on the amount of electricity passed through the electrodes.

4th Symposium on Ion-Selective Electrodes
Mátrafüred, 1984

OPTIMIZATION OF EXPERIMENTAL METHODS FOR DETERMINATION OF SELECTIVITY COEFFICIENTS OF ION-SELECTIVE ELECTRODES BY KNOWN-ADDITION TECHNIQUES

M. ČAKRT

Department of Analytical Chemistry,
Slovak Technical University, Bratislava, Czechoslovakia

ABSTRACT

The multiple known-addition techniques for determination of conditional potentiometric selectivity coefficients are optimized to cover the useful bi-ionic electrode potential response range, to measure reliably the experimental variables, and to choose appropriate working apparatus. Determinations of values of selectivity coefficients for liquid membrane nitrate-selective electrode in the presence of nitrite or chloride ions by using several methods are reported.

INTRODUCTION

Ion-selective electrodes (ISEs) are widely recognized as an important tool for chemical analyses. Despite fundamental advances in their technology, a measure of their ability to distinguish between the primary ion and an interfering ion B in a sample solution still remains the main metrological characteristic of each kind of electrodes. Although various approach were suggested, the electrode selectivity behaviour is usually expressed in terms of selectivity coefficients as they follow from Nikol'skij - Eisenman equation or some of its modification (1). Experimental values of these coefficients are of great importance, especially when are compared with ones based on theoretical models, in developing theories on electrode performance prediction, and even in assesing the suitability of the use of potentiometry with ion-selective electrodes to solve a particular analytical problem. From the analytical chemistry point of view, however, there is still a need for such values that would be applicable for numerical corrections of results (2). It is a good analytical practice to determine the blank-correction values by procedures that are as close as possible to analytical techniques used.

Multiple known-addition methods are often used to determine sample concentrations with ISEs (3-5).Let us suppose that a standard solution of the primary ion A at the analytical concentration c_A is added stepwise in volume increments V_i to a known initial volume V_o of a sample solution which contains both A and interfering ion B at concentrations $c_{A,x}$ and c_B. The resulting concentrations are

$$[A] = (c_{A,x}V_o + c_A \sum_i V_i)/(V_o + \sum_i V_i) \qquad (1)$$

$$[B] = c_B V_o/(V_o + \sum_i V_i) \qquad (2)$$

Assuming that the measured electrode potential in a mixture of univalent ions A and B can be expressed by the equation

$$E_i = E^* + S \log\{[A] + k_{A,B}^{pot} [B]\} \qquad (3)$$

(where E_i is the potential measured vs. an external reference electrode after i-th addition; E^* is the constant that includes the standard electrode potential, the junction potential, and activity coefficient terms; S is the slope of the electrode potential response; and $k_{A,B}^{pot}$ is the bi-ionic coefficient of selectivity), substitution of Eqns.(1) and (2) in Eqn.(3) gives after Gran-type transformation

$$(V_o + \sum_i V_i)10^{E/S} = 10^{E^*/S} (c_{A,x}V_o + c_A \sum_i V_i + k_{A,B}^{pot} c_B V_o) \qquad (4)$$

Because the right-hand side of Eqn.(4) is a linear function of the independent variable $\sum_i V_i$, the values

$$\Phi_i = (V_o + \sum_i V_i)10^{E_i/S} \qquad (5)$$

can be calculated and plotted agaist $\sum_i V_i$. A straight line is obtained, which on extrapolation intercepts the abscissa axis at the point, where $\Phi=0$ and $\sum_i V_i = V_x$. The weighted linear regression is to be used to determine value of V_x, from which concentration $c_{A,x}$ can be calculated:

$$c_{A,x} = - V_x \, c_A/V_o - k^{pot}_{A,B} \, c_B \tag{6}$$

The necessary correction term in Eqn.(6) can be determined similarly by using, instead of the sample solution, the model sample solution containing only the interference ions B at the concentration c_B. The multiple known-addition procedure thus may be used for determination of values of selectivity coefficients. This approach lead to development of six procedures that allow to determine selectivity coefficients of ISEs under bi-ionic conditions (6). Here, the four of them are used for the determination of selectivity coefficients of nitrate ion-selective electrode. Results are compared with the corresponding values obtained by some other methods, and optimization of the working procedures is considered.

EXPERIMENTAL

Instrumentation

The nitrate ion-selective electrode of the type that described Šenkýř and Petr (7) and the Radelkis double-junction silver/silver chloride reference electrode type OP-8203 were used in connection with an Orion model 801A digital pH/mV meter. The solutions to be measured were thermostated to 20 ± 0.2 °C and stirred magnetically.

Chemicals and solutions

Twice distilled water and analytical reagent grade chemicals were used unless stated otherwise. A total ionic strenght of solutions to be measured was adjusted to 1.0 (in mole per liter scale) by using recrystalized sodium sulphate. The liquid ion-exchanger was prepared as follws: 10 mg of ethylviolet (C.I. 42600) are dissolved in 100 ml of water, 10 ml of sodium nitrate solution (10 mol/l) is added and the solution is shaked 5 min with 100 ml of the freshly purified and distilled nitrobenzene. The organic layer is then separated, washed with water, and filtered.

RESULTS AND DISCUSSION

The multiple known-addition methods used for the determinations of the selectivity coefficients of nitrate ISE in the presence of chloride or nitrite ions are summarized in Table 1. In order to optimize working conditions for the reliable determinations, the following aims were considered: (1) minimization of an electrode potential response distortion caused by changes in properties of the electrode active membrane, and (2) maximization of the signal-to-noise ratio whilst keeping the working conditions consistent with the previous requirement.

It is assumed that when a contribution of the last term on right-hand side of Eqn.(3) is great when compared with the $[A]$, the electrode potential response may be altered both in E^* and S values due to changes in composition of the electrode active membrane. Therefore, the maximum allowable value of an interference potential (8,9)

$$\Delta E = S \log(1 + k^{pot}_{A,B}[B]/[A]) \qquad (7)$$

must not exceed the absolute value of about 17.5 mV (at 20 °C) that corresponds to the equal weight of both terms, $[A]$ and $k^{pot}_{A,B}[B]$. When the contribution of the last term in Eqn.(3) is small compared to $[A]$, i.e., when the electrode potential response is close to the Nernstian one, the error in the determination of the selectivity coefficient for a given precision of measurements is to be expected large. Thus, the minimum, but still significant value of the interference potential was taken rather arbitrarily as 1 mV.

Adjustable experimental variables of the known-addition procedures are the initial volume V_o and concentrations of ions A and B in the solution, c_A and c_B, respectively. Their values has to be chosen with regard to the expected value of the selectivity coefficient and to a reliability of the measurement of the V_x value whilst not exceeding the allowable limits of the interference potential. The considerations on the optimization of the procedures can be simplified (6) by introducing the dimensionless parameter P:

$$P = k^{pot}_{A,B}\, c_B/c_A \qquad (8)$$

Appropriate values of P are estimated from maximum and minimum allowable values of interference potential, V/V_o, and from required value of V_x/V. The relatioships between these variables are given in Table 1 together with formulae for calculations of the values of selectivity coefficients. The ratio of the analytical concentrations c_B/c_A is then calculated from an expected value of $k^{pot}_{A,B}$ and from the estimated value of P.

The results summarized in Table 2 were obtained under following conditions:

Method I

1 to 25 ml of a standard sodium nitrate solution ($c_A = 10^{-2}$ mol/l) is added in ten increments to 20 ml of 10^{-2}mol/l NaCl or 10^{-3} mol/l $NaNO_2$ (P = 0.05).

Method II

5 to 20 ml of mixed solution of 10^{-3}mol/l $NaNO_3$ with NaCl (4×10^{-3}mol/l) or $NaNO_2$ (4×10^{-4}mol/l) is added in ten increments to 20 ml of 10^{-3}mol/l $NaNO_3$ (P = 0.02).

Method III

0.2 to 4 ml of 5×10^{-1}mol/l solution of NaCl or 5×10^{-2} mol/l solution of $NaNO_2$ is added in ten increments to 20 ml of 5×10^{-3}mol/l $NaNO_3$ (P = 5).

Method IV

0.2 to 5 ml of mixed solution of 5×10^{-1}mol/l NaCl (or 5×10^{-2}mol/l $NaNO_2$) is added in ten increments to 20 ml of 5×10^{-4}mol/l $NaNO_3$ (P = 5).

Method (a)

Electrode potential was measured in 10^{-4}mol/l $NaNO_3$ and in series of eight solutions with the same concentration of nitrate and with 3×10^{-2} to 10^{-2}mol/l of NaCl or $NaNO_2$.

Method (b)

Electrode potential was measured in series of solutions at continuous variations of the concentrations of $NaNO_3$ from 0 to 10^{-5}mol/l and of NaCl or $NaNO_2$ from 10^{-5} to 0 mol/l.

The value of the electrode response slope S strongly affects the values of V_x that are determined on extrapolation. The dilution method was used after the end of each determination to find the true value of S. Other possible way would be to evaluate the measurements by multiple linear regression. Because the value of the slope S influences also the linearity of the Gran-type plots, a simple check can be done by tests on their linearity, e.g. by using coefficients of correlation.

The construction of the electrode under investigation allowed to compare a behaviour of the electrode with renewal of its active membrane with its response without any renewal. Results of these paralel determinations were in the same range of magnitude. Therefore, it may be concluded that no significant changes in the electrode behaviour were caused by its long-time exposition to the interfering ions under bi-ionic conditions of the determinations.

REFERENCES

1 W.E.Morf, The Principles of Ion-Selective Electrodes and of Membrane Transport, Akadémiai Kiadó, Budapest, 1981

2 F.Oehme, G-I-T Fachz.Lab. 19, (1975) 593

3 M.Mascini, Ion-Selective Electrode Rev., 2, (1980) 17

4 M.Bader, J.Chem.Educ., 57, (1980) 703

5 G.Horvai, E.Pungor, Anal.Chim.Acta, 113, (1980) 295

6 C.Macca, M.Čakrt, Anal.Chim.Acta, 154, (1983) 51

7 J.Šenkýř, J.Petr, Chem.Listy, 73, (1979) 1097

8 J.Bagg, O.Nicholson, R.Vinen, J.Phys.Chem., 75, (1971) 2138

9 G.J.Moody, N.S.Nassory, J.D.R.Thomas, Talanta, 26(1979) 873

10 K.Srinivasan, G.A.Rechnitz, Anal.Chem., 23, (1969) 1203

11 M.S.Okunev, N.V.Khitrova, O.I.Kornienko, Zhur.Anal.Khim., 37, (1982) 5

12 R.B.Dean, W.J.Dixon, Anal.Chem., 23, (1951) 636

Table 1. Relationships between the experimental variables of known-addition methods

	Method I	Method II
Initial solution, V_o	c_B	c_B
Added solution, V_i	c_A	c_A; c_B
$k^{pot}_{A,B}$	$-c_A V_x/c_B V_o$	$-c_A V_x/c_B(V_o+V_x)$
$V/V_o = f(P; \Delta E)$	$P/(10^{\Delta E/S}-1)$	$1/\left[(10^{\Delta E/S}-1)/P-1\right]$
$P = f(V_x; V_o)$	$-V_x/V_o$	$-V_x/(V_o+V_x)$
	III	**IV**
Initial solution, V_o	c_A	c_A
Added solution, V_i	c_B	c_A; c_B
$k^{pot}_{A,B}$	$-c_A V_o/c_B V_x$	$-c_a(V_o+V_x)/c_B V_x$
$V/V_o = f(P; \Delta E)$	$(10^{\Delta E/S}-1)/P$	$1/\left[P/(10^{\Delta E/S}-1)-1\right]$
$P = f(V_x; V_o)$	$-V_o/V_x$	$-V_o/V_x - 1$

V stands for $\sum_i V_i$

Table 2. Selectivity coefficients of liquid membrane nitrate ISE determined by different methods

Method	$10^3 k^{pot}_{NO_3,Cl}$	$10^2 k^{pot}_{NO_3,NO_2}$
I	4.73 ± 0.52	4.65 ± 0.62
II	5.24 ± 0.60	4.32 ± 0.58
III	5.45 ± 0.93	4.75 ± 0.88
IV	5.37 ± 0.83	4.85 ± 0.85
(a)	3.5	4.3
(b)	5.1	3.6

(a) based on method III, ref.(10)

(b) based on "isopotential" method, ref.(11)

Results are expressed as confidence intervals at α= 0.05 for five determinations, ref.(12)

QUESTION

Participants of the discussion: M.Čakrt and G.Horvai

Question:

Some of your last slides showed your results. You had selectivity coefficients like 4 ± 0.8. Is this the standard deviation or three times the standard deviation ?

Answer:

The results were average of five measurements and were expressed as confidence intervals at 0.05α.

4th Symposium on Ion-Selective Electrodes
Mátrafüred, 1984

STABILITY CONSTANTS OF Ca-TRIS COMPLEXES: COMPETITION Ca^{2+}/Mg^{2+}

M.F. G.F.C. CAMÕES

CECUL, Department of Chemistry
Faculty of Sciences, University of Lisbon, Portugal

ABSTRACT

Potentiometric measurements of Ca^{2+} activity in dilute aqueous solutions, with an Orion 93-20 selective electrode, showed Nernstian behaviour ($10^{-1}, 10^{-2}, 10^{-3}$ M), as well as high Ca^{2+} selectively, even towards Mg^{2+}.

Buffering of pure aqueous Ca^{2+} solutions with Tris-Tris HCl 0,005 equimolar buffer (Tris = Trihydroxymethylaminomethane) brings lower $a_{Ca^{2+}}$ values, revealing complexation. Measurements in mixed Ca^{2+}, Mg^{2+} buffered solutions, show nevertheless $a_{Ca^{2+}}$ values higher than expected, what is explained by preferential complexation of Mg^{2+}. NMR observations support a proposed structure for the complex species. There is evidence that apart from the 1:1 complex, appreciable amount of 1:2 complex may be present. Calculation of the corresponding equilibrium constants is attempted.

INTRODUCTION

Recent studies (1,2) show that Tris interacts with Ca^{2+} and Mg^{2+} in aqueous solution. This may lead to difficulties in the interpretation of potentiometric data, when the activity of Ca^{2+} is assessed with an ion-selective electrode, in Tris-Tris HCl pH buffered media, as is often the case in Environmental Chemistry, Biochemistry and Medicine. In this work, potentiometric measurements of Ca^{2+} activities were done in order to prove the Ca^{2+}-Tris interaction, both in the presence and in the absence of Mg^{2+}. These results coupled with NMR spectroscopic measurements, allowed structural considerations, suggesting also possible stoichiometries.

EXPERIMENTAL

- Calibration of the electrode assembly

Ca^{2+} selective electrode - Orion 93-20 / saturated calomel electrode was carried out in aqueous $CaCl_2$ solutions, 10^{-1}, 10^{-2}, 10^{-3}, 10^{-4} M.

To test the selectivity towards Mg^{2+}, potentiometric measurements were made in aqueous $MgCl_2$ solutions, 10^{-1}, 5×10^{-2}, 10^{-2} M.

In order to investigate metal - Tris interaction, readings of potential were taken for the following solutions:

- aqueous $CaCl_2$ solutions 10^{-1}, 10^{-2}, 10^{-3} M, in 0.005 equimolar Tris HCl buffer.

- aqueous $CaCl_2$ solutions 10^{-2} M, in $MgCl_2$ 10^{-1}, 5×10^{-2}, 10^{-2} M.

- $CaCl_2$ 10^{-2} M and $MgCl_2$ 10^{-1} M, 5×10^{-2}, 10^{-2} M binary solutions in 0.005 equimolar Tris-Tris HCl buffer.

Experiments were conducted at room temperature. Reagents were of Analytical Grade with out further purification. Standardization of solutions was not performed.

Equipment - PW9419 Philips digital ion activity meter; NMR - JEOL FX-90

RESULTS AND DISCUSSION

Calibration curve - activities of Ca^{2+} ions in pure aqueous $CaCl_2$ solutions were calculated /3/, the emf values were plotted vs.-log a, Fig. 1. The value for 10^{-1} M $CaCl_2$ was rejected since there was evidence of $CaCl_2$ precipitation,(Table I).

Selectivity towards Mg^{2+} - Potentiometric measurements in $MgCl_2$ solutions show that. the calcium ion selective electrode does not respond to Mg^{2+} at least for the concentration range studied, (Table 2).

$CaCl_2$ in Tris-Tris HCl aqueous solutions - Lower values of Ca^{2+} activity in Tris-Tris HCl indicate the formation of a Ca(II)--Tris complex, (Table 3, Fig. 2).

$CaCl_2$ + $MgCl_2$ aqueous solutions - The presence of Mg^{2+} increases the ionic strength, lower $a_{Ca^{2+}}$ being expected since there is no Mg^{2+} response,(Table 4, Fig. 3).

$CaCl_2$ + $MgCl_2$ in Tris-Tris HCl buffer - A higher $a_{Ca^{2+}}$ for the solution more concentrated in Mg^{2+} (10^{-1} M) can be blamed on a lower value of ionic strength due to complexation of Mg^{2+}. Preferential complexation towards Mg^{2+} is expected, since it is has a concentration 10 x higher than Ca^{2+},(Fig. 3).

The value obtained for 5×10^{-2} M Mg^{2+} solution is explained in terms of simultaneous complexation towards Ca^{2+} and Mg^{2+}.

Considering the 10^{-2} equimolar Ca^{2+}, Mg^{2+} solution, $a_{Ca^{2+}}$ decreased 8% due to the presence of Tris-Tris HCl. In the absence of Mg^{2+} it would have decreased 40%. This leads to the conclusion that Mg^{2+} undergoes preferential complexation.

The Ca-Tris Complex - Evidence of complexation has been reported in previous work (1). Stoichiometric (3:2) and structural considerations made at the time, are inconsistent with further observations. Analysis of the potentiometric and NMR data (Fig. 4) indicates structures as

```
        |
   HO — C
        |  \
   HO — C — C — NH2                      -NH2          (A)
        |           \                        \
            |        Ca                       Ca
                     :                    |  /
           — C — O...H                  -C- O
             |                           |  |
and                                         H

        |                                    |
   HO — C                                    C - OH
        |  \                              /  |
   HO — C — C — N — Ca — N - C  — C — OH               (B)
        |  /      |       |     \  |
   HO — C         H       H        C - OH
        |                          |
```

Similar conclusions are to be drawn for Mg^{2+}.

These results are in agreement with work published by Siegel (2) where much higher Ca^{2+}: Tris concentration ratios were used. For the experimental conditions used in this work, a higher value is assessed for the stability constant ($K_A = 7.1 \times 10^3$). A 1:2 complex is also considered (B).

REFERENCES

1 M. Filomena G.F.C. Camões, E. Osório
"International Symposium on Electroanalysis", U.W.I.S.T. - Cardiff (1983).

2 H. Siegel *et al.*
"Inorganica Chimica Acta", 66, /1982/ 147.

3 Robinson, R.A.; Stokes, R.H., "Electrolyte Solutions", Butterworths, London / 1959.

Table 1 - Results of Ca^{2+} potentiometric measurements in pure aqueous $CaCl_2$ solutions

Ca^{2+} / M	E/mV	γ	$a_{Ca^{2+}}$ / M	log $a_{Ca^{2+}}$
10^{-1}	- 73.8	0.32	3.2×10^{-2}	-1.49
10^{-2}	- 98.2	0.55	5.5×10^{-3}	-2.26
10^{-3}	-128.1	0.79	7.9×10^{-4}	-3.10
10^{-4}	-154.6	0.92	9.2×10^{-5}	-4.05

Table 2 - Potentiometric readings of Ca^{2+} selective electrode in pure Mg^{2+} solutions

$[Mg^{2+}]$ / M	E/mV
10^{-1}	-178.1
5×10^{-2}	-176.1
10^{-2}	-180.6

Table 3 - $a_{Ca^{2+}}$ in Tris-Tris HCl buffer solution

$[Ca^{2+}]$ / M	E/mV	$a_{Ca^{2+}}$
10^{-1}	- 82.6	1.86×10^{-2}
10^{-2}	-106.4	3.31×10^{-3}
10^{-3}	-132.4	4.90×10^{-4}

Table 4 - $a_{Ca^{2+}}$ in Ca^{2+}, Mg^{2+} mixed solutions

$[Ca^{2+}]$ / M	$[Mg^{2+}]$ / M	E/mV	$a_{Ca^{2+}}$
10^{-2}	10^{-1}	-113.7	1.93×10^{-3}
10^{-2}	5×10^{-2}	-112.1	2.17×10^{-3}
10^{-2}	10^{-2}	-109.4	2.64×10^{-3}

Table 5 - $a_{Ca^{2+}}$ in Ca^{2+}, Mg^{2+}, Tris-Tris HCl

$[Ca^{2+}]$ /M	$[Mg^{2+}]$ /M	E/mV	$a_{Ca^{2+}}$
10^{-2}	10^{-1}	-113.4	1.97×10^{-3}
10^{-2}	5×10^{-2}	-112.6	2.09×10^{-3}
10^{-2}	10^{-2}	-110.6	2.42×10^{-3}

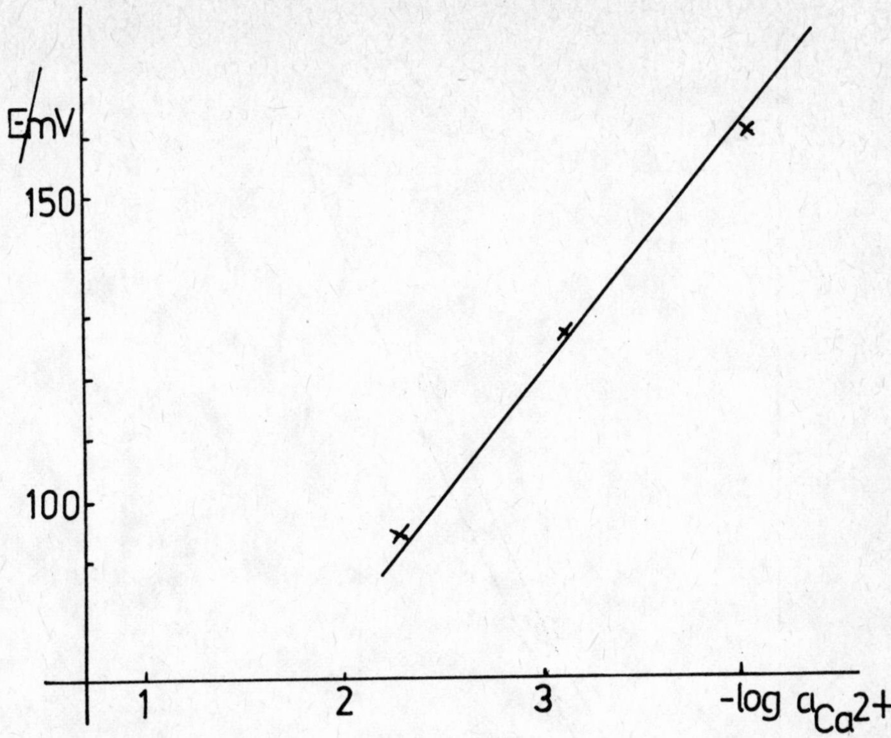

Fig. 1 - Calibration curve for the Ca^{2+} selective electrode assembly.

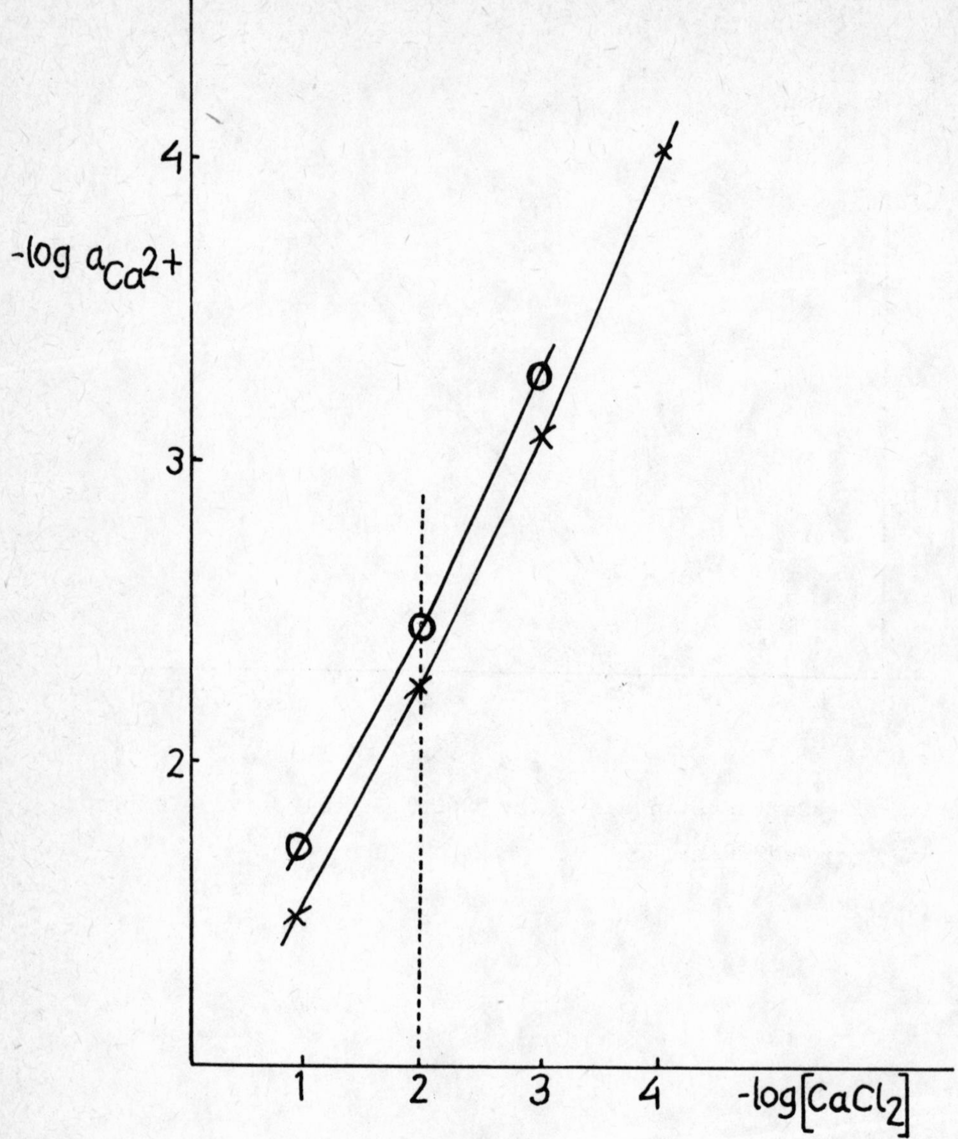

Fig. 2 - Activity of Ca^{2+} in:
x-pure aqueous $CaCl_2$ solutions
o-$CaCl_2$ in Tris

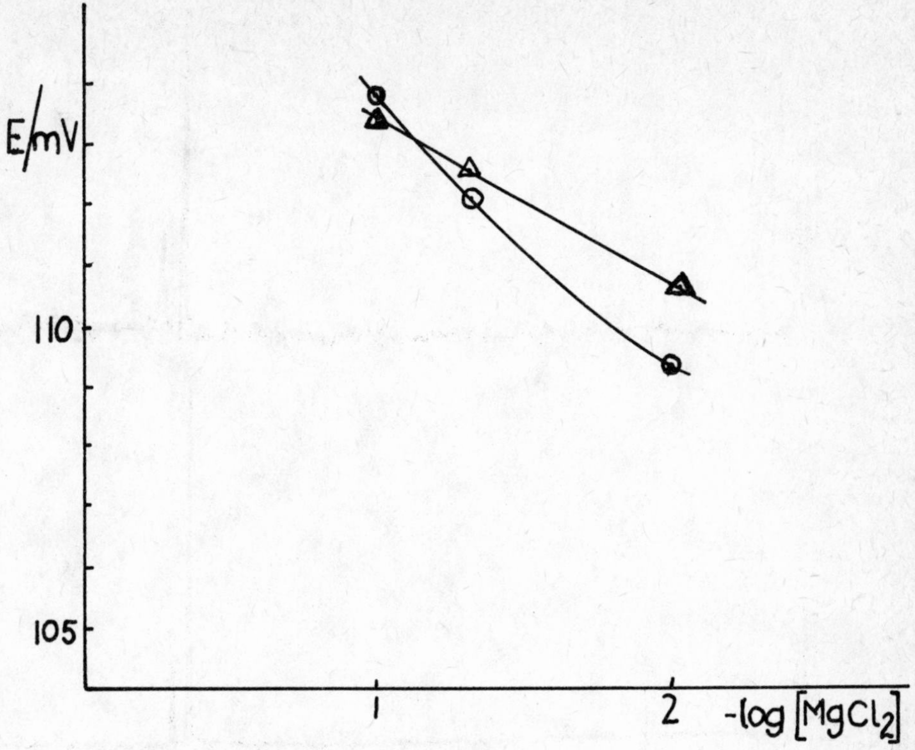

Fig. 3 - Activity of Ca^{2+}, as electrode potentials, in 10^{-2}M $CaCl_2$ solution, owing to:
o-addition of $MgCl_2$
Δ-addition of $MgCl_2$ in Tris

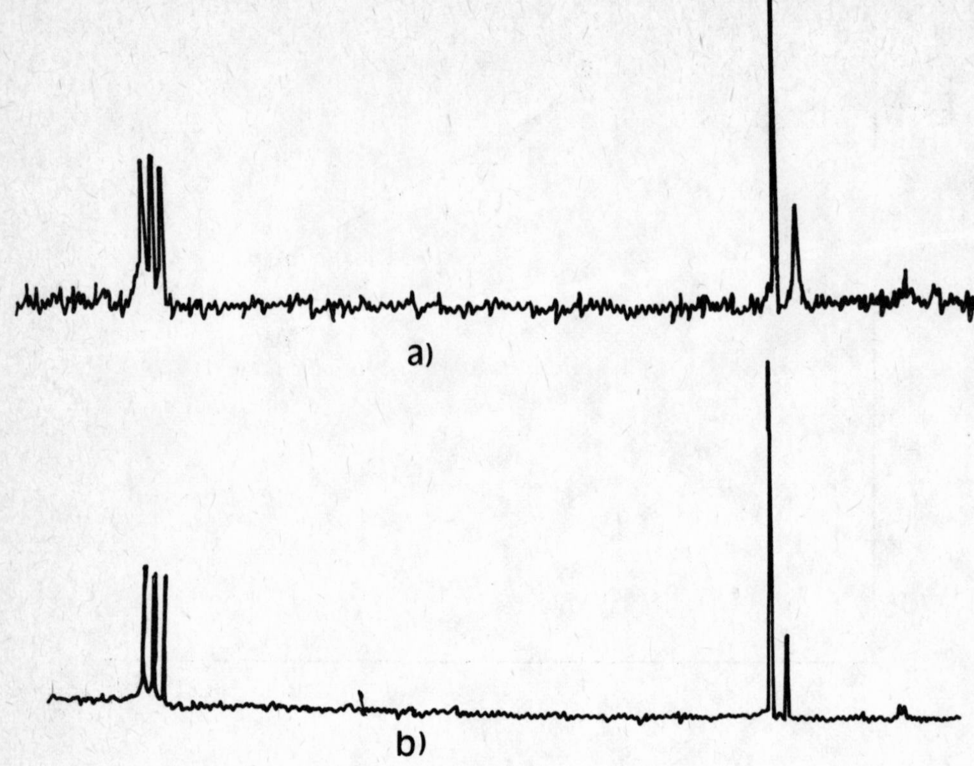

Fig. 4 - ^{13}C NMR spectra in

a) Tris-Tris HCl solution

b) Ca^{2+} + Tris-Tris HCl solution

QUESTIONS AND COMMENTS

Participants of the discussion: M.F.Camões, G.Nagy and J.D.R.Thomas

Question:

What is your advice, following your studies, in relation to the use of TRIS as a buffer system in the clinical field ?

Answer:

For the clinical field my answer is: do not use TRIS. For the environmental field, if you want to do in situ measurements, if you want to do monitoring or discrete sample analysis, my advice is: get rid of organic matter.

Comment:

We have observed that TRIS has an effect on the silver/silver chloride electrode.

Answer:

In this particular study we did not use the silver/silver chloride electrode in the solution, but we used liquid junction. However, we published a paper some 10 years ago on the influence of TRIS on the liquid junction potential in which variations up to 40 mV can occur.

4th Symposium on Ion-Selective Electrodes
Mátrafüred, 1984

CONCENTRATION OF SULPHIDE SPECIES IN CARBONATE BUFFERED SOLUTIONS

M.F. G.F.C. CAMÕES, M.F.G. SILVA and M.I.S. PEREIRA

CECUL, Department of Chemistry
Faculty of Sciences, University of Lisbon, Portugal

ABSTRACT

The H_2S acidity constant, $K_2 = 2,3 \times 10^{-15}$, was calculated from potentiometric measurements of sulphide ion concentration, $[S^{2-}]$, in 1M sodium bicarbonate solutions, with formal Na_2S concentrations, x, ranging from 10^{-3} to 1, at pH 9, kept constant by the addition of concentrated NaOH solution. This value was compared with K_2 calculated from $[S^{2-}]$ measured at different pH values, fixed by the concentration ratios, xM Na_2S / 1M $NaHCO_3$.

Sodium bicarbonate produced solutions of low buffer capacity, thus not fixing the distribution of the species H_2S, HS^-, S^{2-}, H_2CO_3, HCO_3^-, CO_3^{2-}. Consequent changes in the ionic strength, not taken into account, may be responsible for some uncertainty of the results, particularly of those relevant for the calculation of acidity constant K_1.

INTRODUCTION

Knowledge of the speciation of sulphur compounds in aqueous solutions, particularly as H_2S, HS^- and S^{2-}, is of major concern in various situations, such as biological processes (1) and electrodeposition of sulphide films (2), where it is essential to identify the species which control the processes.

Since any relatively small change of pH may result in a drastic change of the concentration ratio, buffering is required, what is usually attempted with carbonate buffer.

The relative amounts of S^{2-}, HS^- and H_2S in Na_2S solutions, are pH dependant, Fig. 1.

$$H_2S \rightleftarrows HS^- + H^+ \qquad K_1 = \frac{[HS^-][H^+]}{[H_2S]} \qquad (A)$$

$$HS^- \rightleftarrows S^{2-} + H^+ \qquad K_2 = \frac{[S^{2-}][H^+]}{[HS^-]} \qquad (B)$$

$$\alpha_0 = \frac{[H_2S]}{C_T} \qquad \alpha_1 = \frac{[HS^-]}{C_T} \qquad \alpha_2 = \frac{[S^{2-}]}{C_T}$$

$$C_T = [H_2S] + [HS^-] + [S^{2-}]$$

In acid solutions H_2S predominates, at intermediate pH almost all sulphide is present as HS^-. Only in highly basic solutions S^{2-} is the dominant species (3).

Experimental measurements of $[S^{2-}]$ and pH, using ion selective electrodes were done in 1M sodium bicarbonate solutions, with sodium sulphide concentrations varying from 10^{-3}M to 1M.

The search for literature data concerning H_2S acidity constants showed different values for the second dissociation constant K_2, ranging from 1.1×10^{-12} (4) to 8.7×10^{-18} (5). Most of these values lead to $[HS^-]$ and $[H_2S]$ not compatible with the total sulphide concentration.

In this work, values of K_1 and K_2 (T = 20°C), valid for the working conditions, were assessed.

Owing to the low buffer capacity found for the carbonate buffer solution (6), Fig. 2, comparison was made with a different buffer, saturated aqueous borax solution, acting in the same pH region (7).

EXPERIMENTAL

Instrumentation

Sulphide Ion Specific Electrode - CA 306-S/Ag-Metrohm

Reference calomel electrode - double junction: satd. KCl, satd. KNO_3

Readings of potential for the sulphide electrode / calomel electrode assembly were taken from the Electrometer-Keitley 610 C. pH measurements were done with the Philips PW9414 digital ion activity meter, or alternatively with a Metrohm pH meter.

Reagents - all reagents used were of analytical grade with no further purification.

Solutions - xM $Na_2S\ 9H_2O$ ($\simeq 10^{-3}$, 5.10^{-3}, 10^{-2}, 5.10^{-2}, 10^{-1}, 5.10^{-1}, 8.10^{-1}, 1) in

i) 1M $NaHCO_3$

ii) satd. $Na_2B_4O_7$, $10H_2O$

Owing to the high water contents of the solid sodium sulphide, standardisation was done with a reference $AgNO_3$ solution, introducing the proper correction to x values.

Calibration - Every set of experimental $[S^{2-}]$ measurements was preceded by adequate calibration of the electrode system, in various solutions of Na_2S, $9H_2O$ in NaOH 1M. The pH meter assembly was calibrated by the two-point standard method.

RESULTS AND DISCUSSION

Determination of K_2 in 1M $NaHCO_3$ - values of $[S^{2-}]$ were obtained from the introduction of measured potentials in the calibration curves, E/V vs log $[S^{2-}]$ / M. Since calibration was done in 1M NaOH, all measurements occurred at constant ionic strength, that is to say, at constant activity coefficient. Calculations were done assuming that $[HS^-] \simeq C_T$ = xM Na_2S. This assumption holds better at pH $\simeq$ 9. For this reason, experimental measurements were done at pH 9, kept constant by the addition of concentrated NaOH solution, Table 1. Calculations led to an average value, $K_2 = 2.3\ (\pm\ 0.6).10^{-15}$, $pK_2 = 14.64$, which compares with an average literature value

$pK_{2_{min.lit.}} = 11.96$ $\qquad$ $K_2 = 1.1\ .\ 10^{-12}$

$pK_{2_{max.lit.}} = 17.06$ $\qquad$ $K_2 = 8.7\ .\ 10^{-18}$

$pK_{2_{avg.lit.}} = 14.51$ $\qquad$ $K_2 = 3.1 \cdot 10^{-15}$

Experimental K_2 values at different pH, fixed by the ratio of concentrations xM Na_2S/1M $NaHCO_3$-K_2 values,(Table 2) were calculated as previously, making similar assumptions. Despite the relative scatter, values are all of the order of 10^{-15},with an average of $2.6 \cdot 10^{-15}$, within the accepted interval.

Acidity constant k_1 - Once a value of K_2 was taken as pertinent to the working conditions, corrected values of $[HS^-]$ were calculated from (B) and $[H_2S]$ from the mass balance relation C_T. Attempts to calculate K_1, this way, did not give reproducible results, Table 3.

While experimental values of K_2 show reasonable self consistency, those of K_1 differ by several orders of magnitude. A reasonable explanation, is that

i) when pH increases to values higher than 9, which is favoured by higher xM Na_2S, ionic species S^{2-} and CO_3^{2-} are present in non-negligible amounts,(Figs. 1 and 3) giving high values of ionic strength. Activity coefficients change as a consequence, and S^{2-} calibration curve, in terms of concentration, looses validity. Similar argument may hold for lower pH values, due to the presence of H_2S.

ii) At high xM Na_2S, H_2S gas formed leaves the solution after saturation (obvious smell), remaining at a concentration lower than the one accounted for.

Experiments in borax buffer - Despite the good buffer capacity (7), both potentiometric and voltammetric measurements (8) lacked reproducibility. Conclusions were not elaborated,but additional processes are likely to occur. Nevertheless an average value of the order of 10^{-15} was found for K_2.

REFERENCES

1 Hugo Guterman, Sam Ben-Yaakov and Aharon Abeliovich, Anal. Chem., 55 / 1983 / 1731.

2 A. Damjanovic, L-S. R. Yeh, P.G. Hudson, J. Appl. Electrochem., 12 / 1982 / 343.

3 I.C. Hamilton, R. Woods, J. Appl. Electrochem., 13 /1983/ 783.

4 Handbook of Chemistry and Physics, 61st ed., Chemical Rubber Co., Cleveland, Ohio (1980-81).

5 W. Giggenbach, Inorg. Chem., 10 /1971/ 1333.

6 W. Stumm, J. Morgan, Aquatic Chemistry - An introduction Emphasizing Chemical Equilibria in Natural Waters - Wiley Interscience, 1970.

7 M. Filomena G.F.C. Camões, M.J. Guiomar M. Lito, Portugaliae Electrochim. Acta, 1 /1984/ 145.

8 M.I.S. Pereira, M.F.G. Silva, Non-published results.

Table 1 - H_2S acidity constant K_2, calculated at pH 9 in 1M $NaHCO_3$

xM Na_2S	$[S^{2-}]$ / M	$10^{15}.K_2$	$\bar{K}_2$
7.90×10^{-4}	1.50×10^{-9}	1.70	
3.74×10^{-3}	9.00×10^{-9}	2.40	
5.40×10^{-3}	1.14×10^{-8}	2.12	$2.30 (\pm 0.62) \times 10^{-15}$
7.90×10^{-3}	1.84×10^{-8}	2.33	
3.74×10^{-2}	1.10×10^{-7}	2.93	

Table 2 - H_2S acidity constant, K_2, calculated at pH values conditioned by the concentration ratio xM Na_2S / / 1M $NaHCO_3$.

xM Na_2S	pH	$[S^{2-}]$ / M	10^{15} K_2
6.90×10^{-4}	7.85	1.00×10^{-10}	2.05
3.34×10^{-3}	8.65	5.07×10^{-9}	3.34
6.67×10^{-3}	8.65	9.00×10^{-9}	2.96
3.34×10^{-2}	8.94	7.67×10^{-8}	2.53
5.40×10^{-2}	9.00	1.61×10^{-7}	2.97
6.67×10^{-2}	9.30	1.46×10^{-7}	1.10
3.34×10^{-1}	9.91	3.92×10^{-6}	1.44
5.34×10^{-1}	10.30	2.34×10^{-5}	2.20
6.67×10^{-1}	10.50	9.11×10^{-5}	4.37

Table 3 - Experimental values of H_2S acidity constant K_1

xM Na_2S	$[HS^-]$ / M	$[H_2S]$ / M	K_1
6.90×10^{-4}	6.21×10^{-4}	6.90×10^{-5}	1.30×10^{-7}
6.67×10^{-2}	3.20×10^{-2}	3.47×10^{-2}	4.62×10^{-10}
3.34×10^{-1}	2.11×10^{-1}	1.23×10^{-1}	2.11×10^{-10}
5.34×10^{-1}	5.13×10^{-1}	2.10×10^{-2}	1.22×10^{-9}

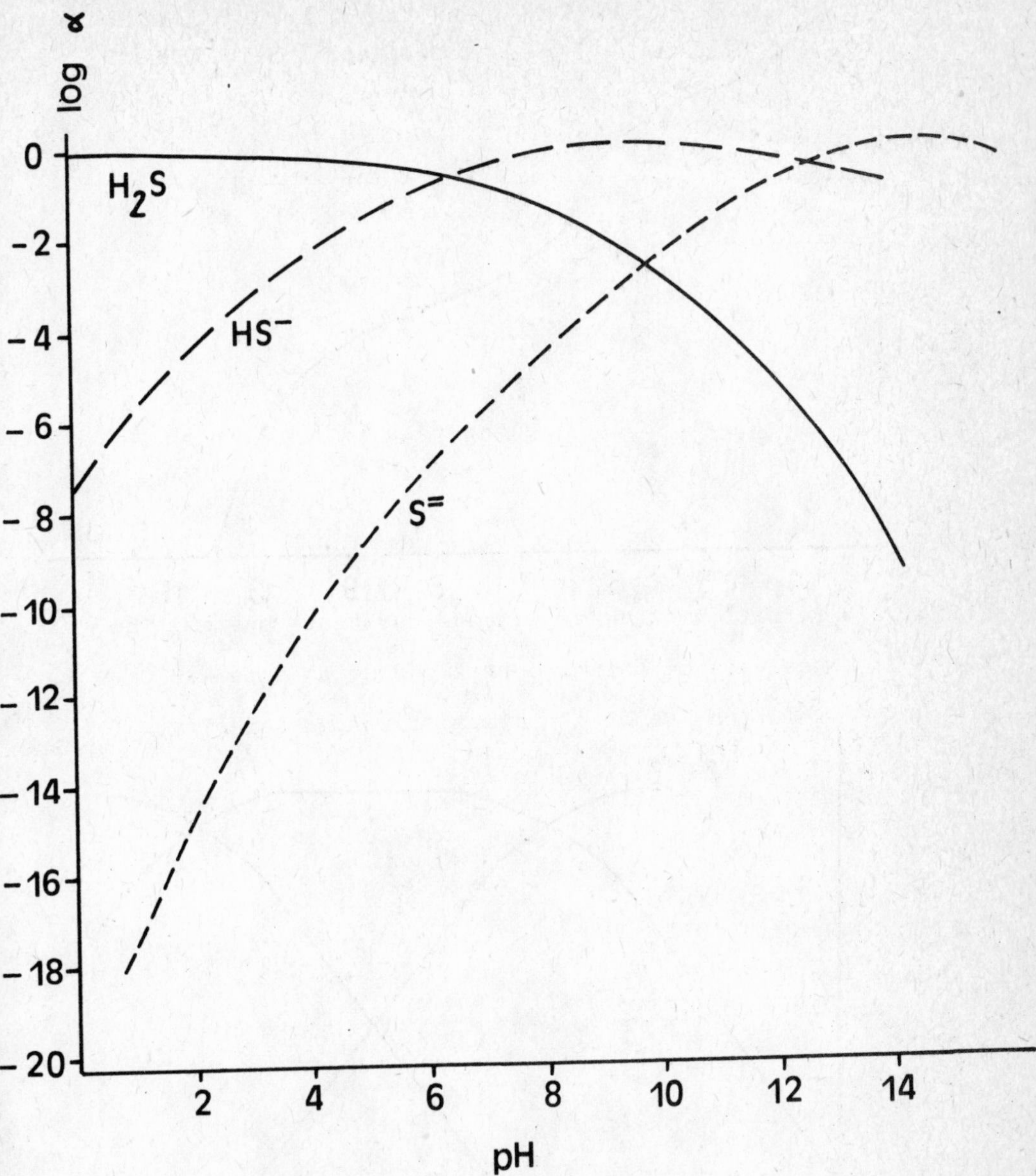

Fig. 1 - Logarithmic fraction of total sulphide concentration present as H_2S, HS^- and S^{2-} from pH0 to pH14.

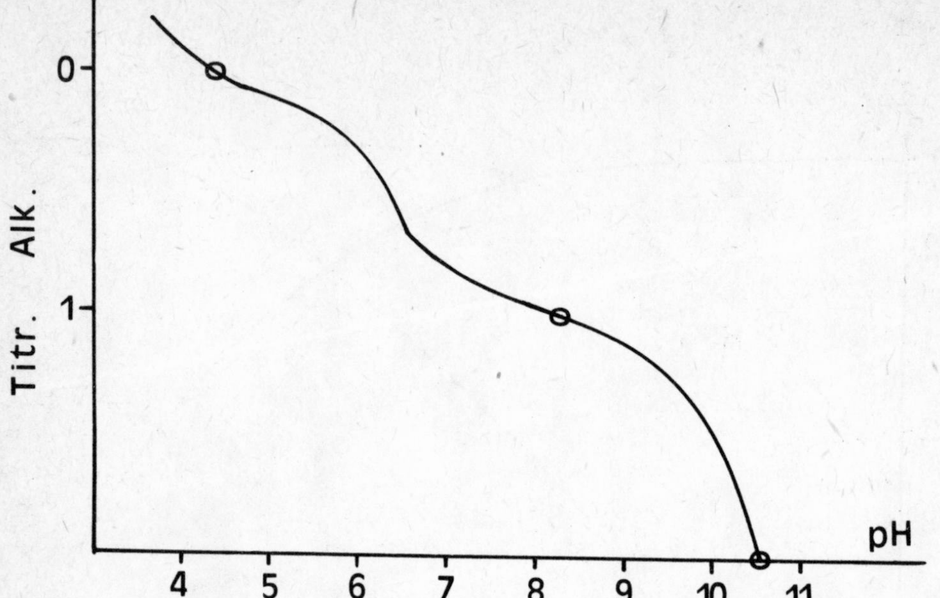

Fig. 2 - Titration curve of aqueous carbonate system. The equivalence point, corresponding to pure $NaHCO_3$ solution, shows a minimum of buffer capacity.

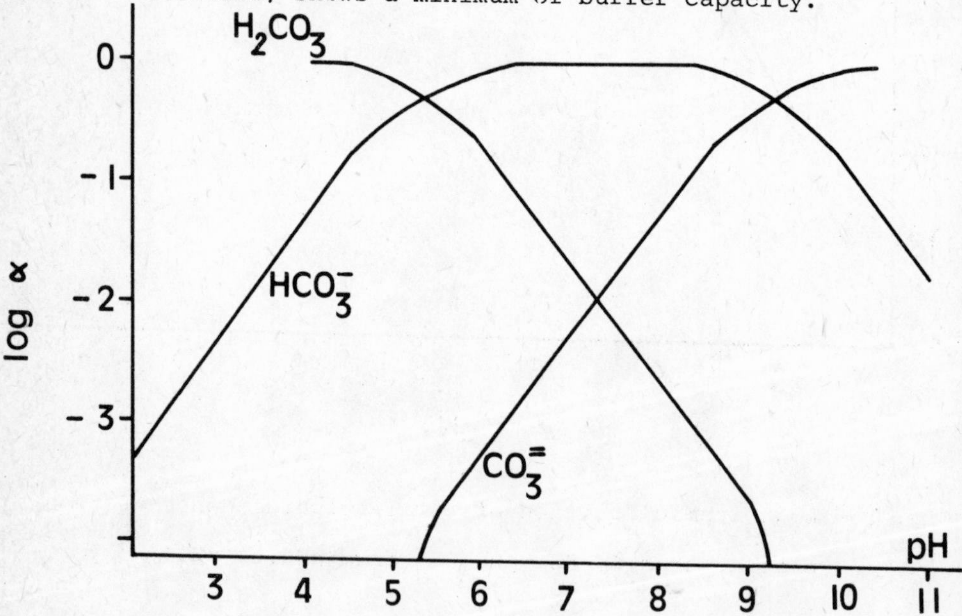

Fig. 3 - Logarithmic fractions of carbonate species present, function of pH.

COMMENT

Participant of the discussion: M.F.Camões and T.Frévert

Comment:

On the protolysis constants of H_2S: The most recent values are given in Analytical Chemistry **49** /1977/ 427. pK_1=6,98 and pK_2=12,97 at 25°.

Answer:

I did not know this particular reference, but the values I know are similar to these data.
I mentioned pK_1 just to show that our procedure was not the best. If we had really wanted to determine a good pK_1, we would have chosen different experimental conditions.

4th Symposium on Ion-Selective Electrodes
Mátrafüred, 1984

THE USE OF ION-SELECTIVE ELECTRODES IN THE ASSAY OF SOME AMINE-TYPE DRUG SUBSTANCES BY TITRATION WITH SODIUM TETRAPHENYLBORATE

B. CSÓTI-GASKÓ, E. BAGÓ-SZMODITS and G. MILCH

Chinoin Pharmaceutical and Chemical Works,
Quality Control Department, Budapest, Hungary

ABSTRACT

The assay of water-soluble drugs with relatively high molecular mass, containing a basic nitrogen has been carried out by direct titration with sodium tetraphenylborate. Potassium, fluoroborate, or tetraphenylborate ion-sensitive indicator electrode was used for the end-point detection. The method is rapid, precise and suitable for routine assays. The relative standard deviation of the method is even in the assay of tablets less than 2 per cent.

INTRODUCTION

Many literature data can be found dealing with the analytical application of sodium tetraphenylborate /2, 3, 4, 5, 6/. More than thirty years ago this reagent, developed by Wittig / 1 / has been first used for the detection and gravimetric assay of potassium and ammonium ions. Later the reagent was applied for the assay of some organic cations either by gravimetric, or by indirect titrimetric methods /7, 8, 9, 10,11/. The development of several ion-selective electrodes enhanced the use of sodium tetraphenylborate as a volumetric solution for direct tirations /12, 13, 14, 15, 19, 20, 21, 22, 23,24/.

In 1976 Vytras has published the development of a potassium ion selective electorde with polyvinyl chloride membrane and valinomycin as the active compound /14/. The potentiometric titration curves obtained with this electorde, marketed as Crytur type 19-15 showed a marked change in the equilibrium voltage and were sufficiently steep.

Some years later Selig /16/ proposed the use of a fluoroborate ion-selective electrode with a double junction reference electrode for the potentiometric precipitation titration of some water-soluble organic cations containing a quaternary ammonium group. The fluoroborate- electrode consists of an electrode body and a replaceable sensing module. This contains a liquid internal filling solution in contact with a gelled organophilic membrane containing an ion-sensitive ion exchanger. The marketed electrode of this type is Orion 93-05.

Peinhardt and Siemroth /17/ published the use of a self-made tetraphenylborate ion-sensitive electrode. It was prepared from a nitrate-sensitive electrode, which was filled with the mixture of 0,01 M sodium tetraphenylborate and 0,1 M sodium chloride solutions.

The aim of our work was the titrimetric assay of some amine type drugs produced in Chinoin Works.

EXPERIMENTAL

Instrumentation

Potentiograph Type E 536 /Metrohm/

Universal pH Titrimeter Type OP 204/1 /Radelkis/ or an instrument of similar performance

Potassium ion-sensitive electrode, Crytur type 19-15

Fluoroborate ion-sensitive electrode, Orion type 93-05

Nitrate ion-sensitive electrode, Crytur type 07-25

Ag/AgCl reference electrode with a salt-bridge of M sodium sulphate, Radelkis OP 0821

Chemicals and solutions

All chemicals used were of analytical or pharmacopoeial grade.

Compounds determined by the titrimetric method are as follows:

Rimazolium methylsulphate /1,6-dimethyl-3-carbethoxy-4-oxo-6,7,8,9-tetrahydro-homopyrimidazole methyl sulphate/

Prenoxdiazine hydrochloride /3-/β,β-diphenyl-/5/-β-piperidinoethyl/-1,2,4-oxadiazole hydrochloride/

Oxprenolol hydrochloride /1-/o-allyloxyphenoxy/-3-izo-propylamino-2-propanol hydrochloride/

Selegiline hydrochloride / /-/-N-/1-phenylizopropyl/-N-methyl-N-propynil ammonium chloride/

The precipitates obtained by titration of the various compounds with tetraphenylborate were isolated by filtration, washing with water and air-drying. The base component of the precipitate has been measured by the non-aqueous titration method published earlier by Chatten /18/. The agreement between calculated and found values was very good confirming the postulated stoichiometry, the 1:1 molar ratio of base and tetraphenylborate in the precipitates.

After having performed the assay of the bulk drugs, we have tried to carry out the assay of the above ingredients in their dosage forms.

First we have checked, whether the usual excipients of tablets would interfere with the assay of the active ingredient content titrated with sodium tetraphenylborate or not. In our experiments out of the excipients used only poly/vinyl-pyrrolidone/ consumed a significant amount of the 0,1 M volumetric solution. So that means, that its interfering effect depends on the base-excipient ratio. In most cases the ratio is convenient, but in the case of Jumex tablets the seligiline hydrochloride - poly/vinyl-pyrrolidone/ ratio is 1:2, this is inconvenient and inhibits application of the method.

The presence of other excipients, as for example methyl-cellulose, Aerosil, talc, stearine, starch, magnesium stearate does not influence neither the shape, nor the potential jump in the end-point.

Potassium /14/, fluoroborate /16/ and tetraphenylborate /17/ ion-selective electrodes proved to be nearly equivalent in the above titration, as it is to be seen in the Figure.

The pH range, which is suitable for the precipitation titrations, is different, depending on the type of compounds. Generally quaternary bases can be titrated in a wide pH range from about pH = 3 to 10.Compounds containing tertiary nitrogen can be successfully titrated only in the weakly acidic pH range 3 to 5. It is not possible to perform precise titration below this pH range.

RESULTS AND DISCUSSION

When we performed a long series of titrations, we found, that after about 15 titrations the response time of the electrode increases and the steepness of the titration curve becomes flat. We suppose, that this is caused partly by precipitate particles, which adhere and contaminate the membrane surface, partly because the solvent in the membrane is gradually saturated with the ion-pair formed during the titrations and this requires the periodical regeneration of the electrodes.

Potassium ion-selective electrode can be regenerated by rinsing and soaking in 0,01 M potassium chloride solution and if necessary by changing the internal aqueous reference solution.

Fluoroborate ion-selective electrode and tetraphenylborate ion-selective electrodes can be regenerated similarly using a 0,1 M sodium fluoroborate solution and the mixture of 0,01 M sodium tetraphenylborate and 0,1 M sodium chloride solutions respectively.

The performance of the electrodes is checked by measuring the electrode slope, which can be defined as the change in potential observed when the concentration changes by a factor of ten. A correct electorde operation is indicated by a differènce of about 56-58 mV-s, but a decrease of about 10 to 12 mV-s can be accepted for titration purposes.

The results obtained by the method compared to those obtained by the official methods are to be seen in the Table.

The relative standard deviation of the method is even in the case of tablets less than 2 per cent.

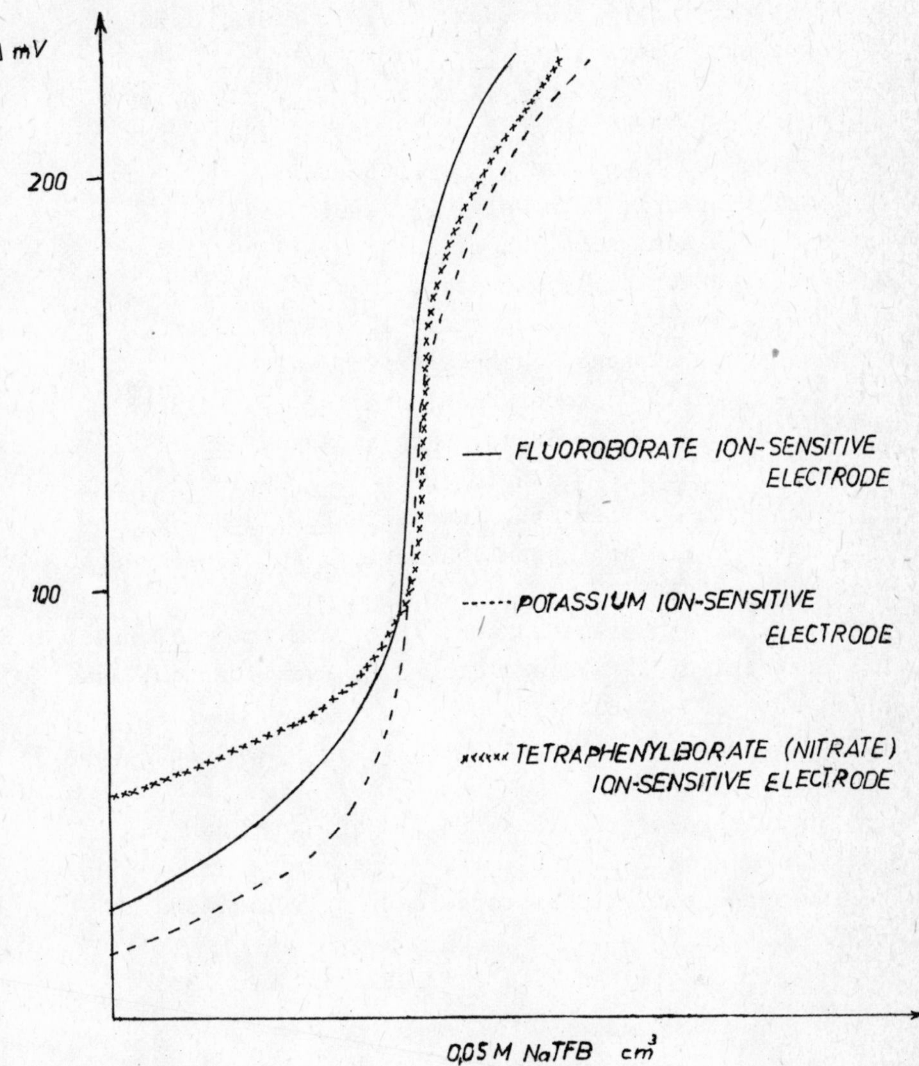

Fig. 1 Titrations with different ion-selective electrodes

REFERENCES

1 G. Wittig, Ann. 563, /1949/ 110

2 L. Buzás, Magy.Kém.Lapja 14, /1959/ 251

3 W. Rüdorff, H.Zannier, Angew. Chem. 64, /1952/ 613

4 E. Siska, E. Pungor, Magy.Kém.Foly. 78, /1972/ 175

5 A. Heyrowsky, Chem. Listy 52, /1958/ 40

6 A. Halász, E. Pungor, Magy.Kém. Foly . 82, /1976/ 640

7 O.E. Schultz, Dtsch.Apoth.Ztg. 21, /1952/ 358

8 J. Bonnard, J. Pharm. Belg. 21,/1966/ 363

9 Z.O. Aklin, Pharm. Acta Helv. 31, /1956/ 457

10 M. Gracza-Lukács, Acta Pharm. Hung. 39, /1969/ 185

11 M. Gracza-Lukács, Gyógyszerészet 18, /1974/ 321

12 M. Gracza-Lukács, Gyógyszerészet 15, /1971/ 330

13 K. Fucamachi, Bunseki Kagaku 24, /1957/ 428

14 K. Vytras, Coll. Czechosl.Chem.Comm. 42, /1977/ 3168

15 M. Gracza-Lukács, Gy. Szász, Gyógyszerészet 25, /1981/ 101

16 W. Selig, Fresenius Z. Anal. Chem. 308, /1981/ 21

17 G. Peinhardt, J. Siemroth, Pharmazie 38, /1983/ 33

18 L.G.Chatten, M. Pernarowski, L.Levi, J.Amer.Pharm.Assoc. Sci.Ed. 48, /1959/ 276

19 K.Vytras, M. Dajkova, V. Mach, Anal.Chim.Acta 127, /1981/166

20 C.E.Efstathiou, E.P. Diamandis, T.P. Hadjiioannou, Anal. Chim. Acta 127, /1981/ 173

21 E.P. Diamandis, E. Athanasiou-Malaki, et al., Anal.Chim. Acta 128, /1981/ 239

22 T.K.Christopoulos, E.P.Diamandis, T.P.Hadjiioannou, Anal.Chim. Acta 143, /1982/ 143

23 E.P. Diamandis, T.K.Christopoulos, Anal.Chim.Acta 152, /1983/ 281

24 K.Selinger, Chemia Analit. 27, /1982/ 51, 223, 383

Table 1. Obtained results by sodium tetraphenylborate titration

Name of drug	n	Mean	Rel.St. Dev.	By Control method
Rimazolium methyl sulphate	7	100,2 %	± 0,33	100,0 %[1]
Its dosage form:				
Probon tablets	7	306,0 mg	± 0,40	302,0 mg[1]
Prenoxdiazine hydrochloride	7	99,8 %	± 0,32	100,2 %[2]
Its dosage forms:				
Libexin tablets	5	98,7 mg	± 1,60	98,0 mg[4]
Libexin combinatum tablets	5	203,0 mg	± 1,65	200,0 mg[4]
Oxprenolol hydrochloride	7	99,3 %	± 0,35	100,0 %[3]
Its dosage forms:				
Trasicor 20 tablets	5	20,3 mg	± 1,90	20,0 mg[5]
Trasicor 80 tablets	5	80,5 mg	± 1,30	80,0 mg[5]
Selegiline Hydrochloride	7	99,8 %	± 0,34	99,6 %[2]

Control methods:

1 By spectrophotometric method according to Chinoin Standard
2 By titration in nonaqueous medium according to Chinoin Standard
3 By titration in nonaqueous medium according to B.P.80.
4 By ion-pair extraction according to Chinoin Standard
5 By spectrophotometric method according to B.P. 80.

QUESTION

Participants of the discussion: G.Milch, and J.D.R.Thomas

Question:

You mentioned that you had tetraphenylborate electrode and in parentheses you had NO_3 ISE. What does this mean ?

Answer:

It was a nitrate sensitive electrode made by Crytur, filled with a mixture of 0.01 mol/l sodium tetraphenylborate and 0,1 mol/l NaCl solution, and it worked as a tetraphenylborate electrode.

4th Symposium on Ion-Selective Electrodes
Mátrafüred, 1984

A NEW, BISMUTH SENSOR FOR RUŽIČKA-TYPE "ALL SOLID STATE" ELECTRODE

M.V. DJIKANOVIĆ and M.S. JOVANOVIĆ*

Chair for Chemistry, Faculty for Metallurgy,
University of Titograd,
YU-81000 Titograd, Yugoslavia

*Institute for Analytical Chemistry,
Faculty for Technology and Metallurgy,
University of Beograd,
YU-11001 Beograd, Yugoslavia

ABSTRACT

The aim of the present work was to synthesize a sensor material for bismuth ions which in a mixture with silver sulphide, could be rubbed into the matrix of Ružička type selective electrode. Such a sensing material should offer not only Nernstian response and short response time, but also the possibility of determining both, Bi^{3+} and S^{2-} ions by direct and titrimetric procedures, due to its dual response.

INTRODUCTION

When speaking of all solid state ion-selective electrodes, the names of professor Ružička and his coworkers, are of an outstanding importance, due to the development of their universal Selectrode /1/. For such type of selective electrode, only sensors for silver and halides /2/, copper /3/, lead /4/ and cadmium /5/ are described in the literature, all of them being the mixtures of sulphides of the appropriate metal and of silver.

Our idea was to prepare a sensor for bismuth on the same principle. Bismuth sulphide is characterized by very small solubility product /$K_{Bi_2S_3} = 10^{-97}$/, while relating to its standard potential and resistance against corrosion, the metal itself is very similar to copper. So, the supposition is logical that Bi_2S_3/Ag_2S sensing powder, should be of the same good characteristics, as is the CuS/Ag_2S mixture. Our further theoretical considerations were extended to the standard potential value of carbon-based bismuth selective electrode, basing the calculations on the equation given by Sato /6/. Also, when preparing sensing sulphide mixture by precipitation, we took into account the necessity for non-stoichiometry of sensor composites, pointed out by Pungor and coworker /7/, responsible for semiconducting properties of the electroactive phase.

Examinations of the behaviour of bismuth selective, Ružička type electrode, were made as follows:

- construction of a calibration graph in a series of standard Bi^{3+} solutions;
- extrapolation of the E/pBi function obtained to unit activity in order to determine the standard potential value and to compare it with the computed one;
- determination of selectivity coefficients for cations;
- investigation of the possibility to determine the unknown concentration of bismuth in solution by direct potentiometry;
- potentiometric EDTA titrations of bismuth solutions, involving bismuth selective ISE.

As any sensing phase, should possess dual, cation and anion response, all the above mentioned investigation steps, were extended also to solutions containing sulphide ions.

Finally, all the investigations were made using a home made version of Ružička universal type selective electrode, /GPE signed/ differing from the Radiometer product in the following:

- polyethylene is used instead of Teflon as hydrophobizing polymer for the graphite matrix of the electrode;
- the solid inner contact of the matrix with the conducting cable, is made by a carbon brush and its copper spring, instead of stainless steel or metallic silver /8/.

EXPERIMENTAL

Instrumentation

A digital, PM 64 Radiometer pH-meter, and a motor driven burette, of 2.50 cm^3 volume, ABU 12 of Radiometer, too, were applied throughout the investigations.

Preparation of sensing material

Sulphide mixtures of bismuth and silver obtained by precipitation, were prepared in mol ratios from 1 to 20 and vice versa. The calculated amount of bismuth nitrate was dissolved in water in the presence of an appropriate amount of sulphuric acid to avoid any hydrolysis, the necessary amount of silver nitrate being added afterwards. Hydrogen sulphide was then passed through the pre-warmed and stirred solution until it became sulphide saturated. The mixture of the precipitated sulphides of bismuth and silver was kept for an hour in this solution afterwards, followed then by the decantation of the precipitate. The mixed sulphide precipitate was then purified by keeping it in redistilled water, two times for an hour, filtered through a G-4 crucible and washed with CS_2 finally. After keeping the precipitate for 24 hours at 105°C, it was pulverised in an agate mortar and was ready for rubbing into the matrix of a GPE electrode.

All the chemicals used were of analytical grade.

Construction of calibration graphs

Because of the well known tendency of bismuth ions to undergo hydrolysis, it appeared to be necessary to keep sulphuric acid concentration at 1 mol/dm^3 in any of the bismuth solutions ranging from 10^{-2} to 10^{-6} mol/dm^3. Almost all of the sensing Bi_2S_3/Ag_2S mixtures showed a slope very near to Nernstian, but with different response times. The mixture with a molar ratio, of 10:1 appeared however to be the best with respect to both slope and response time. The electrode activated with such a sensing powder was used immediately after activation, or after a 24 hour soaking in 0,1 mol/dm^3 EDTA solution, or in 0,001 molar bismuth solution. It was concluded that conditioning of the electrode is not necessary and all the following examinations were made with a freshly prepared one. Potential reproducibility was checked after different periods of time, the results obtained being presented in Table 1; the values are the mean of three independent measurements.

Though there are reports in the literature /9/ that Ružička type carbon substrated electrodes had not been applied for sulphide determinations, we tested Bi_2S_3/Ag_2S 10:1 sensing mixture also in sulphide solutions. Having in mind the pH/pS dependence and the necessity to avoid oxidation of sulphide by air, solutions from 10^{-2} to 10^{-6} molar concentration of sulphide were adjusted to pH slightly above 12, having 1 mg/cm^3 of ascorbic acid added at the same time. It was found that only sulphide mixtures richer in the silver component, gave linear E/logC function having,however, a sub-Nernstian slope.

Applications of Bi_2S_3/Ag_2S /10:1/ electrode

Determinations of bismuth

1. On the basis of well-defined calibration graph of the electrode in solutions of bismuth, we tried first to determine $C_{M_{Bi^{3+}}}$ in samples of only approximately known concentrations. It appeared, however, that direct potentiometry can not be recommended if very exact results are requested.

2. On the contrary, when determining bismuth solutions by EDTA titrations using the above mentioned Bi-selective electrode as the indicator, the results obtained were excellent /Table 2/.

3. Determinations of selectivity coefficient showed the electrode to be non-selective in the simultaneous presence of either copper or antimony. The selectivity coefficient in the presence of cadmium, $k^{pot}_{Bi,Cd}$ was found to be $1,4.10^{-3}$.

Determinations of sulphide

Determinations of sulphide, due to bad calibration graph of the electrode in S^{2-} solutions, were limited only to titrimetry. These solutions, as mentioned above, were of pH 12 being protected against oxidation by ascorbic acid /10/. Instead of applying silver nitrate as conventional titrant, accepting the idea of Slanina et al. /11/, a standard Na_2PbO_2 solution was used instead. Well defined, S shaped titration curves were obtained, the results being shown in Table 3.

RESULTS AND DISCUSSION

A Bi_2S_3/Ag_2S /10:1/ sensing powder activated Ružička type selective electrode gives stable potentials in series of standard bismuth solutions. Response times are from a few tens of seconds to almost ten minutes /most diluted solutions/. The electrode shows a slightly sub-Nernstian slope - 19 mV/decade of concentration in the region of 10^{-2} to $10^{-4,5}$ mol/dm^3.

The electrode is not suitable for determinations of unknown concentrations of bismuth by direct potentiometry, but is excellent as an indicator in EDTA titrations. Due to relative error not exceeding 0,50%, this method preferred to the classical titration using Xylenol-orange as indicator.

The electrode is not selective in the presence of either copper or antimony, but has a remarkable selectivity in the presence of cadmium /$k^{pot}_{Bi,Cd}=1,4.10^{-3}$/.

Though it is not possible to determine sulphide by direct potentiometry using this electrode, because the E/log C function has a sub-Nernstian slope,it is possible to apply succesfully potentiometric titration using standard Na_2PbO_2 solution as titrant. Here, the relative error is of the same magnitude as in the case of EDTA titrations of bismuth.

Finally, it can be concluded, that Bi_2S_3/Ag_2S /10:1/ sensing powder enlarges the family of already known and commercially available, mixed sulphide sensors for the activation of Ružička type ion-selective electrodes.

REFERENCES

1. J.Ružička, C.G.Lamm and J.Chr.Tjell, Anal.Chim.Acta, <u>59</u>, /1972/ 403
2. J.Ružička and C.G.Lamm, Anal.Chim.Acta, <u>53</u>, /1971/ 206

3. E.H.Hansen, C.G.Lamm and J.Chr.Tjell, Anal.Chim.Acta, 62, /1972/ 151
4. E.H.Hansen and J.Ružička, Anal.Chim.Acta, 72, /1974/ 365
5. J.Ružička and E.H.Hansen, Anal.Chim.Acta, 63, /1973/ 115
6. M.Sato, Electrochim.Acta, 11, /1966/ 361
7. E.Pungor and K.Tóth, Analyst, 95, /1970/ 625
8. V.M.Jovanović, M.Radovanović and M.S.Jovanović, 4th Symposium on Ion-Selective Electrodes, Mátrafüred 1984 - Akadémiai Kiadó, Budapest, in press
9. D.Midgley and D.E.Mulcahy, Ion-Selective Electrodes Rew., Vol.5 /1983/ 185
10. R.Bork and H.J.Puff, Z.Anal.Chem., 240, /1968/ 381
11. J.Slanina, E.Buysman, J.Agterdenbos and B.F.Griepink. Microchim.Acta, /1971/ 657

Table 1. Potential/time dependence of Bi_2S_3/Ag_2S selective electrode in a series of standard bismuth solutions

Time	$-\log C_{Bi}$:	2	3	4	5	6
After rubing	E mV SHE	212	195	187	180	178
" 2 hours	"	212	195	186	181	179
" 5 "	"	212	194	184	177	175
" 24 "	"	211	193	182	175	173
" 48 "	"	212	195	183	175	174

Table 2. Results of EDTA titrations of bismuth solutions obtained using Bi_2S_3/Ag_2S sensitized selective electrode

Nr.of anal.	Taken Bi mg	Found Bi mg	Stand. Dev.	Mean Dev.	Relative error in %
9	4,3885	4,3356	0,0279	0,0093	0,50

Table 3. Results of titrations of sulphide solutions using Na_2PbO_2 as titrant and Bi_2S_3/Ag_2S sensitised electrode as indicator

Nr. of anal.	Taken S mg	Found S mg	Stand. Dev.	Mean Dev.	Relative error in%
9	7,2135	7,2728	0,071	0,0078	0,51

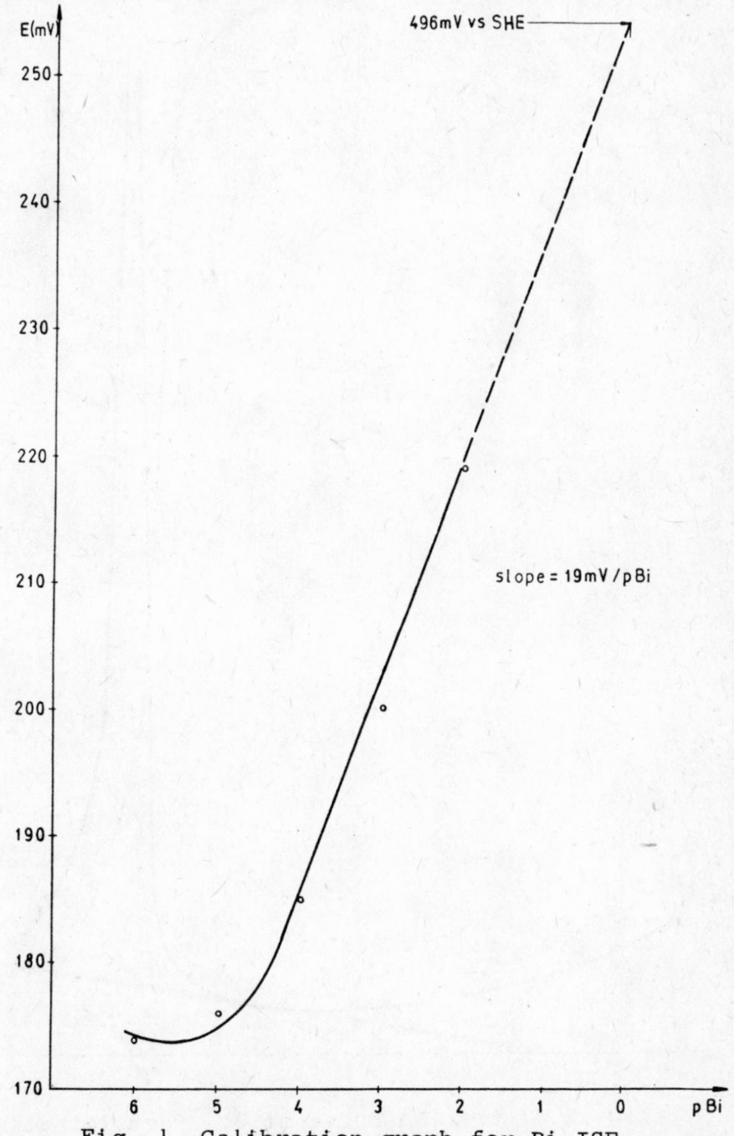

Fig. 1 Calibration graph for Bi ISE

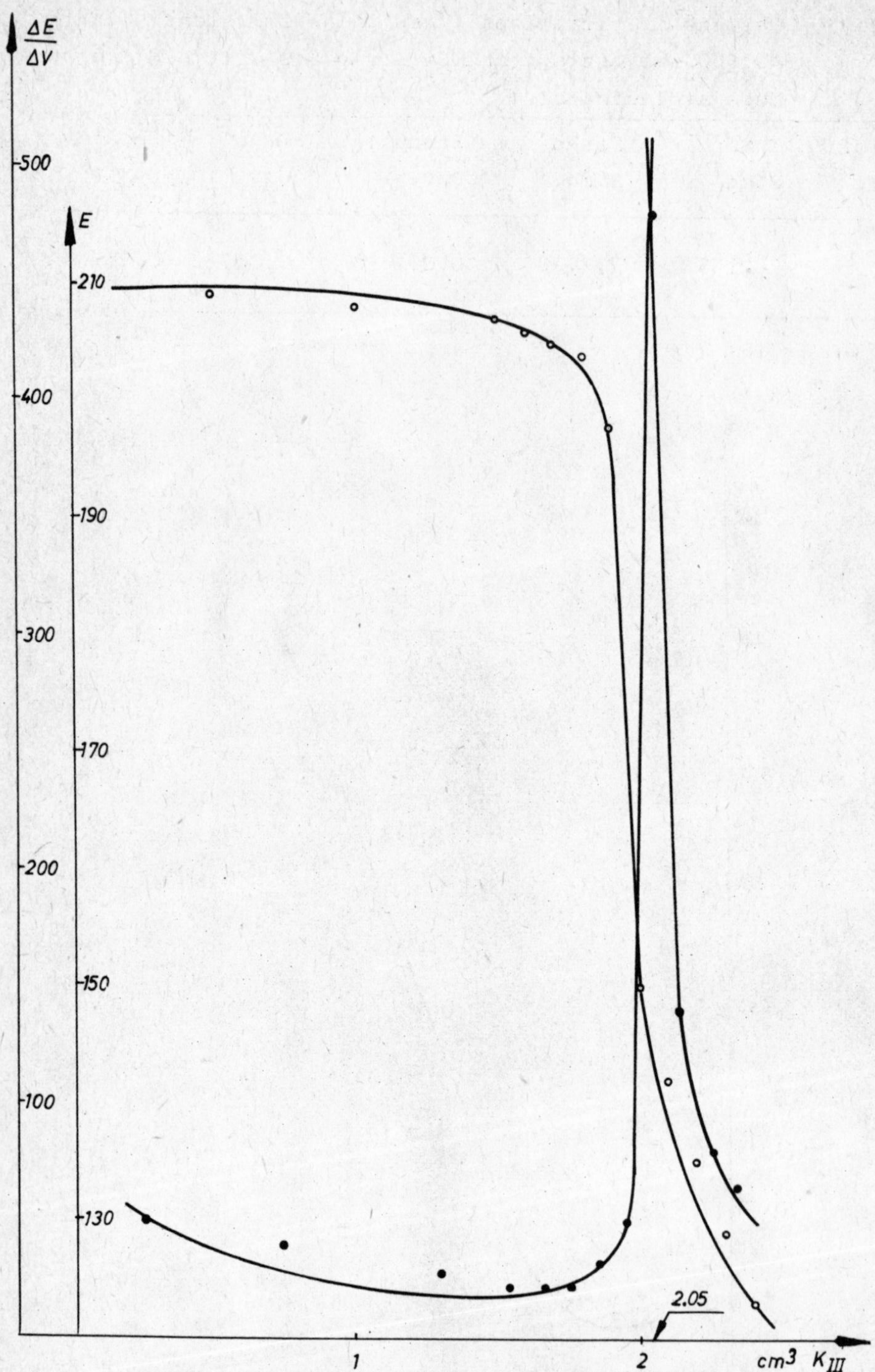

Fig. 2 Titration curve of Bi^{3+} determinations

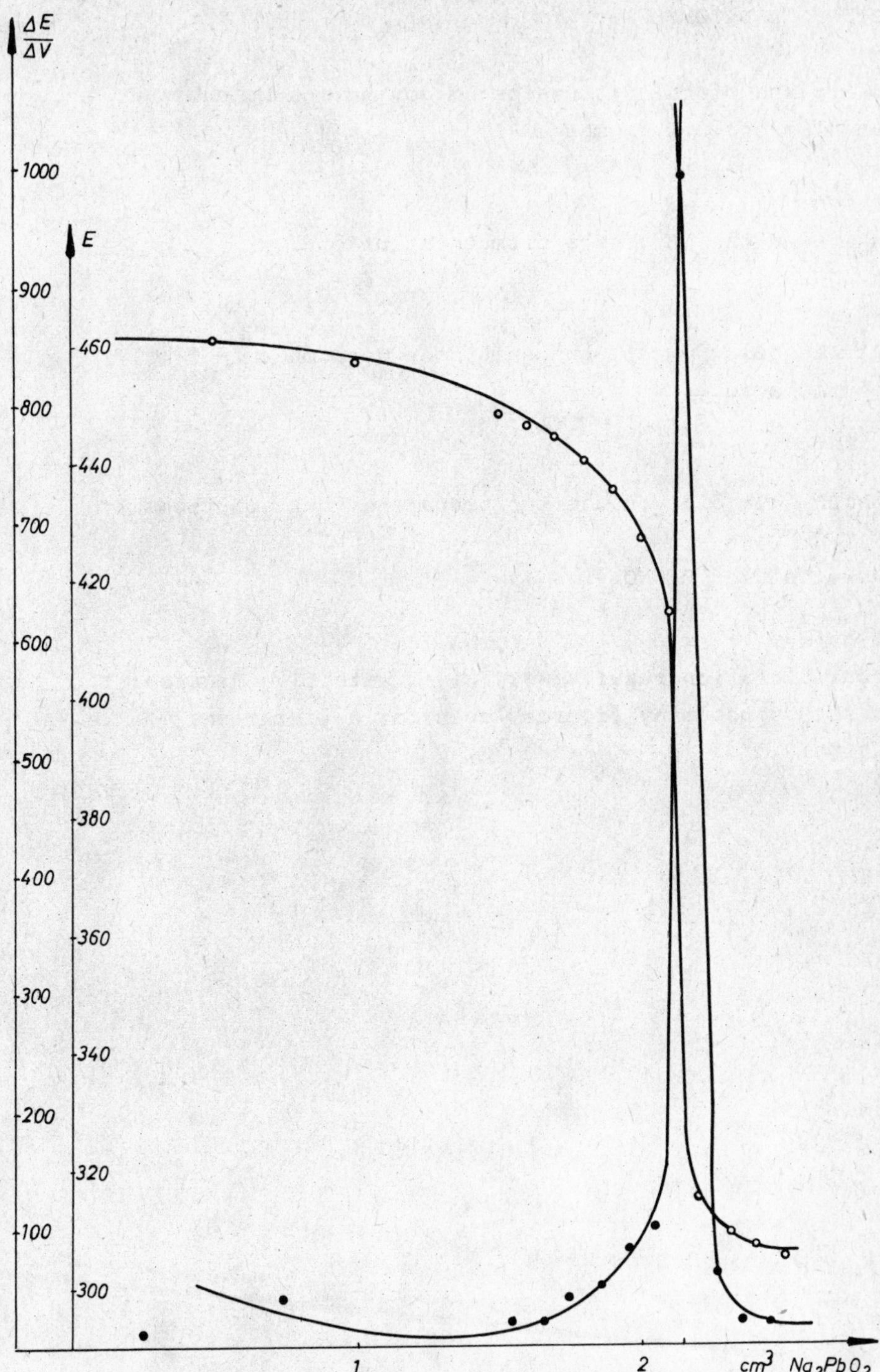

Fig. 3 Titration curve of S^{2-} determinations

QUESTIONS AND COMMENTS

Participants of the discussion: M.Jovanović, M.Neshkova, E.Pungor and J.D.R.Thomas

Question:

What was the pH of the bismuth solution ?

Answer:

It was below 0. All the solutions were made in 1 mol/l sulphuric acid.

Question:

Which salt did you use for preparing your solutions ?

Answer: We used $Bi/NO_3/_3$.

Comment:

Your titration results were given with five figures. I think it is too many figures, even for a titration.

4th Symposium on Ion-Selective Electrodes
Mátrafüred, 1984

SURFACE ANALYTICAL METHODS APPLIED TO ION-SELECTIVE ELECTRODES

M.F. EBEL, H. EBEL, J. WERNISCH, M. GRATZL*, L. PÓLOS*, K. TÓTH* and E. PUNGOR*

Institute for Applied and Technical Physics, Technical University, Vienna, Austria

*Institute for General and Analytical Chemistry, Technical University, Budapest, Hungary

ABSTRACT

Three surface analytical methods with different depth of information (XRFA, SEM, XPS) have been employed to investigate a corrosion process which is responsible for the cyanide response of mixed membranes of silver iodide/silver sulphide.

INTRODUCTION

Surface analytical methods are well applicable for the investigation of processes which take place at the outmost surface and cause progressive in-depth changes towards a stabilized situation. Employing three X-ray methods a range from 1 nm to 0.01 mm can be investigated as shown by Fig.1: XPS (X-ray photoelectron spectroscopy), SEM (scanning electron microscopy) and XRFA (X-ray fluorescence analysis). All these methods offer a quantitative analysis which is of great importance for surface analysis studies. It has to be considered as a great disadvantage if reference samples or standards are necessary. Treating problems as for example surface reactions of ISEs, it is not possible to predict surface compositions and thus to provide reference samples of comparable compositions and concentrations. XPS allows a quantitative analysis without reference samples. As this is of great importance the algorithm which allows a quantitative analysis without reference samples is discussed in this paper.

X P S

The principle of the method which is basing upon the photoelectric effect and its features have been given at the 3rd Symposion on ISE (Mátrafüred) /1/. The applicability for quantitative analysis has also been mentioned. In the meanwhile, we developed an algorithm which leads to better results as it does not depend anylonger on uncertain parameters. The concentration of an element on the surface of aspecific sample can be evaluated from XP-measurements without using reference samples only employing so-called fundamental parameters /2/ as the following equation shows

$$n_i^k = const\cdot[A+B\cdot E_i^k+C\cdot(E_i^k)^2+D\cdot(E_i^k)^3]\cdot\left(\frac{E_i^k}{1000}\right)^{0.7}\cdot$$

$$\frac{\sigma_i^k}{\sigma_C^{1s}}\cdot\frac{1}{A_i}\cdot\left(1-\frac{\beta_i^k}{2}\cdot\frac{3\cdot\cos^2\Theta-1}{2}\right)\cdot C_i\cdot\exp\left(-\frac{d}{\Lambda_L\left(\frac{E_i^k}{1000}\right)^{0.7}\cdot\cos\varepsilon}\right)$$

n_i^k area beneath the photoelectronline of element i and subshell k excited by a defined X-radiation (minus background)

E_i^k photoelectron energy from subshell k and element i excited by a defined X-radiation, in keV. $E_i^k = h\nu - E_{Bi}^k$

$h\nu$ quantum energy (Mg Kα = 1253.6 eV, Al Kα = 1486.6 eV)

E_{Bi}^k electron binding energy of element i and subshell k, in keV

σ_i^k/σ_C^{1s} photoabsorption cross section of quantum energy $h\nu$ in the element i and subshell k, given in multiples of the C1s photoabsorption cross section

A_i atomic weight of element i

β_i^k asymmetry parameter for the subshell k of element i

Θ angle between incident X-rays and direction of detection of the photoelectrons

C_i concentration of the element i (wt%)

A,B,C,D constants which are specific for the spectrometer

ε angle between take-off direction of the photoelectrons and the normal to the sample surface

d/Λ_L reduced thickness of the overlayer (given for photoelectrons with a reference energy of 1 keV. This quantity has to be measured by a separate experiment /3/ and can be found without great experimental set-up for contamination overlayers consisting of carbon-hydrogens /4/.

If the sample consists of m different elements there are m equations for

$$n_i^k\Big|_{i=1....m}$$

together with

$$\sum_{i=1}^{m} C_i = 100$$

For evaluation of the unknown C_1 ... C_m concentrations the following solution is used

$$N_i^k = \text{const}\cdot[A + B\cdot E_i^k + C\cdot(E_i^k)^2 + D\cdot(E_i^k)^3]\cdot\left(\frac{E_i^k}{1000}\right)^{0.7}\cdot$$

$$\frac{\sigma_i^k}{\sigma_C^{1s}}\cdot\frac{1}{A_i}\cdot\left(1 - \frac{\beta_i^k}{2}\cdot\frac{3\cdot\cos^2\Theta - 1}{2}\right)\cdot C_i\cdot\exp\left(-\frac{d}{\Lambda_L\left(\frac{E_i^k}{1000}\right)^{0.7}\cdot\cos\varepsilon}\right)$$

which gives the ratio

$$\frac{n_i^k}{N_i^k} = \text{const}\cdot C_i$$

and

$$\sum_{i=1}^{m}\frac{n_i^k}{N_i^k} = \text{const}\cdot\sum_{i=1}^{m} C_i = 100\cdot\text{const}$$

$$\Downarrow$$

$$\text{const} = \frac{1}{100}\cdot\sum_{i=1}^{m}\frac{n_i^k}{N_i^k}$$

The unknown concentrations C_i in weight % are given by

$$C_i[\text{wt\%}] = \frac{100}{\sum_{i=1}^{m}\frac{n_i^k}{N_i^k}}\cdot\frac{n_i^k}{N_i^k}$$

Using this correlation evaluation can be done by calculators like EPSON HX 20 or HP 41 (program available on request).

S E M and X R F A

There exist comparable considerations how to employ and develop an algorithm for quanititative analysis without referende samples. They are basing similar as the XPS-evaluation on a fundamental parameter approach.

EXPERIMENTAL

These three methods were employed to investigate the corrosion process on the surface of the silver iodide-based membrane electrodes with silver sulphide added which allow indirect cyanide measurements /5,6/. Corrosion was effected at room temperature with 10^{-3} M potassium cyanide solution at different pH values: pH 7, pH 10 and pH 7 with Britton-Robinson buffer and with increasing etching time.

RESULTS and DISCUSSION

Figs. 2 to 4 show the results of the quantitative surface investigations performed on the corroded mixed electrodes. For corrosion time zero an identical composition of all untreated electrodes could be found. At the beginning (short corrosion time) the iodine content of silver iodide/ silver sulphide membranes decreases. The XPS-results show a final stabilized value (after 15 minutes of etching) of the iodine and sulphur concentration at the outmost surface which are identical for all three investigated pH-values. However, the final stabilized values for iodine and sulphur concentrations by SEM differ and they show a dependence on the pH-value and buffer capacity. To understand this, Ar-ion etching was employed on electrodes after etching time of 120 minutes (no selective sputtering could be observed). This treatment caused a change in the surface composition and the XPS-results approach the SEM values.

The results of XRFA with Ag $L\alpha$ and J $L\alpha$ were in good agreement with SEM results, whereas the results with $K\alpha$ radiation gave no change in composition. Thus, it can be observed that the surface layer of the membrane continuously loses the outmost silver particles until a constant AgJ/Ag_2S ratio is achieved at the surface. The leached out particles could be identified as silver sulphide as elemental SEM-micrographs have shown. The corrosion process causes a diffusion layer with a depth comparable to the grain size of silver sulphide which is independent of the corrosion time. Therefore, memory effects and increasing response time do not influence the electrode function, which is in good agreement with literature. Furthermore, the results confirm a steadily renewing of the surface. The surface layer composition is depending upon pH and buffer capacity which is in good agreement with electrochemical behaviour.

ACKNOWLEDGEMENT

Three of the authors (M.F.E., H.E. and J.W.) would like to appreciate the financial support of the "Fonds zur Förderung der wissenschaftlichen Forschung in Österreich" (Projekt Nr. 4272) for purchasing the photoelectron spectrometer.

REFERENCES

1 M.F.Ebel, Communications of the 3rd Symposium on Ion-Selective Electrodes, Mátrafüred 1980, p.80, Hungarian Academy of Sciences.

2 M.F.Ebel, H.Ebel and K.Hirokawa, Spectrochim. Acta, Vol. 37 B, 6 /1982/ 461.

3 M.F.Ebel, G.Zuba, H.Ebel, J.Wernisch and A.Jablonski, Spectrochim. Acta, Vol. 39 B, 5 /1984/ 637.

4 M.F.Ebel, M.Schmid, H.Ebel and A.Vogel, J.Electron Spectrosc. Relat. Phenom., 34, /1984/ 313.

5 J.W.Ross in Proceedings on Ion-Selective Electrodes, NBS, US Special Publication No. 314 /1969/ p. 80.

6 E.Pungor and K.Tóth, Analyst 96, /1970/ 1123.

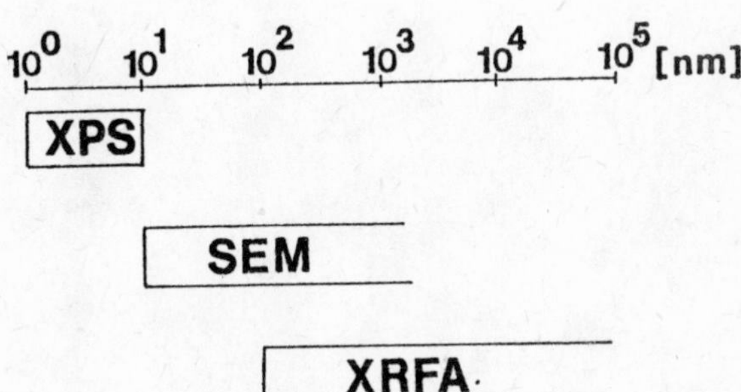

Fig. 1 Survey of the in-depth information of the applied techniques

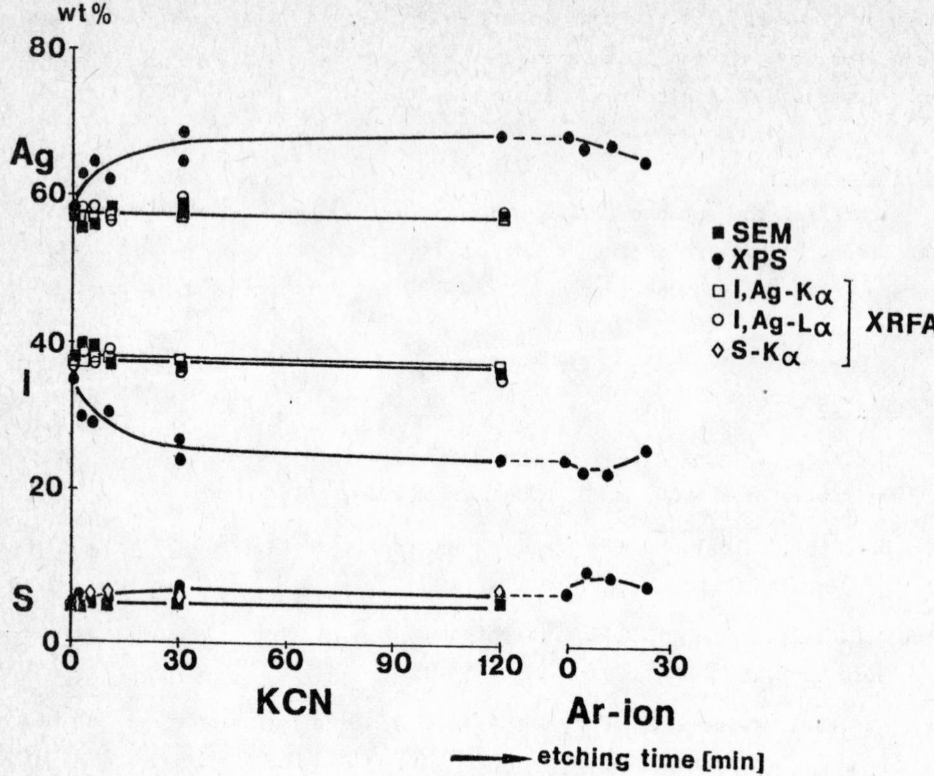

Fig. 2 Results of the quantitative surface analysis of the corroded mixed membrane electrodes for pH 7, giving the elemental composition for silver, iodine and sulphur obtained by XRFA, SEM and XPS in dependence on corrosion time. The XPS-measurements also give results performed after an additional Ar-ion treatment.

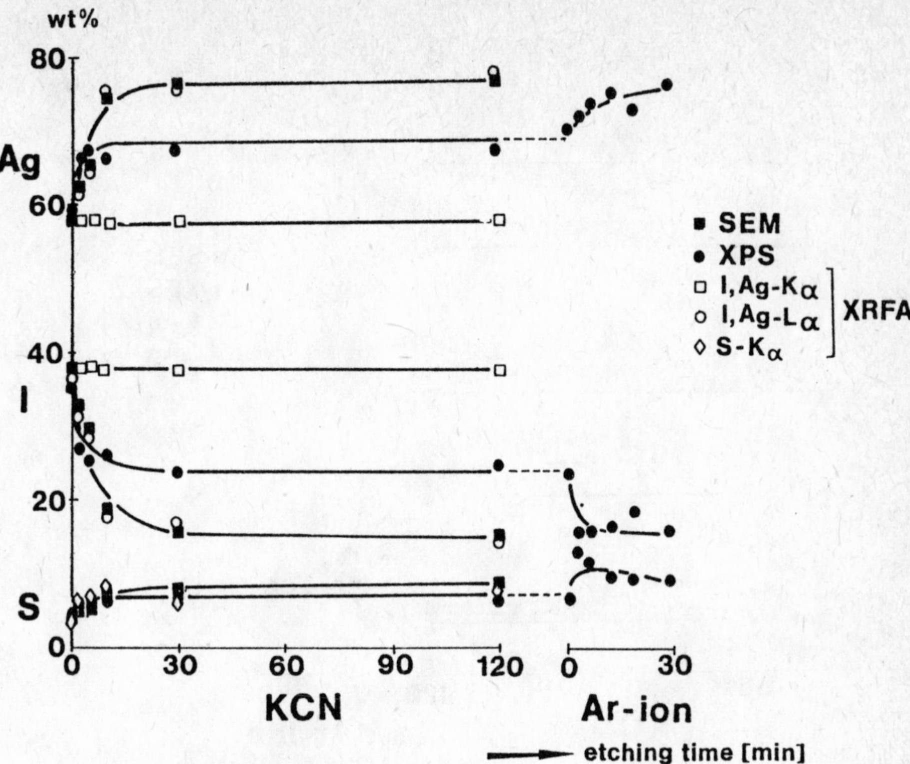

Fig. 3 As Fig. 2, membrane electrode results for pH 10

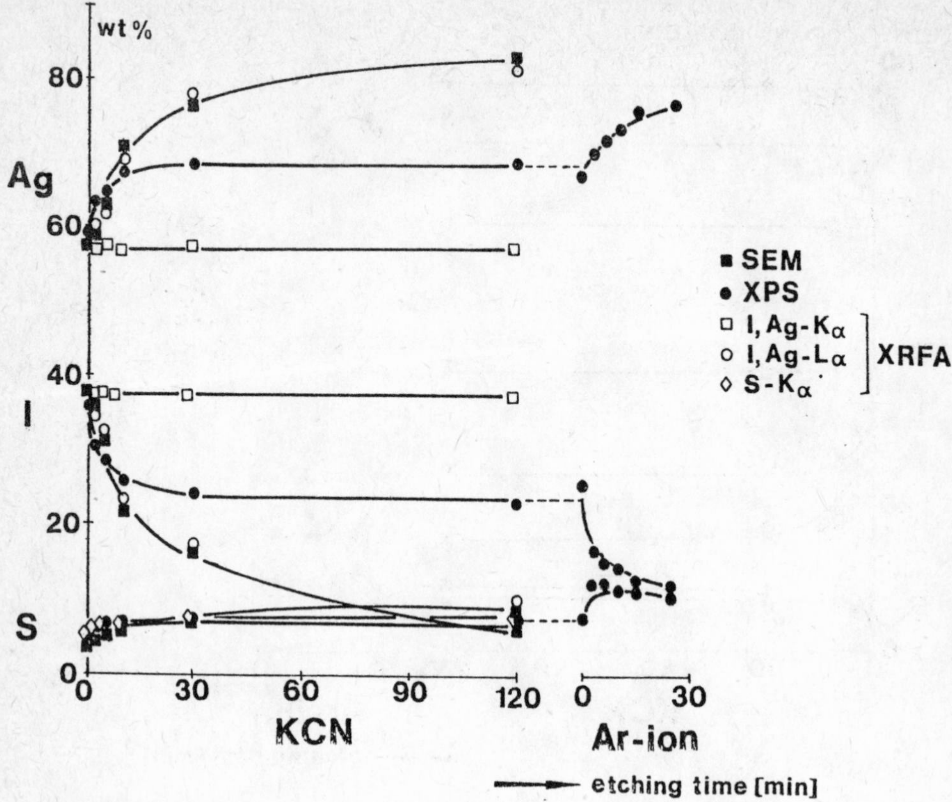

Fig. 4 As Fig. 2, membrane electrode results for pH 7 with a Britton-Robinson buffer

QUESTIONS AND COMMENTS

Participants of the discussion: R.P.Buck, M.Ebel, E.Pungor and J.D.R.Thomas

Question:

With the three methods you indicated, XPS, SEM and XRF, you have been able to cover a range as you mentioned. We have found with analysis of pressed discs for surface components that the methods are so sensitive that they pick up minor components such as carbon contained in the steel of the press in which the disc in pressed. Did you find any problem of the presence of spurious materials present ?

Answer:

Of course we looked for these minor components but they had no influence on the determination of the components we investigated. The problem is always there with the XPS method.

Question:

Do you use these methods to find oxidation products at the electrode surface rather than major constitutional species ? I think we tend to overlook the sensitivity limit sometimes as being determined by chemical reactions with oxygen and I would look for sulphate all the time.

Answer:

XPS is a method capable of telling about the oxidation state, as it not only provides qualitative and quantitative information but tells you about the chemical surrounding.

Comment:

Dr.M.Ebel has done a lot of measurements with the copper electrode, after treatment with oxidizing agents and the two oxidation states of sulphur could be distinguished very well. And as a function of the degree of oxidation, the ratio of the two peak heights changed. It was also possible to see how deep the oxidized layer was, and it came out that it was a monolayer only.

Comment:

We can determine how thick the oxidized layer is by a variant of the method.

4th Symposium on Ion-Selective Electrodes
Mátrafüred, 1984

IN SITU MONITORING OF HYDROGEN SULFIDE IN WATER AND SEDIMENT OF LAKE KINNERET (ISRAEL)

W. ECKERT and T. FREVERT

Chair of Hydrology, University of Bayreuth
P.O. Box 3008, D-8580 Bayreuth, GFR

ABSTRACT

The development of H_2S was monitored *in situ* in the hypolimnetic waters of Lake Kinneret (Israel) using a combined glass/Ag^o,Ag_2S (pH_2S) probe specially deviced for measurements in lakes and sediments. As dissolved oxygen (D.O.) and temperature were also detected the results provide an accurate insight into the vertical structure of D.O. and H_2S concentrations, their alternative occurrence during the thermal stratification of the lake, and the usefulness of the probe for detecting H_2S in environments which are difficult to be accessed by current analytical methods.

INTRODUCTION

An essential hydrochemical and limnological aspect of the aquatic ecosystem of a thermal stratified lake is the microbial hydrogen sulfide release in the anaerobic hypolimnion.

Monitoring the development of H_2S is impossible by use of current analytical sampling methods requiring a strict anaerobic procedure. In this study a novel direct-potentiometric and combined electrode is introduced which can be used for detecting H_2S *in situ* throughout the overlying water and the pore waters of the sediment.

Lake Kinneret proved to be a favourable system to investigate the probe. The lake is situated in the middle part of the Jordan Rift Valley at $32^o42'$ - $32^o55'$ N latitude and $35^o31'$ -

$35^{o}39'$ E longitude. Its surface area - 210 m below sealevel - is 170 km^2 at a length of 22 km and a width up to 12 km. The maximum depth is 43 m with a mean depth of 26 m (1). Lake Kinneret is thermally stratified from May until December. Within this period the thermocline (vertical thermal discontinuity) is find at 15 to 20 m. The mean temperatures of the surface water range throughout the year from 15 to 29°C (2).

Outstanding event in the annual cycle of the lake is a bloom of the dinoflagellate alga *Peridinium cinctum fa. Westii* (from March to June (3)), when water transparency decrease from 3 to 0.6 m. From July to September a bloom of the H_2S-oxidizing phototrophic sulphur bacterium *Chlorobium phaeobacteroides fa. Chlorobiaceae* can be observed (4). In order to investigate the development of the thermocline as well as of the redoxocline (vertical discontinuity of redox intensity), beside of H_2S also the variables D.O. and temperature were detected.

EXPERIMENTAL

Instrumentation

For the H_2S determination we used the glass/silver sulfide covered silver ring electrode without liquid junction (Ag-275-85-6329, INGOLD, Frankfurt/M.) introduced by Frevert and Galster (5, 6). The measurement cell is described as (Fig. 1):

$$Ag^{o}, AgI/0.2\ M\ HI/glass/soln./Ag_2S, Ag^{o}$$

and is responsive to $a_{H_3O^+}$ and $a_{S^{2-}}$.

The half cell potentials are given as:

$$\Delta\varepsilon_{Ag} = \Delta\varepsilon^{o}_{Ag} - \frac{N}{2}\,pK_{sp} + \frac{N}{2}\,pS^{2-} \qquad (a)$$

$$\Delta\varepsilon_{glass} = \Delta\varepsilon^{o}_{glass} - N\ pH \qquad (b)$$

where $N = RTF^{-1}$.

Consequently, the half cell combination will be:

$$\Delta E = const - N\ pH - \frac{N}{2}(pK_1 - pK_2) - \frac{N}{2}\,pH_2S + N\ pH$$

or (for pH ≤ 9)

$$\Delta E = A - \frac{N}{2}\,pH_2S, \qquad where\ pH_2S \equiv -lg\ a_{H_2S}$$

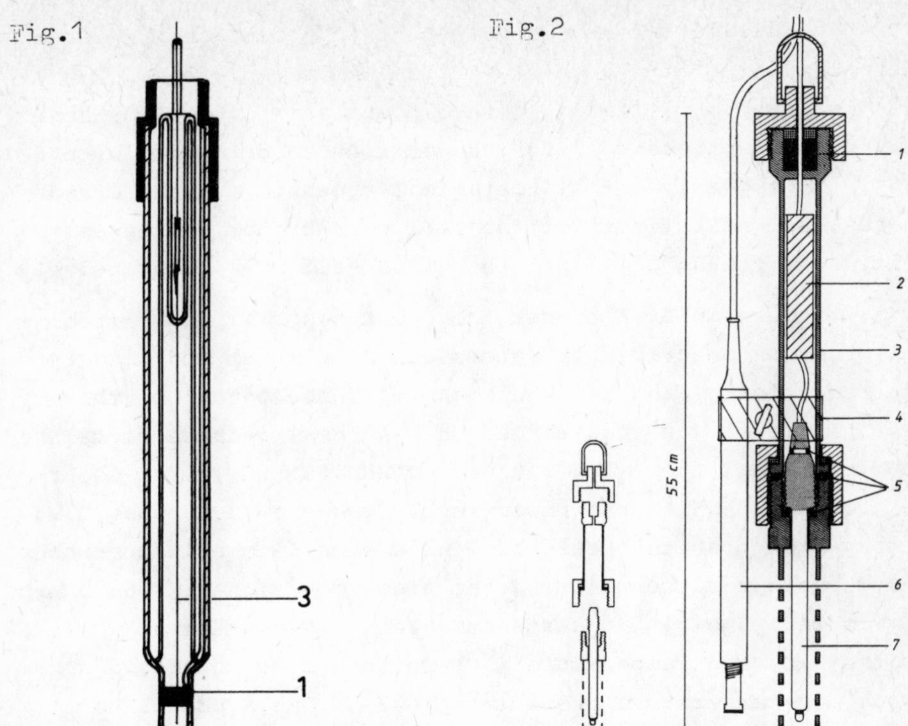

Fig. 1. Combined glass/(Pt)Ag^o,Ag_2S measurement cell (pH_2S probe); 1 - (Pt)Ag^o,Ag_2S ring, 2 - glass half cell, 3 - internal reference (Ag^o,AgI in 0.2 M HI, pH ≃ -0.7)

Fig. 2. Combined D.O./pH_2S probe for the *in situ* detection of dissolved oxygen and hydrogen sulfide (modified from Peters et al., 1984); 1 - silicon rubber sealing, 2 - amplifier (IWA 11, KUNTZE Ltd., Düsseldorf, W. Germany), 3 - metal tube (stainless steel), 4 - clip, 5 - O-rings, 6 - D.O. detector (ORBISPHERE Ltd. Type 2714), 7 - pH_2S probe

The solidified internal electrolyte and the chosen internal reference system of the glass half cell (Ag^o/AgI in 0.2 M HI), leads to low e.m.f. readings (zero point at ~ 0.6 M H_2S) and suits this electrode cell for the continuous detection of H_2S *in situ* irrespective of hydrostatical pressure and temperature changes. The selectivity of the cell corresponds to the selectivities of the Ag^o/Ag_2S and the glass half cell, respectively.

For the use in L. Kinneret the electrode cell was fitted into a probe as described by Peters et al. (7) but modified for lake conditions (Fig. 2). The high impedance signal of the glass half cell was transformed into a low impedance signal by a built-in amplifier (IWA 11, Dr. A. KUNTZE Ltd., Düsseldorf) the necessary additional power supply was provided by a 24 V DC battery pack from the boat side. We used a 60 m silicon rubber sealed cable for transmitting the electrode signal (read on a pH/mV meter pH Digi 88, Wissensch. Techn. Werkstätten, WTW). For the D.O. and temperature (^oC) detection we connected a combined D.O.-temperature probe (model 2714, ORBISPHERE LABORATORIES, Geneva) with the H_2S probe in order to detect all variables at the same site at the same water depth.

All measurements were carried out weekly at the maximum depth of the lake - more frequently during stratification time. The three variables were measured continuously by slowly dropping the probe into the water and eventually into the sediment (~ 5 cm depth).

Calibration of the sensors

A detailed description of the calibration procedure is given by Peters et al. (7).

The electrode cell is screwed into a 100 ml reaction vessel fitted for anaerobic use. After adding 100 ml of a pH-buffer solution (MERCK, pH 5 at 25^oC) the solution is de-aerated with purified N_2. Then 1, 10, 100 and 1000 µl of an iodometrically controlled standard solution (0.1 M $Na_2S \cdot 9\ H_2O$ p.a.) is stepwise added corresponding to the H_2S concentrations $10^{-6.0}$, $10^{-4.96}$, $10^{-3.96}$ and $10^{-2.96}$ M H_2S ($\simeq S_{tot}$).

The mV-readings are due to Nernst equation,

$$pH_2S = pH_2S^o + \frac{(x - mV^o)}{S} \qquad \begin{aligned} x &= \text{reading } (-mV) \\ S &= \text{slope } (mV/pH_2S) \\ mV^o &= \text{reading (mV) at } pH_2S^o \end{aligned}$$

The corresponding data with this study have been:

$$pH_2S = 2.96 + \frac{((-mV)-74)}{30}$$

The precision of the mV readings and the response time is ≤ 1 mV and ≤ 60 sec (at $pH_2S \leq 5$), respectively; no potential drift could be observed. The lower detection limit with this calibration technique is probably given by residual D.O. concentrations oxidizing H_2S (8).

The calibration of the D.O. and temperature probe was made according to the instructions of the manufacturer. The D.O. sensor covered a concentration range from 0.02 to 199.9 ppm and was proved insensitive to H_2S concentrations ≤ 10 ppm.

RESULTS

All H_2S and D.O. concentration profiles are summarized in the isopleth diagram as shown in Fig. 3.

The different temporal phases which can be distinguished due to the lake development are pointed out by some representative profiles shown in Fig. 4 A - L.

When beginning our measurements on May, 12th (Fig. 4 A) we could detect D.O. down to the sediment-water interface while H_2S was still absent.

In the forthgoing time we observed a continuous D.O. depletion in the deep hypolimnion as well as at the thermocline. H_2S could, however, be detected in sediment upper layer (Fig. 4 B). The indicated tendency to an inverse D.O. stratification became obvious with the next profiles (Fig. 4 C - E). Note the extensive anoxic but H_2S free layers above the sediment and in the upper hypolimnion water as concomitant to a steady H_2S increase in the sediment. On 29th of May (Fig. 4 D) we detected H_2S for the first time in the deep hypolimnion water.

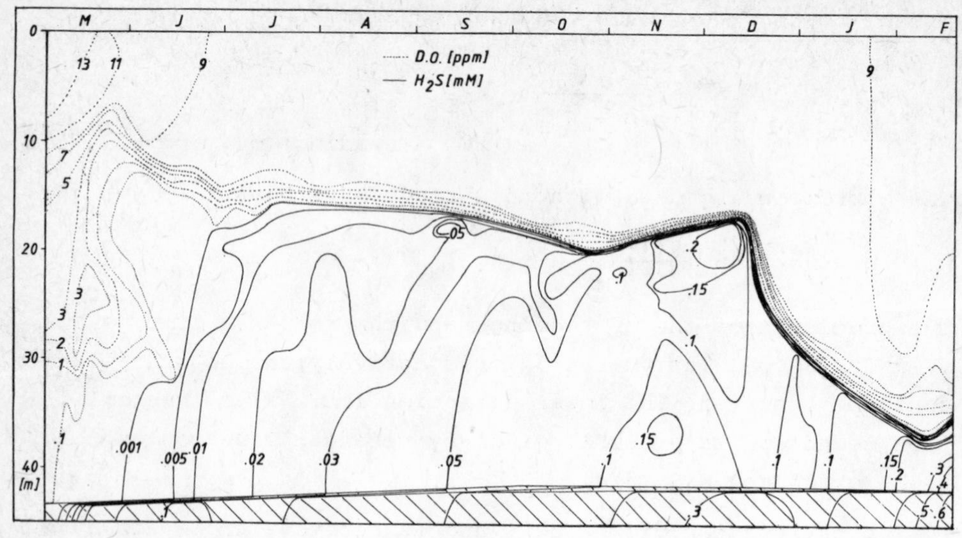

Fig. 3. D.O., H_2S isopleth diagram at maximum depth of L. Kinneret (42 m) from May '83 - Feb. '84.

In a next step the H_2S bearing water layer raised to a depth where still aerated water prevailed (Fig. 4 E). The latter prevented very likely its on-going raise. Only after total depletion of the hypolimnetic D.O. (Fig. 4 F), H_2S raised to the thermocline where a small H_2S concentration peak could be observed.

During the time of summer stratification from July to November two stages can be distinguished: from July until beginning of September we observed a highly stabilised thermocline at 15 m depth and low H_2S concentrations in the *upper* hypolimnion water which contrasted clearly to the increasing concentrations in the *lower* hypolimnion and the sediment (Fig. 4 G). From mid-September until the beginning of November we observed a slow drop of the thermocline with increasing H_2S concentrations at the thermocline (Fig. 4 H, I).

In spite of the permanent cooling down of the water during November the thermocline did not drop under 20 m until mid-December (Fig. 4 J). Note the extreme sharp interface between oxygenised epilimnion and H_2S bearing hypolimnion water and,

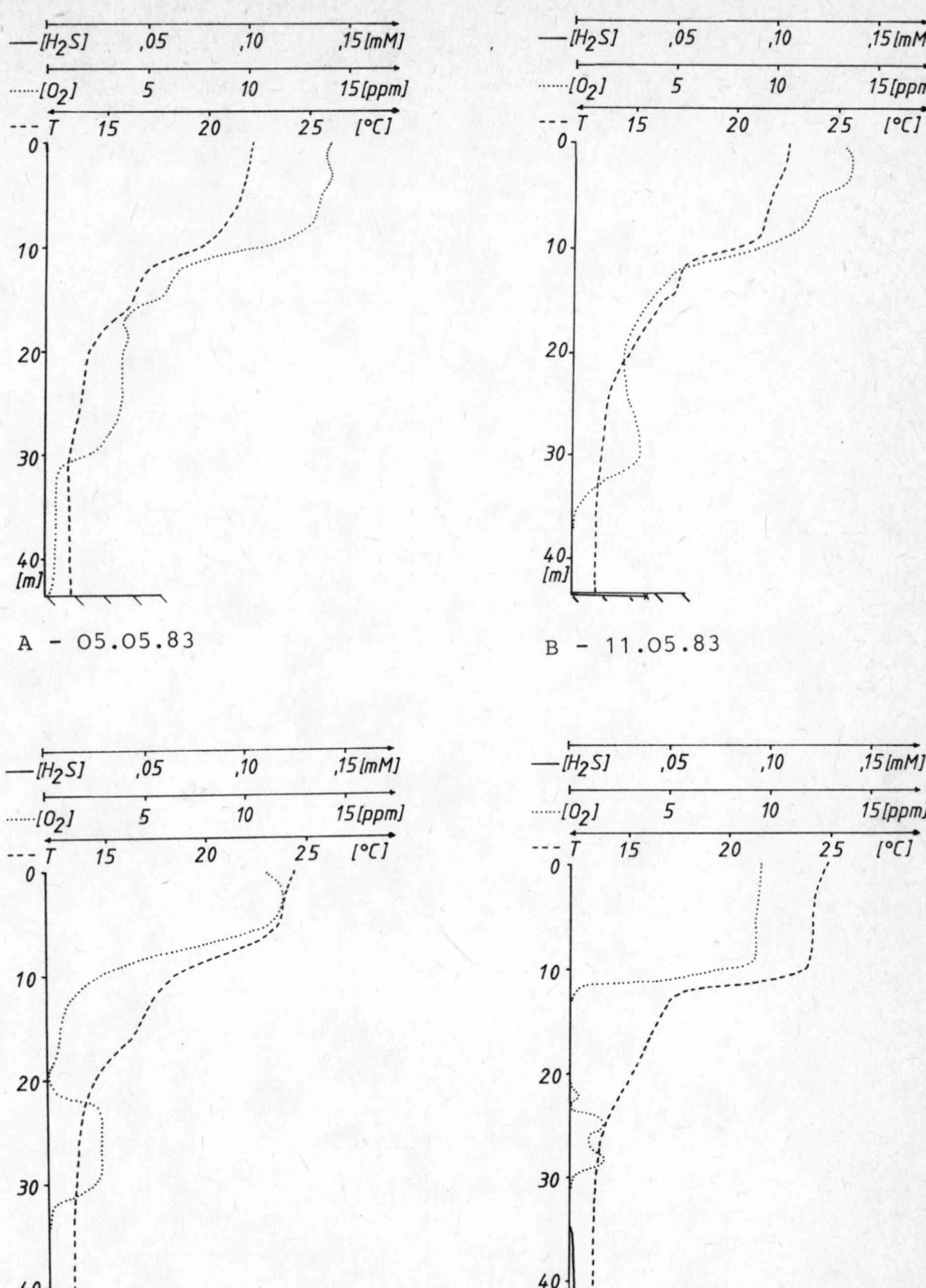

Fig. 4 (A - L). D.O., H_2S concentration profiles versus depth (42 m)

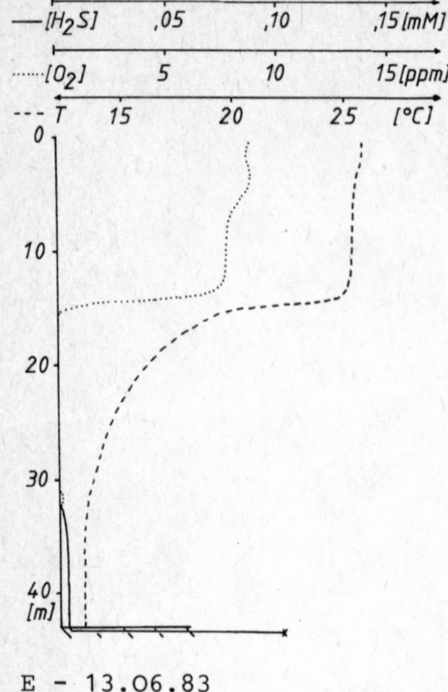

E - 13.06.83

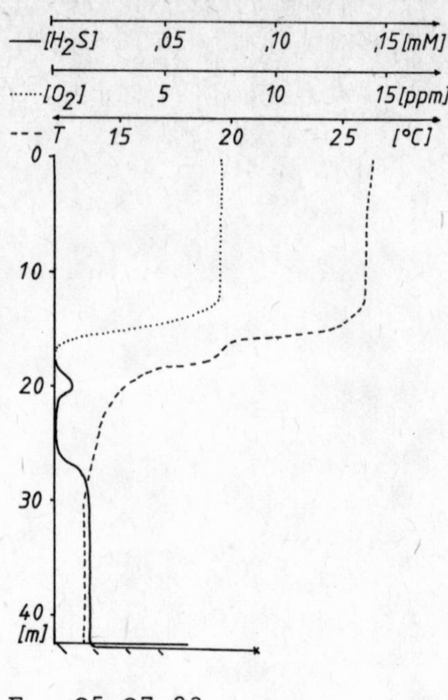

F - 05.07.83

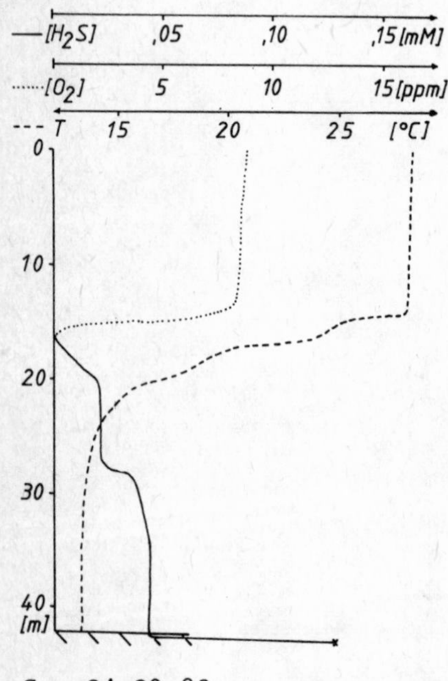

G - 04.09.83

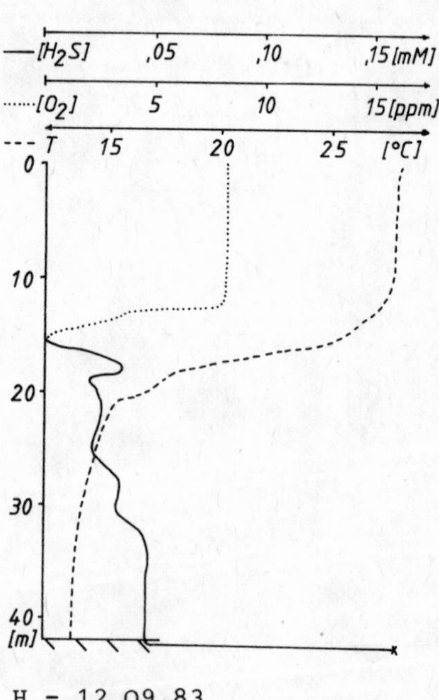

H - 12.09.83

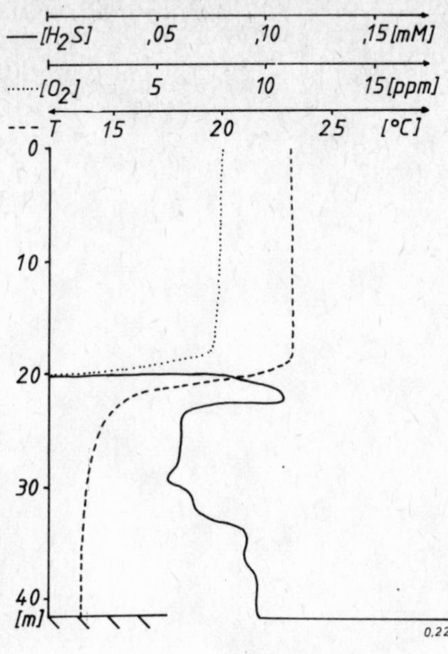

I - 23.10.83

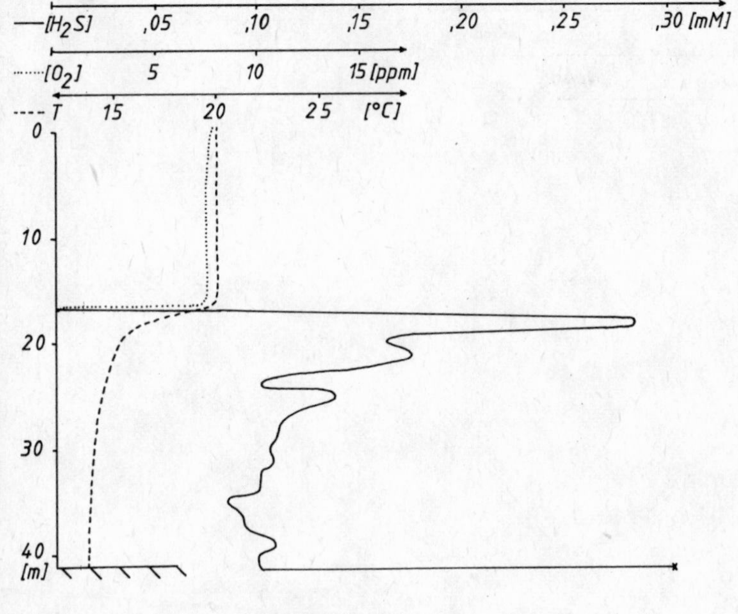

J - 12.12.83

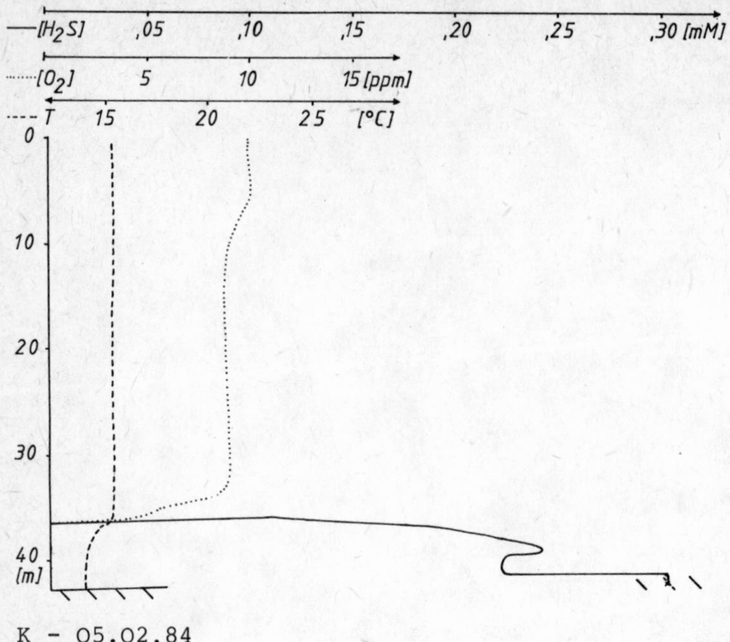

K - 05.02.84

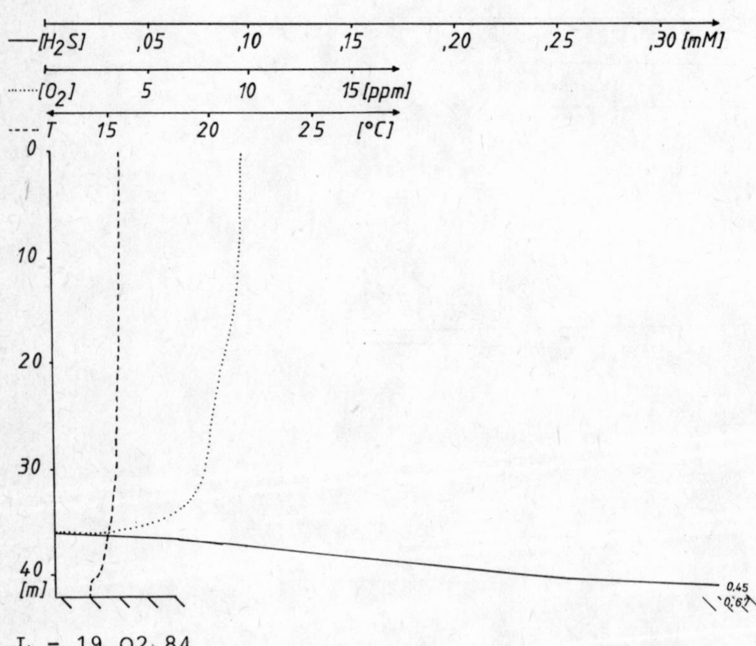

L - 19.02.84

particularly, the increasing H_2S concentration peak at the thermocline.

Subsequent storm events together with the continuous cooling down of the lake water caused a stepwise dropping of the thermocline until 36 m at the beginning of February (Fig. 4 K). After each sinking step the H_2S concentration at the thermocline decreased rapidly starting again to increase until the next thermal shift.

Our last profile taken on 19th of February showed a redoxocline at 38 m *above* the thermocline at 40 m with highest H_2S concentrations in the lake water as well as in the sediment upper layer.

DISCUSSION

Our results during the stratification time May - June (Fig. A - E) have to be considered as influenced by the *Peridinium* bloom. The epilimnetic D.O. concentrations during this period derived from the photosynthetic activity of the algae with a maximum at 3 m water depth (3). The high input of biomass from the declining bloom and its accumulation on the thermocline layer caused very likely an intensified D.O. depletion which is well known from mesotrophic or eutrophic lakes (9). The residual hypolimnetic D.O. is probably depleted by the activit of enclosed aerobic organisms as indicated by the shape of the D.O. concentration peak in Fig. 4 D.

Considering in development of the redoxocline in Lake Kinneret our measurements lead to the assumption that the H_2S release is initiated by the sediment. The always highest H_2S concentration in the sediment and the extensive anoxic but nevertheless H_2S free waters (Fig. 4 C - E) as well as the suppression of H_2S by the residual hypolimnetic D.O. (Fig. 4 E) cannot be explained otherwise.

The vertical H_2S distribution seems further to be correlated to the occurrence of phototrophic sulfur bacteria, whose favourable ecological conditions were established when H_2S raised to the light penetrated water (10), (Fig. 4 F). In 1983 a bloom of *Chlorobium phaeobacteroides* lasted from July until the beginning of September (unpubl. data). The gradual H_2S concentration gradients and the lower H_2S contents in the upper hypolimnion waters (Fig. 4 G) could, therefore, be due to the photooxidation of H_2S by the microorganisms. The declining bloom of *Chlorobium* could result in a H_2S concentration peak at the thermocline as indicated by the concentration profiles in Fig. 4 H.

We took into consideration that the H_2S increase as observed in the subsequent time (Fig. 4 I - L) could be due to a biogenic H_2S production or, likewise, to a change of pH caused by biogenic CO_2 release. We calculated a corresponding pH shift by means of the protolysis coefficient α_o (11). Assuming that the hypolimnetic pH value corresponds to a carbonate buffered solution ($\sim$ 3 mequiv.$\cdot l^{-1}$) near pH 7.0 (1), i.e. $\alpha_o^{H_2S}$ = 0.53, pH 7.0, T = 14^oC, an additional drop of pH of about one unit would be necessary in order to explain the observed pH_2S value which, however, is very improbable to occur. More likely is a common effect of both, H_2S release and pH value shift.

In summary we may conclude that the detailed spatial and temporal insight into the release of H_2S in Lake Kinneret was only possible using the *in situ* electrochemical sensor technique even if such direct-potentiometric techniques may be less precise than titration or spectrophotometrical methods, mainly at higher H_2S concentrations ($S_{tot} > 10^{-3}$ M, $pH_2S < 3$).

ACKNOWLEDGMENTS

The study was granted by the MINERVA Foundation, Rehovot/ Heidelberg, which is gratefully acknowledged. We wish to thank the Israel Oceanographic and Limnological Research Ltd. Haifa (Kinneret Limnological Laboratory, Tiberias) for the steady support of our investigations.

REFERENCES

1 C.Serruya (ed.), Lake Kinneret, Dr.W.Junk, W.V.Publishers, The Hague, Boston, London /1978.

2 S.Serruya, Verh. Int. Verein. Theor. Angew. Limnol., 19, /1975/ 73.

3 T.Berman and U.Pollingher, Limnol. Oceanogr., 19, /1974/ 326.

4 B.Z.Cavari, O.Uriel, M.Gophen and T.Berman, 1st Int. Congr. Bacteriol. Jerusalem (Israel), 11, /1973/ 147.

5 T.Frevert and H.Galster, Schweiz. Z. Hydrol., 40, /1978/ 199.

6 T.Frevert, Schweiz. Z. Hydrol., 42,2 /1980/ 255.

7 K.Peters, G.Huber, S.Netsch and T.Frevert, gwf-Wasser-Abwasser, 125,8 /1984/ 386.

8 T.Frevert, Int. Symp. Elektrochem., RSC, Cardiff (U.K.), /April 1983/ 5.

9 R.Wetzel, Limnology, Saunders, Philadelphia, London, Toronto /1977.

10 N.Pfennig, Ann. Rev. Microbiol., 21, /1967/ 464.

11 W.Stumm, J.J.Morgan, Aquatic Chemistry, Wiley-Interscience, New York /1981.

4th Symposium on Ion-Selective Electrodes
Mátrafüred, 1984

CRYSTALLOGRAPHIC STUDIES ON CO-PRECIPITATED OR SOLID-STATE MIXED $CdS-Ag_2S$, $PbS-Ag_2S$ AND $CuS-Ag_2S$ ION-SELECTIVE ELECTRODE MATERIALS

M. FARKAS-JAHNKE, F. PÁL*, K. TÓTH* and E. PUNGOR*

Research Institute for Technical Physics,
Hungarian Academy of Sciences,
Budapest, Hungary

*Institute for General and Analytical Chemistry,
Technical University, Budapest, Hungary

INTRODUCTION

Mixed sulphide systems, i.e. $CdS-Ag_2S$, $PbS-Ag_2S$ and $CuS-Ag_2S$ systems have been proved to be appropriate as electrode membrane materials selective for Ag^+, Cd^{2+}, Pb^{2+} or Cu^{2+} [1], but pure CdS, PbS and CuS has no response to the relavant ions. The procedures employed to obtain mixed sulphide systems were different; they were prepared either by simultaneous precipitation or mixing together the components of the system, but in all cases the final step of the procedure had been a heat treatment. [2-12]

Crystal structure of the mixed sulphides has been studied by van der Linden et al. [2] Heijne et al. [13] and Ikeda et al. [14] in order to elucidate the operation function of such an electrode system and find the role of the presence of Ag_2S. In $CuS-Ag_2S$ system they found jalpaite /$Ag_{1.55}Cu_{0.45}S$/ in addition to CuS and Ag_2S, while in $PbS-Ag_2S$ and $CdS-Ag_2S$ systems no new phase formation could be identified. They supposed, however, that solid solutions are formed by dissolving Ag_2S into the PbS or CdS lattice.

Recently room temperature procedures have been elaborated for the preparation of mixed sulphides suitable for active materials of heavy metal ion selective electrodes [15].

The aim of the present work was to study the correlation between the crystalline structure and the potential response of the above mentioned three mixed sulphides prepared without heat treatment.

EXPERIMENTAL

For the systematic investigations different CdS-Ag_2S, PbS-Ag_2S and CuS-Ag_2S mixed sulphides, made by co-precipitation or by mechanical mixing and milling the separately precipitated sulphides were used.

Structure investigations carried out using room temperature and in some cases high temperature X-ray diffraction techniques /room temperature patterns made in a Guinier-de Wolff-type camera and high temperature patterns made in a Guinier-Lenné camera using Cu K_{α} radiation were measured by an automatized Zeiss photometer/ were correlated to electrochemical data.

The electrodes made of the mixed sulphides were measured in all cases in a potentiometric cell

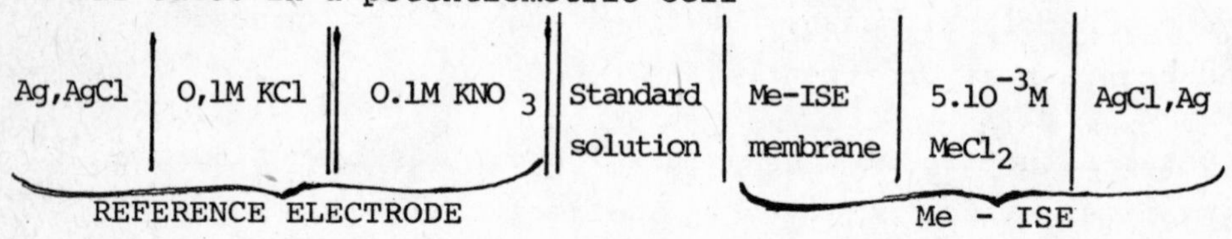

by a precision digital pH meter /RADELKIS OP-208/; /Me = Cd^{2+}, Pb^{2+} or Cu^{2+} /. The reference electrode was a double junction Ag,AgCl /0,1 M KCl/ electrode /RADELKIS OP-821 type/ immersed into 0.1 M KNO_3 solution. As standard solutions 10^{-2}- 10^{-5} M $Me/NO_3/_2$ were used; its ionic strength was adjusted to 0.1 by adding the appropriate amount of KNO_3.

RESULTS AND DISCUSSION

The CdS-Ag_2S system

While X-ray patterns of separately precipitated CdS show the very diffuse lines of quasy-amorphous cubic phase, /Fig.1a/ the separately precipitated Ag_2S is well crystallized and can be identified as argentite. Milling does not influence the structure of Ag_2S, but decreases the half-width of CdS lines which means an increase of the size of CdS crystallites from about 3 nm to 5 nm. At the same time diffuse but well resolved maxima are visible at positions of the 100 and 101 reflections of the hexagonal CdS /Fig.1b/. For comparison the diffracted intensities measured on X-ray patterns taken from powdered hexagonal CdS crystals is shown on Fig. 1c.

X-ray patterns of co-precipitated CdS-Ag_2S samples of mole fraction 6:1, 4:1, 2:1, 1.5:1, 1:1, 1:1.5, 1:2 and 1:6 contain the line system of crystalline Ag_2S together with the somewhat diffuse lines of hexagonal CdS. The intensity ratios of CdS and Ag_2S lines is lower than required for the nominal concentration; it could be explained either by the presence of nuclei sized CdS particles, or by assuming interdiffusion of the two species. Among the samples listed above the one with mole fraction CdS-Ag_2S 1:2 showed the best electrochemical response when used as electrode material.

Milling has a very interesting effect on samples mixed from separately precipitated CdS and Ag_2S. On the X-ray patterns made of these samples the width of the CdS lines decreased remarkably already after a short period of milling. The crystallite size of CdS determined from these half-width values depends from the composition and from the duration of milling; if the composition was CdS:Ag_2S

6:1, the CdS lines were much more broadened on patterns made of separately precipitated and thereafter mixed and milled constituents, as on the patterns of co-precipitated CdS-Ag_2S of the same composition. On the other hand if the composition of the samples was 1:2 mole fraction CdS:Ag_2S, samples made from separately precipitated cadmium sulphide and silver sulphide after milling them together for 9-12 hours gave exactly the same X-ray diffraction pattern as the co-precipitated samples of the same composition /Fig.2/ When measuring the electrochemical response of electrodes made of these materials, strong correlation between the optimal composition and milling time, and the good electrochemical data was found.

The PbS-Ag_2S system

When precipitated separately, PbS show the somewhat diffuse lines of galenite. When precipitated together PbS and Ag_2S in a composition of mole fraction PbS-Ag_2S 1:2, the patterns contain lines of both systems /PbS- galenite, Ag_2S- argentite/ but the PbS lines are less diffuse, the Ag_2S lines more diffuse than on patterns made from the separately precipitated constituents; this means an increase in the PbS crystallite size and either a decrease in the size of Ag_2S crystallites, or a high degree of deformation of the lattice. Milling has the same effect on the structure; when increasing the milling time the initially very sharp Ag_2S lines became more and more diffuse, and after 10 hours of milling the sample has a structure similar to that of the co-precipitated sample with the same composition.

Electrochemical response of the electrodes made of the different samples show a good correlation to the structure results.

The CuS-Ag_2S system

CuS precipitated separately crystallizes as hexagonal covellite, the crystallite size estimated from the half width of the diffraction lines is several time ten nm.

In this system only co-precipitated symples were investigated till now. X-ray patterns of CuS-Ag_2S co-precipitated in a composition with mole fraction 1:1 /which was a suitable electrode material contain besides the Ag_2S lines only some superstructure lines, and no traces of a separate copper-sulphide phase. When investigating this sample with high-temperature X-ray technique, the pattern containing AgS lines and superstructure lines show phase transformation to an unknown high temperature phase and then to a known copper-silver mixed sulphide, jalpaite /Fig 3/ /α- III $Ag_{1,55}Cu_{0,45}S$/, which remains also after cooling the sample to room temperature again. That means that the lattice, giving the superlattice reflections must have been an Ag_2S lattice periodically modified by the built in copper.

CONCLUSIONS

The main conclusion of the present investigation is, that the good electrochemical data of an electrode can be correlated to certain crystallographic properties of its material. These properties are characteristic to the system in question, and are not the same for different systems /i.e. structure of good CdS-Ag_2S electrode materials is not the same as that of good PbS-Ag_2S electrode materials/ but if in the same system any of the procedures produced a good material, the same crystal structure has to be achieved with any other procedure for preapring well working electrodes.

REFERENCES

1. K.Cammann: Working with ion-selective electrodes, Spinger, Berlin-Heidelberg New York, 1979, 61-77.
2. W.E. van der Linden and R.Oostervink, Anal. Chim. Acta, 108, /1979/ 169.
3. M.Mascini and A.Liberti, Anal. Chim. Acta, 64, /1973/ 63.
4. I.Sekerda and J.F.Lechner, Anal. Letter, 9, /1976/ 1099.
5. S.Ikeda, N.Matshuda, G.Nakagawa and K.Ito, Denki Kagaku, 48, /1980/ 16.
6. J.D.Czaban and G.A.Rechnitz, Anal. Chem, 45, /1973/ 471.
7. H.Hirata and K.Higashiyama, Z.Anal. Chem., 257, /1971/ 104.
8. A.F.Zhukov, A.V.Vishnyakov, Ya. L. Kharif, Yu.I.Urusov, F.K.Volymets, E.J. Ryzhikov and A.V. Gordievski, Zh. Anal. Khim., 30, /1975/ 1761.
9. E.L.Colin and H.B.Patrick, Brit U.K. Pat. GB 2,061, 525, 13 May 1981.
10. Ju.G.Vlasov, Yu.E.Ermolenko and V.V.Kolodnikov, Zh.Anal. Khim. 36, /1981/ 889.
11. H.Hirata and K.Higashiyana, Talanta, 19, /1972/ 391.
12. A.V. Gordievskii, V.S.Shterman, Ya.A.Syrchenkov, N.J.Savvin, A.F.Zhukov and Yu.I.Urusov, Zh.Anal.Khim. 27, /1972/ 2170.
13. G.J.M. Heijne and W.E. van der Linden. Anal. Chim. Acta 93, /1977/ 99.
14. S.Ikeda, N.Matsuda, K.Nakagawa, K.Ito, Solid State Ionics 3-4, /1981/ 197.
15. F.Pál, K.Tóth and E. Pungor; in preparation.

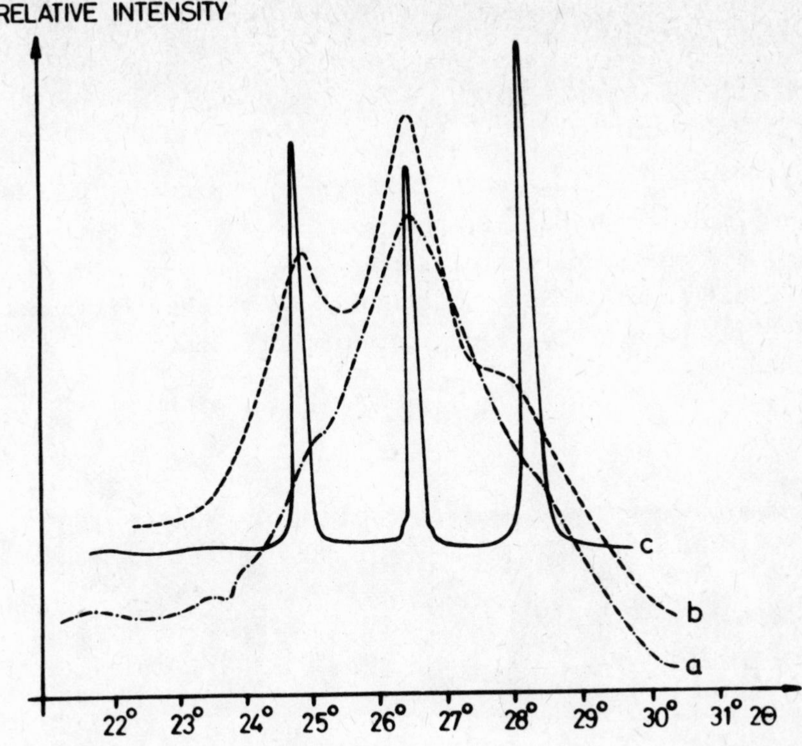

Fig.1. Diffracted intensities measured on X-ray patterns taken in a Guinier de Wolff-type camera with monochromatized CuK$_\alpha$ radiation from

a./ CdS as precipitated

b./ CdS precipitated and milled for 24 hours in a Fritsch-mill

c./ Powdered hexagonal CdS crystals

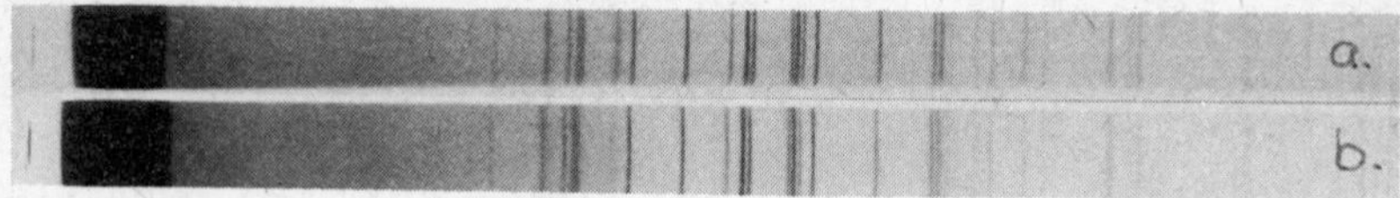

Fig.2. X-ray patterns taken from CdS-Ag_2S mixed sulphides /mole fraction 1:2/ prepared by different methods
a./ co- precipitated sample
b./ sample made by milling 12 hours the mixture of separately precipitated constituents

Fig.3. X-ray pattern made "in situ" during heating the co-precipitated CuS-Ag_2S sample /mole fraction 1:1/ up to 230 °C, and cooling it down to room temperature

QUESTIONS

Participants of the discussion: P.Becker, H.Ebel, M.Farkas-Jahnke, M.Neshkova, K.Tóth and Yu.G.Vlasov

Question:

I did not understand properly, what the difference between the two samples studied was. Could you comment on this ?

Answer:

In one case the mixed copper sulphide-silver sulphide precipitate was prepared by coprecipitation, whereas in the other the precipitates were prepared separately and after mixing, they were ground together for a fairly long period of time.

Question:

Have you found a big difference between the electrochemical behaviour of the two electrode types ?

Answer:

Yes, there is a difference. The mixed precipitate obtained by coprecipitation has a crystalline structure. But the separately precipitated CdS has a quasi-amorphous structure, whereas Ag_2S is crystalline. It was experimentally found that as an effect of grinding the mixture of the separately precipitated CdS and Ag_2S, after a sufficiently long time the whole system becomes crystalline. As long as this is not the case, the electrode characteristics are poorer than those of the electrodes made of the coprecipitated mixture.

Question:

Do you expect the formation of a new phase in the case of the coprecipitated silver sulphide-copper sulphide system ?

Answer:

Yes.

Question:

Usually there is a periodicity of crystallinity when you grind the mixture for a long period of time. Did you also observe this ?

Answer:

We did not observe this. After a 20 hour grinding, the electrode material exhibited appropriate electrode function, while after that it deteriorated. We did not study what happens after that.

Question:

What sort of mill was used ?

Answer:

A ball-type mill.

Question:

Was the sample dry or wet ?

Answer:

Dry.

4th Symposium on Ion-Selective Electrodes
Mátrafüred, 1984

PRECIPITATION TITRATION OF SO_4^{2-} USING A Pb^{2+} SELECTIVE ELECTRODE

D. FREMSTAD

Division of Applied Chemistry, The Foundation for Scientific and Industrial Research at Norwegian Institute of Technology, Trondheim, Norway

ABSTRACT

Through the use of a powerful laboratory computer, evaluation of potentiometric precipitation titrations based on model fitting of the complete titration curve has been implemented for routine analytical work. The algorithm has been tested for the determination of sulphate using a lead-selective indicator electrode. The convergence properties of the algorithm, which is iterative because of the non-linear nature of the model, has been tested and found to be satisfactory for automated unattended operation.

INTRODUCTION

During the last years, the analytical laboratory at SINTEF has been engaged in work concerning the development of ion-selective sensors, and also in the development of analytical methods and automated instruments for the use of such sensors in routine analysis. Much of this work has been concerned with the development of analytical tools for the aluminium industry, since this industry is an important one in our country.

In this connection, a computerized instrument system, marketed under the name "SINTALYZER", has been developed. It is based on the use of ion-selective electrodes and is capable of performing several different analyses, of which the determination of SO_4^{2-} by precipitation titration with Pb^{2+} is one.

The capability of unattended operation is economically important in analytical instruments of this kind. The analytical methods used must therefore be independent of operator intervention, and must, at the same time, provide sufficient information for the control and maintainance of the quality of the analytical results.

TITRATION ALGORITHM

The evaluation of titration curves by fitting a theoretical model of the titration process to the experimental points has long been known. A thorough treatment of this evaluation method applied to acid-base titrations has been given by Pehrsson et al. /1,2/

For the precipitation titration of SO_4^{2-} with Pb^{2+} using a Pb-sensitive indicator electrode, we start with the Nernst equation

$$E = E^{o} + \frac{RT}{nF} \ln[Pb^{2+}] \tag{1}$$

where the symbols used have their usual meaning. For computational purposes eq. (1) can be re-written in the form

$$E = E^{o} + \frac{RT}{nF} \ln(\frac{B}{V}) = E^{o\prime} + \frac{RT}{nF} \ln(B) \tag{2}$$

where

V is the solution volume,

B is the amount (in mole units) of Pb^{2+}

and

$$E^{o'} = E^{o} - \frac{RT}{nF} \ln(V) \qquad (3)$$

The algorithm is developed under the condition that the solution volume remains virtually constant throughout the titration. In practice, this is ensured by using solution volumes of about 50 ml and keeping the titration volume below 0.25 ml. Equation (2) is then valid throughout the titration range.

The solubility product K_s of $PbSO_4$

$$K_s = [Pb^{2+}] \times [SO_4^{2-}] \qquad (4)$$

can be re-written in the form

$$K_s = \frac{AB}{V^2} \qquad (5)$$

where

B and V are defined as above, and

A is the amount (in mole units) of SO_4^{2-} present in the volume V.

We then obtain an expression for the amount of free SO_4^{2-} present in the volume V:

$$A = \frac{V^2 K_s}{B}$$

This expression is valid for that part of the titration curve where precipitation of $PbSO_4$ has started to take place. In the algorithm, which is written for stepwise addition of the titrant, this restriction is taken into account by neglecting the first experimental point.

Utilizing the fact that the sulphate titration is symmetrical, we then obtain the following expression for the amount of Pb^{2+} added at the equivalence point:

$$T_E = T_B - B + \frac{V^2 K_s}{B} \tag{6}$$

where

T_B is the (cummulative) amount of Pb^{2+} added, and
B is the corresponding amount of free Pb^{2+}.

Assuming that the value of $E^{o'}$ is precisely known, the amount of free Pb^{2+} can be calculated from the experimental EMF-value, using eq. (2):

$$B = \exp\{\frac{nF}{RT}(E - E^{o'})\} \tag{7}$$

Assuming, more realistically, that our estimated value of $E^{o'}$ is in error by an amount ΔE^o, $E^{o'}$ estimated = $E^{o'} + \Delta E^o$, we would calculate an incorrect value of B. Denoting this incorrect value by B', we get

$$B' = \exp\{\frac{nF}{RT}(E - E^{o'})\}\exp\{-\frac{nF}{RT}\Delta E^o\} \tag{8}$$

Here, the first exponential is B, and, denoting the second (error) exponential by 1/f, we have

$$B = B'f \tag{9}$$

Combining eq. (6) and eq. (9), we obtain

$$T_E = T_B - B'f + \frac{V^2K_s}{B'f} \tag{10}$$

For calculational purposes we define the combined constant k_s by

$$k_s = \frac{V^2K_s}{f} \tag{11}$$

so that, finally

$$T_E = T_B - B'f + \frac{k_s}{B'} \tag{12}$$

In eq. (12), T_E is the equivalence point, and T_B and B' can be calculated from the autoburette and voltmeter readings, respectively. The constants T_E, f and k_s are determined by an iterative least squares procedure. To this end, eq. (12) can be re-written

$$0 = z = -T_E + T_B - B'f + \frac{k_s}{B'} \tag{13}$$

where the first equality is valid for a perfect model fit. Varying T_E, f and k_s by the amounts ΔT_E, Δf and Δk_s, we can then write

$$\Delta z = \frac{\partial z}{\partial T_E}\Delta T_E + \frac{\partial z}{\partial f}\Delta f + \frac{\partial z}{\partial k_s}\Delta k_s$$

or

$$\Delta z = -\Delta T_E - B'\Delta f + \frac{1}{B'}\Delta k_s \tag{14}$$

Eq. (14) is linear in all variables, and z can be minimized using a linear least squares procedure.

Before the analyses are started, the value of $E^{o'}$ must be estimated from calibration runs, using eq. (2). The calibration need not be accurate. The iteration procedure set out below will converge if $E^{o'}$ is estimated with an accuracy of about +/- 100 mV.

An initial approximation of k_s must also be provided. Again, only a rough estimate is needed. The iteration procedure will converge even if this estimate is in error by 3-4 orders of magnitude.

For each sample, the iteration procedure then is as follows:

1. Set f=1 and estimate T_E by locating the two experimental points between which the change in measured EMF is the largest.

2. Compute the coefficients in eq. (14) for each experimental point and obtain the corresponding estimates of Δz from eq. (13).

3. Calculate ΔT_E, Δf and Δk_s by means of the least squares procedure.

4. Update T_E, f and k_s.

5. Resume the iteration at step 2.

The iteration is assumed to have converged when two successive values of T_E in step 4 above differ by less than a predetermined amount. In our program this value is set equal to 0.01% of the current value of T_E.

INSTRUMENT AND IMPLEMENTATION

The above algorithm is easily implemented on a modern laboratory computer. In our implementation, the computer system DECLAB MINC, manufactured by Digital Equipment Corporation, was used. The computer controls three independent analytical stations, each station consisting of a high-impedance digital voltmeter to which the electrode pair is connected, an automatic burette, and an optional sample changer for unattended operation. The complete hardware system is shown in Fig. 1.

The MINC computer had a LSI11/23 processor with floating point hardware and memory management, 256 kb main memory, and two RL01 disks with 5Mb capacity each. The operating console was a VT125 low resolution graphic terminal.

The precipitation titration routine forms part of a larger program system which is written partly in FORTRAN IV and partly in the assembly language MACRO-11. The program executes under Digital Equipment Corporation's low-end real-time operating system RT-11.

The titration algorithm uses fixed titration increments entered by the analyst at the beginning of a run. The number of titration points to be taken per titration curve is entered at the same time.

The real-time part of the program contains a DMV-handling routine that can be set to accept or reject EMF readings using different stability criteria. For the precipitation titration, the best results were obtained when an equilibration time of 4-6 seconds was allowed between titrant addition and EMF recording.

At the end of the titration procedure, the analytical result is printed out together with the final values of the parameters k_s and f. The approximate constancy of k_s and f constitutes one means of checking the quality of the analytical results after a series of samples have been analysed without operator intervention. More importantly, the experimental points and the model analytical curve with the equivalence point marked are displayed on the graphic screen at the end of each titration. For unattended operation, this display can be transferred to a dot-addressable matrix printer, thus leaving a record of each analysis analoguous to Fig. 2.

The titration curves shown in this paper are dot-by-dot reproductions of curves displayed on the graphic terminal. The figures were produced on a DECWRITER LA100 matrix printer.

EXPERIMENTAL

The analytical curves shown in Figs. 2-7 were obtained using an ORION Pb^{2+} -selective electrode, a RADIOMETER double junction reference electrode, a RADIOMETER PHM84 digital voltmeter and a RADIOMETER ABU80 automatic burette. The titrator solution was 0.1m $Pb(NO_3)_2$ in all experiments.

The titrant solution and the Na_2SO_4 standard solutions were prepared from Merck chemicals of analytical grade.

The sulphate titration, as applied to flue gas absorption solutions from aluminum works, has been discussed elsewhere. /3/

RESULTS AND DISCUSSION

Fig. 2 and Fig. 3 illustrate the effect of varying the number of titration steps. Both curves were obtained titrating 6.5 μmoles of Na_2SO_4 standard in solution volumes of approximately 50 ml, containing 80 vol% ethanol and 20 vol% H_2O. The number of titration steps was 15 in Fig. 2 and 6 in Fig. 3. The results obtained were 6.53 and 6.42 μmoles, respectively, indicating that the algorithm does not require a close spacing of experimental points. A series of tests using 5, 6, 8, 10, 15 and 20 titration increments gave an average result of 6.53 μmoles with a standard deviation of 0.12 μmoles.

The effect of increasing the solubility product is shown in Figs. 4-6. Again, 6.5 μmoles of Na_2SO_4 was titrated in solution volumes of 50 ml. The ethanol contents were 80%, 60% and 40% respectively. The results obtained were 6.65, 6.75 and 7.25 μmoles Na_2SO_4.

Finally, Fig. 7 shows the determination of sulphate in a primary gas absorption solution from an aluminium plant. The solution volume was again 50 ml and the sulphate content was 16.5 μmoles.

The examples above were chosen in order to illustrate the functioning of the algorithm under less than ideal conditions. The results indicate that the method will give acceptable results for routine analytical work even when applied to weak solutions with titration curves having a gradual slope around the equivalence point.

The utilization of the algorithm in automated titration equipment gives rise to some special considerations. In most automatic titrators, the burette capillary will remain immersed in the solution throughout the titration. If the time allowed for electrode equilibration between titrant additions is too long, even a small diffusion of titrator ions from the capillary into the solution can have a marked effect on the analytical curve, specially in the region of steepest slope. It is important, therefore, that the electrodes used should have short response times. Together with the number of titration steps, the time allowed for electrode equilibration also determines the sample throughput of the instrument. Optimum values should therefore be determined for each analysis to which the algorithm is applied.

REFERENCES

1 L.Pehrsson, F.Ingman and A.Johansson, Talanta, 23/1976/769

2 L.Pehrsson, F.Ingman and S.Johansson, Talanta, 23/1976/781

3 K.Nagy, E.Kleven, T.A.Fjeldly and D.Fremstad, Z.Anal.Chem. 262/1979/362

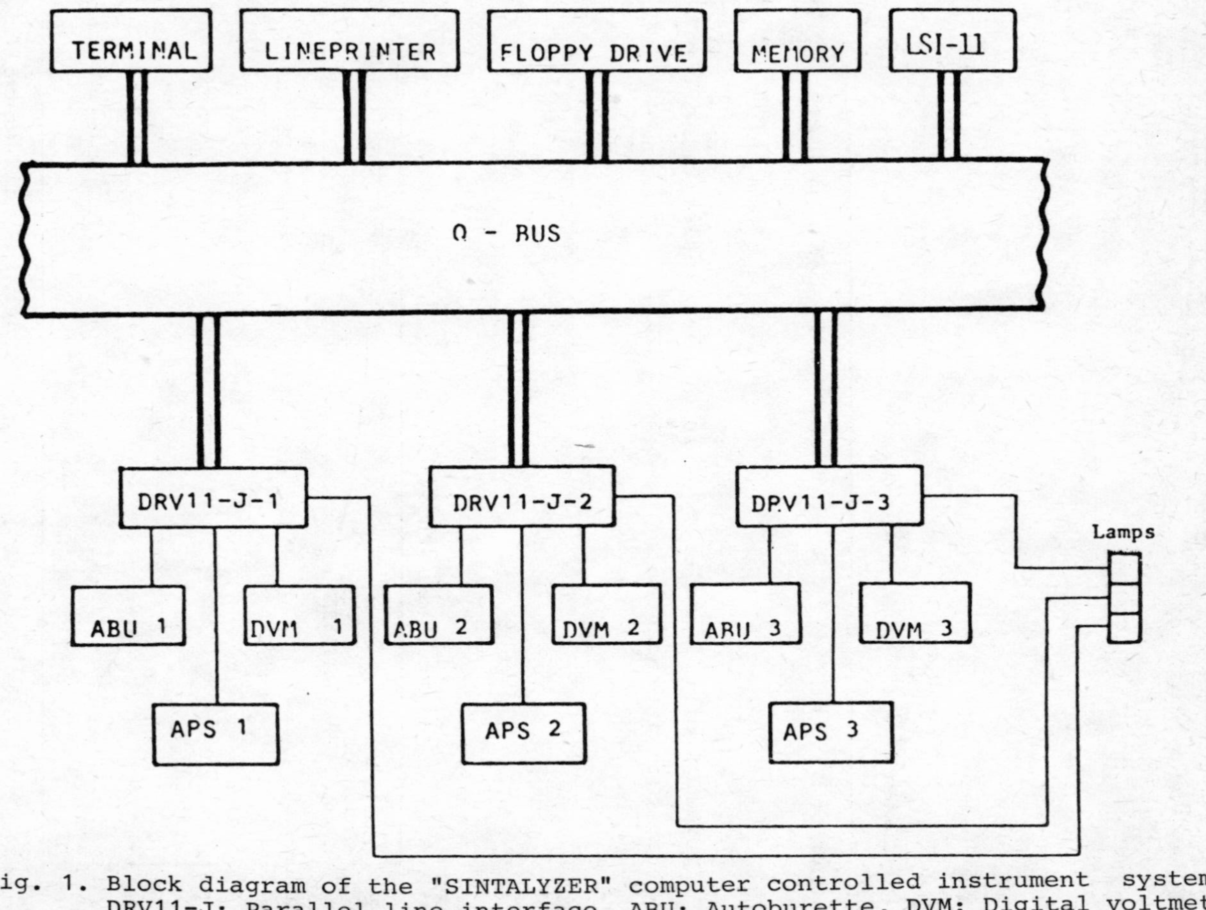

Fig. 1. Block diagram of the "SINTALYZER" computer controlled instrument system. DRV11-J: Parallel line interface, ABU: Autoburette, DVM: Digital voltmeter, APS: Sample changer

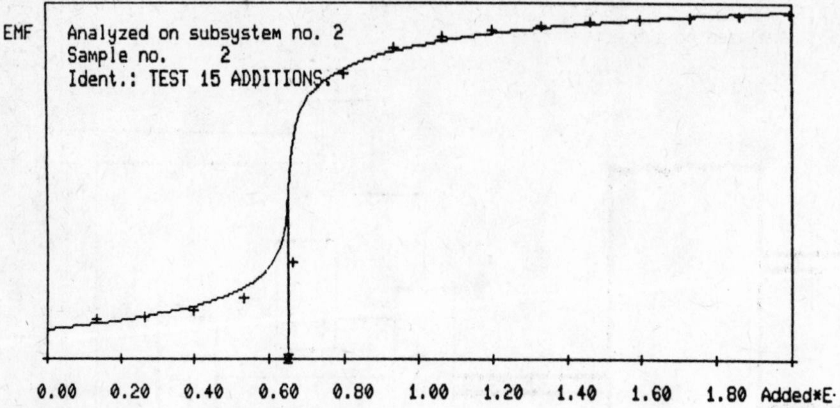

Fig. 2

Titration of 6.5 µmoles SO_4^{2-} in 50 ml solution using 15 titration increments.

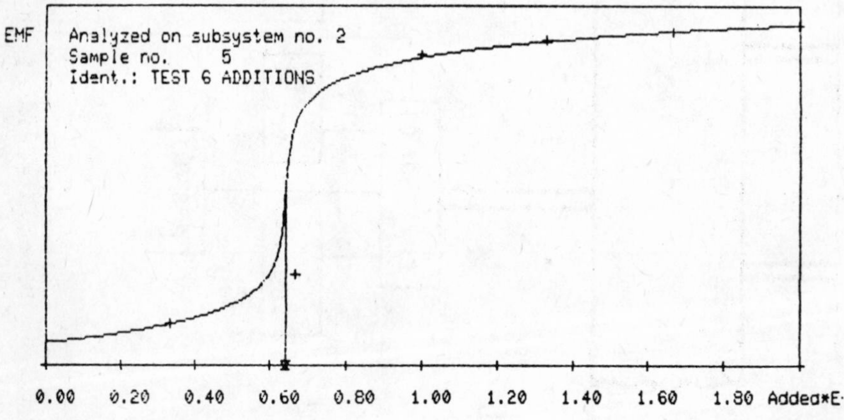

Fig. 3

Titration of 6.5 µmoles SO_4^{2-} in 50 ml solution using 6 titration increments.

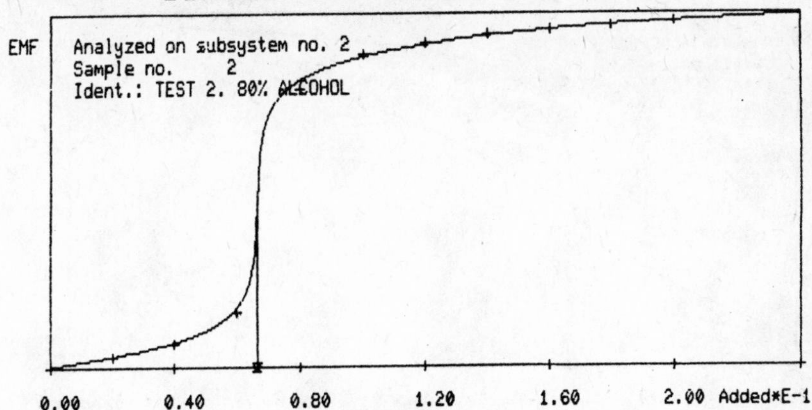

Fig. 4

Titration of 6.5 µmoles SO_4^{2-} in 50 ml solution containing 80 vol% ethanol.

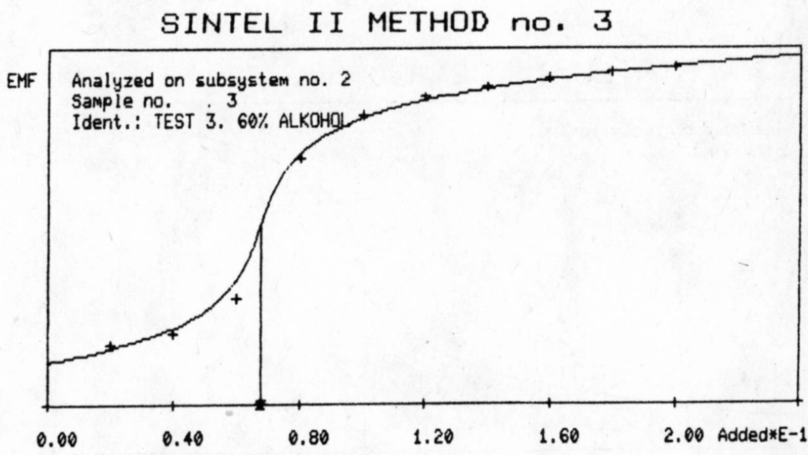

Fig. 5

Titration of 6.5 µmoles SO_4^{2-} in 50 ml solution containing 60 vol% ethanol.

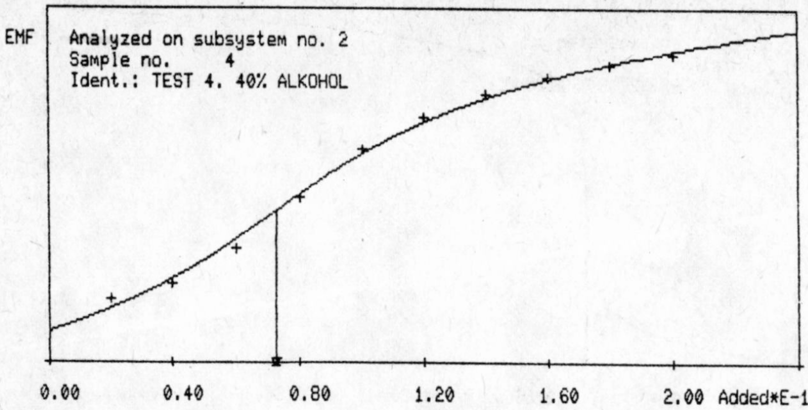

Fig. 6

Titration of 6.5 µmoles SO_4^{2-} in 50 ml solution containing 40 vol% ethanol.

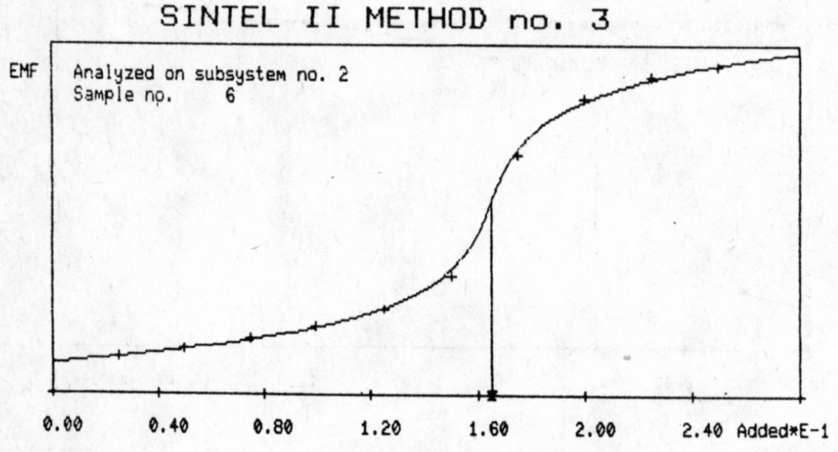

Fig. 7

Titration of SO_4^{2-} in primary gas absorbate from an aluminium plant.

QUESTIONS

Participants of the discussion: H.Müller, K.Nagy and E.Pungor

Question:

What was the membrane material of the novel solid-state lead-selective electrode ?

Answer:

We have not used any new membrane material but we used the amplifying electronics. The novelty of the electrode is that we have impedance transforming electronics in the electrode.

Question:

How did you accelerate the formation of lead sulphate ?

Answer:

We added lead sulphate to the solution, this way the precipitation is faster.

4th Symposium on Ion-Selective Electrodes
Mátrafüred, 1984

EFFECT OF PRIMARY AND INTERFERING IONS ON SURFACE PROCESSES OF PRECIPITATE-BASED ION-SELECTIVE ELECTRODES

E.G. HARSÁNYI, K. TÓTH and E. PUNGOR

Institute for General and Analytical Chemistry,
Technical University of Budapest, Gellért tér 4,
Hungary

INTRODUCTION

The role of the adsorption of primary ions on the surface or solid electrodes in affecting the actual electrode potential was mentioned in the literature already about fifty years ago /1/. For electrodes of the second kind /Pb-PbS, Ag-AgCl/ Nernstian response was found by Boltunov /2/, but only for the primary ions. He explained this phenomenon by the adsorption of primary ions on the surface of the electrodes.

In the literature of ion-selective electrodes the adsorption and desorption of primary ions was mentioned for the first time in the work of Ross /3/ in connection with the silver sulphide based electrode. Buffle and coworkers /4/ studied the fluoride adsorption at the surface of a fluoride electrode. Hulanicki and coworkers /5/ determined the amount of iodide ions adsorbed on the surface of silver iodide electrode. Blaedel and Dinwiddie /6/ mentioned the role of copper ion adsorption in determining the lower detection limit of the copper/II/ sulphide electrode.

We have investigated in detail the adsorption-desorption processes in the course of studies on the AgI /7/, CuS /8/, AgCl /9/ and Ag_2S /10/ ion-selective electrodes. In these works the adsorption and desorption of the primary ions at the electrode surface were identified by the use of a microcell arrangement. Our results showed that with electrodes previously soaked in deionized water, during calibration in the microcell, the concentration of the test solution decreased, that is, the actual concentrations were lower than the nominal ones. In the case of precipitate-based electrode membranes this phenomenon seemed to be general. The concentration decrease in the solutions correlated well with a Freundlich-type adsorption isotherm. The maximum amount of primary ion adsorbed could also be calculated by the aid of the Langmuir adsorption isotherm equation, and generally it was found at a primary ion concentration of about 10^{-4} mol/l. Below this concentration the deviations from the ideal Nernstian response might be connected with adsorption--desorption as well as dissolution processes.

In the present paper the adsorption-desorption processes observed in the presence of interfering ions will be discussed.

EXPERIMENTAL

The experimental microcell set-up used in earlier adsorption measurements and also in this work consisted of an electrode membrane of relatively large surface area and a solution of small volume /200-300 µl/ /Fig.1/. In such circumstances the whole solution volume could be treated as a phase boundary after reaching the final equilibrium state. Ten minutes proved to be satisfactory in most cases to reach equilibrium after immersing the electrode into the solution. The concentration change of the small volume of solution was measured by atomic absorption spectrometry using the impulse nebulization technique with 50 µl portions. By this method the cation concentration changes could be followed.

The potentiometric calibration curves in the microcell were plotted from data obtained in the tenth minute. The electrode potential change was also recorded as function of the time during the ten minute measuring time.

As a model material the copper/II/sulphide homogeneous electrode was used. A natural covellite crystal of 0,3 cm^2 surface area was used as electrode membrane and 200 µl solution volumes were used beside conventional solution volumes.

As interfering ions silver, cadmium, and lead were studied. The ionic strength of the solutions was adjusted with 0,1 mol/l potassium nitrate.

The interference of silver in the determination of copper using a CuS ISE is obvious, but for comparing the interference processes the study of this kind of cases proved to be useful.

In the case of the silver interference, since the Ag_2S has a lower solubility product than the CuS /$K_{SO\ Ag_2S}=6,62x10^{-50}$, $K_{SO\ CuS}= 1,28x10^{-36}$/, a precipitate exchange is to be expected.

Calibration curves for Cu and Ag ions with the CuS electrode in 50 ml solution volume are shown in Figure 2 /curves A and B/. At low silver ion concentrations /10^{-6}-10^{-5} mol/l/ the silver calibration curve runs fairly close to the copper curve. Above 10^{-3} mol/l silver ion concentration the calibration curve has a Nernstian slope for silver ions. At 10^{-3} mol/l silver ion concentration the electrode is supposedly converts to Ag_2S and functions as a silver electrode. For comparison, a silver ion calibration curve with a Ag_2S electrode was also taken and plotted in the Figure /curve C/. Above 10^{-3} mol/l silver ion concentration the two curves are very close.

To determine how the interfering effect is changing with the concentration of the interfering ion, measurements were carried out in the microcell in 200 µl solution volumes containing copper ions of fixed concentration /10^{-5} mol/l/and silver ions of increasing concentrations /10^{-6}-10^{-1} mol/l/. The electrode potential vs. pc_{Ag^+} curve is shown in Figure 3. The actual copper and silver concentrations measured by AAS are also given in Table 1. Below 10^{-4} mol/l silver ion concentration the electrode potential values correlate with the actual copper ion concentrations in the test solution. Above this concentration a rapid potential increase can be observed parallel with a drastic increase in the copper concentration.

On the basis of the results a precipitate exchange reaction on the electrode surface is assumed at and above 10^{-3} mol/l silver ion concentrations. At and below 10^{-4} mol/l silver ion concentrations only the double layer might be changed under the given experimental conditions. The silver ions adsorb simultaneously with the copper ions and gradually replace the adsorbed copper ions.

We can calculate a coefficient value from the experimental data which is formally a "selectivity" coefficient, but it can be called rather a "non-selectivity" coefficient. The calculated values at different increasing silver ion concentrations were between 3 and 10^{13}, using separate solution technique. The latter correspond to the theoretical value obtained from the ratio of the solubility products.

The nature of the interference of cadmium and lead ions in copper determination with CuS electrode was also studied.

In Figure 4 the calibration curves for Cu, Cd, and Pb taken with the CuS based electrode are given. The curves relate to 50 ml sample solution volumes. Both Cd and Pb ions show a potential increase as the concentration increases indicating an interference.

In an other experiment at a fixed copper ion concentration /10^{-5} ml/l/ the cadmium and lead concentrations were changed using the microcell arrangement.

The electrode potential change as the function of $pc_{Cd}{}^{2+}$ is given in Figure 5. The Figure shows also the copper calibration curve. The actual copper ion concentrations measured by AAS in the microcell, given in the Table 2 prove that the electrode responds up to 10^{-2} mol/l Cd concentration to the actual copper ion concentration being in the solution, which is different from the nominal value. The difference between the nominal and actual copper ion concentration may due to the adsorption of the copper ions from the solution to the surface of the electrode. At higher cadmium concentrations /$c_{Cd} \geq 10^{-3}$/ the desorption of copper ions increases the actual copper concentration in the test solution.

In the case of lead interference studied in the microcell /Figure 6, Table 3/ similar results could be obtained. The electrode responds to copper actually present at the interface up to 10^{-3} mol/l lead concentration. Above this concentration the adsorption of lead on the surface of the electrode may be assumed.

The release of copper ions from the membrane material by precipitate exchange cannot be supposed to be of a significant extent because the actual copper concentration does not increase even at 1 mol/l interfering ion /Cd, Pb/ concentration in contrast to the silver interference. It must also be mentioned that the CuS electrode functions well as a copper electrode after measuring with it in concentrated lead or cadmium solutions.

In view of these results one can understand that the measured selectivity values are usually higher than the theoretical ones. Selectivity values obtained for interference of Cd and Pb on copper are given in Table 4.

For the nature of Cd and Pb interference on the copper response the model worked out by Lindner, Tóth and Pungor /11/ proved to be valid. At constant primary ion concentration a rapid increase in the interfering ion concentration results overshoot-type response curves in the adhering solution layer. Such transient signals at the additions of lead to 10^{-5} mol/l copper solution are given in Figure 7 measuring in 50 ml solution volume. The explanation for this kind of curves is that the primary ion concentration is increasing at the interface after addition of the interferant due to desorption and then decreases due to diffusion to the direction of the bulk.

REFERENCES

1. N.Isgarischew, Z.für Elektrochem. 32, /1926/ 281
2. Yu.A.Boltunov, Zsurn.Obs.Himii 23, /1937/ 2831
3. J.W.Ross in R.A.Durst /Ed./ Ion-Selective Electrodes, NBS Special Publication No. 314. Ch. 12. 1969
4. J.Buffle, N.Parthasarathy, W.Haerdi, Anal.Chim.Acta 68, /1974/ 253
5. A.Hulanicki, A.Lewenstam, M.Maj-Zurawska, Anal.Chim.Acta 107, /1979/ 121
6. W.J.Blaedel, D.E.Dinwiddie, Anal.Chem. 46, /1974/ 873
7. E.G.Harsányi, K.Tóth, L.Pólos, E.Pungor, Anal.Chem. 54, /1982/ 1094
8. E.G.Harsányi, K.Tóth, E.Pungor, Anal.Chim.Acta 152, /1983/ 163
9. E.G.Harsányi, K.Tóth, E.Pungor, Anal.Chim.Acta /in press/
10. E.G.Harsányi, K.Tóth, E.Pungor, Y.Umezawa, S.Fujiwara, Talanta 31, /1984/ 579
11. E.Lindner, K.Tóth, E.Pungor, Anal.Chem. 54, /1982/ 202

Table 1.

Solution composition	Actual Cu^{2+} conc.	Actual Ag^{+} conc.
10^{-5} mol/l Cu^{2+}	$6,3.10^{-6}$ M	-
10^{-5} M Cu^{2+} + 10^{-6} M Ag^{+}	$6,3.10^{-6}$ M	1.10^{-6} M
10^{-5} M Cu^{2+} + 10^{-5} M Ag^{+}	$6,3.10^{-6}$ M	$4,4.10^{-6}$ M
10^{-5} M Cu^{2+} + 10^{-4} M Ag^{+}	$1,06.10^{-5}$ M	$6,9.10^{-5}$ M
10^{-5} M Cu^{2+} + 10^{-3} M Ag^{+}	$6,5.10^{-5}$ M	-
10^{-5} M Cu^{2+} + 10^{-2} M Ag^{+}	1.10^{-3} M	-
10^{-5} M Cu^{2+} + 10^{-1} M Ag^{+}	$6,3.10^{-3}$ M	-

Table 2

Solution composition	Actual Cu^{2+} conc.
10^{-5} M Cu^{2+}	$2,3.10^{-6}$ M
10^{-5} M Cu^{2+} + 10^{-5} M Cd^{2+}	$5,6.10^{-6}$ M
10^{-5} M Cu^{2+} + 10^{-4} M Cd^{2+}	$9,4.10^{-6}$ M
10^{-5} M Cu^{2+} + 10^{-3} M Cd^{2+}	$1,6.10^{-5}$ M
10^{-5} M Cu^{2+} + 10^{-2} M Cd^{2+}	$2,6.10^{-5}$ M
10^{-5} M Cu^{2+} + 10^{-1} M Cd^{2+}	$3,8.10^{-5}$ M
10^{-5} M Cu^{2+} + 1 M Cd^{2+}	$2,7.10^{-5}$ M
10^{-5} M Cu^{2+} + 10^{-2} M Pb^{2+}	$1,5.10^{-5}$ M
10^{-5} M Cu^{2+} + 10^{-1} M Pb^{2+}	$1,4.10^{-5}$ M
10^{-5} M Cu^{2+} + 1 M Pb^{2+}	$1,7.10^{-5}$ M

Table 3.

Solution composition	Actual Cu^{2+} conc.
10^{-5} M Cu^{2+}	$3,1.10^{-6}$ M
10^{-5} M Cu^{2+} + 10^{-5} M Pb^{2+}	$1,3.10^{-5}$ M
10^{-5} M Cu^{2+} + 10^{-4} M Pb^{2+}	$1,35.10^{-5}$ M
10^{-5} M Cu^{2+} + 10^{-3} M Pb^{2+}	$1,5.10^{-5}$ M

Table 4.

	$K_{Cu^{2+},Cd^{2+}}$	$K_{Cu^{2+},Pb^{2+}}$
Theoretical value	$9,1x10^{-8}$	$1,4x10^{-8}$
Experimental value /mixed solution technique/	$5x10^{-3}$	$3x10^{-3}$
Experimental value /separate solution technique/ at 0,1 mol/l concentrations	$1,4x10^{-3}$	$4,3.10^{-3}$

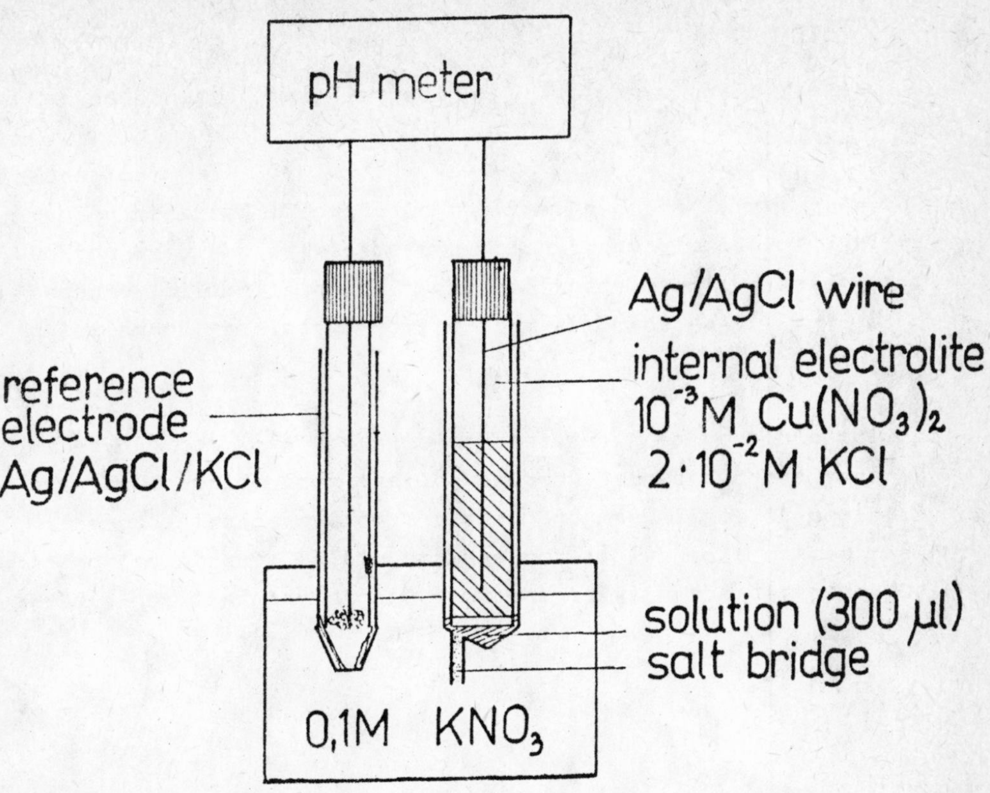

Fig. 1 Microcell arrangement for the electrode potential measurement

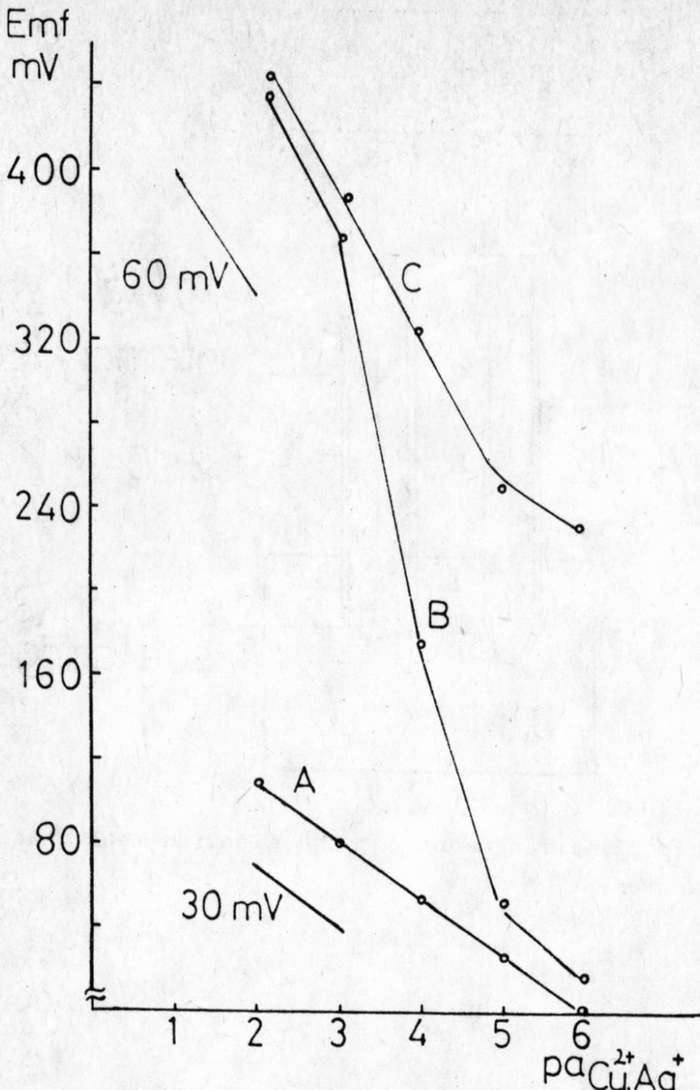

Fig. 2 Calibration curves taken in 50 ml solution volumes
A: Cu^{2+} ion calibration using CuS electrode;
B: Ag^{+} ion calibration using CuS electrode;
C: Ag^{+} ion calibration using Ag_2S electrode

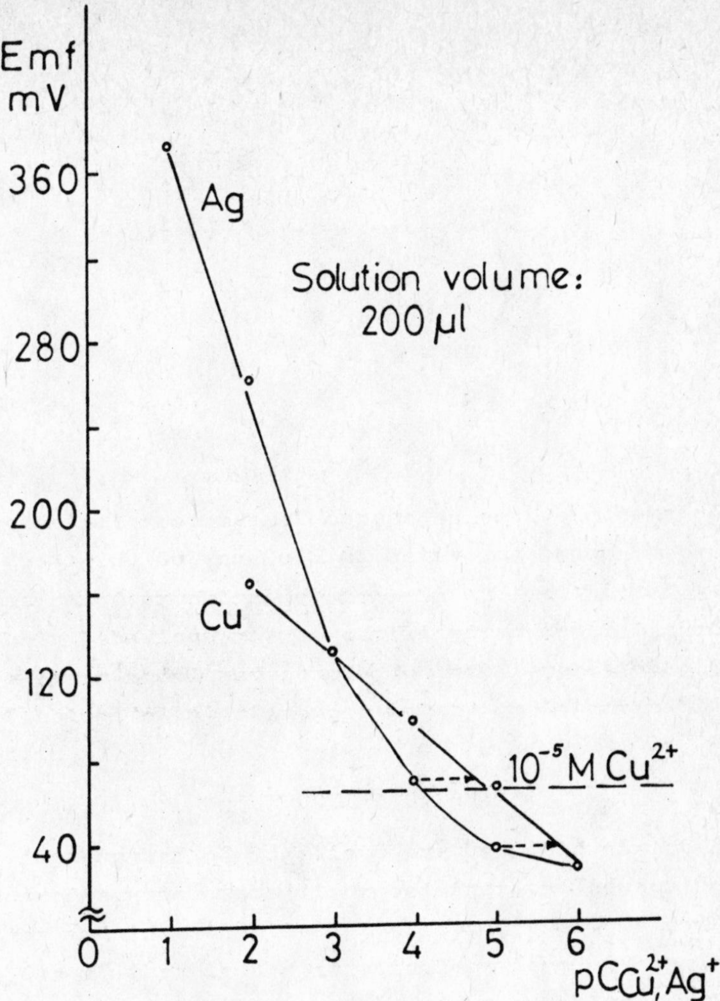

Fig. 3 Electrode potential change of 10^{-5} mol/l copper solution by adding increasing silver ion concentrations /10^{-6}-10^{-1} mol/l/. For comparison a copper ion calibration curve is drawn also. Solution volume 200 µl

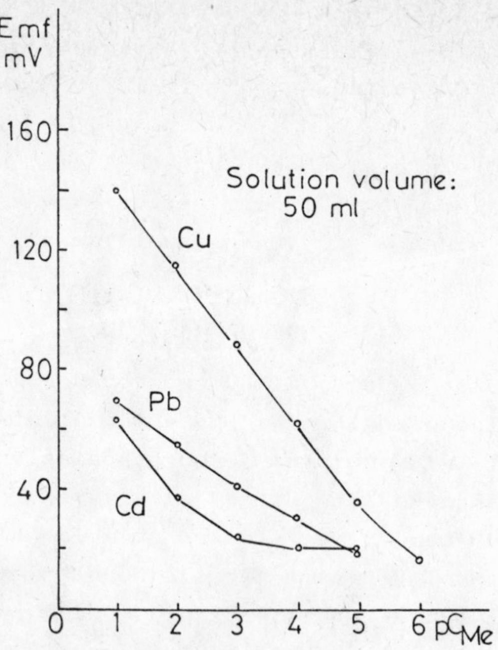

Fig. 4 Calibration curves for Cu^{2+}, Pb^{2+} and Cd^{2+} ions taken with the CuS electrode. Solution volume 50 ml

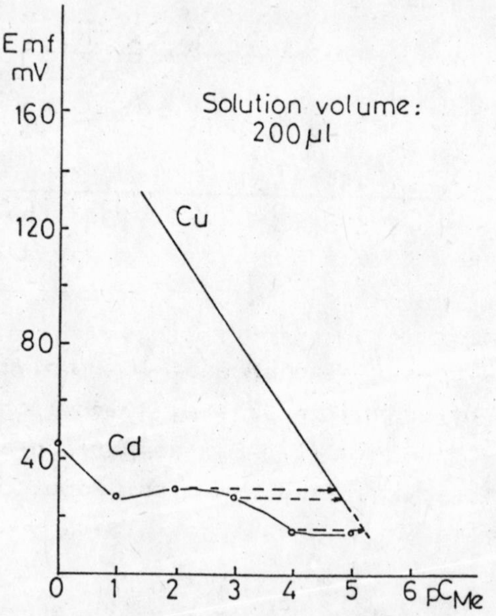

Fig. 5 Electrode potential change of 10^{-5} mol/l copper solution by adding increasing cadmium ion concentrations /10^{-6}-1 mol/l/. The Cu^{2+} calibration curve is also given. Solution volume: 200 µl

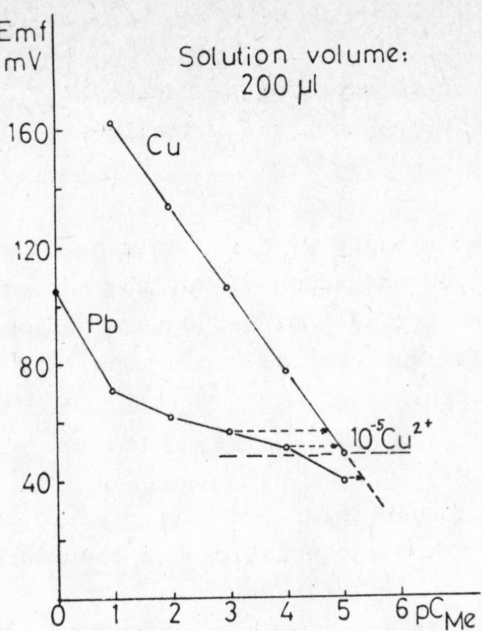

Fig. 6 Electrode potential change of 10^{-5} mol/l copper solution by adding increasing lead ion concentrations /10^{-6}-1 mol/l/. The Cu^{2+} calibration curve is also given. Solution volume: 200 µl

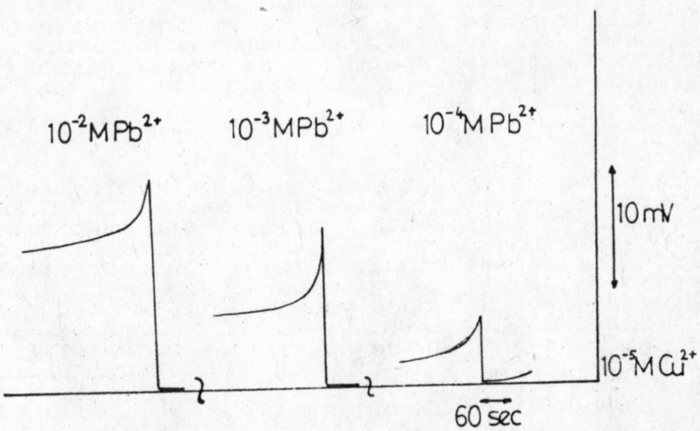

Fig. 7 Overshoot-type curves obtained by adding increasing concentration of Pb^{2+} ions to a solution of 10^{-5} mol/l Cu^{2+} ion concentration. Solution volume: 50 ml

QUESTIONS AND COMMENTS

Participants of the discussion: P.Becker, R.P.Buck, M.Ebel, E.Gráf-Harsányi, M.Neshkova and R.Stella

Question:

You have shown a table with two columns, the first containing the nominal concentration of Cu, the second the values which we found as a result of AAS determination. How did you prepare the so called "nominal" solutions, and how did you check the concentration ?

Answer:

The standard copper solutions were prepared freshly by serial dilution. The concentration was checked by AAS.

Question:

What was the pH of the solution ? Have you made blank experiments, too ?

Answer:

We used 0.1 mol/l KNO_3 solution to adjust the ionic strength. The pH was between 4 and 5. No buffer was used.

Blank measurements were also carried out with the system, without the electrode, and this was corrected for.

Question:

You have two different explanations, two different mechanisms concerning the interference by silver and cadmium. And you had the same experimental proof, I mean, the concentration was always increasing. But once you said it to be due to the exchange process and in the other case to desorption. Could you please comment on that ?

Answer:

The same type of measurement was carried out when investigating the silver and cadmium interference. The silver interference seemed to be a clear process, as silver sulphide has a much lower solubility product than copper sulphide, so we expected a precipitate exchange. And in this case we expected that the exchanged copper will desorb, and due to this the copper concentration will increase remarkably.

But in the other case a slight increase in copper concentration was measured, but this may be due to the copper adsorbed previously to the surface being desorbed. It has been shown earlier, that if the copper sulphide solid electrode is in contact with a solution containing the primary ion, adsorption is always taking place, a copper ion excess can always be measured on the electrode surface.

It has not been cleared up yet whether there occurs a precipitate exchange as well, because it is of a small extent, much smaller than in the case of silver.

If we calculate a non-selectivity constant for silver with the separate solution technique, it gives a value which can be expected from the solubility products, but in the other case not, which means that the processes involved are different at the electrode surface.

Comment:

On the problem of checking the nominal concentrations when you go down with the dilution: I am not sure that a cross--control with another method may be so powerful to prove the concentration, because you need another calibration and if you want to have good results, you have to go down with the concentration, so the problem reproduces itself. I think serial dilution may be more precise than another technique used to check the concentration.

Answer:

When I measured the actual concentration after contact with the electrode, and plotted the potential values against the actual concentration. I could get a Nernstian curve, even in this low concentration range.

Comment:

I only wanted to say that sometimes it is not worthwile to

do this checking. Often, you can take the concentration that results from dilution.

Comment:

We have started an XPS and SEM study of the problems of the copper sulphide electrode. We have only qualitative results, which are in agreement with what dr.Gráf has found. We are going to study the problem further to get quantitative results.

Comment:

I think radio-tracer technique is the best method to see all the changes that are going on at the electrode surface, because you simply see where the radioactivity goes and you have to deal with it because you are not allowed to throw anything away that is radioactive.

Answer:

The work I have reported on is only an approach towards understanding better the processes at the electrode surface. I think there are neglections in this approach, and it is very difficult in the low concentration range to decide what kind of processes are taking place. We have also made radio-tracer measurements with the copper electrode. But in the low concentration range we had problems because isotope exchange may take place between the solution and the electrode, which may influence the process under investigation.

Question:

Would you just summarize which data agree with theory and which do not ?

Answer:

In the case of cadmium and lead interference, the selectivity coefficient does not correlate with the theoretical one, whereas with the silver interference, it correlates.

4th Symposium on Ion-Selective Electrodes
Mátrafüred, 1984

CALIBRATION CHARACTERISTICS AND MASS LOSS OF REACTIVE ION-SELECTIVE ELECTRODES AS FUNCTION OF OPERATION TIME

M. GRATZL, L. GRYZELKÓ, J. KŐMIVES, K. TÓTH and E. PUNGOR

Institute for General and Analytical Chemistry,
Technical University, Budapest, Hungary

ABSTRACT

The membranes of silver halide based cyanide, thiosulphate and ammonia sensing electrodes react with the ions measured, hence they are chemically corroded during their operation. Membranes of different compositions /pure silver halide as well as silver halide/silver sulphide mixtures of molar ratios 10:1, 1:1 and 1:10, and occasionally pure silver sulphide/ have been investigated. The calibration characteristics, life times and surface morphologies of new /unused/and corroded /used/ electrode membranes are presented. On the basis of these data the working mechanisms of the different reactive ion-selective electrodes can be better explained. Accordingly, not only the halide content but generally also the silver sulphide particles are gradually leached out of the mixed membranes. Hence, the pure silver halide and common mixed membrane - based reactive ion-selective electrodes are nearly equivalent in performance. A decrease in the proportion of the silver halide in the membranes significantly below a molar ratio of 1:1, however, leads to worse characteristics.

INTRODUCTION

Certain precipitate based ion-selective electrodes sense, beside their primary ions also components which react with the membrane material in a fast and reversible chemical reaction, generating highly soluble products. For silver halide based membranes such reactions may take place with some complexing agents:

$$\underline{AgX} + 2L^{m} \rightleftharpoons AgL_2^{m+1} + X^- \qquad /1/$$

where X means I, Br or Cl, and L may be CN^-, $S_2O_3^{2-}$ or NH_3. As stationary diffusion profiles of the soluble species may establish within a second in stirred solutions, the activity of the primary ion X^- quickly reaches a stable level at the membrane surface, if the sample contains ligand L^m /see Fig.1./. This level is related to the activity in the sample as follows:

$$a'_x = \frac{D_L}{2D_x} a_L \qquad /2/$$

where a′ and a mean activites at the electrode surface and in the solution bulk, respectively, and D is diffusion coefficient. Thus, if ion X^- is not present in the sample then the ligand activity can be measured with the appropriate silver halide-based membrane electrode:

$$E = E_x^o - S \lg a'_x = E_L^o - S \lg a_L \qquad /3/$$

where $E_L^o = E_x^o - S \lg \frac{D_L}{2D_x}$, and the other symbols carry their usual meanings.

As the basis of the operation of these electrodes is the stationary chemical reaction between the membrane material and the component to be measured, they may be called "reactive ion-selective electrodes". The more thoroughly investigated electrode belonging to this group of sensors is the AgI based cyanide electrode /1-5/ because of its outstanding practical importance, but many properties of the AgI, AgBr and AgCl based cyanide and thiosulphate electrodes and the AgCl based ammonia electrode have also been reported /3,6,7/.

Due to reaction /1/ the sensing surface of the reactive electrodes is steadily renewed during their operation. This causes, however, also a certain loss of membrane material during use, which leads to limited life time. In order to diminish this effect, mixed silver halide/silver sulphide membranes of molar ratios 1:1 /8/ or 7:3 /9/ have been prepared. The supposed formation of a porous Ag_2S matrix covering the sensing surface /10-13/ was assumed to reduce the rate of corrosion in the case of mixed membranes.

In this work this hypothesis was re-examined by investigating different pure and mixed membrane reactive electrodes, and on the basis of the results new practical conclusions were drawn.

EXPERIMENTAL

The membrane materials were prepared by the appropriate precipitation titrations /14/; the mixed materials were obtained by co-precipitation. The membranes with a diameter of 7 mm and thickness of about 2 mm were pressed at 4×10^{-6} Pa /~40 atm/ using washed and dried precipitates. Bayer Silopren Paste E Type 3035 adhesive was used to fix the membranes into the electrode body. The corrosion of the membranes were carried out in closed vessels and stirred solutions, which were frequently renewed /at least after 15 min/.

In the indicator electrodes 10^{-3}mol/l $AgNO_3$/Ag junction, and a Radelkis double junction reference electrode were used. For potential measurements a Radiometer pHM-84 pH-meter, while for the surface morphology studies a Jeol JSM 50/A scanning electron microscope were employed.

RESULTS AND DISCUSSION

Table 1. represents the calibration characteristics of several cyanide electrodes involving homogeneous membranes, of which the Ag_2S based sensor does not belong to the reactive electrode group as its working mechanism is different /15,16/. The electrodes referred to by Table 1. had been used to measure cyanide for a long period of time before the data reported were determined.

Table 2. represents the calibration characteristics of different heterogeneous /mixed/ membrane cyanide electrodes. Data referring to new /unused/ as well as to corroded /= used during given periods/ electrodes are presented.

Table 3. summarizes the losses of membrane material of different pure and mixed membrane based reactive electrodes, due to corrosion by the sample solution.

The data displayed in Table 1. and 2. prove that the corrosion by cyanide does not affect the quality of the electrodes, except for the mixed membrane electrodes containing a large amount of Ag_2S /AgI/Ag_2S molar ratio of 1:10/. In this latter case the electrode function gets quickly very poor during operation.

In the course of the investigations no significant memory effects could be observed, except again the case of 1:10 mixed membranes.

According to Table 3. the life time of mixed membrane electrodes seems to be not longer but rather slightly shorter than that of the respective pure membrane electrodes /if membranes of equal masses are taken into account/. This trend is supported by the mass losses of pure as well as 10:1 and 1:1 mixed membranes. The 1:10 membranes represent again an exception: their loss of mass is much smaller, and decreases further during use.

These results contradict the hypothesis that a steadily thickening porous Ag_2S matrix covers the common mixed membranes during operation /10-13/, because its expectable consequences /changing calibration characteristics, memory effects getting more and more significant during use, increased life time/cannot be observed in practice. Rather a more realistic picture on the corrosion mechanism of mixed membranes should be drawn for explaining these experimental facts.

As the Ag_2S is a mechanically harder material than AgI /17, 18/, the Ag_2S microcrystals are surrounded by the softer AgI and thus, separated from each other. Hence, when AgI is being dissolved from the membrane by some complexing agent, some Ag_2S particles also leave the membrane that are not "stuck" any more by AgI to the membrane bulk. So, the outermost membrane layer steadily loses AgI as well as Ag_2S, which means that a porous matrix of Ag_2S can hardly be formed on the surface. Rather it is being steadily renewed during operation, which means that no significant differences in calibration characteristics and life times can be expected between pure silver halide, and 10:1 or 1:1 silver halide/silver sulphide mixed membrane based electrodes, or between new /unused/ and "old" /corroded/ reactive sensors. The lack of memory effects can also be easily interpreted by the above dissolution mechanism.

The exceptional behaviour of 1:10 AgI/Ag_2S membranes is due to the fact that at such a low AgI concentration most Ag_2S particles are fastened to the membrane by many other Ag_2S particles, too. Thus, the Ag_2S microcrystals remain on the membrane surface even after the dissolution of the surrounding AgI, and a porous covering matrix of Ag_2S can be formed. This is responsible for the worsening of the calibration characteristics, the memory effects and also for the slower dissolution of the membrane.

This interpretation of our experimental findings /Tables 1-3/ could be supported by surface analytical studies /photoelectron spectroscopy, scanning electron micrography and X-ray fluorescence analysis /19/ /. Results of scanning electron microscopy are also in perfect accordance with the dissolution mechanism suggested above /see Figs 2-4./.

The structures and working mechanisms of the other reactive ion-selective electrodes /AgI, AgBr and AgCl based CN^- and $S_2O_3^{2-}$, as well AgCl based NH_3 electrodes/ are very similar to the AgI/CN^- case /7,20/. Hence, the results of this work are also essentially valid for the other homogenous and mixed membrane based reactive electrodes /21/.

CONCLUSION

According to the results presented, the commercially available homogeneous and mixed reactive ion-selective electrodes are essentially equivalent in any practical aspect investigated /calibration characteristics, memory effects, life times/, unless the molar ratio of silver halide/silver sulphide in the sensing membrane decreases significantly below 1:1. The porous silver sulphide covering layer assumed in the literature for 1:1 membranes can be formed in reality only in the latter cases, where the advantageous effect of increased life time is spoilt by bad electrode operation. Hence, the membrane composition should be kept between pure silver halide and 1:1 mixed membrane. In this range the sensing surface is steadily renewed during operation with respect to both the halide and the sulphide content.

REFERENCES

1. J.Havas, K.Tóth, I.Szabó and E.Pungor, Proc.Anal.Chem.Conf., Budapest, /1966/ 159.
2. Instruction Manual on Cyanide Ion Activity Electrode /model 94-06/. Orion Research Inc., 11 Blackstone str.,Cambridge, Mass. USA /1967/.
3. E.Pungor and K.Tóth, Analyst, 95, /1970/ 625.
4. I.Sekerka and J.F.Lechner, Water Res., 10, /1976/479.
5. M.Gratzl, F.Rakiás, G.Horvai, K.Tóth and E.Pungor, Anal.Chim. Acta, 102, /1978/ 85.
6. H.Komiya, Japan Analyst, 21, /1972/911.
7. W.E.Morf, G.Kahr and W.Simon, Anal.Chem., 45, /1974/ 1538.
8. J.W.Ross, J.H.Rieseman and M.S.Frant, US Patent No.3874, 16th February /1971/.
9. T.Aomi, Denki Kagaku, 46,/6 /1978/.
10. D.H.Evans, Anal.Chem., 44, /1972/875.
11. G.P.Bound, B.Fleet, H.von Storp and D.H.Evans, Anal.Chem., 45, /1973/788.
12. M.S.Frant, Plating 58, /1971/686.
13. R.J.Sipson, Metal Finishing J., 18, /1972/265.
14. E.Pungor, J.Havas, K.Tóth and G.Madarász, Hung.Pat.152106 /1964/.
15. P.K.C.Tseng and W.F. Gutknecht, Anal.Chem. 47, /1975/2316.
16. I.Sekerka and J.F.Lechner, Anal.Chim.Acta 93, /1977/139.
17. P.W.Bridgman, Proc.Am.Acad.Arts.Sci., 71, /1973/387 cited in Gmelin, Silber Teil B2 Verbindungen, Springer, Berlin, 238 /1974/.

18. J.W.Mellor, "A Comprehensive Treatise on Inorganic and Theoretical Chemistry", Vol.III., Longmans, Green and Co., London, 441 /1952/.
19. E.Pungor, M.Gratzl, L.Pólos, K.Tóth, M.F.Ebel, H.Ebel, G.Zuba and J.Wernisch, Anal.Chim.Acta, 156, /1984/ 9.
20. M.Gratzl, "Reactive and Air Gap Potentiometric Sensors", candidate's thesis, Technical University, Budapest /1984/.
21. M.Gratzl et al., Unpublished results /1983/.

Table 1. Calibration characteristics of some homogeneous membrane-based cyanide electrodes after long operation
A: calibration in 10^{-4}-10^{-1}mol/l KI + 10^{-1}mol/l KNO_3
B: calibration in 10^{-4}-10^{-1}mol/l KCl + 10^{-1}mol/l KNO_3

	A			B
	Radelkis OP-I J^- electrode	home-made AgI electrode	home-made Ag_2S electrode	home-made AgCl electrode
S /mV/decade/	58.4	56.3	62.0	53.6
r^2	0.999	0.999	0.999	0.997

Table 2. Calibration characteristics of different mixed membrane-based cyanide electrodes as function of operation time
Operation /corrosion/ in 10^{-3}mol/l KCN + 10^{-1}mol/l KNO_3 /pH >10/
Calibrations in 10^{-4}-10^{-1}mol/l KI + 10^{-1}mol/l KNO_3

AgI/Ag_2S molar ratio		time of operation /corrosion/ /min/						
		0	0.5	2	5	15	60	1500
10:1	S /mV/decade/	-	58.4	57.3	54.9	63.0	60.0	60.2
	r^2	-	0,999	0,998	0,997	1.000	0,998	0,999
1:1	S /mV/decade/	-	54,7	49.5	50.3	57.7	52.5	50.5
	r^2	-	0.999	0.992	0.998	0.989	0.867	0.998
1:10	S /mV/decade/	58.0	-	54.2	44.6	20.6	2.9	- 1.7
	r^2	0.986	-	0.998	0.941	0.456	0.207	0.696

Table 3. Losses of membrane material im 25 ml 10^{-2}mol/l KCN + 10^{-1}mol/l KNO_3 solutions /pH >10/, and life times of different cyanide electrodes. The life times are referred to the loss of half mass of the electrode membranes /thickness of ~2 mm/ under usual conditions /operation in stirred 10^{-3}mol/l KCN solutions/.

AgI/Ag_2S molar ratio	loss of mass m /mg/	average m /mg/	remark	life time /hours/
pure AgI	8.1 8.3	8.2		~150
10:1	10.1 7.3 8.5 8.7	8.65 8.17 /1/	/1/ average without the first point	
1:1	7.6 9.2 9.9 8.3	8.75		~140
1:10	1.5 0.8	1.15 /2/	/2/initially unused /new/ membranes	
	0.2 0.1	0.15 /3/	/3/corroded /used/ membranes	

Fig.1. Schematic picture of the stationary diffusion profiles in the vicinity of a reactive ion-selective electrode immersed into stirred solution; d means the thickness of the Nernstian diffusion layer.

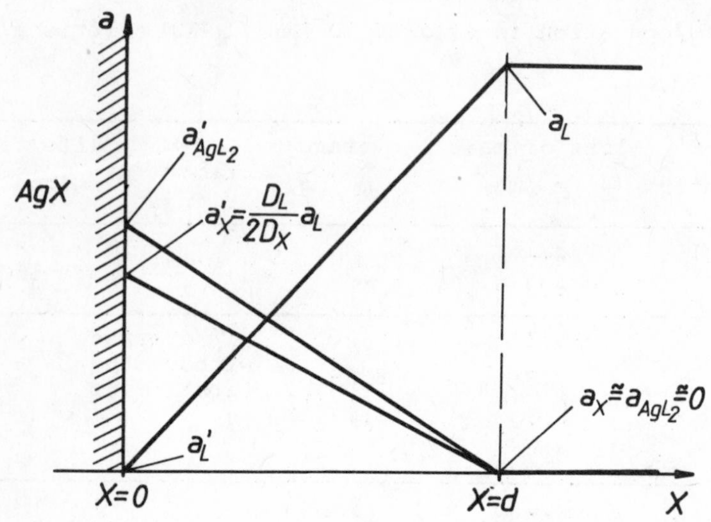

Fig.2. The surface of a used /corroded in KCN/ 1:1 AgI/Ag_2S membrane /scanning electron microscope, 90°,1800 */. The individual Ag_2S particles standing out of the surface can be clearly seen.

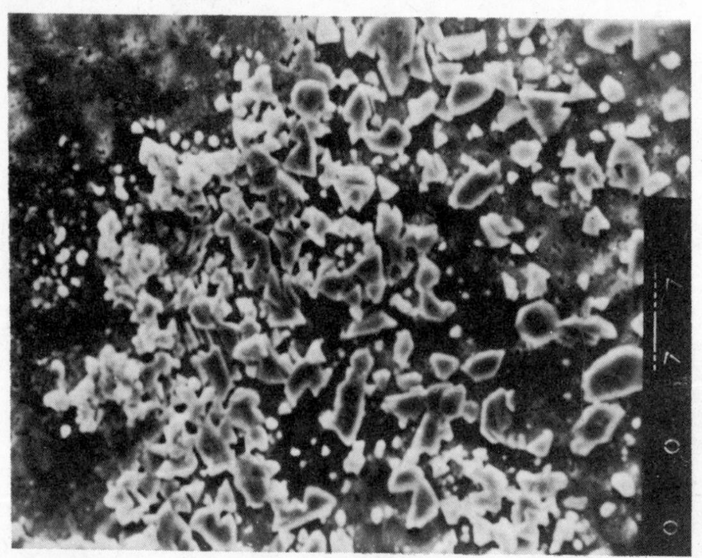

Fig.3. The surface of an unused /new/ 1:10 AgI/Ag_2S membrane /as before, 5400 */. Gaps can be observed between the Ag_2S microcrystals. Small bright dots can be observed on the surface.

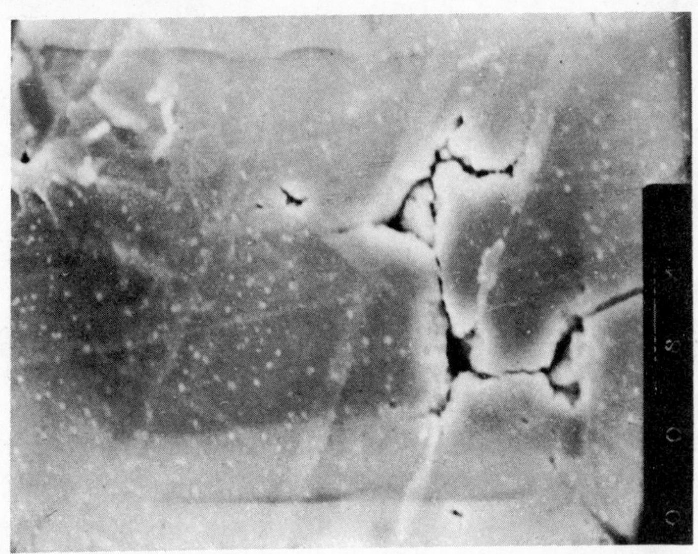

Fig.4. The surface of a used /corroded in KCN/ 1:10 AgI/Ag_2S membrane /as before/. Deep holes have been formed in the place of dissolved AgI. The small bright dots disappeared /hence, they represented probably AgI/.

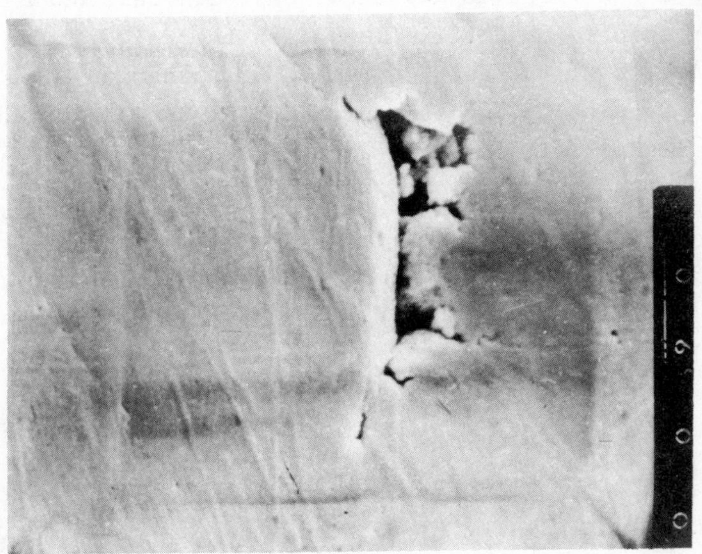

QUESTIONS AND COMMENTS

Participants of the discussion: H.Ebel, M.Gratzl, K.Nagy, M.Neshkova, Z.Noszticzius and E.Pungor

Question:

You mentioned some memory effects. How did you recognize them experimentally ?

Answer:

I spoke about memory effects only in the case of mixed-membrane electrodes which contain a large excess of silver sulphide. These electrodes do not work properly. We have recognized these memory effects from the irreproducibility of the calibration graph. However, this has no practical importance because these electrodes were bad anyway. With good electrodes we do not have to worry about memory effects.

Question:

Could you give an explanation why you mix silver iodide with silver sulphide ?

Answer:

I was not the first to mix them. Our work has shown that the silver sulphide has no influence on the calibration characteristics of the electrodes and the response time, or any other practically important feature of the electrode. Maybe the conductivity is better.

Comment:

It is said that the role of silver sulphide is to reduce the resistance and to make it easier to press, and to avoid the cracks sometimes occurring.

Comment:

The glue in this case is silver iodide, because silver sulphide is harder than silver iodide. So I do not understand why manufacturers use the mixture. The pH meters now available have such a high input impedance that the reduction of resistance due to the presence of silver sulphide is not important.

Comment:

We have also found that the silver iodide pellet is just as good as the mixed pellet, and the difference in resistance is so small, a few kΩ-s that it does not justify the mixing.

4th Symposium on Ion-Selective Electrodes
Mátrafüred, 1984

LOW ELECTRIC RESISTANCE GLASSES FOR MICROELECTRODES

E. HOPÎRTEAN, E. VERESS, E. STEFĂNIGĂ, H. LINGNER and C. SAVICI

Institute of Chemistry, Cluj-Napoca, Romania

ABSTRACT

The experimental data obtained for a number of H^+-sensitive and K^+-sensitive glasses are reported. A series of pH electrode glasses, based on two oxide systems: SiO_2-Li_2O-BaO-La_2O_3 and SiO_2-Li_2O-Y_2O_3-La_2O_3 respectively, have been obtained. The best pH-electrode characteristics were those of the glass containing SiO_2-Li_2O-Y_2O_3-La_2O_3-PbO. The K^+-sensitive glasses based on the oxide system SiO_2-Na_2O-Al_2O_3 have been obtained. The oxide system containing TiO_2besides Al_2O_3 gives glasses with low electric resistance and good electrode performance.

INTRODUCTION

The paper deals with the production of some glasses for use in the manufacture of pH- and K^+- sensitive electrodes. We studied two oxide system in order to obtain glasses for pH-sensitive electrodes having low electric resistance: SiO_2-Li_2O-BaO-La_2O_3 (I) (1-6), and SiO_2-Li_2O-Y_2O_3-La_2O_3 (II) (7-12). It is known that the glass obtained from system (I) is difficult to process and the electrodes thus obtained have a large range of linearity, but they have a large internal electric resistance.The second oxide system gives an electrode glass with a rather low electric resistance but the H-function does not cover the whole strong basic range.

We intended to obtain glasses of these two systems,having optimal pH characteristics, respectively with the largest operation range and with the lowest electric resistance. For this purpose the composition of the two systems were modified by adding one of the following oxides: U_3O_8, Fe_2O_3, V_2O_5, PbO, Cs_2O.

The references(13-15) indicate that the introduction of some oxides such as Al_2O_3, Ga_2O_3, TiO_2, In_2O_3, B_2O_3 in the glass of the pH-sensitive electrodes, determines the setting up of the metallic function. The oxide system SiO_2-Na_2O-Al_2O_3 was used as basic system for the K^+-sensitive glasses.We studied the effect of the oxides TiO_2, UO_2, CeO_2, GeO_2 and Cs_2O introduced in the glasses, upon the electrode behaviour towards K^+ ions.

EXPERIMENTAL

In order to obtain electrode glasses, the following analytical grade reagents were used: SiO_2, Na_2CO_3, Li_2CO_3, BaO, TiO_2, UO_2, CeO_2, GeO_2, Al_2O_3, Fe_2O_3, Y_2O_3, La_2O_3,V_2O_5, U_3O_8, PbO and Cs_2O.

The glass melting was achieved at 1350-1400°C in an electrical oven of the KO II type.

The pH-sensitive electrodes were checked using the following standard solutions:

- 1 M HCl (pH = 0);
- pH standards (NBS) for the 1.67 - 9.23 pH range;
- buffer solutions for the 9 - 12.5 pH range prepared from 0.1 M glycocoll , 0.1 M NaOH and having c_{Na+} = 3 M, achieved with NaCl;
- 1 M NaOH solution prepared in 2 M NaCl.

The K^+-sensitive electrode was checked with KCl solutions covering the 10^0- 10^{-5}M concentration range and containing 10^{-1}M LiCl, 10^{-2}M NaCl, 1 M $Mg(NO_3)_2$ and 1 M $CaCl_2$, respectively.

The determinations were made in thermostat conditions (20 ± 0.1°C), under stirring,witha"pH-loo" type pH-meter. The e.m.f. of the following symmetrical electrochemical cells was measured:

Ag/AgCl reference electrode	Standard solution for pH	pH-selective glass electrode

and, respectively:

SCE with double junction	Standard solution for K^+	K^+-sensitive glass electrode

The internal electric resistance of the electrodes was measured with a TR 2201 type Teraohmmeter.

RESULTS AND DISCUSSION

The results regarding the pH electrode characteristics of the glasses of the oxide system SiO_2-Li_2O-BaO-La_2O_3 are given in Table 1. The particularly high electric resistance of the electrodes made of this glass is to be noted.

The introducing in the glass of one of the oxides Fe_2O_3, Y_2O_3 or V_2O_5 determines the strong decrease of the internal electric resistance of the electrodes, a slight narrowing of the linear response range (the alkaline error at pH = 13 being 0.4 pH units), and the improvement of the qualities of the glass on flame processing. The U_3O_8 addition yields the lowest internal electric resistance of the electrodes, but linear response range extends only up pH = 10. The Cs_2O addition gave the best results. The internal electric resistance of the electrodes decreased with an order of magnitude without affecting the response in strong alkaline medium. The electrode has no alkaline error at pH = 13 and c_{Na^+} = 3M.

The pH characteristics of the glass electrodes obtained from the oxide system SiO_2-Li_2O-Y_2O_3-La_2O_3 are given in Table 2. The increase of Li_2O content in the glass has the expected effects: the strong decrease of the internal electric resistanceof the electrodes and a slight increase of the alkaline error. U_3O_8 shows the same effect as seen with system (I). The best electrode characteristics were those of the glass containing PbO: the lowest internal electric resistance and the widest range of linear response. The glass is practically alkaline error free at pH = 12.

These results indicate two optimal oxide systems: SiO_2-Li_2O-BaO-La_2O_3-Cs_2O and SiO_2-Li_2O-La_2O_3-Y_2O_3-PbO. In order to obtain pH microelectrodes the glass obtained from the oxide system containing PbO was preferred, owing to its high electric conductivity and to the wide pH range of the linear response.

The results obtained with the K^+- sensitive electrode glasses are given in Table 3.

The presence of TiO_2 and CeO_2 oxides together with Al_2O_3 assures a good electrode behaviour towards the K^+ ion. In the glasses containing TiO_2, the partial replacement of Na_2O by Cs_2O or GeO_2 provides membranes with low electric resistance but it induces the decrease of the electrode sensitivity and narrows the linear range of the response. The oxide system containing TiO_2 besides Al_2O_3 gives glasses with low electric resistance and good electrode performance. The working characteristics of the K^+- sensitive electrode obtained from this glass are shown in Table 4. The same electrode was used to indicate the end point for the titration of potassium with calcium tetraphenylborate solution.

The use of this titrant is justified by the good selectivity of the electrode for calcium. The results are shown in Table 5. It is to be noted that potassium determination is also possible in the presence of very high calcium and magnezium concentrations owing to the favourable selectivit coefficient of the electrode in the presence of these ions.

REFERENCES

1. G.A.Perley, Anal.Chem., 21, /1949/ 391.
2. K.Schwabe, Chem.Tech.(Berlin), 6, /1954/ 3o1.
3. B.Lengyel and F.Till,Egypt.J.Chem., 1, /1958/ 99.
4. M.M.Shults, A.I.Parfenov, Chen'De-yuim,T.T.Bondarenko and Yu.Ya.Mekhryushov, Vest.Leningr.Univ.,No.4%1963/155.
5. M.M.Shults, N.V.Peshekhonova and T.V.Lipets, Vestn.Leningrad.Univ., No.4 /1963/, 16o.
6. A.I.Parfenov, M.M.Shults, N.N.Kochergina, V.P.Ivanov and S.B.Evnina, Vestn.Leningr.Univ.,No.4 %1963/ 162.

7. E.P.Arthur and R.W.Nolan, U.S.Patent No.3.238.050 /1966.
8. M.Stepinak and M.Karsulin,Z.Anorg.Allg.Chem., 355, /1967/ 219.
9. A.I.Parfenov M.M.Shults,T.N.Nebrasova and I.P.Polazova, Vestn.Leningr.Univ.,No.4 /1963/ 126.
10. H.H.Cary and W.P.Baxter,U.S.Patent,No.2.462.843 /1945.
11. H.Moore and R.S.de Silva.J.Soc.Glass Technol.,36/1952/51.
12. A.I.Parfenov,A.F.Klimov and O.V.Mazurin, Vestn.Leningr. Univ., No.10 /1959/ 129.
13. B.P.Nikolskii,Zhur.fiz.Khim., 7/1953/724; 10/1973/495; 10 /1973/ 513.
14. M.M.Shults, Vestn.Leningr.Univ.,No.4 /1963/ 174.
15. G.A.Rechnitz and G.Kugler, Z.analyt.Chem.,210/1965/174.

Table 1. pH Characteristics of different glasses based on oxide system (I)

Oxide system	Internal electric resistance at 20°C (MOhms)	Linear response range at [Na+]=3M (pH)	Sensitivity (mV/pH)	Alkaline error at pH=13 (pH)
SiO_2-Li_2O-BaO-La_2O_3	2000	0 - 13.6	57.3	-
SiO_2-Li_2O-BaO-La_2O_3-Fe_2O_3	350-400	0 - 12.5	57.4	0.25
SiO_2-Li_2O-BaO-La_2O_3-Y_2O_3	350-450	0 - 12.0	56.6	0.40
SiO_2-Li_2O-BaO-La_2O_3-V_2O_5	150-200	0 - 12.0	57.2	0.40
SiO_2-Li_2O-BaO-La_2O_3-U_3O_8	75-125	0 - 10.0	57.8	2.00
SiO_2-Li_2O-BaO-La_2O_3-Cs_2O	150-200	0 - 13.5	57.2	-

Table 2. pH Characteristics of different glasses based on oxide system (II)

Oxide system	Internal electric resistance at 2o °C (MOhms)	Linear response range [Na^+]= =3M (pH)	Sensitivity, mV/pH	Alkaline error at pH=12 (pH)
SiO_2-Li_2O(25)-La_2O_3-Y_2O_3	150-200	0-11.5	58.1	0.30
SiO_2-Li_2O(28)-La_2O_3-Y_2O_3	50-100	0-11.0	57.6	0.40
SiO_2-Li_2O(30)-La_2O_3-Y_2O_3	10-30	0-10.5	57.1	0.60
SiO_2-Li_2O(25)-La_2O_3-Y_2O_3-U_3O_8	75-125	0-10.0	57.8	0.90
SiO_2-Li_2O(30)-La_2O_3-Y_2O_3-U_3O_8	5-15	0-10.o	57.0	0.90
SiO_2-Li_2O(30)-La_2O_3-Y_2O_3-PbO	5-15	0-12.0	57.0	-

Table 3. The electrode characteristics of K^+-sensitive glasses

Oxide system	Internal electric resistance, 2o °C (MOhms)	Sensitivity (mV/pK)	Linear response range (pK)
SiO_2-Na_2O-Al_2O_3	3 - 1o	51.5	0-3.8
SiO_2-Na_2O-Al_2O_3-TiO_2	2 - 5	54.5	0-4.0
SiO_2-Na_2O-Al_2O_3-UO_2	5- 10	50.0	0-3.0
SiO_2-Na_2O-Al_2O_3-TiO_2-CeO_2	20 - 30	52.5	0-3.8
SiO_2-Na_2O-Al_2O_3-TiO_2-Cs_2O	5 - 10	43.0	0-3.0
SiO_2-Na_2O-Al_2O_3-TiO_2-GeO_2	5 - 10	42.5	0-3.0

Table 4. Selectivity data of K^+-sensitive glass based on SiO_2-Na_2O-Al_2O_3-TiO_2

Interfering ion (M)	Linear response range (pK)	Sensitivity mV/pK	Selectivity coefficient, $K_{K,M}$
-	0 - 4.0	54.5	-
Na^+ $1o^{-2}$M	0 - 2.8	50.0	1.58x$1o^{-1}$
Li^+ $1o^{-1}$M	0 - 3.5	50.0	3.16x$1o^{-3}$
Ca^{2+} 1 M	0 - 2.8	52.5	1.58x$1o^{-3}$
Mg^{2+} 1 M	0 - 3.0	54.5	$1o^{-3}$

Table 5. Results of the potentiometric titration of the same sample of K^+ in the presence of different salts. Titrant:calcium tetraphenylborate 10^{-1}N (F = 1.213)

Potential jump at the equivalence point (mV)	Equivalence volume (ml)	Error[x] (%)	The foreign salt present in the sample
32	2.06	-	-
13	2.02	- 1.94	NaCl 10^{-2}M
30	2.05	- 0.48	$CaCl_2$ 10^{-1}M
28	2.05	- 0.48	$CaCl_2$ 3 M
33	2.06	0	$MgCl_2$ 10^{-1}M
22	2.07	+ 0.48	$MgCl_2$ 3 M

[x]Errors have been calculated by using as true value V_e= 2.06 ml, obtained in the absence of foreign salt.

4th Symposium on Ion-Selective Electrodes
Mátrafüred, 1984

LOW RESISTANCE LIQUID MEMBRANE ION-SELECTIVE ELECTRODES

G. HORVAI, T.A. NIEMAN* and E. PUNGOR

Institute for General and Analytical Chemistry, Technical University, Budapest, Hungary

*School of Chemical Sciences, University of Illinois, Urbana, IL, USA

ABSTRACT

Low resistance neutral carrier electrodes have been prepared by varying the geometrical dimensions and the composition of the membranes. A quaternary ammonium tetraphenylborate additive efficiently decreased the membrane specific resistance. The role of the additive in membrane behaviour has been studied by transport experiments. A model is suggested to explain the observations.

INTRODUCTION

Neutral carrier ISEs are usually fabricated by dissolving the neutral carrier in an organic solvent and then incorporating this solution either into a porous membrane or into a polymer film. Microelectrodes may be constructed without a supporting material. Optimised membrane compositions have been reviewed recently /1/. Membranes made with these compositions and incorporated into commercially available electrode bodies, like those available from Philips, have a resistance in the megohms range. This is statisfactory for most applications. In a recent work /2/ we needed ISEs with much lower resistance. From the data shown in this paper it will be apparent that the problem can be solved. The success of the work was partly due to lowering the specific resistance of the membranes by suitable additives. The role of these additives in the conduction mechanism of the membranes is an intriguing problem closely related to topics of vivid current interest like the polarization of ITIES. A preliminary study of the ionic transport across the low resistance membranes is also reported here.

EXPERIMENTAL

Materials and instruments

Valinomycin was purchased from Sigma. Aliquat 336 /"tricaprylilmethylammonium chloride"/ and high molecular weight PVC were produced by Aldrich. Bis/2-ethylhexyl/sebacate /DOS/ and 2-nitrophyenyl octyl ether /ONPOE/ were produced by Fluka for use in ion-selective electrodes. Sodium tetraphenylborate /NaTPhB/ was AR from Mallinckrodt. Tetrahydrofuran /THF/ from MCB, Baker and Fluka was stabilized with BHT. Sodium chloride, potassium chloride and ammonium chloride were analytical reagents. All chemicals were used as obtained. Deionized water was purified either on a Milli-Q system /Millipore/ or double distilled in quartz apparatus.

pH-mV meters from Corning /model 130/ and Radelkis /OP-208/1/ were used. Electrode resistances were measured by a bipolar pulse conductance instrument using 0.2 or 2 ms pulse width or with a Radelkis OK-102/1 conductometer. Silver-silver chloride reference electrodes were used.

Transport experiments were made with a simple galvanostat assembled in our laboratory. AAS measurements were made on Varian Techtron AA6 with a custom made sample introduction device.

Preparation of membranes and electrodes

Membranes were cast after dissolution of all components in THF. NaTPhB and Aliquat were dosaged in THF solutions. Compositions of the individual membranes are given in Table 1. The 21 mm diameter membranes were glued to clear vinyl tubing of 5/8" i.d. and 7/8" o.d. with VLP vinyl repair fluid /P.D.I.Inc., St. Paul, Minn./. The tube and the glue were found to be inert. The membranes were about 0.1 mm thick.

RESULTS AND DISCUSSION

Membrane resistance can be decreased by increasing the surface area of the electrode, by decreasing the thickness of the membrane and by decreasing its specific resistance. In this work all three methods were combined. The geometrical dimensions were pushed to the practically acceptable limits. The specific resistance has been decreased by various additives /Table 1./.

Tetraphenylborate salts have been in use for some time in ISEs to avoid interference from lipophylic anions. They also decrease the specific resistance of the membrane. With inorganic tetraphenylborate salts the limits of this decrease are set by the maximum allowable 1:1 molar ratio of tetraphenylborate to valinomycin and the limited solubility of valinomycin in the membranes. Electrodes with the ratio above 1:1, like electrode 5 in Table 1, do not show potassium-selective behaviour.

However, if tetraphenylborates of lipophylic cations are used, the molar ratio of the salt to valinomycin can be raised higher than 1:1 without impairing the selectivity and sensitivity of the electrode. Electrode 4 of Table 1 shows K^+/Na^+ selectivity identical to electrodes 2 and 3 which represent well established electrode compositions. On the other hand, electrode 4 has a slightly lower slope /56 mV/decade/ than electrodes 2 and 3 /58 mV/decade/ and it is more prone to drifting.

Although one could envisage optimization of the behaviour of the low resistance electrodes, for example by testing similar other additives in different concentrations, this was not the aim of the present work. Rather it was attempted to investigate why the low resistance electrodes maintain the same selectivity as the membranes without additives. This question arises because the ion-selective behaviour has often been attributed to a membrane transport number near unity with respect to the primary ion. Comparison of electrodes 2,4 and 6 in Table 1 shows, however, that the increase in membrane AC conductivity is due to the additives, and independent from the presence of valinomycin. Hence it follows that the additives must carry most of the AC current in membrane 4, so that the transport number for potassium may not be close to 1. To see if the DC behaviour is similar or not we made a series of transport experiments across the electrode membranes. 9.4 μA was passed for one hour across the membrane. Subsequently the 2.0 ml filling solution of the electrode was analysed by AAS.

Two setups were used:

A:+Ag,AgCl 10^{-4}M KCl,10^{-2}M NaCl Membrane 2.0 ml 10^{-4}M NH_4Cl AgCl, Ag-

B:-Ag,AgCl 10^{-4}M NH_4Cl Membrane 2.0 ml 4.10^{-4}M KCl AgCl,Ag +

With cell A the appearance of Na^+ and K^+ on the cathode side was studied; cell B was used to study the K^+ decrease on the anode side. Since parallel measurements show a large variability one can only observe some trends at this time. These can be summarized as follows.

With electrode composition 2, i.e. no additives, most of the current in cell A is transported by sodium and potassium. Potassium is transported preferentially.

With electrode composition 4, i.e. with the low specific resistance potassium selective electrode, much less sodium and potassium is transported to the cathode side /cell A/ than expected from the total charge passed. On the other hand, the decrease in the potassium concentration on the anode side /cell B/ is close to the expected value. These observations can be reconciled with the assumption that on the anode side only cations from the aqueous phase are transported across the membrane/ solution interface, i.e. tetraphenylborate does not leave the membrane, while on the cathode side some potassium /and sodium, if present/ leaves the membrane, but at the same time also Aliquat cations arrive at the surface and probably form organic-soluble Aliquat chloride with chloride ions from the cathode chamber. Consequently the membrane composition would change during the transport experiment. One does, indeed, observe about 50 percent decrease in membrane conductance after the transport experiment.

It is very remarkable that electrode 6, which contains the additives but no valinomycin, did not transport in cell A virtually any sodium or potassium to the cathode side. This observation may shed some light on the role of valinomycin in the electrodes. Valinomycin facilitates the entrance into and the transport across the membrane. Most notably it regulates the relative rate of entrance of sodium and potassium. It is, however, not necessary for the potassium-selective behaviour that the potassium transference number be unity across the whole membrane.

Figure 1. shows a graphical representation of the suggested membrane transport processes.

The above hypotheses need to be accepted at this time with due caution and further experimental confirmation is needed. Nevertheless, the techniques used here appear to be promising. It is also noted that the same experimental technique could be used to study transport phenomena of ITIES where the narrow potential window renders purely electrochemical studies difficult.

ACKNOWLEDGEMENT

AAS measurements were kindly made for us by dr E.Harsányi.

REFERENCES

1 D.Ammann, W.E.Morf, P.Anker, P.C.Meier, E.Pretsch, W.Simon, Ion-Selective Electrode Rev., 5, /1983/ 3

2 G.Horvai, T.A.Nieman, to be published.

Table 1

Composition and resistance of representative membranes

/5/8" diameter, 0.1 mm thickness/

No.	Valinomycin µmoles	Aliquat µmoles	NaTPB µmoles	DOS mg	ONPOE mg	PVC mg	Resistance kohm
1	-	-	-	34	-	16.5	4900
2	0.35	-	-	27	7	16.5	250
3	0.35	-	0.15	27	7	16.5	48
4	0.35	0.59	0.75	27	7	16.5	10
5	0.35	-	0.75	26	7	16.5	21
6	-	0.59	0.75	26	7	16.5	10

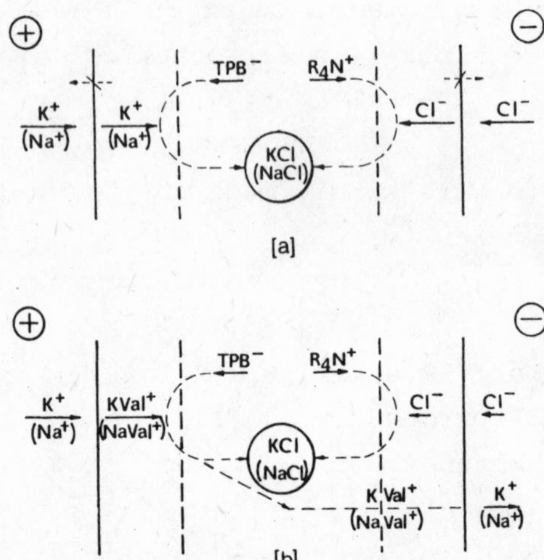

Fig. 1. Suggested transport processes across the low resistance membranes.
The membrane is between the two solid vertical lines. The dashed vertical lines separate the bulk of the membrane phase from the surface-near region. KCl may accumulate in water droplets forming in the membrane.
a./ membrane without valinomycin
b./ membrane with valinomycin

QUESTIONS and COMMENTS

Participants of the discussion: R.P.Buck, G.Horvai, J.Janata, M.Neshkova, Z.Noszticzius, W.Simon and J.D.R.Thomas

Question:

You mentioned that KCl is formed in the membrane, dissolved in water droplets. Could you explain, how this happens ?

Answer:

I was referring to potassium entering the membrane on one side and chloride entering it on the other side. And as they migrate within the membrane due to diffusion or a current, they probably meet somewhere. And if they meet they can form KCl. I also showed a dashed line which wanted to represent that tetraphenylborate may play a role in the transport of potassium within the membrane. Another dashed line showed the possible transport of chloride ions by quaternary ammonium. This is only an assumption, I cannot say anything certain about this type of transport at the moment. But at the time when potassium and chloride, either in free form or in tetraphenylborate salt or quaternary ammonium salt meet, potassium chloride can form. And, as the permeability of this membrane to water vapour is relatively high, water can condense in that, or if there are droplets originally in the membrane, the condensed water will increase these droplets.

Question:

You mentioned that potassium and chloride ions cannot pass through the membrane. The previous question implied the question of how the mechanism you propose can prevent them from leaving the membrane.

Answer:

I said that tetraphenyl borate cannot leave the membrane, neither can the quaternary ammonium leave the membrane across the membrane - solution interface. Potassium can enter and leave the membrane and so can chloride. But, under the conditions applied they both enter but they do not leave. I did not say they cannot, I only said they do not leaves during the time of the experiment, which is one hour. So, either this time was not enough for the cation to appear on the other side, or it did not move at all towards the other side.

Question:

You had a table where you showed the composition of the membrane without valinomycin. And also, there is a sentence in your abstract which says that the selectivity against sodium is virtually unchanged. Does that mean that in the absence of valinomycin you have an electrode just as good as with valinomycin ?

Answer:

No. I meant that the selectivity of the membrane with valinomycin and the organic salt additive is the same as that of a membrane containing valinomycin but no organic salt additive.

Comment:

Concerning the water droplets within the membrane. I liked your statement very much, as in normal ion-selective solvent--polymeric membranes there is water at a concentration of about 10^{-2} mol/dm^3. We postulated in 1976 that this water is responsible for the compensation of charge, that is for conserving electroneutrality within the membrane. We called it water clusters.

Answer:

In our case, with the quaternary ammonium and the tetraphenyl borate present, we do not speak of any charge within these water clusters or droplets, we do not really know that they are droplets. They are so small that we cannot see them under a light microscope.

Question:

I would like to ask for some more detail of the polarizing

technique you use. What is the polarizing current ?

Answer:

The current was 10 μA and the time of a measurement was usually one hour. From this you can calculate the charge transported.

Question:

How did you decide that potassium was the only ion entering the membrane ?

Answer:

I had pure potassium chloride solution on one side of the membrane and I measured the change in potassium concentration on this side before and after passing the current. I also knew the volume, so I could calculate how much potassium left the solution. And I also calculated from the charge passed through, how much potassium had to pass across the interface. And the two values were roughly the same.

Question:

How much was the difference in concentration ?

Answer:

When I used a volume of 2 ml, the solution was 4×10^{-4} mol/l before and 2.4×10^{-4} mol/l after the experiment. The difference is 1.6×10^{-4} mol/l.

Remark:

There was a point that worried me in your talk, which had to do with identifying which species crossed the interfaces - a priori. You can measure these single interface processes, we have measured them. You can change the voltage. Potassium goes in most easily, then sodium. It is not true that tetraphenylborate or aliquat are locked in there, it is just a question of how much voltage you apply to force the current. The sequence goes as the free energies.

Answer:

I completely agree. Under the conditions I used, with the concentrations in the range 10^{-2}-10^{-4} mol/l, tetraphenylborate did not leave the membrane. This is a conclusion from what I have measured and not an a priori statement that it cannot leave the membrane at all.

Comment:

You have to have a quite high voltage to get the current you mentioned 10 μA.

Answer:

I did not measure it, but the maximum was 7V, as this was the maximum output of the device I used.

4th Symposium on Ion-Selective Electrodes
Mátrafüred, 1984

POTENTIOMETRIC STUDY OF THE CHLORIDE TOLERANCE OF TREES

H. HÖDREJÄRV

Tallinn Technical University, Tallinn, Estonian SSR

ABSTRACT

A direct potentiometric determination of chloride in the leaves of trees,with the silver - silver chloride electrode, can be used in the investigation of salt tolerance of trees. Due to the matrix effect,the standard addition method has been suggested.

INTRODUCTION

The major identifiable pollution source of chlorides is road salt.Salt is used during the winter to clear the streets from snow - a usual practice in the northern regions.Beside contamination of ground and well water,a serious problem is the salt tolerance of the plants growing in the soils with raised salt concentration.For example,in Boston,USA,an average of 107 tons.km^{-2}.$year^{-1}$ of salt was applied in the urban district and between 39...63 t.km^{-2}.y^{-1} in its suburban areas (1).In the regions where the snowing period rises to 3...4 months a year,the salt amounts used are about 1000 t.km^{-2}.y^{-1}. At the same time the chloride input from the atmosphere is usually about 2 t.km^{-2}.y^{-1} (2).The chloride content of soils in the towns with high salt application is about 30..80mg.kg^{-1} (DW) as the mean of a year.

An excess of soluble salts is harmful for plant growth. Generally the growth decreases with salinity,except the true halophytes,which are resistant to 10..20 g.kg^{-1}NaCl in the soil,and for which salt is necessary to stimulate growth.For the glycophytes as low as 0.1 $g.kg^{-1}$ NaCl can already cause damage (tomatoes,peas,beans).The salt tolerance has been investigated by L.Bernstein (3) for fruit crops.By his investigations severe damage for the apple tree takes place when the chloride content of leaves exceeds the range 1.3..1.8% and for the grapes 2.4..2.8% Cl(DW).Some trees seem to be very sensitive to the chloride ion.In addition to the general osmotic growth inhibition,characteristic leaf burn symptoms develop when chloride accumulates to harmful levels in the leaves.The osmotic effect and chloride accumulation are not equally dominant in all kind of trees.There are individual trends to counteract the influence of the salt,depending on several factors.Information is lacking on the salt tolerance of the trees growing in urban districts.However,this characteristic is important along such factors as disease and frost resistance and must also be taken into account.Leaf analysis would be one way to find out the chloride accumulation ability of the trees.Among the various methods currently used in such investigations,the electrochemical techniques provide us with the sensitivity,accuracy and speed necessary in a variety of situations.The determination of chloride ion concentration has been an object of several investigations.Classical examples in this field are the methods worked out by H.Laitinen (4),J.M.Kolthoff and P.K.Kuroda (5),E.Pungor et al. (6) etc.,who used for that purpose potentiometric or amperometric titration and potentiometric direct determination with the ion selective electrode.However,very little is known about the possibilities of using these methods for the analysis of biological samples including the ISE practice.From the point of environmental control,which always must consider the necessity of a great number of analyses to give statistically correct information, direct potentiometry with ISE is the best,because it is not time consuming.

EXPERIMENTAL

The leaves of the trees were collected in spring and autumn at the 1.3...1.8 m height in amounts of 100 g of fresh leaves. The leaves were dried in air to air-dry. 10 g of air-dry leaves were ground in a laboratory mill and dried at 110°C. About 0.4..0.8 g of the leaves were digested with 25 ml of distilled water at 80°C for an hour, and left after this overnight. The solution was separated by centrifugation. The leaves were double washed with distilled water, which was after this combined with the prime centrifugate. 10 ml of ethyl alcohol was added and the solution was diluted to 100 ml with acetate buffer (2 M acetic acid and 2M sodium acetate). The pH of the final solution was 5.

The concentration of the chloride ions was determined in the solution at 20 $\pm$ 0.2^{o}C using a potentiometric method with the silver-silver chloride electrode, selective to chloride ion. The electrode was made by coating electrolytically a silver rod (ø=2 mm; l=6 mm) with silver chloride. The rod was cathodically cleaned in an 1 N H_2SO_4 solution before the coating. The coating was carried out in an 0.1 N HCl solution with the silver rod as the anode and platinum electrode, S=1 cm^2, as the cathode. The coating time was 10 min. The current density was about 10 mA.cm^{-2}. The potentiometer with an accuracy of 0.2 mV was used. The commercial silver-silver chloride reference electrode was connected with the solution by a KNO_3-agar-agar salt bridge. The calibration of the electrode was carried out using a leaf extract with a known chloride content. An amperometric titration method was used as the checking method (7).

RESULTS AND DISCUSSION

The results of the investigation of chloride content in the leaves of trees are shown in Table 1. As a rule, the mean concentration ($\bar{x}$) of the chloride is greater in the town A, which, first of all, is situated on the sea coast, and, secondly, is about twice bigger (0.5 million inhabitants) than the other town - which is an inland one. In general, the chloride concentration in the dry (110°C) leaves was in the range of 0.1.. ..3.0% Cl. The highest concentration was found in limes: $\bar{x}$=0.69% (max 3.03%). The highest mean concentration was found in elms: $\bar{x}$= 0.74% (max 2.35%) and the lowest mean in oaks: $\bar{x}$= 0.11% (max 0.24%). In the other species the concentrations were found as follows: birch - $\bar{x}$=0.31% (max 1.70%); horse-chestnut $\bar{x}$= 0.61% (max 2.11%); maple $\bar{x}$=0.69% (max 1.91%); ash-tree $\bar{x}$=0.60% (max 2.00%); white willow $\bar{x}$=0.30% (max 0.67%). The concentration of chloride is rising in summer.

As the result of this investigation, differences in the salt accumulation for different kinds of trees have been found. The trees, growing in urban conditions, have considerably more chloride in the leaves than the same kinds of trees in "clean" areas. In autumn, 1983, the chloride content was checked in the area far from the pollution sources. The results were as follows: lime 0.21% Cl; birch 0.07% Cl; oak 0.08% Cl; maple 0.14% Cl; elm 0.06% Cl; horse-chestnut 0.07% Cl; ash-tree 0.14% Cl.

The applicability of the procedure developed for chloride determination in leaves depends, in general, on two factors: the efficiency of the chloride ion transport into the solution by a treatment of the leaves with water, and the interpretation of chloride response of the electrode. As the result of the water treatment of the leaves, we have a colloidal solution rich of organic matter of plant origin. Some part of it is well soluble, the other part is in a swollen form and can bind a lot of solvent. The chloride ion transport from the plant tissues into the solution must take place since the chloride ions accumulate within the vacuoles of the leaves and are released easily after the grinding process. Since the organic matter can be partly a highly charged polyelectrolyte, its presence, as well as the presence of colloidal particles, can cause remarkable trouble and it must be taken into consideration in the interpretation of the results. The problem is well known in the biological sample analysis as a matrix effect. We have checked the matrix influence by an independent analytical method: the amperometric titration with a platinum rotating electrode in an 1 N HNO_3 solution. The correlation between these two methods was satisfactory, when the standard addition method was used for the ISE potentiometry. The calibration plot is linear in the concentration range of 1.1..3.8 pCl (1.1..4.3 for NaCl standards)(Figure 1). The ratio of the slopes of the standard addition plots: solutions of the leaves of several species of trees to the standard NaCl solutions (in acetate buffer), is used as an index of the matrix effect. The ratio was for all investigated trees in the range of 1.1..1.3. (Figure 2).

Selectivity problems are usual for silver electrodes. The selectivity coefficient

$$K = \frac{a_x}{a_{Cl}},$$

where the a_x and a_{Cl} are the activities (isopotential) of the ions, was used. The selectivity constants for some ions are as follows: $K_{Br} = 0.4$; $K_I = 16$; $K_{CN} = 400$. In solutions containing bromide and iodide the life time of the electrode is shorter. The accuracy of the used method in the concentration range of 0.1..0.7% Cl in the leaves is 3..1% (P=0.05) and enables the investigation of the salt tolerance of the trees.

REFERENCES

1 J.Caesar, R.Collier, J.Edmond, F.Frey, G.Matisoff, A.Ng, R.Sallard, Environmental Science and Technology, 10, /1976/ 697.

2 V.Karise, Eesti Loodus, N.4, /1965/ 234.

3 L.Bernstein, Agriculture information, Bull.N.292, US Department of Agriculture, /1980/ 1.

4 H.A.Laitinen, W.P.Jennings, T.D.Parks, Ind.Eng.Chem., Anal.Ed., 18, /1946/ 355.

5 J.M.Kolthoff.P.K.Kuroda, Analytical Chem., 23, /1951/ 306.

6 E.Pungor, K.Tóth, P.Gábor-Klatsmányi, Hung.Scient.Instr., 49, /1980/ 1

7 H.H.Hödrejärv, A.J.Vaarmann, L.A.Uibo, Proc.of Tallinn Technical University, N.479 /1980/ 55.

Table 1. Chloride content in the leaves of urban trees in the towns A and B (% DW) /$\bar{\bar{x}}$ arithmetic mean; x_{min}-x_{max} concentration range; n number of samples; s spring and a autumn/.

year		1980 $\bar{\bar{x}}$ (x_{min}-x_{max})	1981	1982	1983
		LIME (Tilia corolata)			
A	s	0.74(0.1-2.5)	0.56(0.1-1.3)	0.43(0.1-1.5)	0.57(0.1-2.3)
n=60	a	1.22(0.2-3.0)	0.79(0.2-1.7)	0.52(0.2-1.6)	1.05(0.1-2.0)
B	s	0.29(0.1-0.5)	0.51(0.1-1.3)	0.35(0.1-1.0)	0.38(0.1-0.9)
n=21	a	0.89(0.2-2.1)	0.59(0.1-2.4)	0.74(0.2-1.5)	0.75(0.3-2.0)
		BIRCH (Betula pendula)			
A	s	0.20(0.1-0.5)	0.13(0.1-0.2)	0.25(0.1-1.1)	0.15(0.1-0.3)
n=8	a	0.71(0.3-1.7)	0.27(0.1-0.4)	0.30(0.2-0.6)	0.43(0.2-0.7)
B	s	0.09(0.1-0.2)	0.08(0.0-0.1)	0.10(0.1-0.1)	0.07(0.1-0.1)
n=5	a	0.12(0.1-0.2)	0.13(0.1-0.1)	0.21(0.1-0.4)	0.13(0.1-0.2)
		OAK (Quercus robus)			
A	s	0.10	0.08	0.08	0.11
n=1	a	0.17	0.14	0.11	0.23
B	s	0.07(0.0-0.1)	0.10(0.0-0.2)	0.13(0.1-0.2)	0.09(0.0-0.4)
n=13	a	0.18(0.1-0.5)	0.12(0.1-0.5)	0.15(0.1-0.2)	0.25(0.1-0.3)
		HORSE-CHESTNUT (Aesculus hippocastanum)			
A	s	0.43(0.1-0.8)	0.49(0.2-1.0)	0.25(0.1-1.1)	0.42(0.1-1.0)
n=24	a	1.3(0.4-1.6)	1.0(0.2- 1.5)	0.58(0.2-1.5)	0.72(0.2-1.2)
B	s	-	0.26(0.1-0.5)	0.12(0.1-0.2)	0.19(0.1-0.3)
n=4	a	-	0.47(0.2-0.6)	0.84(0.5-1.1)	0.54(0.3-0.8)
		MAPLE (Acer platanoides)			
A	s	0.57(0.1-1.0)	0.50(0.3-0.8)	0.44(0.3-0.8)	0.63(0.3-0.9)
n=10	a	1.43(1.1-1.7)	0.60(0.2-1.2)	0.61(0.1-1.5)	0.73(0.3-1.2)
B	s	-	0.73(0.6-0.9)	0.20(0.2-0.2)	0.42(0.1-0.7)
n=3	a	-	0.80(0.2-1.2)	0.32(0.1-0.5)	0.54(0.3-0.8)
		ASH-TREE (Fraxinus excelsior)			
A	s	0.78(0.7-0.8)	0.47(0.5-0.6)	0.46(0.1-0.6)	0.35(0.3-0.4)
n=4	a	-	0.74(0.5-0.7)	o.46(0.4-0.5)	0.61(0.5-0.7)
B	s	0.73(0.2-0.4)	0.30(0.1-0.4)	0.12(0.1-0.2)	-
n=3	a	0.90(0.4-1.4)	0.59(0.2-1.0)	0.33(0.3-0.4)	-

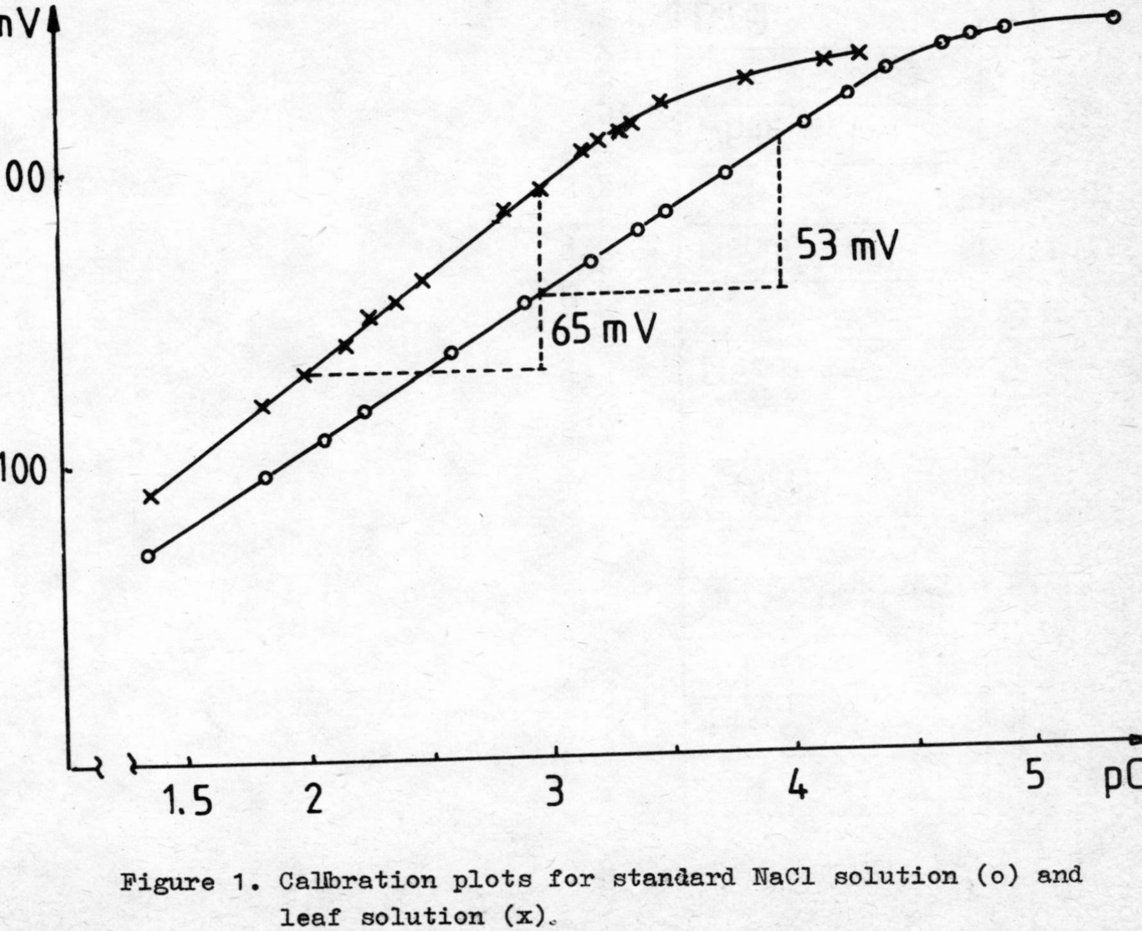

Figure 1. Calibration plots for standard NaCl solution (o) and leaf solution (x).

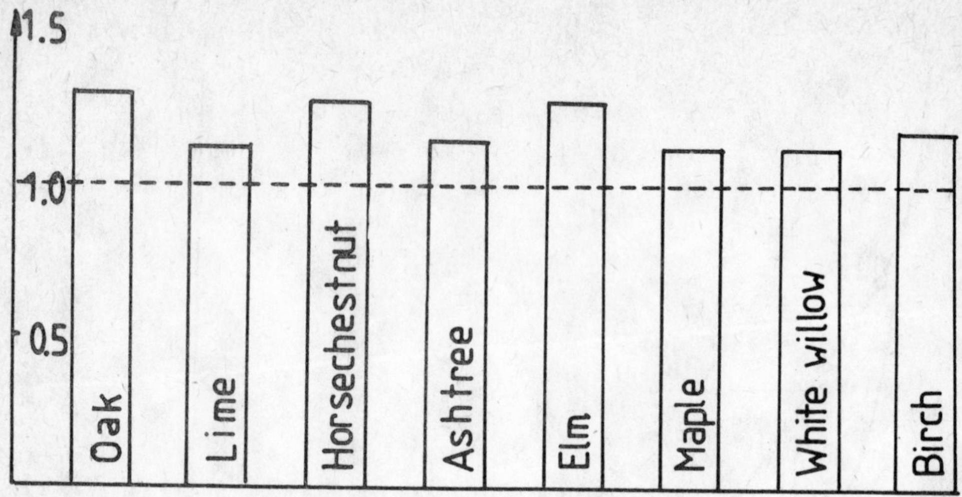

Figure 2. The ratio of slopes of standard addition plots: solutions of leaves vs. standard NaCl solution.

QUESTION

Participants of the discussion: H.Hödrejärv and R.Stella

Question:

You measured chloride in the leaves of different trees. Do you wash the leaves before doing the analysis ? I ask this because Tallinn is on the sea, and there might be some chloride on the leaves coming from airborne salt.

Answer:

We did not wash the samples because it is possible to wash out the chloride from the leaves too. But we did analyses in different distances from the sea. And the effect of sea-spray is not so strong, Baltic sea has not a high salt concentration. We did not find any significant change with the distance.

[illegible] decided [illegible] to [illegible]
[illegible] also [illegible]
[illegible] the [illegible]
[illegible]

[illegible] the [illegible]
[illegible]
[illegible]
[illegible] situation [illegible]

4th Symposium on Ion-Selective Electrodes
Mátrafüred, 1984

THE BEHAVIOUR OF COPPER ION-SELECTIVE ELECTRODES IN PRESENCE OF SOME INTERFERING ANIONS

A. HULANICKI, A. LEWENSTAM and T. SOKALSKI

Department of Chemistry, University of Warsaw,
Warsaw, Poland

ABSTRACT

The potential response of copper selective electrodes containing $CuS+Ag_2S$, CuS or Cu_2S as active membrane components was investigated in chloride, bromide and/or thiosulphate media. Their behaviour may be explained on the basis of the diffusion layer model and complex chemical reactions. Elimination of chloride interferences in copper determination was suggested.

INTRODUCTION

There is a number of questions in the field of ion-selective electrodes which were discussed from the beginning of application and investigation of electrodes at the end of sixties. One of them is the erratic behaviour of copper electrodes in presence of some anions, e.g. chloride [1], being evidenced by the increase of electrode slope, the lack of response reproducibility and the prolonged response times. Until now there is no definite answer concerning the mechanism of such behaviour nor procedure for practical interference elimination. To get some more informations about those phenomena we have investigated various copper electrode response in solutions containing chloride or bromide ions in the presence of thiosulphate.

EXPERIMENTAL

Potential measurements were performed using following

instruments: pHM-64 - Radiometer, and OP-206 - Radelkis.

As indicator electrodes were used: Orion 94-29A ($CuS+Ag_2S$), with CuS membrane - own-made [2], and Cu_2S single crystal [3].

As reference electrode was used s.c.e. K-401 - Radiometer.

All reagents were of analytical grade and bidistilled water from quartz still was used. The measurements were carried out at 25 $\pm$ 1°C.

RESULTS AND DISCUSSION

The copper ion function of all three electrodes was studied in chloride, bromide and thiosulphate solutions. The results at some selected concentration (Fig. 1) represent differences characteristic for the differences of membrane composition. In chloride as well as in bromide media stable potentials were reached after relatively long times of several hours, except the Orion electrode, for which the equilibration time was approx. 10 min. The electrode response in thiosulphate media became stable after a few minutes. The slow and poorly reproducible response in halide solutions is a consequence of formation of amorphous sulphur in the redox reaction [2]. Slowing down this reaction in reducing thiosulphate medium may lead to formation of crystalline sulphur [4].

The behaviour of the Orion electrode differs from other copper electrodes because in presence of 1 M chloride or bromide a zero-slope region was observed for copper concentrations larger than $10^{-3.5}$ M. This is due to the formation of a new electroactive phase composed of AgCl or AgBr. Such precipitate obviously cannot be formed in presence of thiosulphate because of strong silver complexation. Also in 2 or 3 M bromide the formation of $AgBr_2^-$ or $AgBr_3^-$ prevents the precipitation of the insoluble silver salts. The formation of an electroactive phase in solutions containing 1 - 3 M chloride and 1 M bromide [5] but not in 3 M bromide was confirmed in separate experiments (Fig. 2).

An interesting feature is the changing electrode slope in presence of complexing anions. The stability of copper(II) complexes with halides as ligands is rather low and can be practically neglected. This is not the case with copper(I) complexes. Thiosulphate forms relatively strong complexes [6] with copper(II) ($\beta''_2 = 10^{12.3}$), with copper(I) ($\beta'_2 = 10^{12.9}$) as well as with silver ($\beta_2 = 10^{13.4}$). There is a good approximation to assume that in the considered concentration range the complexes with two ligands are formed only.

On the basis of our studies [5] it may be assumed that in investigated conditions the following reactions predominate in halide $[X^-]$ media, depending on the membrane composition:

$$Ag_2S + 2Cu^{2+} + 4nX^- \rightleftharpoons 2AgX_n^{1-n} + 2CuX_n^{1-n} + S \quad (1)$$

$$CuS + Cu^{2+} + 2nX^- \rightleftharpoons 2CuX_n^{1-n} + S \quad (2)$$

$$Cu_2S + 2Cu^{2+} + 4nX^- \rightleftharpoons 4CuX_n^{1-n} + S \quad (3)$$

and in thiosulphate media, respectively:

$$Ag_2S + 2Cu(S_2O_3)_2^{2-} + 4S_2O_3^{2-} \rightleftharpoons 2Ag(S_2O_3)_2^{3-} + 2Cu(S_2O_3)_2^{3-} + S \quad (4)$$

$$CuS + Cu(S_2O_3)_2^{2-} + 2S_2O_3^{2-} \rightleftharpoons 2Cu(S_2O_3)_2^{3-} + S \quad (5)$$

$$Cu_2S + 2Cu(S_2O_3)_2^{2-} + 4S_2O_3^{2-} \rightleftharpoons 4Cu(S_2O_3)_2^{3-} + S \quad (6)$$

Those equations indicate that in fact the processes occuring in all studied solutions are qualitatively of the same character, but they differ significantly in their quantitative course, because of the differences in stability of respective complexes.

The potential of all mentioned electrodes may be described in interfering ligand media by the same general type of equations. Their derivation in the case of halide interferences is somewhat more complicated because of weak and stepwise complexation and also because of halide precipitation in the case of CuS + Ag_2S membrane. This is not the case when thiosulphate is used as a complexing medium. The full derivation for membranes containing CuS has been pre-

sented elsewhere [5] , but for the Cu_2S membrane may be obtained in a similar manner.
These equations are the following:
for the $CuS-Ag_2S$ membrane:

$$E = E^o + \frac{RT}{2F} \ln \frac{BC}{\beta''_2 K_2{}^{1/2}} + \frac{RT}{F} \ln \frac{C_{Cu}}{[C_L - 2(B + C)\cdot C_{Cu}]^2} \qquad (7)$$

for the CuS membrane:

$$E = E^o + \frac{RT}{2F} \ln \frac{4B^2}{\beta''_2 K_1} + \frac{RT}{F} \ln \frac{C_{Cu}}{[C_L - 4B \cdot C_{Cu}]^2} \qquad (8)$$

for the Cu_2S membrane:

$$E = E^o + \frac{RT}{2F} \ln \frac{4B^2}{\beta'_2{}^2} + \frac{RT}{F} \ln \frac{C_{Cu}}{[C_L - 4B \cdot C_{Cu}]^2} \qquad (9)$$

where B stands for the ratio of diffusion coefficients of di- and monovalent copper complexes, C - is the ratio of diffusion coefficients of divalent copper and silver complexes, C_L and C_{Cu} - are total concentrations of thiosulphate and copper, respectively, K_1 and K_2 - are equilibrium constants for reactions (5) and (4) , respectively, β''_2 and β'_2 - are cumulative stability constants for $Cu(S_2O_3)_2^{2-}$ and $Cu(S_2O_3)_2^{3-}$, respectively, and the other symbols have their usual meaning.

The validity of our considerations was checked by comparison of calculated and experimental data. For this purpose the equations (7), (8) and (9) in the general form:

$$E = \text{const.} + S \log \frac{C_{Cu}}{[C_L - A \cdot C_{Cu}]^2} \qquad (10)$$

has been numerically linearized and its parameters calculated. The electrode slope in all cases approaches that for a monovalent cation, i.e. 59 mV/pCu. Slightly smaller slope was found only for the CuS electrode. This may be explained by the significantly larger conductivity of covelite [7] . The parameter A varies for different membrane materials, but is within limits, which can be expected taking into account the probable values of the diffusion coefficients. The values of const. were compared with the corresponding values calculated on the basis of available equilibrium constants and approximate ratios of diffusion coefficients. The agreement (Table 1) seems to justify the validity of proposed models.

The presented investigations have also important practical aspects. In real analytical samples chloride in variable concentrations occurs often as an interferent in determination of copper. Addition of thiosulphate as a matrix modifier eliminates the effect of chloride and enables undisturbed determination of copper with the additional advantage of having the analytical curve with a slope close to 59 mV/pCu.

Addition of large excess of thiosulphate should be advantageous, because then $C_L \gg C_{Cu}$, and the equations (7), (8) and (9) are approximated to a simple linear dependence between measured potential and log C_{Cu}. This has however also drawbacks because the excessive concentration of thiosulphate increases the solubility product of the membrane material, makes impossible determination of small concentrations of copper in the sample solution and causes excessive corrosion of the electrode. This is especially harmful for the CuS membrane, for which the conditional solubility product in 0.04 M $S_2O_3^{2-}$ equals $K_{so} = 2 \cdot 10^{-16}$. This is also seen from the early curvature of the calibration curve on Fig. 1c. Therefore depending on the expected concentration of copper either 0.4 M $S_2O_3^{2-}$ should be used ($C_{Cu} \geqslant 10^{-3}$ M) or 0.04 M $S_2O_3^{2-}$ ($C_{Cu} = 10^{-3} - 10^{-2.5}$ M) (Fig. 3).

CONCLUSIONS

The processes occuring at the membrane of a solid-state electrode and responsible for the potential response in presence of interfering ions may be simple or complex. In the simple processes only the ion-exchane at the membrane surface and diffusion towards the membrane from the bulk should be considered. Such case is represented e.g. by the bromide interference on the chloride ion selective electrode [8] . However it must be remembered that in such cases also the properties of the new solid phase influence the response.

In many instances the processes are more complex when components of the solution sample react with one or more reactants in the system. In such cases we deal with so-called different mechanism interferences (DM-interferences) [9] . To this group belongs the discussed case of copper selective electrode in solution containing various complexing ligands. The complexity of the system results from the fact that we encounter here, besides the exchange reaction between the solid sulphides and metal ions in solution, also redox reaction in which the sulphide ion is oxidized to elemental sulphur by copper(II), complexation reactions of copper(II), copper(I) and silver ions and in some instances precipitation reactions in which the insoluble silver salts are formed. Those processes in turn are dependant on the transport phenomena, assuming that the chemical processes are fast and not rate determining.

Detailed study of the behaviour of this electrode in thiosulphate media leads also to important practical implications enabling practical elimination of chloride interferences in analytical problems.

REFERENCES

1. J. Gulens, Ion Selective Electrode Rev., 2, (1980) 117
2. E. Ghali, B. Dandapani and A. Lewenstam, J. Appl. Electrochem., 12, (1982) 369
3. T. Hepel, Anal. Chim. Acta, 123, (1981) 151, 161; 142, (1982) 217
4. T. Biegler and D. A. Swift. J. Appl. Electrochem., 9, (1979) 545
5. A. Lewenstam, T. Sokalski and A. Hulanicki, Talanta (in press)
6. W. Buckel and R. Hilsch, Z. Physik, 128, (1950) 324
7. L. G. Sillén and A. E. Martell, Stability Constants of Metal-Ion Complexes, Spec. Publ. No 17 and No 25. The

Chemical Society, London, 1964 and 1971
8. A. Hulanicki and A. Lewenstam, Anal. Chem., 53, (1981) 1401
9. G. den Boef and A. Hulanicki, Pure Appl. Chem., 55, (1983) 553

Table 1. Comparison of experimental and calculated values of parameters of equation (10)

	const., mV		S, mV/pCu		A	
	calc.	exp.	calc.	exp.	calc.	exp.
CuS	⟨-187;-73⟩	⟨-62;-57⟩	59.2	⟨50;55⟩	⟨1.8;4⟩	1.3
$CuS+Ag_2S$	⟨-193;-79⟩	⟨-183;-136⟩	59.2	⟨52;59⟩	⟨1.8;4⟩	3.5
Cu_2S	⟨-247;-189⟩	⟨-209;-188⟩	59.2	⟨50;58⟩	⟨1.8;4⟩	2.5

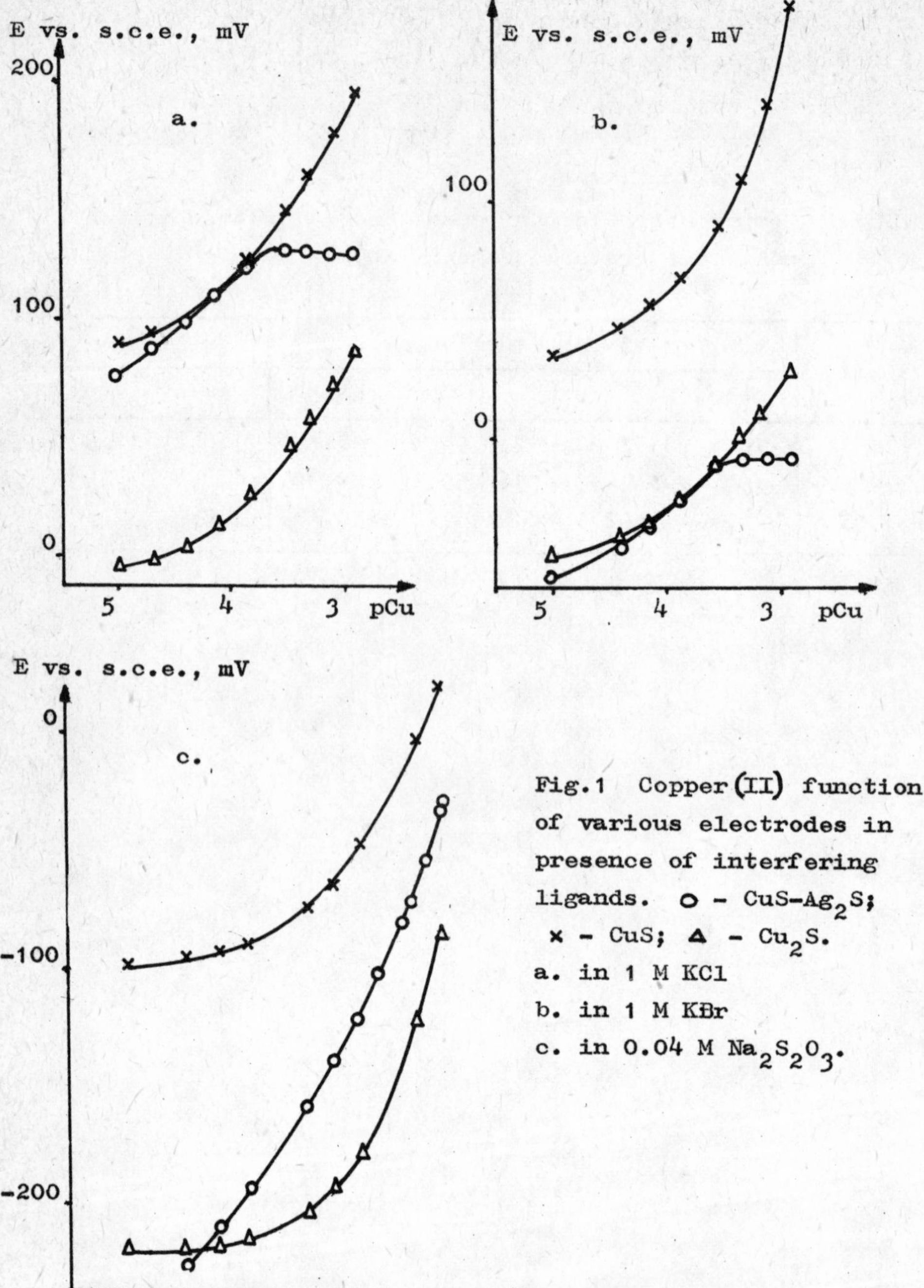

Fig.1 Copper(II) function of various electrodes in presence of interfering ligands. o - $CuS\text{-}Ag_2S$; × - CuS; Δ - Cu_2S.

a. in 1 M KCl

b. in 1 M KBr

c. in 0.04 M $Na_2S_2O_3$.

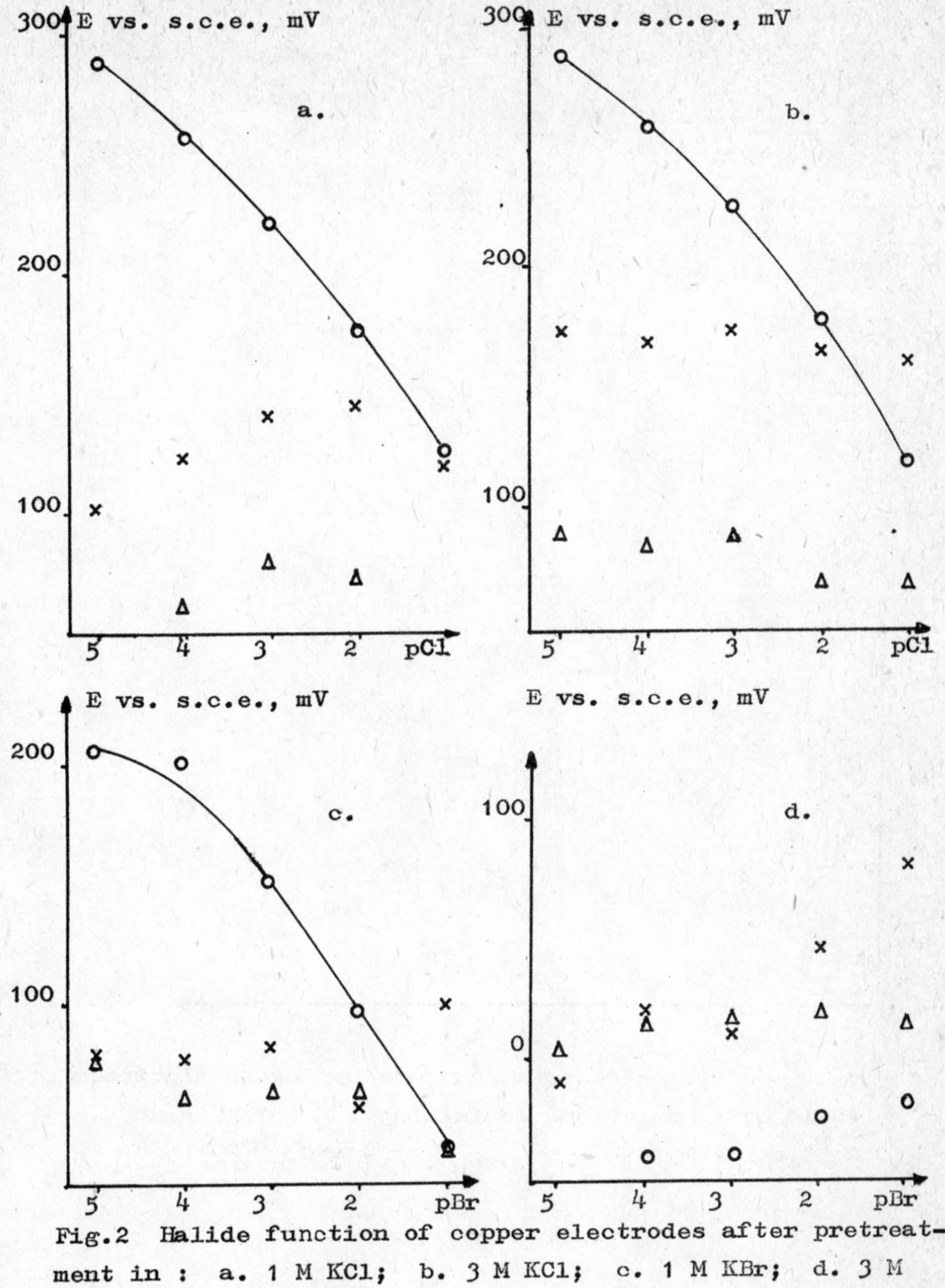

Fig.2 Halide function of copper electrodes after pretreatment in : a. 1 M KCl; b. 3 M KCl; c. 1 M KBr; d. 3 M KBr. o – $CuS-Ag_2S$; × – CuS; Δ – Cu_2S.

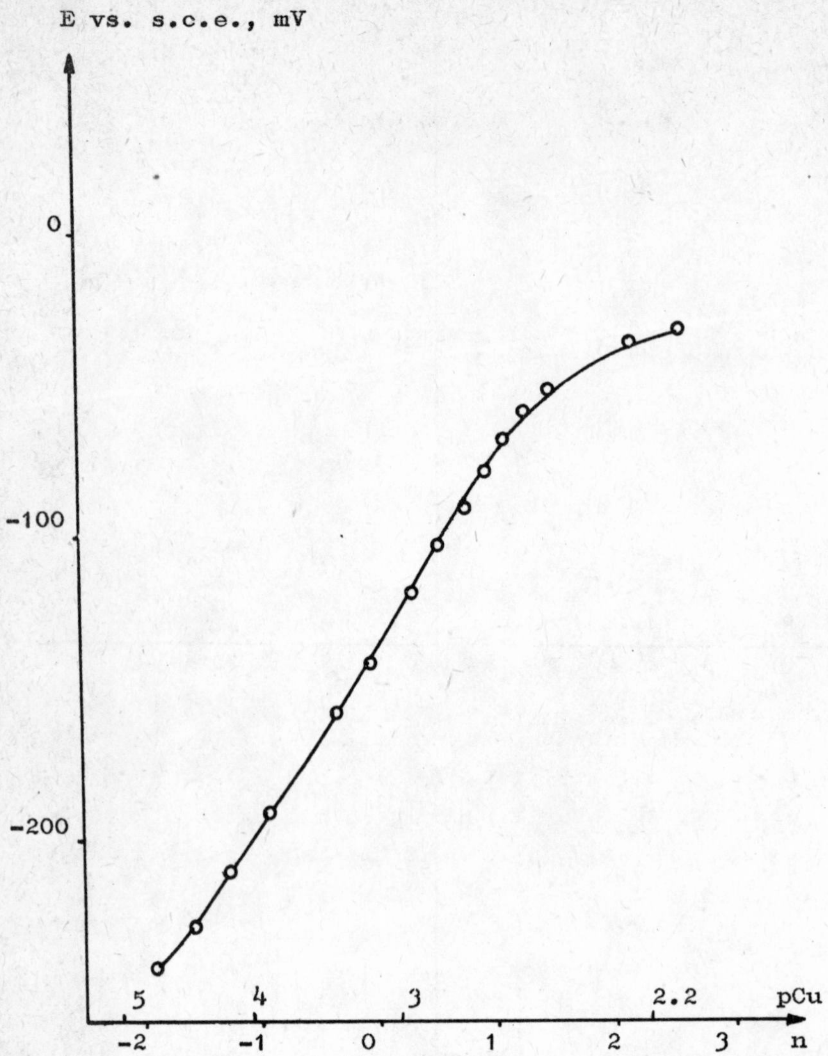

Fig.3 Corrected calibration curve of Orion electrode ($CuS-Ag_2S$) in solutions containing 1 M chloride and 0.04 M thiosulphate. $n = C_{Cu}(C_L - A \cdot C_{Cu})^{-2}$.

4th Symposium on Ion-Selective Electrodes
Mátrafüred, 1984

INVESTIGATION OF THE INFLUENCE OF INERT MEMBRANES ON THE SELECTIVITY OF LIQUID STATE ISE BY POTENTIOMETRY AND CYCLIC VOLTAMMETRY

B. HUNDHAMMER, S. BECKER and H.J. SEIDLITZ

Technical University "Carl Schorlemmer"
Leuna-Merseburg, Department of Chemistry
DDR - 4200 Merseburg, Otto-Nuschke-Str.

ABSTRACT

The application of voltammetric techniques to study the ion transfer across the interface of immiscible electrolyte solutions is an additional tool in interpreting the selectivity behaviour of liquid state ISE. Voltammetric investigations are not restricted to the liquid-liquid interface, but it may be also applied if two liquid phases are separated by an inert hydrophilic membrane.
It is the aim of the paper to compare the potentiometrically obtained results with the observed voltammetric behaviour of such interfaces with respect to the influence of the inert membrane.

INTRODUCTION

Ion transfer across the interface of immiscible electrolyte solutions has been studied by several electrochemical techniques (for a summary see (1,2,3)). Those experiments yield very useful information about the thermodynamics and the kinetics of the ion transfer from water to the organic phase which can be used to elucidate the response mechanism of liquid state ISE. Key thermodynamic parameter is the standard Gibbs energy of ion partition between water and the membrane solvent. It has been shown that standard Gibbs energies of partition can be obtained from voltammetrically determined half-wave potentials if the experimental ΔE scale with an arbitrary reference point is transformed to a potential scale with $\Delta\varphi = 0$ V as zero point (4). This point can be fixed according to Parkers tetraphenylarsonium tetraphenylborate assumption (TPAsTPB-assumption) (5,6,7) if TPAsTPB is employed as supporting electrolyte in the organic phase (4). The standard Gibbs energies of anion partition have been evaluated by this method in the system water/nitrobenzene, water/1,2-dichloroethane (4), water/acetophenone (8) and water/nitrobenzene-chlorobenzene (9).

If the standard Gibbs energies of partition for the individual ions are known, the potential response and the selectivity behaviour of liquid state ISE can be predicted taking salt partition and ion exchange into account (10).

Voltammetric investigations of the ion transfer are not restricted to the liquid-liquid interface, but they can be applied to systems in which the two liquid phases are separated by an hydrophilic membrane. This arrangement resembles a liquid state electrode using cellulose dialysis membrane in order to fix the membrane phase (11). The membrane is supposed to be inert with respect to the response of the electrode. On the other hand, it is known that cellulose membranes exhibit ion-exchange properties or they may exclude large particles from penetrating the membrane. In this paper the influence of cellulose dialysis membrane (PT 150) upon the standard Gibbs energies of ion partition in the model system water/nitrobenzene will be discussed. The results of the voltammetric investigations are compared with those of potentiometric measurements.

EXPERIMENTAL

The voltammetric measurements were carried out with a four electrode potentiostat with automatic IR-compensation by means of positive feedback. The IR-compensation was set to the nearest point before the potentiostat starts oscillating. The potentiostat PS 3 (Forschungsinstitut Meinsberg) was employed. The voltammograms were recorded with an X-Y-recorder (Endim 620.02). The potentiometric measurements were carried out with the pH-meter MV 87 (VEB Präcitronik Dresden).

All solutions were prepared from salts of analytical grade with twice-distilled water. Crystal violet tetraphenylborate (CVTPB) and TPAsTPB were employed as supporting electrolytes in nitrobenzene, lithium sulfate serving as supporting electrolyte in the aqueous phase. For the potentiometric experiments the organic phase was prepared by shaking equal volumes of 0.1 M aqueous solutions of a salt of the primary ion and a 10 mM solution of CVTPB in nitrobenzene for two hours. The aqueous phase was used as an internal solution in the liquid state electrode. The experiments were carried out at laboratory temperature (20 $\pm$ 2 ^{o}C).

RESULTS AND DISCUSSION

Fig. 1 compares the cyclic voltammogram for the transfer of ClO_4^- across the water-nitrobenzene interface with the cyclic voltammogram if the two phases are separeted by a cellulose dialysis membrane. In Fig. 1 the ΔE potential scale is transfered to a $\Delta\varphi$ potential scale according to the TPAsTPB-assumption. It can be seen that the peak potentials of the current peaks are not to be changed in the presence of the cellulose dialysis membrane. This behaviour indicates that there is no change in the standard Gibbs energies of partition between water and nitrobenzene or in other words, there is no interaction of the perchlorate ion with the membrane. This also holds for the other anions studied (NO_3^-, I^-, SCN^-). This observation is well in accordance with the results obtained potentiometrically (Fig.2). From the potential shift of the calibration curves of the ions to each other relative standard Gibbs energies of ion partition can be calculated if one of the ions is chosen as reference. Table 1 compares standard Gibbs energies obtained in this way with those obtained by cyclic voltammetry at the water-nitrobenzene interface. The agreement between the relative standard Gibbs energies obtained by the two different methods shows that the cellulose dialysis membrane can be considered to be inert with respect to the anion transfer across the water-membrane interface.

Besides the standard Gibbs energies of partition the diffusion coefficient can be evaluated from the voltammogram. The plot of i_p for the transfer of ClO_4^- vs the square root of the sweep rate is shown in Fig.4. The apparent diffusion coefficients are approximately by one order in magnitude less than the diffusion coefficients in water. They are found to be independent of the thickness of the membrane. The current peaks due to the transfer of the anions from water to nitrobenzene change to a wave if the two phases are separated by the membrane. This observation may be explained by the assumption of a diffusion layer with constant thickness.

The transfer of tetrabutylammonium ions ($TBuA^+$) from water to nitrobenzene is observed in the cyclic voltammogram at the liquid-liquid interface (Fig.3a). If the membrane is introduced there is no indication of the transfer of $TBuA^+$ (Fig.3b). Since there is no reason to assume that $TBuA^+$ is excluded from the transfer due to the size of the ion (r_{TBuA^+} = 3.8 Å, the diameter of the membrane pores is about 100 Å), this observation can be explained if ion-exchange at the membrane is taken into consideration. The exclusion of $TBuA^+$ from the transfer is also observed by potentiometric measurements. Fig.5 shows the potential dependence of a nitrate-selective electrode upon the concentration of TBuACl. Without membrane the response of the electrode is typically for the cation caused by the exchange of CV^+ against $TBuA^+$ in the organic phase. If the membrane is present the electrode response indicates that $TBuA^+$ is excluded from the ion-exchange reaction. Thus we conclude that the employment of special membranes will be one possibility to improve the selectivity of ISE.

Futher studies on this topic are in progress in our laboratory.

REFERENCES

1 J.Koryta, Electrochim.Acta, 24 /1979/ 293.

2 J.Koryta, Hung.Sci.Instr., 49 /1980/ 25.

3 J.Koryta, Electrochim.Acta, 29 /1984/ 445.

4 B.Hundhammer and Theodros Solomon, J.Electroanal.Chem., 157 /1983/ 19.

5 O.Popovych, Crit.Rev.Anal.Chem., 1 /1970/ 73.

6 A.J.Parker, Chem.Rev., 69 /1969/ 1.

7 B.G.Cox, G.R.Hedwig, A.J.Parker and D.W.Watts, Aust.J. Chem., 27 /1974/ 477.

8 Theodros Solomon, Hailemichael Alemu and B.Hundhammer, J.Electroanal.Chem., 169 /1984/ 303.

9 Theodros Solomon, Hailemichael Alemu and B.Hundhammer, J.Electroanal.Chem., 169 /1984/ 311.

10 B.Hundhammer, H.J.Seidlitz, S.Becker, S.K.Dawan and Theodros Solomon, J.Electroanal.Chem., in press.

11 J.Ross, Science, 156 /1967/ 1378.

Table 1. Relative standard Gibbs energies of ion partition between water and nitrobenzene based on ClO_4^- as reference ion

Ion	$G^{o\ w\ nb}_{p(rel)}$ / kJ mol^{-1} obtained potentiometrically	$G^{o\ w\ nb}_{p(rel)}$ / kJ mol^{-1} obtained voltammetrically
ClO_4^-	0	0
I^-	12.1	11.5
SCN^-	8.8	8.6
NO_3^-	17.0	17.0
Br^-	22.5	22.9

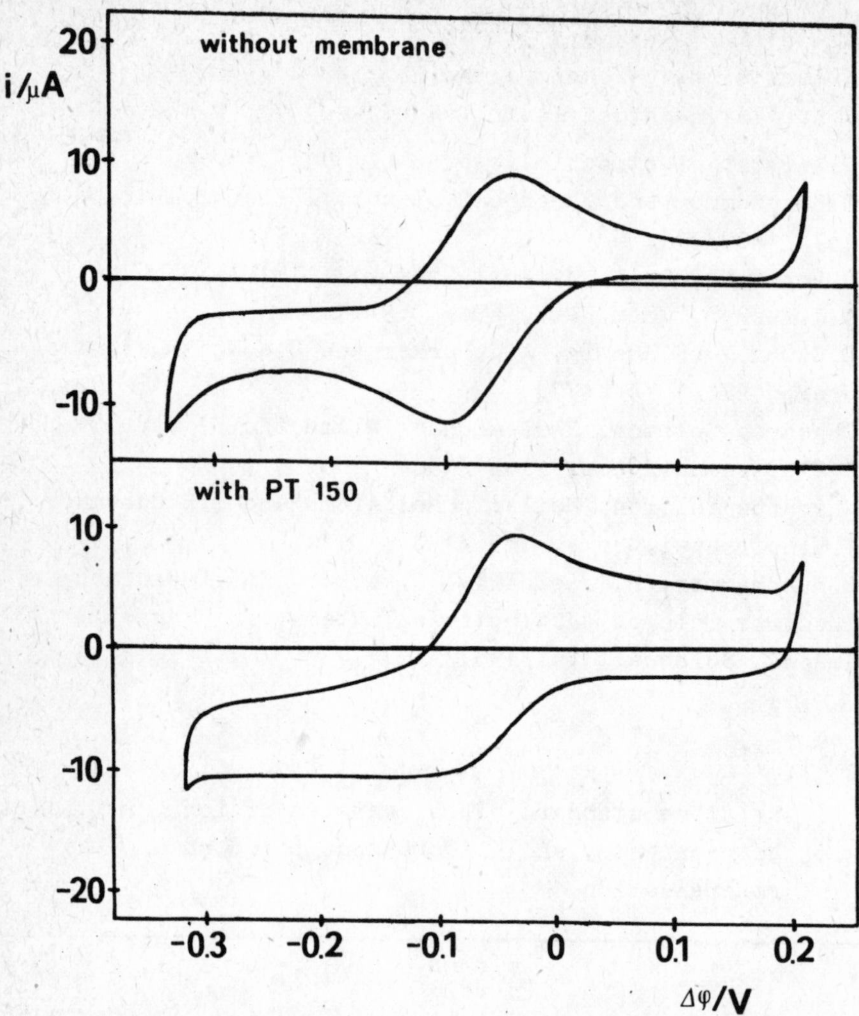

Fig. 1. Comparison of the cyclic voltammogram of the transfer of ClO_4^- across the water-nitrobenzene interface with the corresponding cyclic voltammogramm in the presence of a cellulose dialysis membrane PT 150. Supporting electrolytes: 10 mM CVTPB in nitrobenzene

10 mM LiF in water

Sweep rate 10 mV s^{-1}; $c_{ClO_4^-}$ = 0.25 mM.

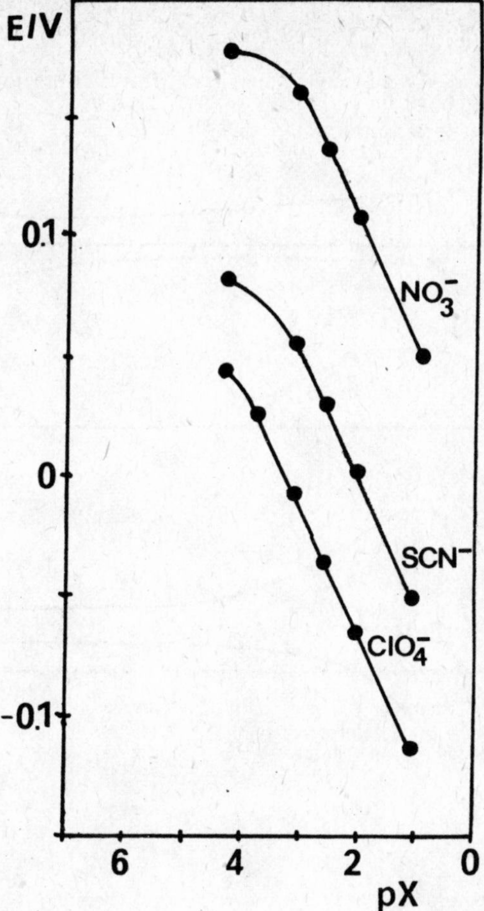

Fig.2. Calibration curves of liquid state ISE with PT 150.

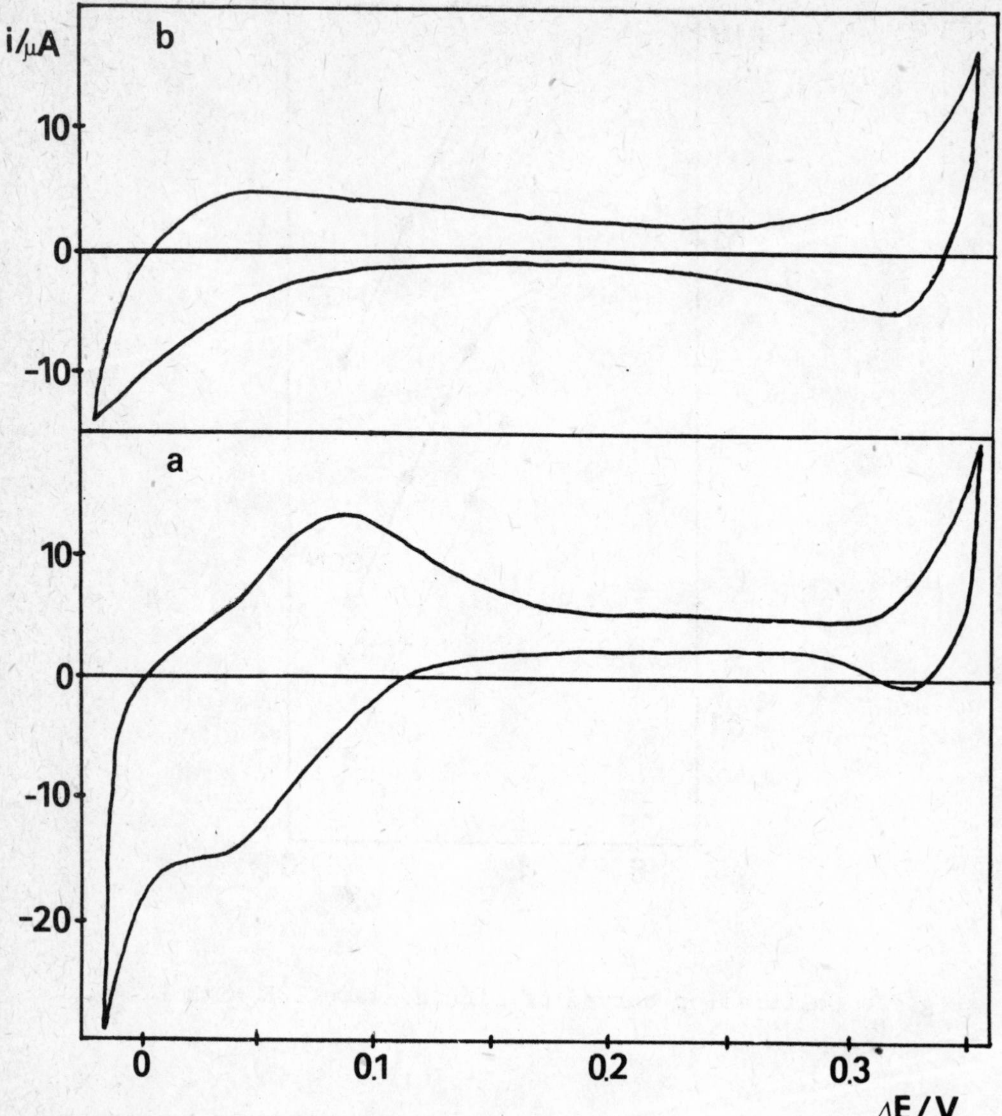

Fig.3. Cyclic voltammograms for the transfer of $TBuA^+$
a) without membrane
b) with PT 150
Conditions as in Fig.1.

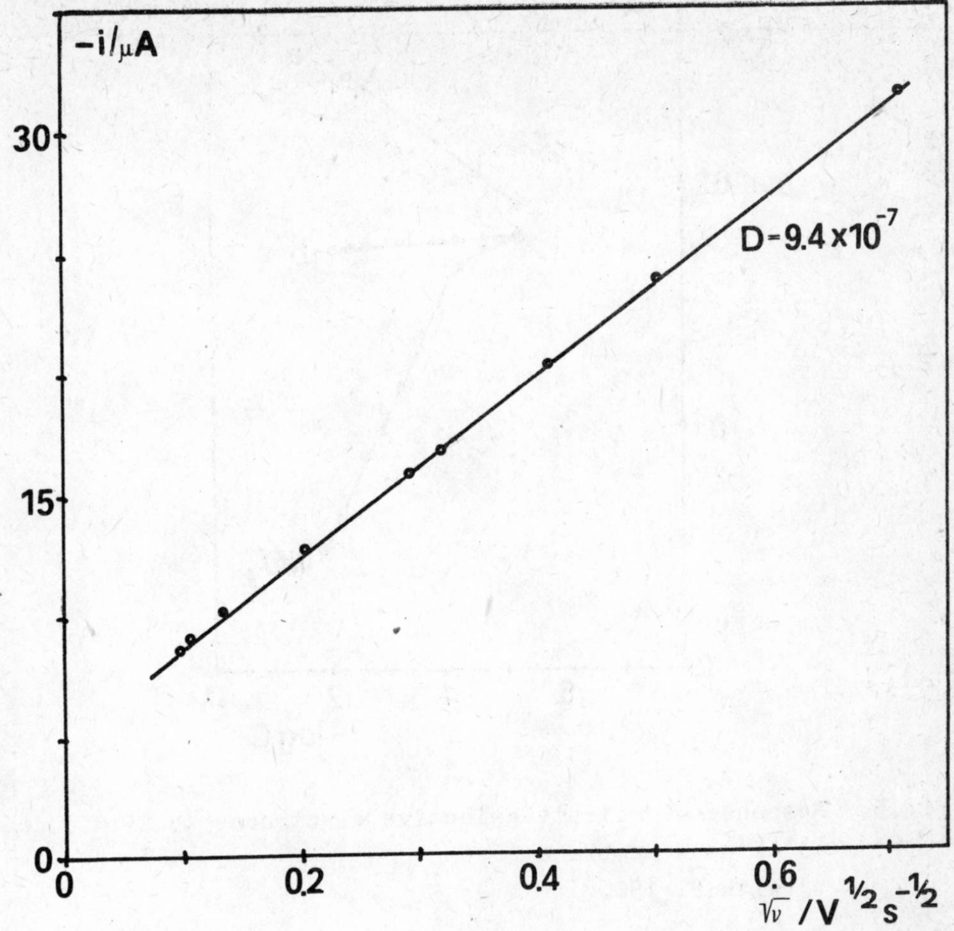

Fig.4. Dependence of the peak current upon the square root of the sweep rate for the transfer of ClO_4^- (w) to ClO_4^- (nb) with PT 150.

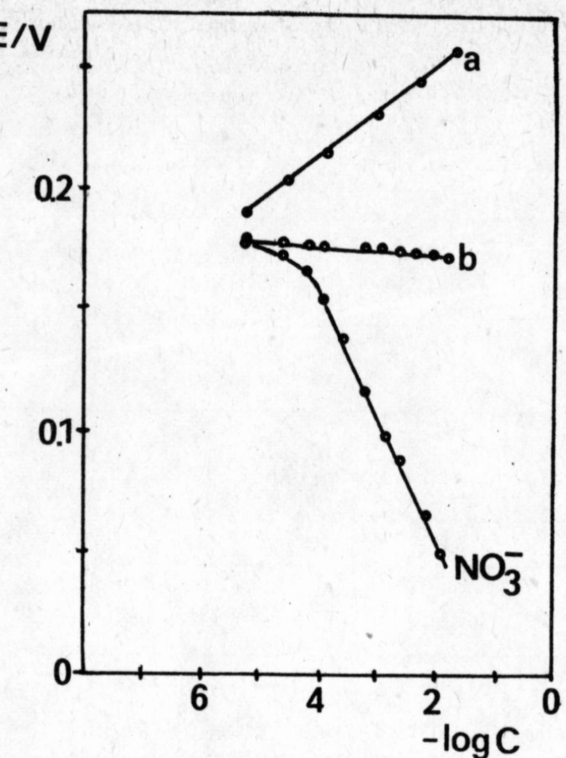

Fig.5. Response of nitrate-selective electrodes to TBuACl.
a) without membrane
b) with PT 150.

4th Symposium on Ion-Selective Electrodes
Mátrafüred, 1984

FLOW-THROUGH TITRATION OF AMINO ACIDS USING pH-SENSITIVE GLASS ELECTRODE

N. ISHIBASHI, T. IMATO, C. AZEMORI and Y. ASANO*

Faculty of Engineering, Kyushu University, Hakozaki, Higashiku, Fukuokashi, 812, Japan

*Denki Kagaku Keiki Co., Ltd., Musashinoshi, 180, Japan

ABSTRACT

Rapid analysis of amino acids based on the Formol Titration reaction was successfully conducted by using a flow-single point titration method. By an injection technique, amino acids such as glycine, L-valine, L-leucine and glutaminc acid etc. could be determined at sampling rates of 40 to 50 samples/hr. A separate determination of amino acids was also possible in combination with a HPLC technique.

INTRODUCTION

Rapid volumetric analyses by using a continuous flow technique have been developed by many investigators [1-7]. Astrom [8] has applied "Single Point Titration method [9-11]" to the flow injection analysis of acids and bases, using a buffer solution which is ingeniously designed to show the linear pH response to the sample concentration.

In the present paper, the theoretical treatment of the flow-single point titration method using a simple buffer solution (a weak acid with its conjugate base) is made for determination of acids and bases. After the experimental test of the proposed flow-single point titration method, a flow analysis of amino acids based on the Formol Titration reaction was successfully performed by the proposed method in combination with FIA and HPLC techniques.

THEORETICAL CONSIDERATION

We illustrate the principle of the flow-single point titration using one buffer solution (a weak acid and its conjugate base) by an example of the determination of an acid. The two channel flow system is shown in Fig. 1 (a), where a buffer solution containing a weak acid HA with its conjugate base A^- is pumped through one channel at a flow rate of V_{Buff} and water is pumped through another channel at the flow rate of V_S (ml/min). In injection analysis, a sample acid HB is injected into the water stream, but in order to perform theoretical analysis, we treat the case that an acid HB is continuously pumped as shown in Fig. 1 (b). This manifold is applicable to a continuous monitoring for a flowing sample. The potential of the glass electrode governed by the buffer solution HA - A is expressed by Eq. (1).

$$E_1 = E^O + 0.059 \log K_{a,HA} + 0.059 \log (C_{HA}/C_A) \quad (1)$$

where E^O and $K_{a,HA}$ are the standard potential of the electrode and the dissociation constant of acid HA, respectively. C_{HA} and C_A are the original concentrations of HA and A^-, respectively, in the buffer solution. E_1 represents the baseline potential. When the sample solution is mixed with the buffer solution, a following reaction (2) occurs and the expression of the electrode potential becomes Eq. (3).

$$A^- + HB \rightleftarrows HA + B^- \quad (2)$$

$$E_2 = E' + 0.059 \log [(C_{HA}V_{Buff} + C_XV_S)/(C_AV_{Buff} - C_XV_S)] \quad (3)$$

where $E' = E^O + 0.059 \log K_{a,HA}$ and C_X is the concentration of HA formed by the reaction (2). The potential change ΔE is expressed by Eq. (4) in the case for $V_{Buff}=V_S$.

$$\Delta E = 0.059 \log [(1 + (C_X/C_{HA}))/(1 - (C_X/C_A))] \quad (4)$$

When HB is a weak acid of the dissociation constant $K_{a,HB}$, C_X is expressed as follows.

$$C_X = [K_r(C_A+C_{HB})+ C_{HA} - (K_r(C_A + C_{HB})+ C_{HA})^2 -4(K_r -1)K_rC_AC_{HB}] \\ /2(K_r -1). \quad (K_r \neq 1). \tag{5}$$

where $K_r = K_{a,HB}/K_{a,HA}$. When HB is a strong acid, C_X is identical to the concentration of the sample C_{HB}. In the case K_r = 1, the potential change ΔE becomes as follows:

$$\Delta E = 0.059 \log [1 + (C_{HB}/C_{HA})] \tag{6}$$

Some of the calibration curves calculated by using Eqs. (4)-(6) are shown in Fig. 2 for the case that $C_{HA}=C_A$. In the case of $C_X/C_{HA} \ll 1$ and $C_X/C_A \ll 1$, Eq.(4) is reduced nearly to $\Delta E = 0.059 \times 2\ (C_X/C_{HA})$. Namely, the approximately linear relation holds between the concentration of a sample acid and the potential change. As can be seen from Fig. 2, the sensitivity depends on the parameter of K_r which is the relative values of the dissociation constant between the acid used as the buffer solution and the sample acid. When K_r is greater than 10^3, the calibration curves become almost the identical to those in the strong acid like HCl because of almost complete advancement of the reaction of Eq. (2).

In injection analyses, an injected sample is dispersed during flowing through the tube. However, a degree of dispersion at the place of the detector may be constant, if the flow condition is strictly controlled. In such injection analysis, the electrode response will be proportional to that calculated by Eqs. (4) - (6).

EXPERIMENTAL

Apparatus. A flow injection system (FICS-10, Denki Kagaku Keiki Co., (DKK), Japan) equipped with a peristaltic pump (Gilson minipuls pump II) and an injector was used. A flow-through type glass electrode detector, which has a hold up volume less than 10 µl, was used. The electrode potential was measured with an Ion-Meter (IOC-10, DKK, Japan) and a recorder (SR651, Watanabe Sokki Co., Japan). A HPLC system was constructed with a pump (L-4000W,Yanagimoto Co., Japan) an injector (100 ul) and a separation column. The chromtographic separation of amino acids was performed by means of the ODS column (4 mm i.d., 250 mm length, ODS-120A, Toyo Soda Co., Japan), by using water as an eluent.

Reagents. All reagents were of analytical grade and used without purification. A solution of formaldehyde was diluted with deionized water and neutralized by NaOH solution before use, since the received solution contains a small amount of formic acid.

RESULTS AND DISCUSSION

Test of the proposed method.

Figure 3 shows the flow diagram for injection analysis of acids and bases and the typical response peak signals for hydrochloric acid samples. A good linear relationship exists between peak heights and concentrations of hydrochloric acid as expected from Fig. 2. By comparing peak heights obtained to the theoretical calibration curves, the sample HCl is found to be diluted by about 1.3 times at this manifold. Reproducibility of peak heights for triplicate injection is good and the sampling rate of about 50 samples/hr is possible at this flow rate.

Application to the Formol Titration.

The Formol Titration [12,13] has been used for the determination of amino acids since Sorensen [14] applied it for quantitative analyses of amino acids. The reaction of the Formol Titration is as follows, if we take glycine as an example.

$$H_2NCH_2COOH + n\ HCHO \longrightarrow HOH_2CNHCH_2COOH \text{ or } (HOH_2C)_2NCH_2COOH$$

The N-hydroxymethyl glycine or N-di(hydroxymethyl) glycine formed by the reaction behaves as more stronger acid than the parent amino acid and can be determined by the FIA using a manifold shown in Fig. 4 (a). An amino acid sample injected into the water stream, is confluenced to the formaldehyde solution stream. The acid produced is confluenced to the stream of the phosphate buffer solution and changes the pH of the buffer solution. Figure 4 (b) shows the response peak signals for glycine. The linear relationship exists between peak heights and concentrations of glycine. Similar calibration curves were obtained for L-valine, L-alanine with the identical sensitivity to that for glycine. This is because that the pK_a values of the acids formed from amino acids [12,13] are sufficiently smaller than that of HPO_4^{2-} used as the phosphate buffer solution of HPO_4^{2-} - PO_4^{3-}. The sensitivity of glutamic acid was found to be about two times larger than that of glycine since the acid formed from glutamic acid by reaction with formaldehyde behaves as the diprotic acid . Based on the successful results for the sample containing a single amino acid species, the total concentration of mixed aminoacids in Sake (Japanese rice wine) was determined by the same method as well as the conventional Formol Titration method. The Sake samples were neutralized to pH 7 with the standard NaOH solution before analysis in order to mask the organic acids. The correlation between the proposed method and the conventional Formol Titration method was good.

Separate determination of each amino acid was conducted by applying the flow Formol Titration to the HPLC analysis. Figure 5 shows the HPLC system and the resulting chromatogram for amino acids. The reagent solution containing formaldehyde and Na_2HPO_4 - Na_3PO_4 buffer solution was added in the post column technique to the amino acids stream eluted from the chromatographic column. The linear calibration curves were obtained in the concentration range from 0 to 1.5 x 10^{-2} M for each amino acid. The peak heights decrease in the order of glycine, L-valine and L-leucine, irrespective of the same concentration. This decrease is due to the broadening of the peak. The FIA and HPLC combined flow titration method described in the present paper can be useful for continuous monitorings of flowing sample solutions by using the manifold shown in Fig. 1 (a). Applications of this flow-titration technique to redox titration and chelatometric titration will be described in elsewhere.

REFERENCES

1. W. J. Blaedel and R. H. Laessig, Anal. Chem., 36, /1964/1617.
2. B. Fleet and A. Y. Ho, Anal. Chem., 46, /1974/ 9.
3. G. Nagy, K. Tóth and E. Pungor, Anal. Chem., 47, /1975/1460.
4. G. Nagy, Zs. Fehér, K. Tóth and E. Pungor, Anal. Chim. Acta, 91, /1977/ 87.
5 G. Nagy, Zs. Fehér, K. Tóth and E. Pungor, Anal. Chim. Acta, 91, /1977/ 97.
6. J. Ruzicka, E. H. Hansen and H. Mosbeak, Anal. Chim. Acta, 92, /1977/ 235.
7. A. U. Ramsing, J. Ruzicka and E. H. Hansen, Anal. Chim. Acta, 129, /1981/ 1.
8. O. Astrom, Anal. Chim. Acta, 105, /1979/ 67.
9. G. Johansson and W. Backen, Anal. Chim. Acta, 69, /1974/ 415.
10. O. Astrom, Anal. Chim. Acta, 88, /1977/ 17.
11. O. Astrom, Anal. Chim. Acta, 97, /1978/ 259.
12. M. Levy, J. Biol. Chem., 99, /1933/ 767.
13. M. L. Anson and J. T. Edsall, Ed., "Advances in protein chemistry" vol. II, Academidc Press, New York, 1945, p289
14. S. P. L. Sorensen, Biochem. Z., 7, /1908/ 45.

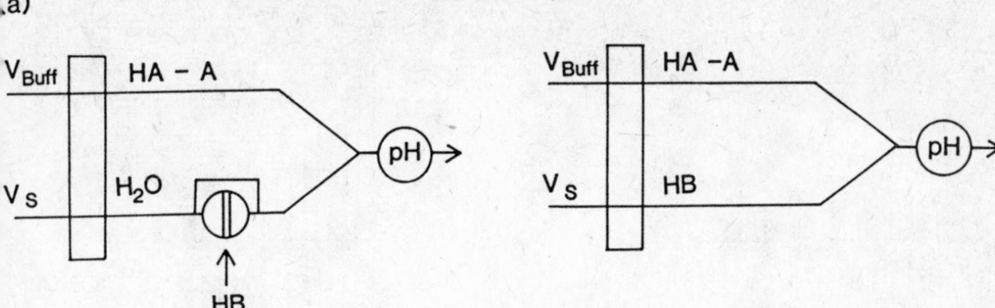

Fig. 1. Flow diagrams of continuous single point titration.
(a) Sample injection method.
(b) Continuous flow method.

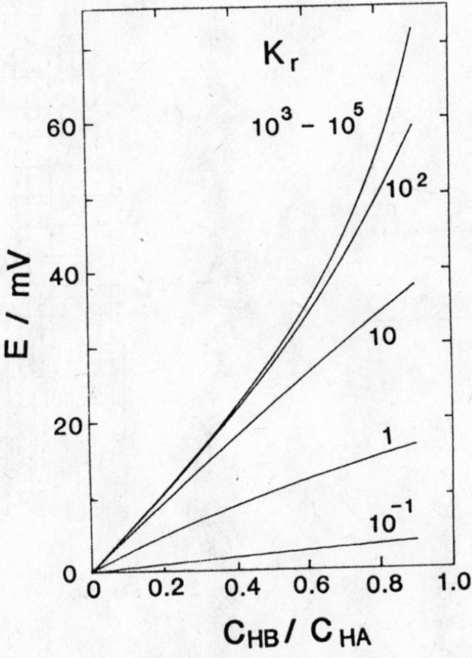

Fig. 2. Theoretical calibration curves.
K_r; ratio of the dissociation constants of an acid of a sample to a weak acid of a buffer solution.
($K_r = K_{a,HB}/K_{a,HA}$)

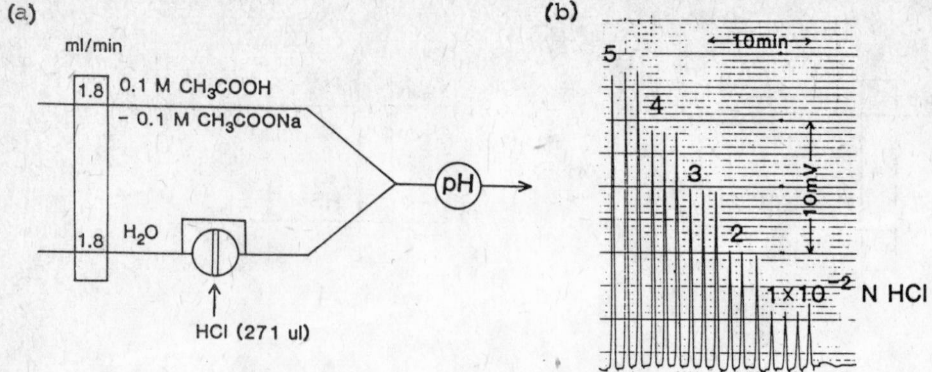

Fig. 3. (a) Flow diagram for determination of acids or bases.
(b) FIA peaks for hydrochloric acid.

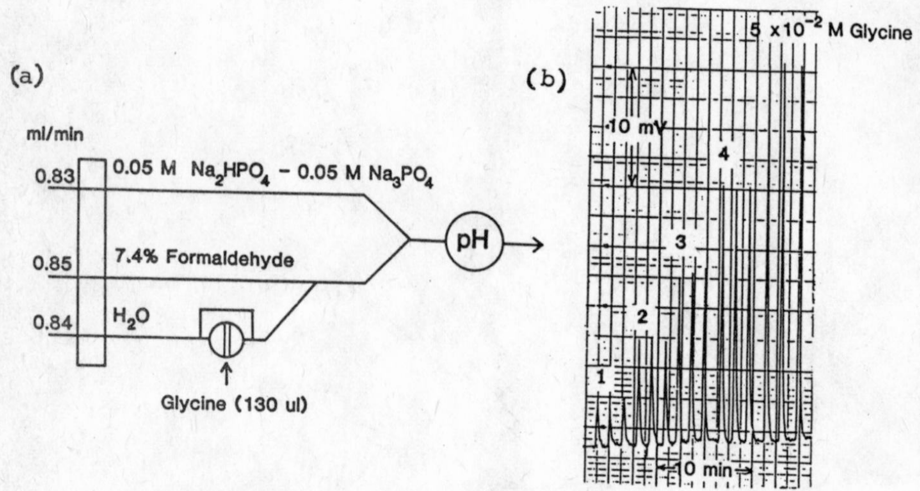

Fig. 4. FIA of amino acids, by using Formol Titration.
(a) Flow diagram.
(b) FIA peaks for glycine.

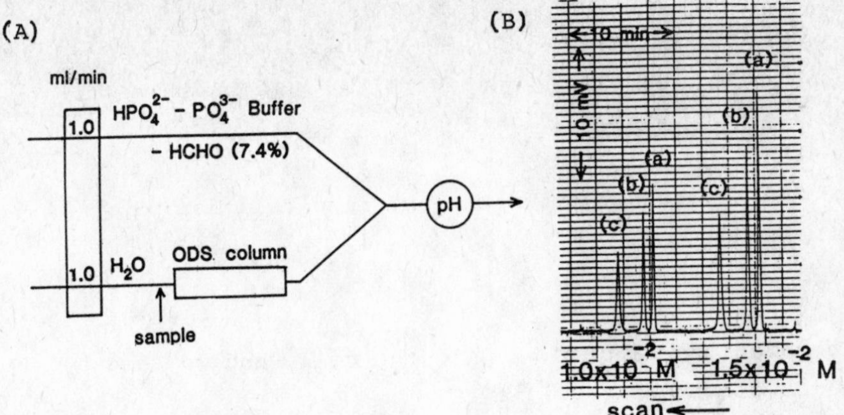

Fig. 5. (A) Flow diagram for HPLC determination of amino acids.
(B) Chromatograms of amino acids.
Sample (100 ul); (a):glycine, (b):L-valine, (c):L-leucine. Concentration of amino acids is indicated in (B).

QUESTION

Participants of the discussion: H.Müller and N. Ishibashi

Question:

What was the construction of your flow cell for pH determination ?

Answer:

We screwed the electrodes into the cell, and we used a flat glass electrode.

4th Symposium on Ion-Selective Electrodes
Mátrafüred, 1984

BEHAVIOUR AND PROPERTIES OF SILVER BASED, MELT COATED SILVER HALIDE SELECTIVE ELECTRODES

V.M. JOVANOVIĆ, M. RADOVANOVIĆ and M.S. JOVANOVIĆ*

Institute for Chemistry, Technology and Metallurgy, Department for Chemical Engineering and Inorganic Technology, Yu-11001 Beograd, Njegoševa 12, Yugoslavia

*Institute for Analytical Chemistry, Faculty of Technology and Metallurgy, University of Beograd, YU-11001 Beograd, Karnegieva 4/II, Yugoslavia

ABSTRACT

Four types of silver based, AgX/X=Cl, Br and I respectively/ melt coated electrodes were prepared and tested in halide and silver solutions. Their resistance towards strong oxidants was checked, and their applicability for determination of chloride in potable waters, iodide in a sample and for titration of cyanide was also investigated.

INTRODUCTION

Since the pioneering work of Kolthoff and Sanders /1/ on fused silver halide electrodes with inner reference, but also on coated wire type in 1937, the interest for halide CWEs was re-born nearly 30 years later. Bishop and coworker /2/ described in their paper the anodical preparation of Ag/AgX electrodes in 1963. Another important report on silver halides obtained from fused materials, was made known by Siemroth and coworkers /3/ at this very place in 1976. Harzdorf /4,5/ thoroughly investigated the behaviour of anodically prepared Ag/AgX electrodes in 1974 and again in 1982, especially relating to oxidants and on non-porosity of the deposit.

Bearing in mind Kolthoff and Sanders's experiences with silver halide disc electrodes obtained from molten state, our intention was to produce silver based, silver halide electrodes which any analyst can prepare in the most simple way under ambient conditions, avoiding any complicated apparatus. A silver strip was coated with silver halide, being immersed into the melt of either KX allowing the metal to react chemically /type A/, or into the already molten AgX obtained in different ways /types B, C and D/ - see Fig. 1.

EXPERIMENTAL

Instrumentation

Digital pH-meter MA5705 "Iskra"-Kranj /Yugoslavia/ or Radiometer PHM62, motor piston-burette Metrohm E-415, Zeiss K201 /GDR/ recorder and reflectance light microscope Reichert Me2862/IV /Austria/, were used for direct measurements, potentiometric titrations and examinations of electrode surface.

ISEs used were as follows: 1. Anionic set of universal Ružička type 'Selectrode' /Radiometer/ F3005, F3006 and F3007, 2. self-made version of universal Ružička type electrodes described previously /6/ designated GPE-Cl, GPE-Br and GPE-I differing from those of Radiometer production as shown /Fig.2/.

Silver strips /99.9999%/ some 15 cm in length, 3 mm wide and 1 mm thick, polished to mirror-like appearance, were applied to be coated with the melt.

Substances to be melted were put either in platinum or in porcelain crucibles and heated on a Bunsen flame without any protection.

Chemicals and solutions

All chemicals used were of analytical purity.

Standard halide solutions except those of pX 1.12 and 2.05, were adjusted to uniform ionic strength using 10^{-3} molar KNO_3 solution. Chlorides and cyanides to be determined were titrated with a 0.0279 molar $AgNO_3$ solution.

Preparation of melt-coated halide electrodes

Type A

About 1 g of each, KX and KNO_3 were pulverised, put into the crucible and heated just above melting point. A silver strip was then kept in the melt during a period of time depending on the halide in question: 15 min.for KCl + KNO_3, 7 min. for KBr + KNO_3 and 3 min. for KI + KNO_3 melt. After slow cooling, the Ag/AgX electrode was washed twice in boiling water. Appearance: slight dark colour of not uniformly thick /several µm/ coverage. Conditioning: a few hours in distilled water prior to use.

Type B

A stoichiometric admixture of $AgNO_3$ and of KX was pulverised, put into a crucible and got just to fuse. The silver strip was kept in the melt a few times for a second or two, just to obtain a coating of visually uniform thickness. After slow cooling, the electrode was washed twice in boiling water. Appearance: Glassy coating approx. 0.1 mm thick and of pink colour for chloride, brown for bromide and black for iodide electrode.

Conditioning: not necessary.

Type C

The electrode of type B, after washing in boiling water, was kept for a few seconds in a saturated solution of H_2S.

Type D

Either of the silver halides was precipitated using equal volumes of 0.1 molar $AgNO_3$ and KX solutions /pH=3/. The precipitate was thoroughly washed in a G-4 crucible, and a volume of approx. 5 cm^3 of saturated H_2S solution was forced to penetrate the precipitate afterwards, in order to obtain only a surface layer of Ag_2S. After keeping overnight at 80°C, the mass was pulverised. The colour of the sensing powder, used also for the activation of self-made GPE matrix, was from mice-grey for chloride, to dark yellow-green for iodide electrode. Preparation of the coverage as for type B. Conditioning is not necessary for any of the black Ag/AgX/Ag_2S electrodes with glassy appearance.

/Non/-porosity test of the electrodes

Although reflectance light microscopy indicated no "open holes" of the coverage of any of the freshly prepared electrodes /Fig.4/, the durability of those of type A did not exceed a few weeks. Also, a parallel drift of a few mV was evident. Accepting Huber's /7/ definition of porosity and Harzdorf's /5/ conclusion that redox influences could be a useful criterion of /non/-porosity, tests shown in Fig. 3 indicated clearly that durable, nonporous electrodes were only those of thick, glassy appearance /types B and D/. We assume that thin layer Type A electrodes whose metallic silver undergoes reaction being in the melt, gets a layer with "open holes" during the usage due to dissolved and torn off AgX film /Fig. 5/. The observed potential drift of type A electrodes is a probable consequence of leaching phenomena pointed out by Morf et al. /8/. On the contrary, type B and D electrodes having glassy appearance, retain their unchanged cover even after prolonged usage /Fig. 6/. We concluded also that thick layer type C electrodes /thin Ag_2S film over AgX/ though reliable in use, are not resistant against redox influences because no Ag_2S film is obtained from molten state.

Application of Coated Wire Halide Electrodes

The especially promising type B and D electrodes were used for further research. Chloride selective electrodes were applied for determination of chloride in drinking waters /Table 1/ substituting Mohr's visual end-point detection as indicated by Moody and Thomas /9/. Iodide selective electrodes can be used for cyanide determinations as explained by Pungor and coworkers /10/ /Table 2/, and in iodide determination /Table 3/ by direct potentiometry.

RESULTS AND COMMENT

As shown by Fig. 7, the behaviour of all the CWHEs is very similar to the electrodes of Radiometer production of the kind, regarding Nernstian slope and the limit of detection. CWHEs/type B and D/ are, however, of much shorter response time.

Determinations of chloride in potable waters, yield higher results if direct potentiometry is applied instead of potentiometric titration.

Determinations of cyanide, accepting the results of depolarisation end point method /11/ as trustworthy ones, yield better results if Ag/AgI or Ag/AgI/Ag_2S instead of silver metal was the indicator.

Also, the determination of iodide with iodide coated-wire electrodes yields results agreeing with those obtained with a F3007 electrode.

REFERENCES

1. I.M.Kolthoff and H.L.Sanders, J.Am.Chem.Soc., 59, /1937/ 416
2. E.Bishop and R.G.Dhanenshwar, Analyst, 88, /1963/ 424
3. J.Siemroth, I.Henning and R.Claus, Proc. 2nd Symposium on Ion-Selective Electrodes, Mátrafüred 1976 Akadémiai Kiadó Budapest 1977, p. 185
4. C.Harzdorf, Z.Anal.Chem., 270, /1974/ 23
5. C.Harzdorf, Anal.Chim.Acta, 136, /1982/ 61
6. V.M.Jovanović and M.S.Jovanović, 7th Yugoslav Congress on Chemistry, Novi Sad 1983, Abstract of Papers II, p. V-33
7. K.Huber, Z.Electrochem., 59, /1955/ 448
8. W.E.Morf, G.Kahr and W.Simon, Anal.Chem., 46, /1974/ 1539
9. G.J.Moody and J.D.R.Thomas in H.Freiser's Ion Selective Electrodes in Anal.Chem., Plenum Press N.Y. 1981 /I/ p.351
10. B.György, L.André, L.Stehli and E.Pungor, Anal.Chim.Acta 46, /1969/ 318
11. M.S. Jovanović, F.Sigulinsky and M.Dragojević, Talanta, 13, /1966/ 1275

Table 1. Determination of chloride level of potable waters using chloride selective electrodes

Water	Cl^- in mg/dm^3				
	found: titrimetry /10 dtmn/			direct potentiometry	
	Mohr	F3005	CWE /B/D/	F3005	CWE /B/D/
I	15.9	15.3	14.9/14.6	9.0	22.4/11.2
II	18.7	17.5	16.4/16.6	22.0	28.2/14.1
III	42.0	40.0	40.2/40.0	44.0	50.1/25.5

Table 2. Determination of 0.001 molar cyanide solution /pH=9.8/ by different argentometric procedures

No. of anal.	taken CN^- mg /Ag ind.el./	found CN^- mg		
		Liebig	DEP	CWE Ag/AgI /B/D/
10	2.75	3.57	3.06	2.66/2.84

Table 3. Determination of iodide concentration in a sample containing 10^3-fold concentration of chloride

concentration of I^- in mol/dm^3		
reported	found, direct potentiometry	
	F3007	CWE Ag/AgI /B/D/
$2.0.10^{-5}$	$2.1.10^{-5}$	$1.7.10^{-5}$

[x]Perhaps, Deposited on Wire Electrode is more proper term.

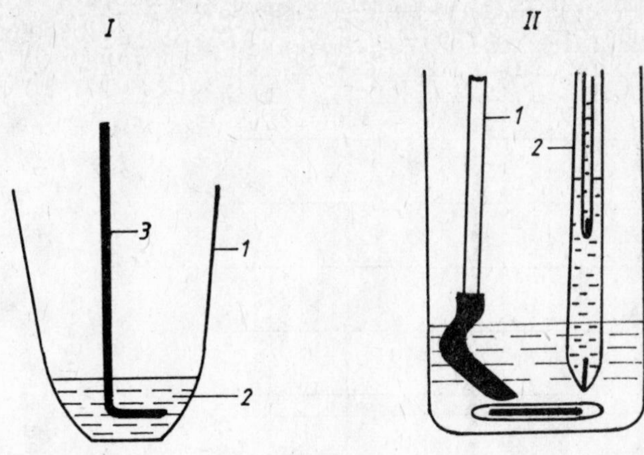

Fig. 1 Coated Wire Halide Electrode

I Preparation: 1-crucible, 2-melt, 3-Ag strip

II Usage: 1-indicator Ag/AgX, 2-double junction RE

Fig. 2 Cross-section of GPE body

1-rubbed in sensor powder, 2-graphite-polyethylene matrix /2:8 mass ratio/, 3-Teflon tube, 4-carbon brush, 5-brush copper spring, 6-coaxial cable

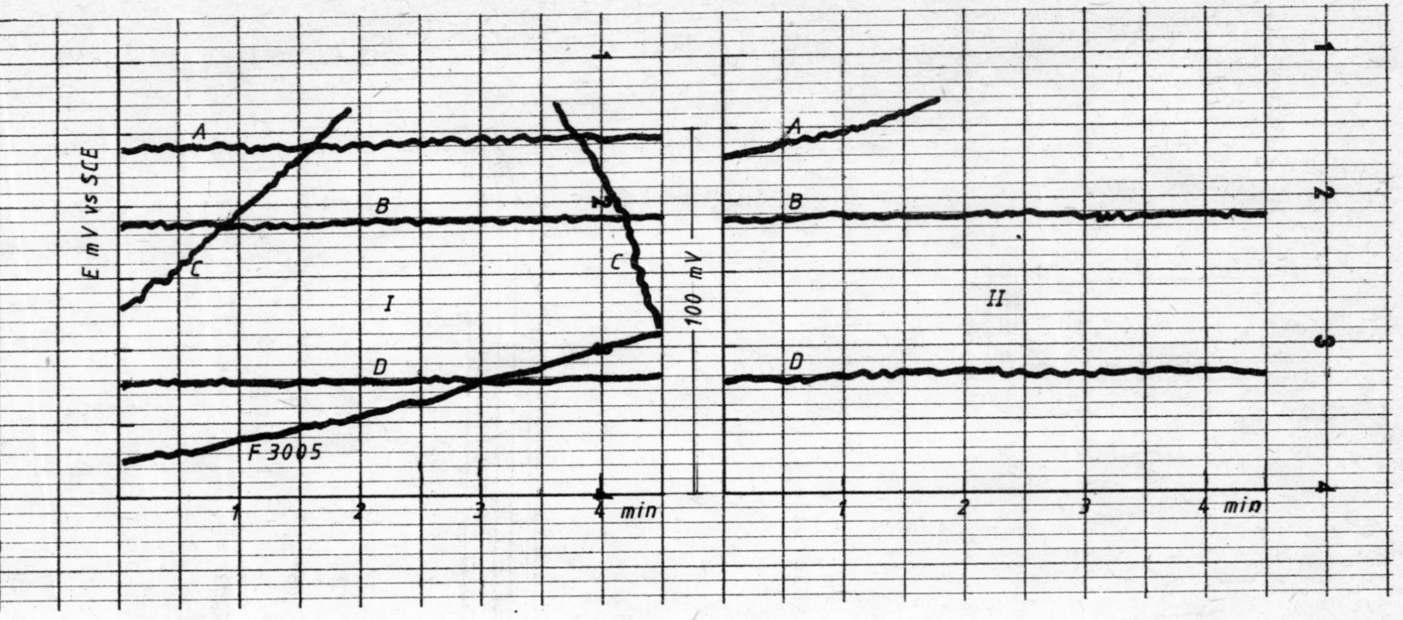

Fig. 3 Potential/time dependence of type A,B,C and D CWHEs and of F3005 in 0.01 mol/l $KMnO_4$ solution /pH=0/ I: freshly prepared, II: after prolonged use in chloride solutions

Fig. 4A Longitudinal section of a freshly prepared type A Ag/AgCl CWE. Note etched surface of silver and a thin AgCl layer.
Enlargement 1:750

Fig. 4B Longitudinal section of a freshly prepared type B Ag/AgCl CWE. Note thick AgCl layer. Enlargment 1:750

Fig. 5 Longitudinal section of a used type A Ag/AgCl CWE. Note only islands of AgCl in the pores of silver. Enlargment 1:750

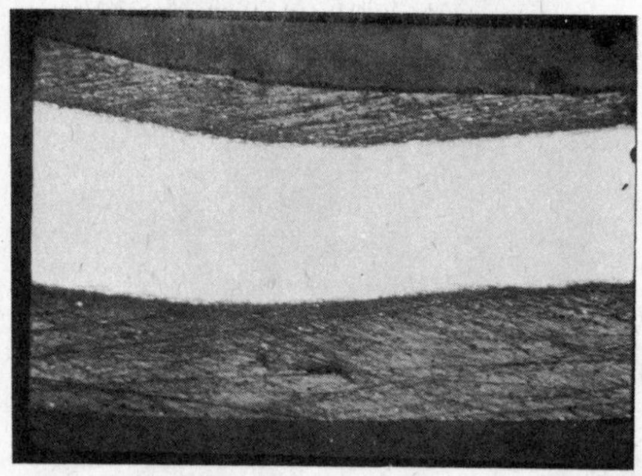

Fig. 6 Longitudinal section of used type B Ag/AgCl CWE. Note the absence of "open holes" in AgCl layer. Enlargment 1:100

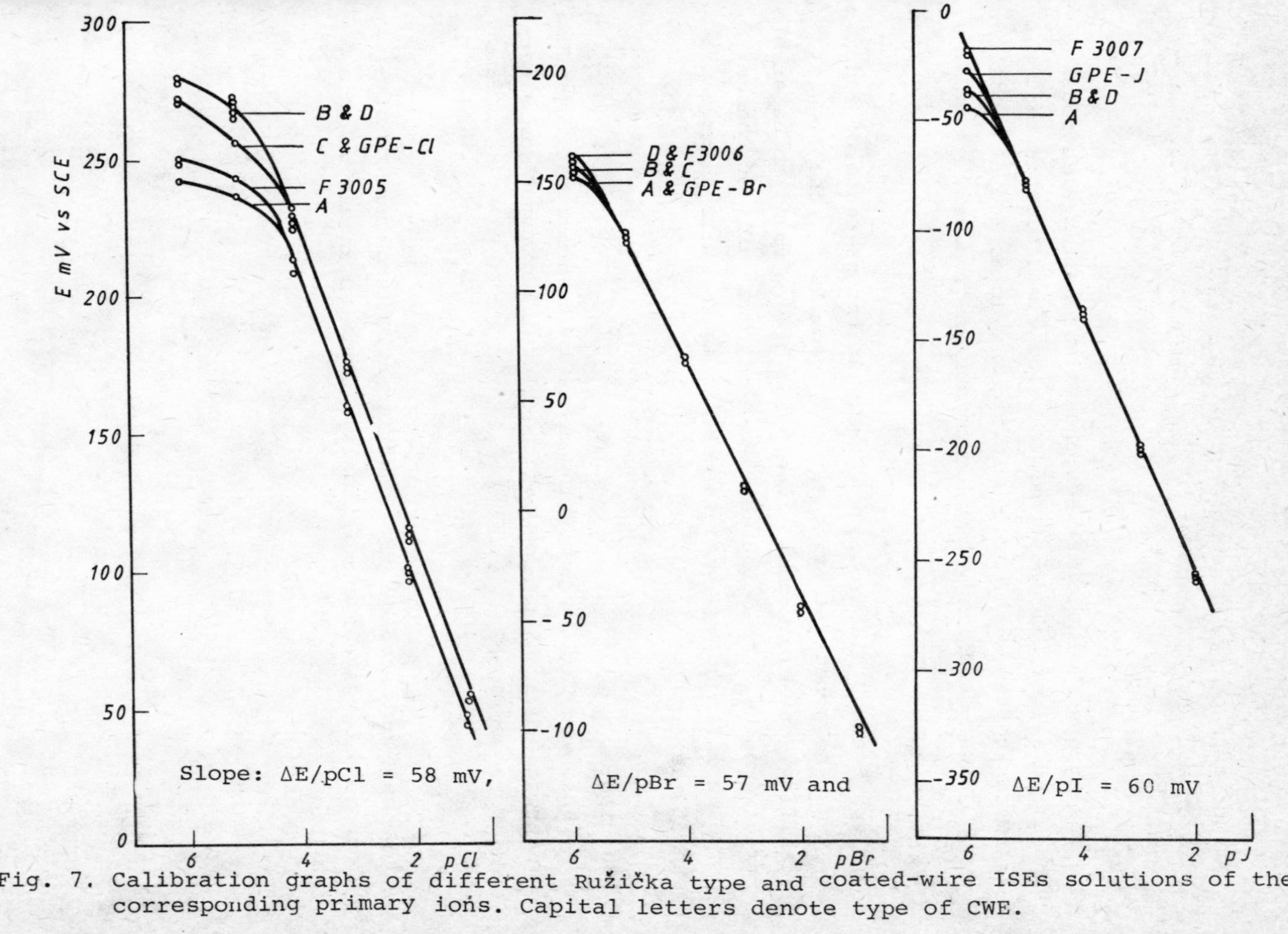

Fig. 7. Calibration graphs of different Ružička type and coated-wire ISEs solutions of the corresponding primary ions. Capital letters denote type of CWE.

QUESTIONS

Participants of the discussion: R.P.Buck, M.Jovanović, V.Jovanović and A.Zhukov

Question:

The main difference between the silver halide electrodes and mixed silver halide-silver sulphide electrodes is that the latter is better when working in redox systems and also, it has a good light-stability. How about your electrodes ?

Answer:

Type B and D electrodes work well in the presence of redox systems, no potential drift was observed. As for the light stability, our electrodes, type B and D were found to be stable for six months, the colour of the electrodes did not change. So, even if light has some effect, it is so small that we cannot see it.

Question:

Could you tell us something about the reproducibility of electrode preparation? How did the E^o value change from electrode to electrode ?

Answer:

We have not studied yet a sufficiently large number of electrodes to give a statistically justified answer to this question.

4th Symposium on Ion-Selective Electrodes
Mátrafüred, 1984

FTIR-ATR-SPECTROSCOPIC DEPTH PROFILING ON ION-SELECTIVE BIS (CROWN ETHER)S BASED MEMBRANES

R. KELLNER, G. GÖTZINGER, K. TÓTH*, L. PÓLOS* and E. PUNGOR*

Institute for Analytical Chemistry, Technical University Vienna, Austria

*Institute for General and Analytical Chemistry, Technical University, Budapest, Hungary

INTRODUCTION

In ion-selective membranes made of polymer-softener mixtures such as PVC and o-nitrophenyl octyl ether /o-NPOE/ or PVC and dioctyl sebacate /DOS/ and ionophores such as valinomycin or bis-crown-ether derivatives /1,2/ the depth /e.g. BME-ligands/ distribution of the ionophores in the surface near area of the membrane may be of interest in order to elucidate the electrode response mechanism.

The analysis of such samples requires the use of non-destructive molecular specific surface techniques which can be operated under normal air pressure conditions. Among the known methods of surface analysis FTIR-ATR /Fourier Transform Infrared Attenuated Total Reflectance/ spectrometry fulfills all the requirements while the other methods /XPS, AES, SIMS, .../ are destructive to the sample due to the need of high vacuum and also to the process of signal generation /3/.

Good optical contact between sample and reflection element provided, ATR-spectroscopy is chararterized by a range of penetration depth of the IR-radiation of approximately 0,2 μm /Ge 60°/ to 3,1 μm /KRS-5, 45°/ in the mid infrared region /3000 cm^{-1} and 800 cm^{-1} respectively//4/ /see Fig.1/.

EXPERIMENTAL

Three types of PVC-membranes have been investigated in this work by ATR-spectroscopy /see Table 1/.

All membranes were cast from tetrahydrofuran /THF/ solutions on a glass plate and analyzed by FTIR-ATR-spectrometry on both sides under different experimental conditions /reflection element Ge, KRS-5, angle of incidence $\Theta_1 = 45^o$, $\Theta_2 = 60^o$, different pressures between the sample and reflection elements/. There were no differences observed in the IR-spectra of the air facing side as compared to the glass facing side of the membranes.

By subtraction of the spectra of sample C /reference without ligand, BME-15/ from the spectrum of sample A the spectrum of the ligand BME-15 could be clearly obtained /Fig.2, left side/. The usual way to get a depth profil with rigid samples is to plot the ratio of the band intensities for two different penetration depth: I_{45^o}/I_{60^o} for KRS-5 /Fig.2, right side/.

The curves obtained by this ratio method show a significant parallel shift as compared to the ideal theoretical value of 1,6 obtained for the softener sample cast directly on the reflection element, because the membrane samples are non rigid.

This parallel shift can be explained by different pressures applied to the samples at placing them at the reflection element. At extremely low pressures the sample is not in perfect optical contact to the reflection element what affects I_{60^o} more than I_{45^o} and produces a high value for I_{45^o}/I_{60^o}. This result shows the high sensitivity of this mode of evaluation to the applied pressure.

The ligand /BME-15/ band at 1730 cm^{-1} shows however, a persistant high I_{45^o}/I_{60^o} as compared to the other values which supports the subsurface enrichment model presented in the Discussion Part.

In order to eliminate the uncertainty produced by the different pressures applied at sample preparation the relative intensities of the individual bands related to the 1610 cm^{-1} softener band in the spectra of PVC+softener and PVC, softener and bis/crown ether/s resp. have been calculated for $\Theta = 45^o$ as well as for $\Theta = 60^o$. The averages of the respective $I_{\hat{\nu}}/I_{1610}$- ratios are displayed in Fig.3 for low and high pressures /normalized spectra/.

Of special interest is the ratio I_{1730}/I_{1610}/crown ether/

softener/. For low pressure and a smaller penetration depth this ratio is smaller than for high pressure and high penetration depth /see Table 2/.

Both effects /high pressure and high penetration depth/ deliver information from subsurface layers whereas low pressure and low penetration depth increase the contribution of the signal from the surface area.

From these results it can be concluded that BME-ligand is present also in the surface layer, but there is an enrichment in layers deeper than $\approx 0{,}7$ µm. This finding is corroborated by the investigations of the surface layer $\angle 0{,}7$ µm by using Ge-reflection elements with $\Theta = 60^{o} - 70^{o}$ /$d_{p-1730} = 0{,}29$ um/ in order to enlarge the data set of Table 2 /Fig.4/.

In addition to membranes A containing 4 mg BME-15 ligand each and showing crystallization of the crown ether under the microscope special samples with a 10-fold lower BME-15 content /0,4 mg each/ have been analyzed using KRS-5 reflection elements /$\Theta = 45^{o}$ and $\Theta = 60^{o}$, high and low pressure/.

As a significant difference, these samples showed no crystallization of BME-15 ligand, but the ratios of BME-15 to softener band were at a rather high level comparable to membrane A samples and were almost intensitive to changes of Θ or pressure /see Table 2/. This is indicative for a homogeneous distribution of the /dissolved/ BME-15 ligand within the analyzed surface area /Fig.4/.

RESULTS AND DISCUSSION

It can be concluded from the measurements, that /see Fig.4/

a/ Membranes B with 0.4 mg ligand:
consist of homogeneously distributed /dissolved/ bis/crown ether/s in the investigated surface area of approx 0,2-2 µm. The high bis/crown ether/s to softener ratio, however, as compared to membranes A points to an enrichment of the crown ether in a surface area of the membrane. In CSE 4 membranes this ratio is larger by a factor of approx 3 only and not 10 as expected by the preparation formula.

b/ Membranes A with 4 mg ligand:
contain in addition to the /homogeneously distributed/ dissolved crown ether part also crystalline crown ether. Under normal pressure, loss conditions these crystals are concentrated in a sub-surface area and covered by the softener /the lower the penetration depth is, the lower is the ratio of BME-15 and o-NPOE-band intensities/. This finding is corroborated by the increasing BME-15 to o-NPOE ratio when increasing the pressure applied to the membrane at the reflection element. By that action more BME-15 crystals are transported into the 0,2-2 μm zone of penetration depth and replace part of the softener.

The technique is promising for carrying out molecular specific studies of the membrane surface composition also under in situ-conditions /in contact with aqueous solutions/5/.

REFERENCES

1 W.E.Morf, W.Simon, in "Ion-Selective Electrodes in Analytical Chemistry" Ed. by H.Freiser, Plenum Press, New York, London Vol.1. Chapter 3.

2 L.Tőke, B.Ágai, I.Bitter, E.Pungor, K.Tóth, E.Lindner, M.Horváth, J.Havas, PCT Int. Appl. WO 8300,149 /Cl. co7J5/ /18/, 20 Jan. 1983 Appl. 81/1,999, 09. Jul.1981.

3 Kellner R., Speciation in Organic Surface and Trace Analysis, in Proceedings of Euroanalysis V Conference, Aug. 24-31, 1984, Cracow, A.Hulanicki ed, Akad. Kiadó Budapest /in press/

4 Harrick N.J., Internal Reflection Spectroscopy New York: Interscience, 1967

5 Neugebauer H., Nauer G., Brinda-Konopik N., Kellner R., Fres. Z. Anal. Chem. 314, 266 /1983/

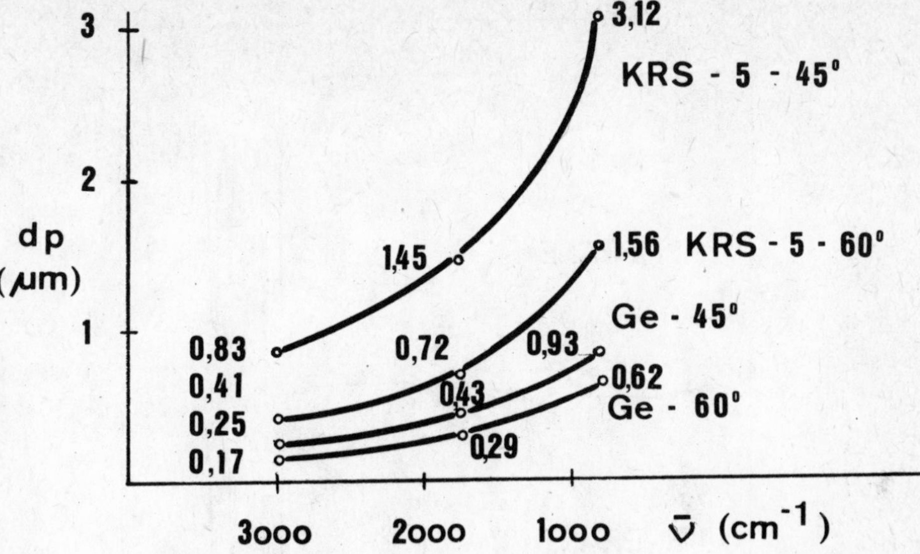

Fig. 1 Penetration depth of IR- radiation at different wavenumbers in ATR- experiments with different reflection elements /Ge: n = 4, KRS-5 : n = 2,4/ and samples with n = 1,5

Table 1. Membrane systems studied

Membrane components	Membranes		
	A	B	C
PVC	60 mg	120 mg	60 mg
o-NPOE	120 mg	120 mg	120 mg
BME-15[x]	4 mg	0,4 mg	---

x

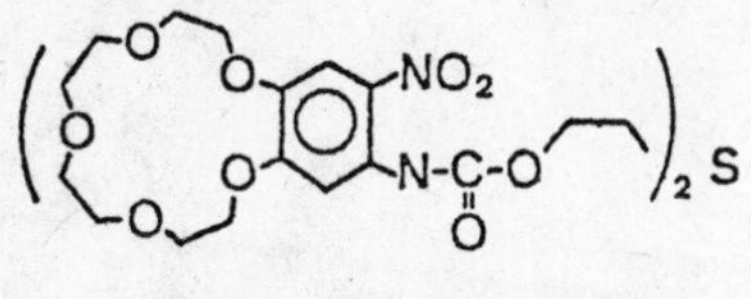

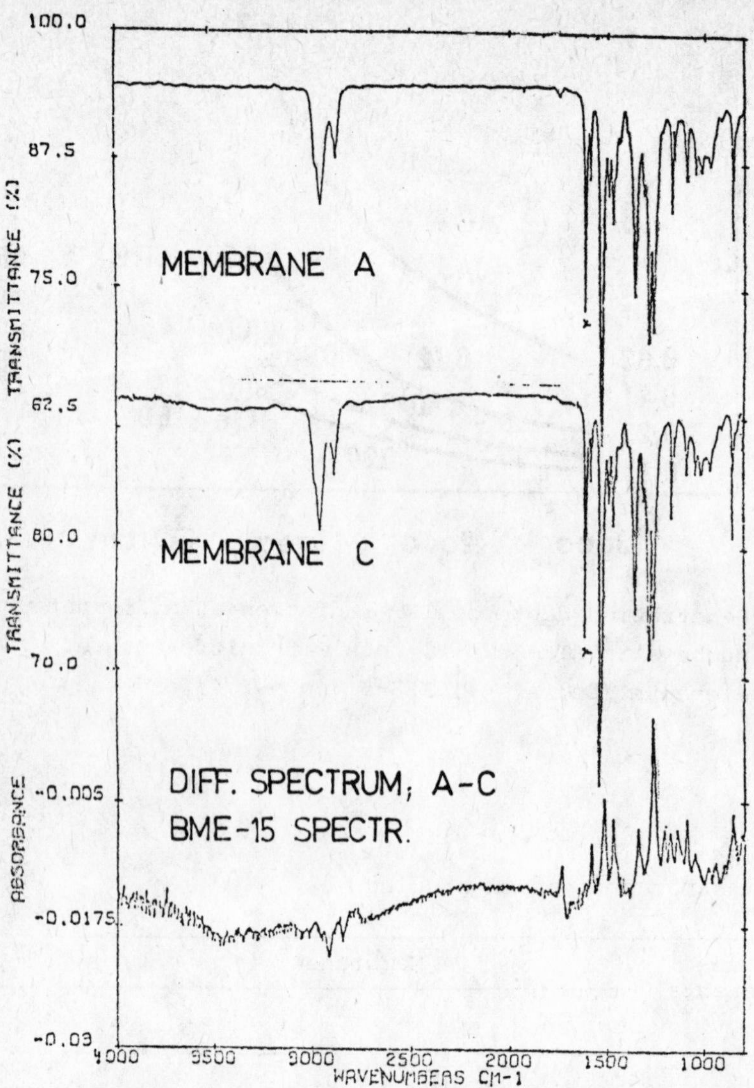

Fig. 2 Absorbance subtraction anf depth profiling applied selective

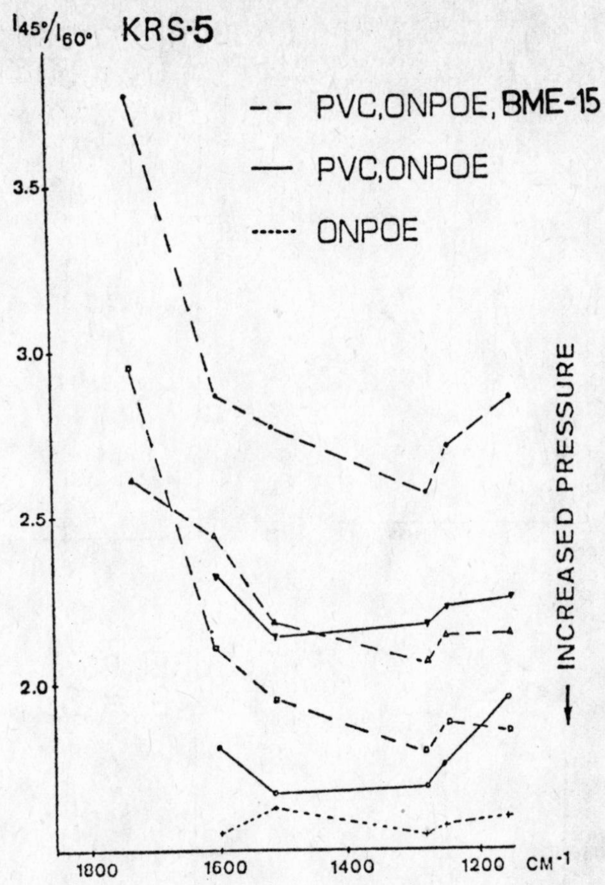

to FTIR-ATR-spectra of PVC, o-NPOE, BME-15 based ion membranes

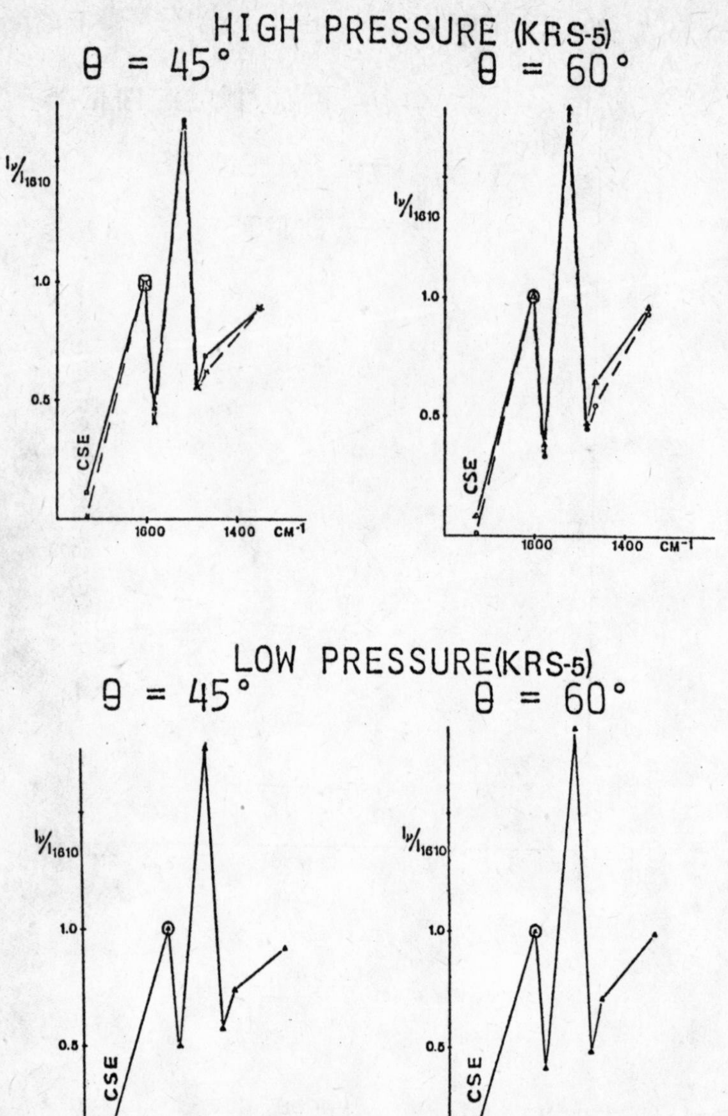

Fig. 3 FTIR-ATR-spectra of PVC, o-NPOE, BME-15 based membrane samples normalized to the 1610 softener band

MEMBRANE A : 4mg BME-15 MEMBRANE B : 0,4 mg BME-15

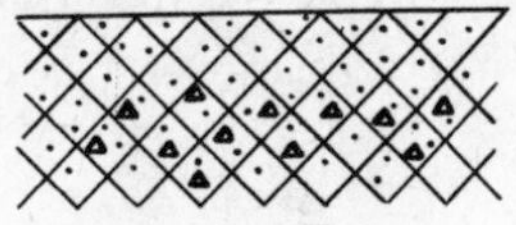

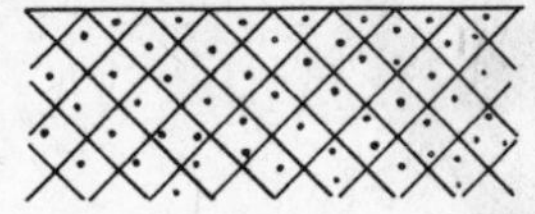

LOW PRESSURE

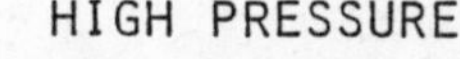

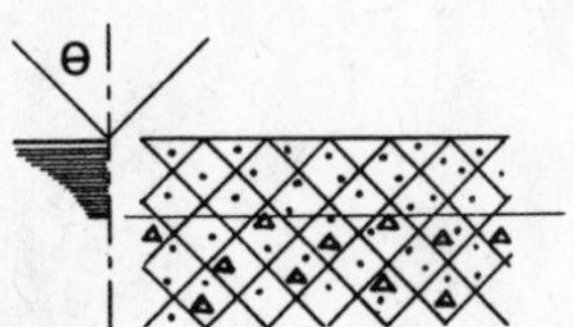

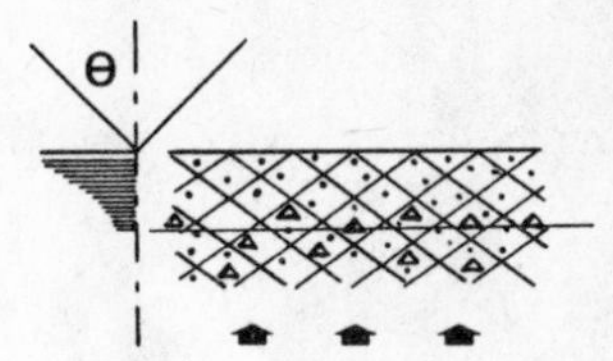

MATRIX

BME-15 DISSOLVED

BME-15 CRYSTALS

Fig. 4 Models for the BME-15 containing membranes investigated

Table 2. Relative intensities at different penetration depth. with membranes containing ionophores in different concentrations

		Relative intensities I_{1730}/I_{1610}			
		$\theta = 60^{\circ}$		$\theta = 45^{\circ}$	
		Membrane B 0,4 mg BME-15	Membrane A 4 mg BME-15	Membrane B 0,4 mg BME-15	Membrane A 4 mg BME-15
KRS-5	low pressure	0,03	0,09	0,04	0,12
	high pressure	0,04	0,17	0,04	0,21
GE	low pressure		0,05		0,07

4th Symposium on Ion-Selective Electrodes
Mátrafüred, 1984

BEHAVIOUR OF SODIUM SENSITIVE GLASS ELECTRODE IN CONCENTRATED STRONG ELECTROLYTE

G. KÓNYA, and L. TOMCSÁNYI

ALUTERV-FKI, Budapest, Hungary

ABSTRACT

The origin of the "sub-Nernstian"slope of a sodium sensitive glass electrode has been investigated in different sodium chloride /sodium hydroxide and sodium hydroxide/ sodium aluminate mixtures of different concentrations. The anion effect on the operation of sodium sensitive glass electrode can be evaluated from these experiments. The knowledge of this effect is important in the industrial application of sodium sensitive glass electrode in monitoring sodium concentration in the Bayer-process.

INTRODUCTION

Sodium sensitive glass electrodes have been frequently used in monitoring sodium ion concentration in solutions of industrial and biological origin. Wilson et al. (1) evaluated the behaviour of different commercially available sodium selective glass electrodes. The conclusion of their work was that the slopes of the potentials of different electrodes were almost Nernstian(56.1-59.6 mV/decade) and remained almost constant during about a four months period. Sodium ion concentration of the solutions were up to 0.1 M. On the contrary, the values of E^o strongly varied (63.7-162.8 mV) and decreased with aging of the electrodes. According to Eisenman (2), the principal factor in aging is probably the hydration of the membrane glass.

Sodium selective glass electrode was first applied in concentrated solutions by Bergner (3) who monitored the sodium ion concentration in pulping liquors. The solutions were diluted by ammonium carbonate buffer to level out differences in ionic strength. The method was accurate with a standard deviation of less than 1 % for samples of 1 to 5 M sodium concentration. Variation of E^o was eliminated by frequent calibration of the cell.

Sodium selective glass electrode was used in undiluted, concentrated solutions of sodium hydroxide and also in sodium aluminate (4). The error of the method was less than 1 rel. % in strong aluminate liquors. According to those experiments the sodium selective glass electrode had "super-Nernstian" response in certain aluminate containing sodium hydroxide solutions.

The aim of this work is to investigate whether the anion effect can be responsible for this phenomenon.

EXPERIMENTAL

Radelkis OP-Na-0711P sodium selective electrodes, Radelkis OP-Cl-0711P chloride selective and home made Hg/HgO electrodes, as well as double junction Ag/AgCl and calomel reference electrodes were used. The electrodes were pretreated by soaking them in the solutions prescribed by the producer.

Radiometer pHM 64, Radelkis OP 208 pH meters and Solatron DVM were used for the emf measurements.

All experiments were performed at 25.0±0.1°C and all chemicals used were of analytical grade.

RESULTS AND DISCUSSION

The emfs in pure NaCl solutions of different concentrations up to 5 M were measured using a sodium selective glass electrode against a chloride ion-selective electrode. Activities were calculated using activity coefficients of

L.Meites (5). The emf vs. log a_{NaCl} was found to be strictly linear in the 0.15-5.0 activity range with a slope of 118.4 mV/decade as it is shown in Fig.1. Almost the same slope was given by Shatkay and Lerman (6).

In the case of sodium hydroxide the emf of the cell of sodium selective glass electrode vs. Hg/HgO electrode was measured. The emf values had quite large standard deviation and the emf was not stable within a reasonable time. A tipical response of the sodium selective glass electrode vs. Hg/HgO electrode is shown in Fig.2. Emf was read at the peak, because its position is thought to be the real response time of the system. After this time a sloping potential curve (0.1-0.2 mV/min) can be observed.

This latter effect is probably due to the aging of sodium selective glass electrode on the effect of hydroxide ions. Therefore, the criteria of Bates et al. (7), namely that the emf should be determined with less than 0.5 mV random error for determination of mean activity coefficient, can not be fulfilled here. The degradation of the surface of the electrode was detected through the measurement of aluminium concentration of soaking solution of high sodium hydoxide concentration. Fig.3. shows the aluminium concentration as a function of soaking time.

Due to the degradation effect of hydoxide ions, the slope of the sodium selective glass electrode vs. Hg/HgO cell was not the same as it is in the case of the sodium selective electrode vs. chloride ISE cell as it is shown in Fig.1. (curve b). The slope is 111.1 mV/decade but the reproducibility was poor.

For the further study of this anion effect 1:1 NaCl - NaOH mixtures were investigated. The emfs of a sodium selective glass electrode vs. Ag/AgCl cell were measured as a function of sodium concentration. As the junction potential and the true activity coefficients can not be calculated accurately in these solutions, the emf - concentration function is evaluated and compared with the NaCl and

NaOH systems in Fig.4. The shape of the electrode potential of sodium selective electrode vs. concentration function contiuously changes with the hydroxide ratio of solutions and is probably influenced by changing of activity coefficients and junction potential too.

Further aging experiments were performed under industrial conditions. Relatively fast changes in E^o were observed, however, the variation of the slope was much smaller in strong aluminate liquors (digesting liquor with high sodium hydroxide and low aluminate content). Aging speed decreases with time and our sodium selective glass electrode can be used for as long as after 6 months, although with different slope and E^o. This is an important fact from an application point of view, because the higher the aluminate content in the liquor is the lower aging can be detected. Therefore a successful application can be expected for aluminate solutions of higher aluminate concentration.

Further studies of anion effect on sodium selective glass electrode are in progress in our laboratory.

REFERENCES

1. M.F.Wilson, E.Haikala and P.Kivalo, Anal.Chim.Acta 74, (1975) 395.
2. G.Eisenman, in C.N.Reilley (Ed.), Advances in Analytical Chemistry and Instrumentation, Vol. 4, Wiley-Interscience, New York, 1965.
3. K.Bergner, Anal.Cim.Acta 87, (1976) 1.
4. I. Bertényi, L. Tomcsányi in E. Pungor (Ed.), Ionselective Electrodes, 3, Akadémiai Kiadó, Budapest, 1981.
5. L.Meites (Ed.), Handbook of Analytical Chemistry McGraw-Hill, Inc., New York, 1963.

6. A.Shatkay and A.Lerman, Anal. Chem., 41, (1969) 514.

7. R.G.Bates, A.G.Dickson, M.Gratzl, A.Hrabéczy-Páll, E.Lindner and E.Pungor, Anal. Chem., 55, (1983) 1275.

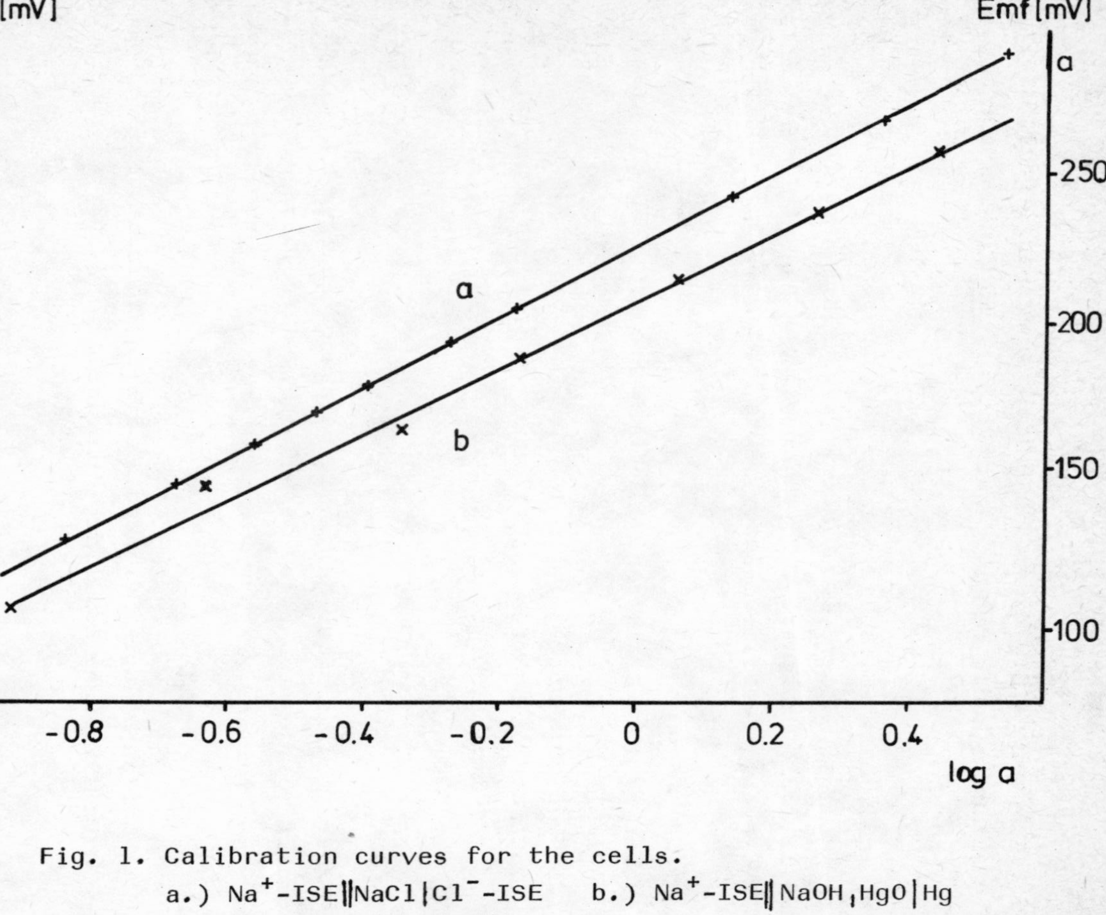

Fig. 1. Calibration curves for the cells.
a.) Na^+-ISE‖NaCl|Cl^--ISE b.) Na^+-ISE‖NaOH,HgO|Hg

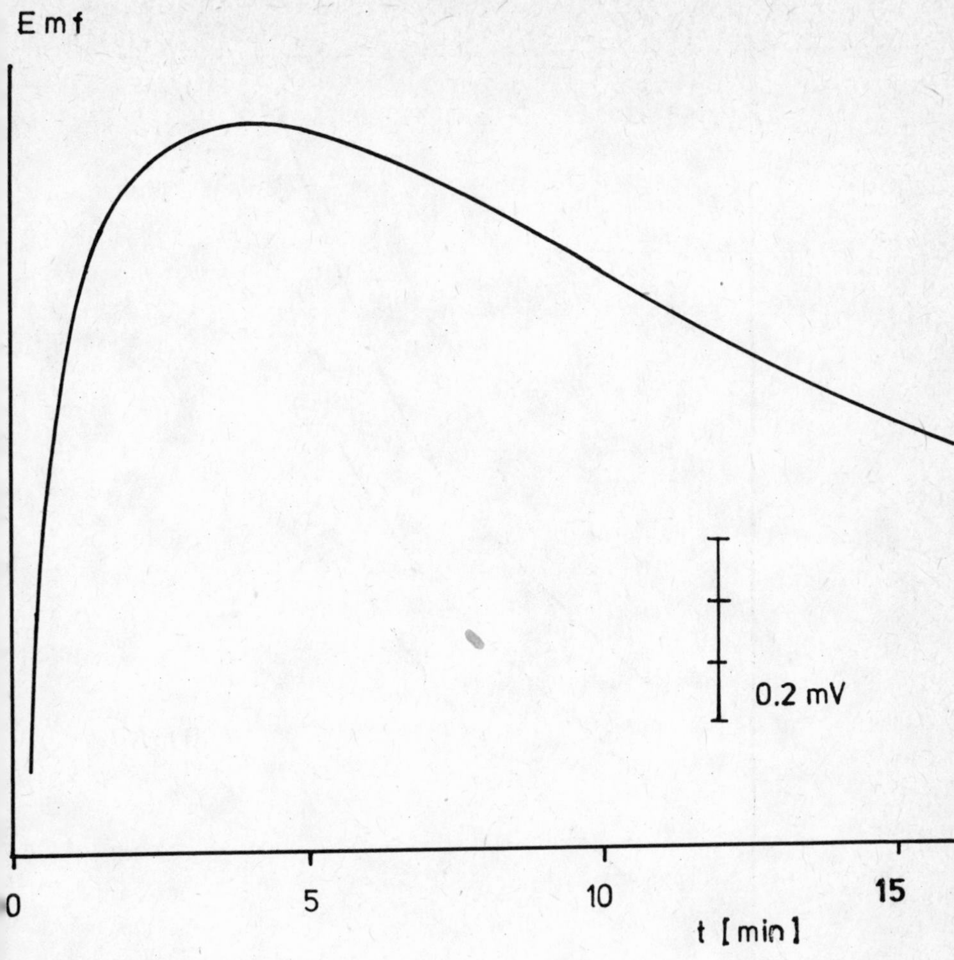

Fig. 2. A typical response of the Na^+-selective glass electrode vs. Hg | HgO electrode cell.

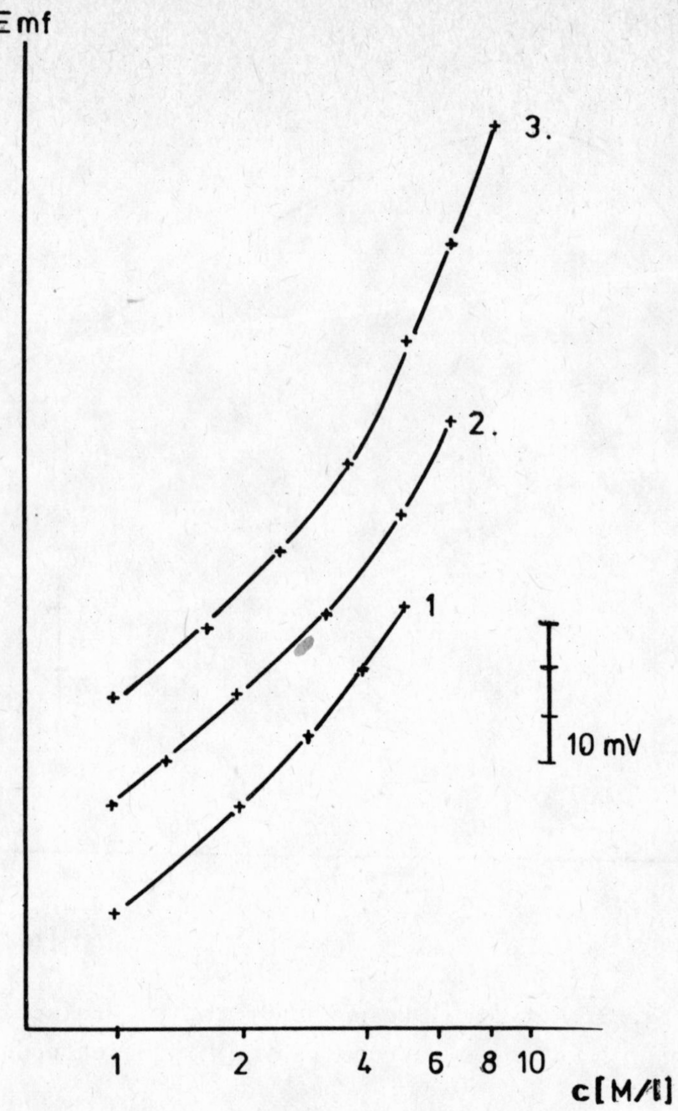

Fig. 3. Calibration curves in different electrolytes.
1.) NaCl 2.) NaCl-NaOH 1:1 mixture 3.) NaOH

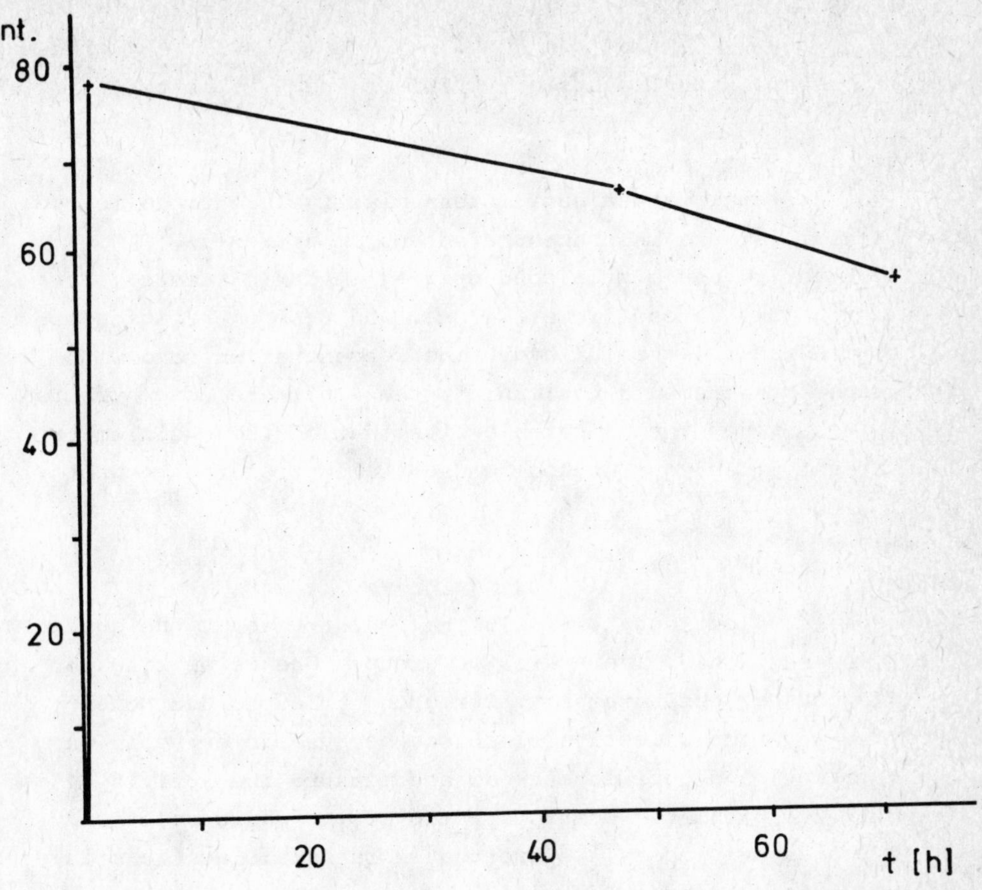

Fig. 4. Aluminium content of the soaking solution
(in arbitrary units).

QUESTIONS AND COMMENTS

Participants of the discussion: E.Pungor and L.Tomcsányi

Question:

I do not think that a sodium ion-selective electrode is the best sensor for use in concentrated sodium hydroxide. It is better to use e.g. a conductivity measuring device, or more favourably an oscillometric one. And from the results one can obtain the free sodium hydroxide concentration, and with some other measurements you can get the aluminate concentration. Is the result you get comparable with that of the oscillometric measurement ? Can your method be used to control the technological process ?

Answer:

There is a long debate in alumina industry about the questions you mentioned. There are two groups. One is working on oscillometry or conductometry. Our group tried to use potentiometry with ion-selective electrode because in our opinion conductometry and oscillometry do not measure the activity, but rather a parameter characteristic of the whole solution.

In the Bayer process, we use the expression: caustic soda, g/l, that is, sodium oxide, g/l although we know that the most important parameter in the Bayer process is the hydroxide ion activity. It is true that the value we get with oscillometry depends primarily on the hydroxide ion concentration, but there is sodium, aluminate, and complexes present in the solution. Therefore we want to provide an activity value. We are at the beginning of the work. We have to measure an activity and transform it into a concentration: caustic soda g/l.

Question:

If you measure sodium ion concentration, how do you derive from that the free sodium hydroxide concentration ?

Answer:

We did not calculate the hydroxide ion concentration. We

try to prepare a practically applicable hydroxide ion selective electrode, but now we only measured the sodium activity.

Comment:

The system in the Bayer process is very complicated when we were measuring aluminium hydroxides containing chloride, sulphate and nitrate, and we found various kinds of complexes. In a complicated system like we have in the Bayer process I do not think that your technique, or oscillometry or conductometry can provide chemically exact results. But the question is, how to relate the oscillometric result to the potentiometric ones.

Answer:

The colleagues working with oscillometry gave us the task of determining the sodium activity, as, at some point of the technology they needed the free hydroxyl ion to aluminate ratio and the sodium ion concentration as well, and they could not get it from the conductivity data they measured. We try to help them in determining the sodium. This is the primary aim of our work.

4th Symposium on Ion-Selective Electrodes
Mátrafüred, 1984

ION-SELECTIVE ELECTRODES BASED ON SnO_2 FILMS

F. KORMOS

Institute of Chemistry, Cluj-Napoca, Romania

ABSTRACT

Antimony doped SnO_2 films with semiconductor properties, on glass substrate are used to elaborate various ion-selective electrodes. The semiconductor properties of the film and the dependence of sensitivity of the electrodes on the electrical conductivity of the film were studied as well. Some cations or anions sensitive tin oxide electrodes were used in potentiometric measurementes in biological samples. The ion-selective SnO_2 electrodes give the same performances as the usual electrodes and have a low price due to the absence of noble metals in construction.

INTRODUCTION

Actually there is a trend to elaborate more accesible and less costly electrodes as compared with the usual ones. For that purpose new electrode materials like graphite (1-3), B_4C_3 (4), film of SnO_2 (3,5) were introduced.

The SnO_2 doped film has semiconductor properties. It is deposited on various inert substrates mainly on glass. SnO_2 film is used since 1962 (6) as an electrode material because it possesses the following advantages (9): mechanical and electrochemical stability, a large useful potential range, low double-layer capacitance, good optical transparency and lack of surface phenomena. The electrodes based on SnO_2 films were elaborated especially for use as redox indicator electrodes in potentiometric or amperometric titrations(6-8).

Ion-selective electrodes for H^+ (1o), S^{2-} (1o), Cl^- (11) were obtained in 1979.

In the present paper we describe ion-sensitive tin oxide electrodes for some cations and anions. The semiconductor properties of the film and the dependence of the sensitivity of the electrodes on the electrical conductivity of the film were studied as well.

EXPERIMENTAL

The SnO_2 film was obtained by mechanical spraying on a glass rod (ø = 3 mm, l = 5 cm) heated at 55o°C, of a solution having the following composition: $SnCl_4$ 3M, $SbCl_3$ $7.1o^{-2}$ M (used as doping substance), ethanol 1o %. This composition corresponds to a doping level of 1,7 mole % Sb^{3+}. The SnO_2 film was prepared at a higher doping level too corresponding to 4,4 mole % Sb^{3+}.

The H^+-selective electrode was obtained by electrolytical coating of the SnO_2 film with a thin antimony layer. The electrolyte had the following composition: 5o ml H_2O , 5o ml $SbCl_3$, 3o g HCl and 3 g tartric acid. 3 V voltage, 25o µA current at 2o°C and 26 h deposition time were used.

The ion-selective electrode based on PVC membrane was obtained introducing the glass rod covered with SnO_2 film in a mixture of 5 (w/w) % PVC, 74,5 (w/w) % tetrahydrofuran , 2o (w/w) % dioctylphtalate and o,5 (w/w) % electrode active substance (potassium tetraphenylborate for cation selective electrodes and cetylpyridinium for anion selective electrodes). The mixture was preheated at 6o-7o°C and cooled before use. The covered rod was air dried 24 h.

The ion-selective electrode based on SnO_2 film is shown in Fig.1. The electrodes were prepared both with SnO_2 films at 1,7 mole % Sb^{3+} doping level and with 4,4 mole % Sb^{3+} respectively.

Analytical grade substances were used to obtain the electrodes and to prepare the standard solutions. The potentiometric determinations were made with Clamann-Grahnert MV 87 type pH-meter at 2o°C under continuous stirring the test solution. For the measurements of the differential capacitance in alternating current (12) the device shown in Fig.2

was used. The supporting electrolyte solution was 1 M KCl at pH = 5,5. The electrical resistance (R) of the SnO_2 film was made using the four point method (12).

RESULTS AND DISCUSSION

The SnO_2 films doped with antimony are n type semiconductors. The number of charge carriers (N) in the SnO_2 films was determined measuring the differential capacitance in alternating current, using the Mott-Schottky equation:

$$\frac{1}{C_s^2} = \frac{1}{C_H^2} + \frac{2}{\varepsilon_0 N \varepsilon} \left(E - E_{fb} - \frac{kT}{q} \right) \qquad (1)$$

where C_s is the interface capacitance, C_H is the Helmholtz layer capacitance, E is the potential applied at the interface, E_{fb} is the flat band potential, ε_0 is the dielectric constant of vacuum, ε is the dielectric constant of SnO_2.

The conductivity of the film (σ) was determined by measuring the resistivity (ρ) and applying equation (2):

$$\sigma = \frac{1}{\rho} = \frac{1}{\delta \cdot R} \qquad (2)$$

where δ -being the film thickness = o,6 μ.

The results of differential capacitance and conductivity measurements are shown in Table 1. It results that a higher doping level assures a higher conductivity.

The electrode function of our H^+-selective was established determining simultaneously the same data also with a metal-antimony electrode. The results are shown in Table 2. The following important differences were observed: the electrode based on SnO_2 film had a better repeatability and a lower temperature coefficient that of Sb, but it has a higher electrical resistance. The electrode is feasible at a lower cost but has the disadvantage that it cannot be used in F^- containing media having a glass support.

The PVC coated SnO_2 electrode with active electrode material potassium tetraphenylborate (KTFB) functions as a cation selective electrode, and the electrode with cetylpyridinium as active electrode material is anion selective. The characteristics of these electrodes based on SnO_2 film at a doping level of 1,7 mole % Sb^{3+} are shown in Table 3.

These PVC membrane electrodes have the following characteristics: electric resistance 5o MΩ, temperature coefficient o,7 mV/°C (between 2o-8o°C) and response time 8o-1oo sec. The available pH range: 4-9 and the reproducibility was ± 5 mV in the first five days. After this time the coating must be repeated otherwise unsatisfactory reproducibilities are obtained.

The selectivity of the electrodes (Table 4) was established according to the K.Srinivasan-G.A.Rechnitz method (13). It was observed that among the cation sensitive electrodes the Tl^+- sensitive electrode has the best selectivity and among the anion sensitive electrodes the most selective electrodes were those of SCN^- and TBF^- sensitives against the ions investigated.

Comparing these results with those obtained with PVC membrane electrodes on SnO_2 film support at a 4,4 mole % Sb^{3+} doping level, it results that their electrical resistance decreased to < 1o MΩ and the response time was reduced to 3o sec. The linearity range and their selectivity remained unchanged but in the case of the K^+, NH_4^+, NO_3^-, ClO_4^- sensitive electrodes respectively an extended measuring range up to 10^{-6}M was observed. These can be explained by the higher conductivity of the semiconductor SnO_2 films with 4,4 mole %Sb^{3+}.

The electrodes based on SnO_2 film were used in direct potentiometric titrations, carried out by the standard addition method (14). The unknown concentration was calculated with the equation (3):

$$C_x = C\left[10^{\Delta E/s}\left(1 + \frac{V_x}{V}\right) - \frac{V_x}{V}\right]^{-1} \qquad (3)$$

where C_x is the concentration of the sample, C is the concentration of the standard solution, V_x is the volume of the sample, V is the volume of standard solution added, E is the potential difference of the sample before and after the standard addition, s is the electrode sensitivity.

Our pH-sensitive electrode was used for the determination of the pH in gastric juice, blood serum and urine. The results were compared with those obtained with the glass or antimony metal electrodes (Table 5). One can see that the results obtained with the SnO_2 based electrode were identical with those obtained with the usual ones.

The high selectivity of the K^+-sensitive PVC membrane electrode (10^{-1} towards NH_4^+ and 2.10^{-3} towards Na^+) permitted the determination of K^+ ion in blood serum and urine. The results were in good agreement with those obtained with flame-photometry (serum 7.10^{-2}M, urine 6.10^{-3}M).

The NO_3^--sensitive PVC membrane electrode was used to determine NO_3^- in drinking water. The results were the same as those obtained by the colorimetric methode.

In conclusion we may assert that coated ion-selective electrodes based on SnO_2 film give the same performances as the usual ones. Their advantages are: the construction is simpler, they are easy to prepare and regenerate, and have a low price due to the absence of noble metals in construction.

Acknowledgement:Professor László Kékedy,Babeş-Bolyai University,Cluj-Napoca,Romania,is thanked for valuable suggestions.

REFERENCES

1 W.S.Selig, J.Chem.Educ.,Am.Chem.Soc.,61,(1984) 81.
2 A.Ansaldi and S.I.Epstein,Anal.Chem.,45,(1973) 595.
3 K.H.Heckner and I.F.Müller,Z.Phys.Chem.,261,(1980) 585.
4 W.J.Bladel and G.W.Schieffer,J.Electroanal.Chem.,80,(1977) 259.
5 D.E.Albertson and H.N.Blount,Anal.Chem., 51,(1979) 556.
6 W.Cooper, Nature, 194,(1962) 569.
7 J.Kane, G.P.Schweizer and W.Kern, J.Electrochem.Soc., 122,(1975) 1144.
8 L.Kékedy and F.Kormos, Talanta, 26,(1979) 584.
9 L.Kékedy, M.Olariu and F.Kormos,Analusis,10,(1982)288.
10 H.A.Laitinen and T.M.Hseu,Anal.Chem., 51,(1979) 1550.
11 H.Cha,K.Y.Choo and H.Kim., Taehan Hwahakhoe Chi, 25,(1981) 127.
12 O.Elliot,O.L.Zellmer and H.A.Laitinen, J.Electrochem.Soc., 117,(1970) 1343.

13 K.Srinivasan and G.A.Rechnitz, Anal.Chem.,41,(1969)12o3.
14 J.Havas, Ion és molekulaszelektiv elektródok biológiai rendszerekben. Akadémiai Kiadó, Budapest 198o.

Table 1. Differential capacitance and conductivity values of differently doped SnO_2 film electrode.

Doping level	R Ω	ρ Ω.cm	σ $\Omega^{-1}.cm^{-1}$	N carriers/cm^3
1,7 % Sb^{3+}	378	2,26.$1o^{-2}$	4,4.1o	2,7.$1o^{19}$
4,4 % Sb^{3+}	95	5,7.$1o^{-3}$	1,7.$1o^{2}$	1,1.$1o^{2o}$

Table 2. The characteristics of the H^+-sensitive electrodes

Functional characteristics	Electrode SnO_2 film	Antimony
Lineal pH range	2,8 - 9,5	2,8 - 9,5
Measurable pH range	2 - 1o	2 - 1o
Temperature range	o - 6o°C	o - 6o°C
Electrical resistance	1 kΩ	1oΩ
Repeatability	± 2 mV	± 5 mV
Response time	1oo sec.	1oo sec.
Temperature coefficient	o,9 mV/°C	1,3 mV/°C
Precision	o,2 pH	o,2 pH
Interference	F^-,Cu^{2+},ox,red.	Cu^{2+},ox,red.

Table 3. Performance data of the PVC membrane electrodes on SnO_2 film

Electrode	Linearity range mol/l	Measuring range mol/l	Sensitivity mV/pC
K^+	$1o^{-1}$- 5.$1o^{-4}$	$1o^{-1}$ - $1o^{-5}$	55
NH_4^+	$1o^{-1}$- $1o^{-4}$	1 - 5.$1o^{-4}$	56
Tl^+	$1o^{-1}$- $1o^{-3}$	$1o^{-1}$-5.$1o^{-3}$	52
NO_3^-	$1o^{-1}$- $1o^{-4}$	$1o^{-1}$- $1o^{-5}$	55
ClO_4^-	$1o^{-1}$- $1o^{-4}$	$1o^{-1}$-5.$1o^{-5}$	57
SCN^-	$1o^{-1}$- $1o^{-3}$	$1o^{-1}$- $1o^{-4}$	54
TFB^-	$1o^{-1}$- 5.$1o^{-3}$	$1o^{-1}$- $1o^{-4}$	53

Table 4. Selectivity data of PVC membrane electrodes based on SnO_2 film

X Interfering ions	Cation sensitive electrodes		
	$k_{K^+,X}$	$k_{NH_4^+,X}$	$k_{Tl^+,X}$
K^+	-	1	10^{-2}
NH_4^+	10^{-1}	-	5.10^{-3}
Na^+	2.10^{-3}	10^{-4}	10^{-3}
Mg^{2+}	10^{-5}	5.10^{-4}	10^{-5}
Tl^+	2.10^{-3}	5.10^{-3}	-

	Anion sensitive electrodes			
	$k_{TFB^-,X}$	$k_{NO_3^-,X}$	$k_{ClO_4^-,X}$	$k_{SCN^-,X}$
ClO_4^-	10^{-3}	1	-	10^{-3}
NO_3^-	10^{-2}	-	1	10^{-2}
SCN^-	10^{-3}	10^{-2}	10^{-3}	-
TFB^-	-	10^{-4}	10^{-4}	10^{-4}
Cl^-	10^{-3}	10^{-2}	10^{-2}	5.10^{-3}

k - is the selectivity coefficient of the electrode

Table 5. The pH of some biological samples determined with several type of H^+-sensitive electrodes

Sample	pH		
	SnO_2 electrode	glass electrode	antimony electrode
Gastric juice	2,50	2,60	2,50
Urine	6,30	6,60	6,40
Serum	8,45	8,15	8,10

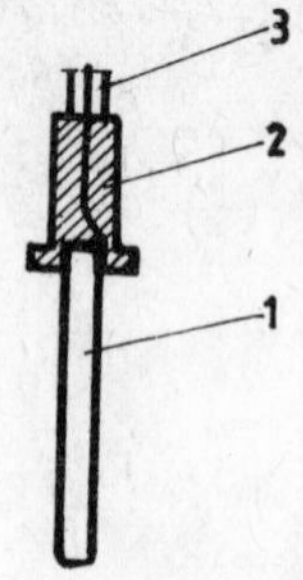

Fig.1. Schematic design of an electrode based on SnO_2 film.
1 - cylindrical glass support, 2 - cap,
3 - contact wire

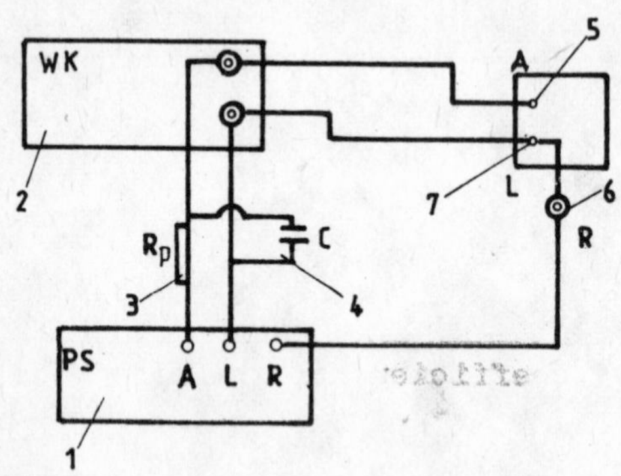

Fig.2. The schematic electrical circuit used for the differential capacitance measurements in alternativ current
1 - potentiostat Tacussel type PRT,
2 - impedance measuring bridge with self balancing, Wayne-Kerr type B 641
3 - resistance
4 - capacity
5 - auxiliary electrode
6 - reference electrode
7 - SnO_2 indicator electrode

QUESTION

Participants of the discussion: F.Kormos and J.D.R.Thomas

Question:

Would you equate your electrode in any way with a coated wire electrode ?

Answer:

Yes, it is a coated wire type electrode.

4th Symposium on Ion-Selective Electrodes
Mátrafüred, 1983

A SYSTEM FOR AUTOMATIC TITRATIONS WITH COMPUTER INTERPRETATION OF DATA

R. KUCHARKOWSKI, W. KLUGE, E. KASPER and A. DRESCHER

Academy of Sciences of GDR,
Central Institute of Solid State Physics
and Material Research
GDR - 8027 Dresden, Helmholtzstr. 20

ABSTRACT

A system for automatic titrations with data output on paper tapes and the complex data interpretation computer program TITDAT are represented. The program, which is suited to all kinds of titrations and indications includes a test mode to fix the optimum interpretation algorithm in dialogue with the terminal and an automatically processing routine mode. Experiences and results are demonstrated by the example of chloride precipitation titration, indicated by chloride ion-selective electrodes, applied to characterize phases in the system Bi_2O_3-BiOCl.

INTRODUCTION

High-precision determinations of main components of compounds and special alloys more and more raise to one of the most important fields of chemical analysis, especially for characterization of electronic materials, where determinations of small differences from theoretical stoichiometry or fixations of phase compositions and phase boundaries are necessary. In such cases the relative error of the analytical result may not exceed 0.1 % (1).

Figure 1 shows the T-x-diagram of the system Bi_2O_3-BiOCl, which is interesting for the preparation of thick-layer pastes. Here we had to distinguish between phases with very similar compositions, e.g. the phases 5 and 6 have a theoretical difference in chloride concentration of 0.17 % (2, 3).

The most favourable method to solve such a problem is the titration analysis, often carried out with potentiometric indication using ion-selective electrodes. In the last part of this paper some experiences and results of chloride determination by this technique are reported.

The accuracy of a titration is generally limited by the correct and reproducible detection of the equivalence point (EP). To realize the required accuracy, the EP must be fixed within a few thousandths of a milliliter, which is only a little part of a drop.
Three principal methods for EP determination are known:

1. End-point titrations: For its application the EP potential must be exactly known and constant over the whole set of samples, including the calibration samples.
2. Graphic methods: These are unsuited for high-precision interpretations and routine titrations.
3. Computer processing of data measured and stored during the titrations: Only these methods are capable to deliver reliable and precise results for different titration and indication variations under alternating conditions of sample compositions, ionic force, electrode response etc.

HARD-WARE AND SOFT-WARE DESIGN

For carrying out computer interpretations of titration data a device for automatic control of the titration process and data output on paper tapes was developed (4,5). Figure 2 shows a scheme of the titration work place: The interface TS, constructed in the Central Institute of Solid State Physics and Material Research, connects the autoburette (ABU 13,

RADIOMETER Copenhagen) and the pH- and mV-meter (PHM 64, RADIOMETER) with a data serializing system including a paper punch (Funkwerk Erfurt) and controls the incremental titration and the data output process. In the case of potentiometric indicated titrations the output of potential-volume measuring values occurs according to the principle of quasi-equidistant potential steps, which delivers the largest data density with respect to the volume axis at the steep part of the titration curve, that means near the EP. Control parameters are the potential step width, the stopping time between two increments and the end-potential.
The data interpretations are carried out with the computer SM4. The complex computer program TITDAT is written in FORTRAN IV and allows to interprete all kinds of titrations and indications. (It is not limited to the potentiometric indication method.)

TITDAT includes the following options:

1. Correction of measuring values:
   The necessity for corrections of data is given, if the volume output of the burette was reset to zero during titration or if erroneous values are to select (punching failures etc.)

2. Reduction of data density:
   If the potential step width was chosen so small that the differences of contiguous values approximate the size of the random error of the measuring values, difficulties at further calculations can arise. Therefore the data density must be reduced.

3. Curve smoothings:
   They are necessary especially for computing of inflection points. TITDAT allows to smooth the original titration curve and/or its 1. derivative. The polynom and the number of measuring values, over which shall be smoothed must be interactive read in at the terminal. (The confir-

med standard is: titration curve: polynom of 2. degree, 5 points; 1. derivative: polynom of 1. degree, 3 points).

4. Computing of inflection points:
Potentiometric indicated titration curves generally have an inflection point at or near by the EP. TITDAT calculates inflection points by approximation of segments of in each case 5 points of the titration data set to a polynom of 3. degree, two-fold derivation of this polynom and search zeros in the 2. derivative. After processing the whole data set of one titration curve in such a way, the zero point with the highest slope of the curve is set equal the EP. This calculation method has the advantages, that in contrast to GRANS linearization method no knowledge about the special titration function is required, and in contrast to the interpolation methods of HAHN, WOLF and KELLER, RICHTER no presumptions concerning the volume step width and distances are necessary and that a large number of measured values can be used to compute the result.

5. Computing of intersection points:
Photometric, amperometric, conductometric and partially potentiometric indicated titrations too result in quasilinear titration curves. The intersection point of extrapolated straight parts of the titration curve generally approximates the EP.
A special option of TITDAT searches those linear parts of the curve, stores the corresponding measuring values, selects outliers within these values, computes the parameters of the concerning regression straight lines and their intersection points.

6. Simulation of end-point titrations:
If potentiometric indicated titrations have a large potential jump at the EP, end-point titrations sometimes yield more reliable results than inflection point calculations.

In these cases the program can compute the imaginable end-point volume for every preset end-point potential by interpolation.

7. Graphic representations:
The original and smoothed titration curves as well as their derivatives can be represented by a digital plotter or on a vector display in automatically calculated or preset scales. Moreover magnifications near the EP and simultaneous graphs of a whole titration set are possible.

8. Statistical interpretations of results:
All results of calculated equivalence volumes are automatically stored on a data file for following computations of standard deviations and confidence intervals including selection of outliers. In such a manner an objective comparison of results obtained by different interpretation algorithms is given.

Figure 3 shows the chartflow of the TITDAT program. It has two principal processing modes: the test mode and the routine mode. During test operations in dialogue with the terminal the required parameters are read in and the switches for program branchings are set and stored, processing of the same data set with other parameters and branchings are possible. After transfer to the routine mode the program runs automatically by means of the stored algorithm for interpretation the whole set of titrations.

EXPERIENCES AND RESULTS

Some experiences and results of the work with this titration data interpretation system shall be demonstrated at the example of chloride titrations with silver nitrate solutions indicated potentiometrically by chloride ion-sensitive electrodes. Figure 4 shows a titration curve and its 1. and 2. derivatives without any curve smoothings. The shape of the 2. derivative demonstrates, that no unique EP can be obtained. The standard smoothing operations however lead to definite results (figure 5). Comparable results are received by reducing the density of measuring values to the half (figure 6), but sometimes by such a way the precision is diminished simultaneously.

Figure 7 represents curves of titrations indicated by a potential-instable ISE. End-point titrations would fail completely, but computing of inflection points with TITDAT leads to a good precision of equivalence volume values. This is demonstrated by the congruence of the 2. derivatives of these titration curves (figure 8).

Figure 9 shows two series of chloride titration curves, received by titrations of equal samples with a normal sensitive ISE (series A) and a very unsensitive ISE (series B). A precise interpretation of titration data of the series B by conventional methods would be difficult or impossible, but TITDAT yields no significant deviation of the mean values and the standard deviations of these two series. The 2. derivatives of all curves are of good congruence (figure 10).

Finally the properties of the described system of titration data interpretations shall be summarized: The advantages are:

- No previous knowledge about the shape of the titration curve, the equivalence point potential and the reproducibility of potential response is necessary.
- After storage of an adequate number of measuring values during titration, all types of titrations and indications can be interpreted with optimum accuracy by one of the TITDAT versions, presupposed that the titration process was carried out in chemical equilibrium.
- A set of titration data can be processed by several algorithms and the results can be statistically compared.

The disadvantage of the system is, that till now exists a chronological difference between the titration process and the output of its results and that the computer cannot directly control the process. This will be overcome by establishment of a process computer.

REFERENCES

1 R. Kucharkowski, A. Drescher and O. Großmann, Z. Chem. 19, /1979/ 281

2 H. Oppermann, G. Stöver, N. Mattern and R. Kucharkowski, 17. Annual Session of the Society for Crystallography and Solid State Chemistry, Leipzig /1983/ (Poster)

3 R. Kucharkowski, H. Oppermann and G. Stöver, 4. Symposium on Solid State Analytics, Karl-Marx-Stadt /1984/ (Poster)

4 R. Kucharkowski, E. Kasper and A. Drescher, 3. Symposium on Solid State Analytics, Karl-Marx-Stadt /1981/ (Poster)

5 R. Kucharkowski and W. Kluge, 5. Working Session Progress in Metallurgic-analytical Chemistry, Rauschenbach (GDR) /1983/ (Poster)

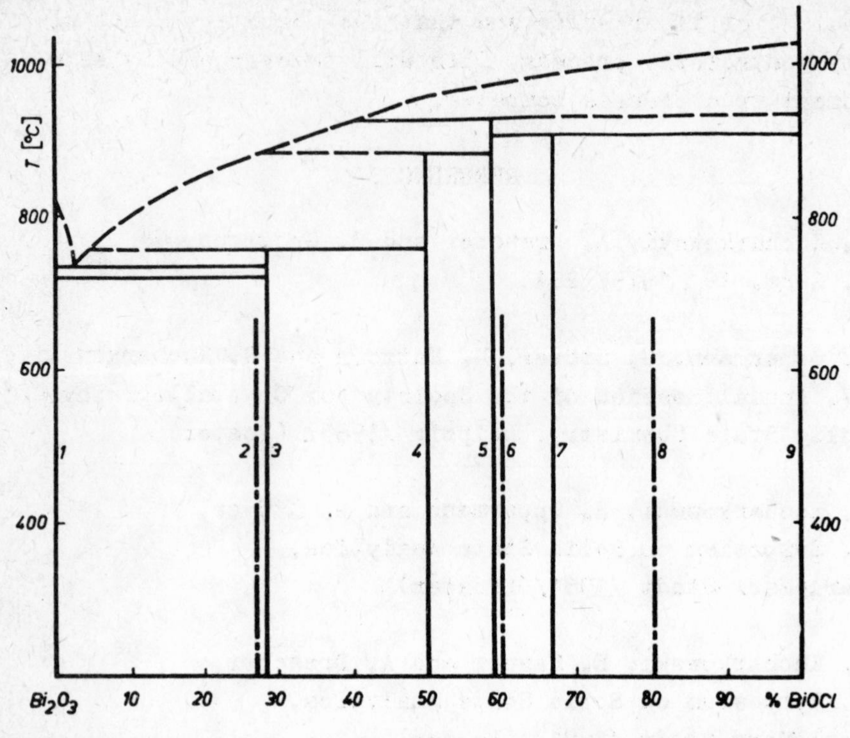

phase num.	composition	mol% BiOCl	m% Cl theoretical	m% Cl analyzed	identified
1	Bi_2O_3	0	0	-	+
2	$Bi_{19}O_{27}Cl_3$	27.3	2.36	-	-
3	$Bi_{12}O_{17}Cl_2$	28.6	2.49	2.51±0.03	+
4	Bi_3O_4Cl	50	4.88	4.86±0.03	+
5	$Bi_{24}O_{31}Cl_{10}$	58.8	6.04	6.04±0.03	+
6	$Bi_7O_9Cl_3$	60	6.21	-	-
7	$Bi_{12}O_{15}Cl_6$	66.7	7.19	/7.38/	+
8	$Bi_6O_7Cl_4$	80	9.41	-	-
9	BiOCl	100	13.61	-	+

Fig.1 T-x-diagram of the system Bi_2O_3-BiOCl

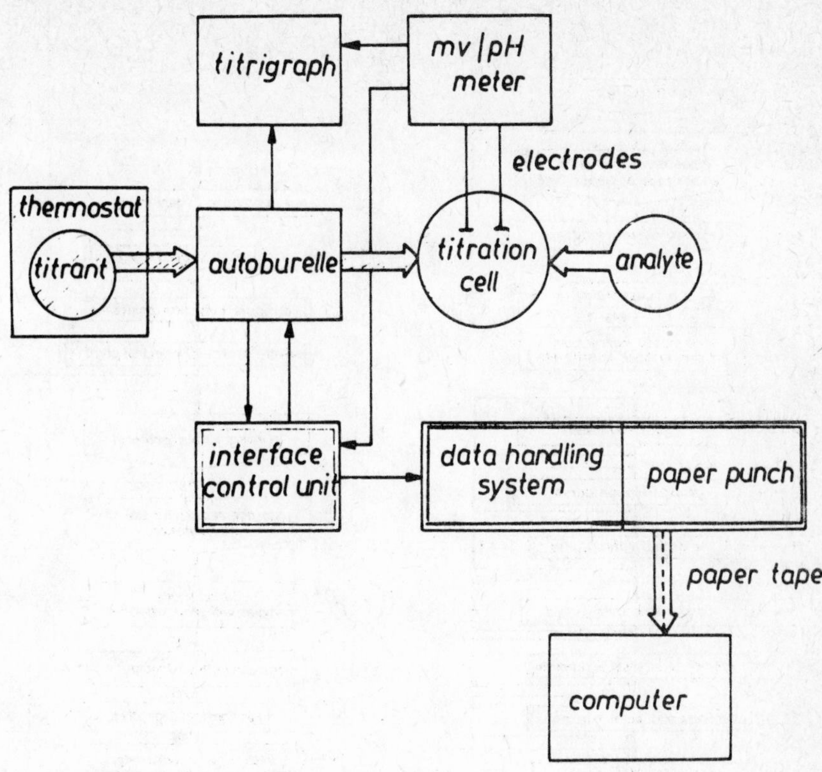

Fig.2 ZFW titration work place

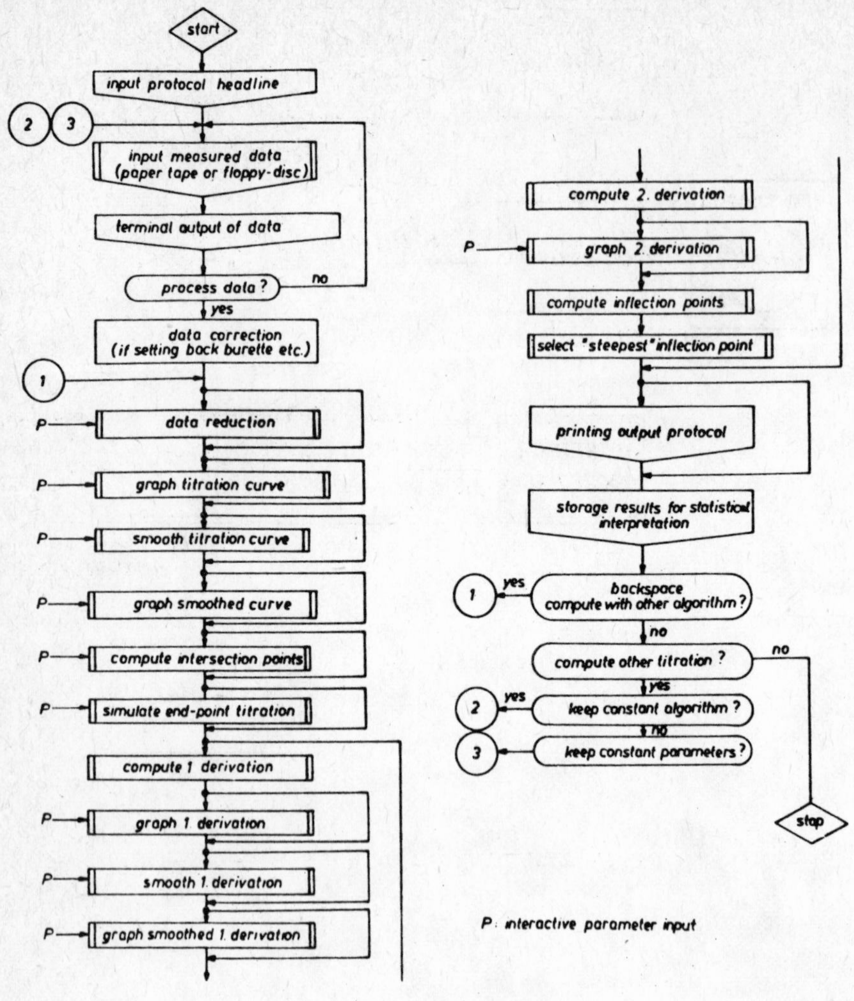

Fig.3 Flowchart of computer program TITDAT

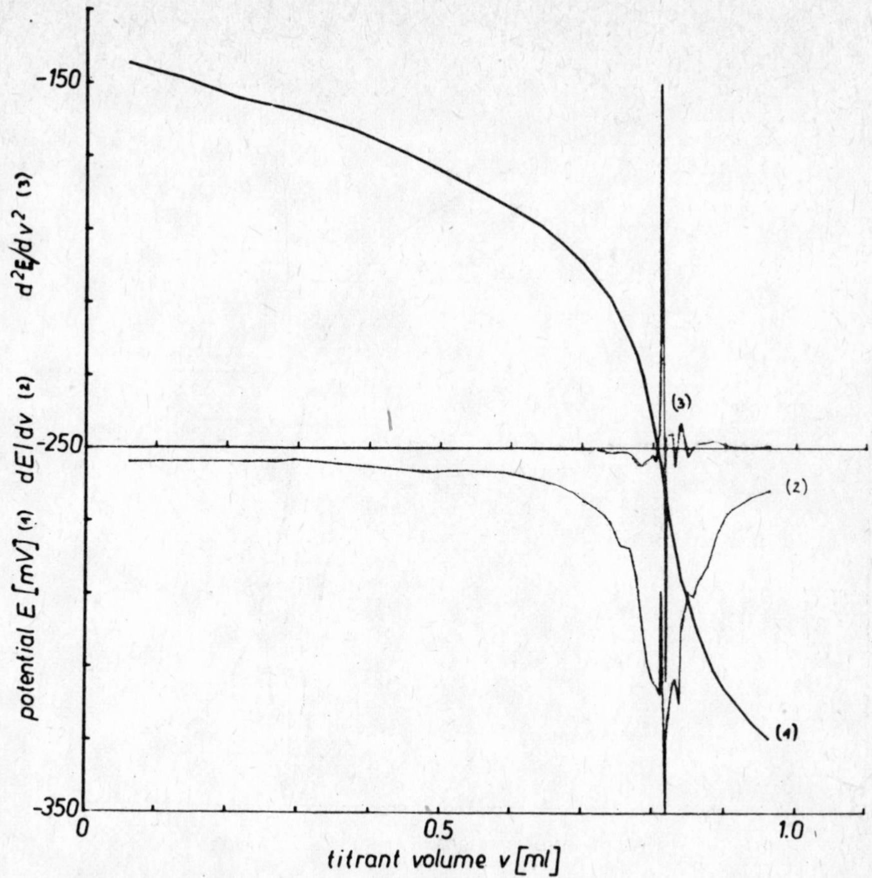

Fig.4 Cl$^-$-titration curve using a Cl$^-$ - ISE, its 1. and 2. derivation /high density of measuring values, without curve smoothings/

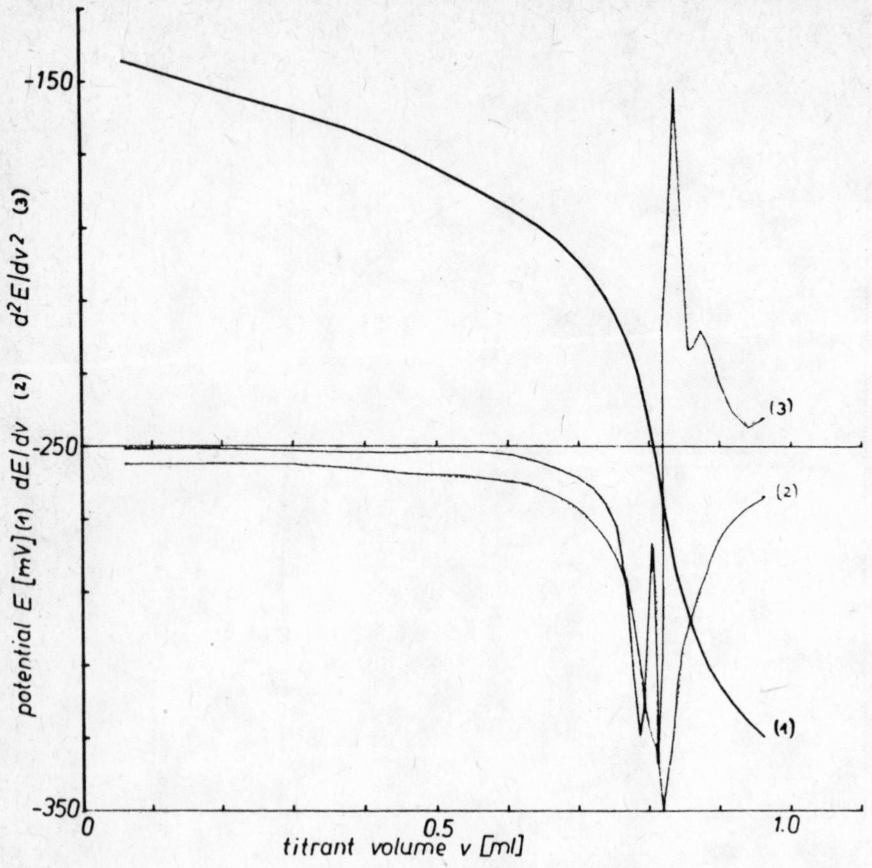

Fig.5 Cl^--titration curve using a Cl^--ISE, its 1. and 2. derivation /high density of measuring values, curve smoothings of titration curve /5/2/ and 1.derivation /3/1//

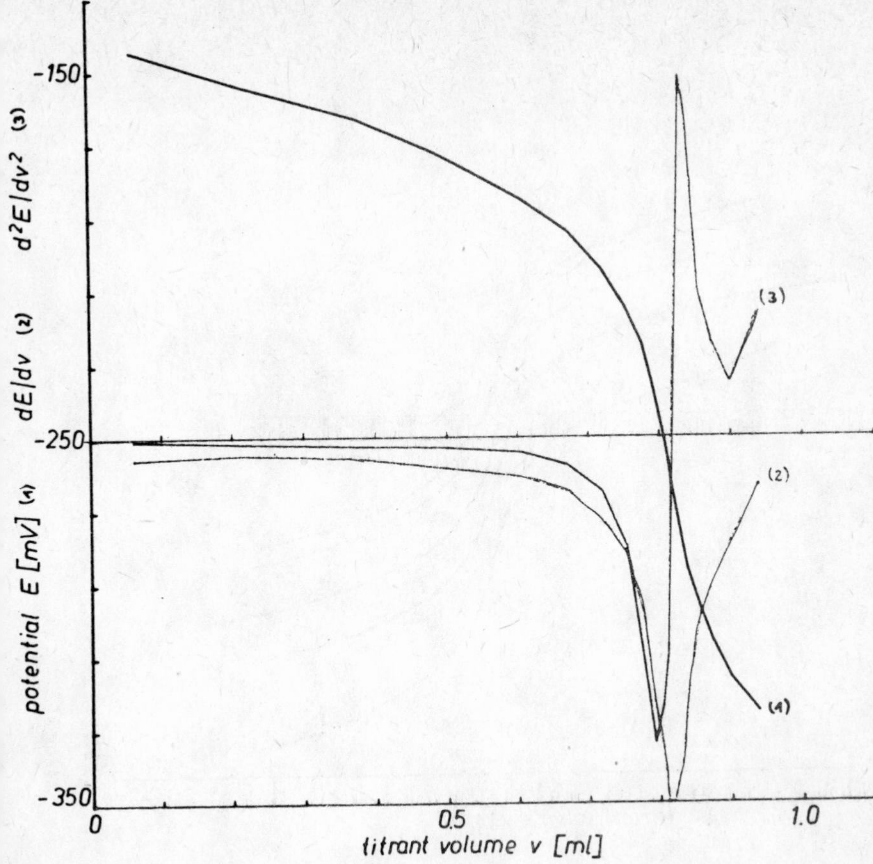

Fig.6 Cl^- titration curve using a Cl^- ISE, its 1. and 2. derivation /normal density of measuring values, without curve smoothings/

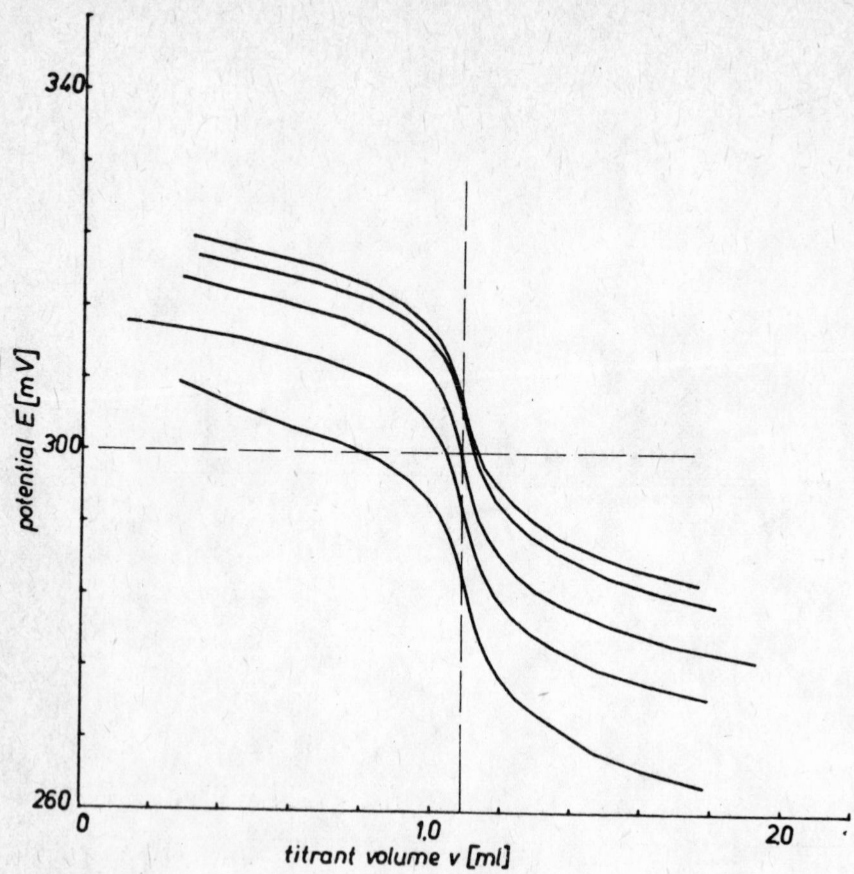

Results of computing inflection points with TITDAT:

1.095 ml
1.099 "
1.090 "
1.094 "
1.090 "

$\bar{v}$ = 1.094 ml

Fig.7 Titrations of equal Cl^- samples using a potential-inconstant Cl^- ISE

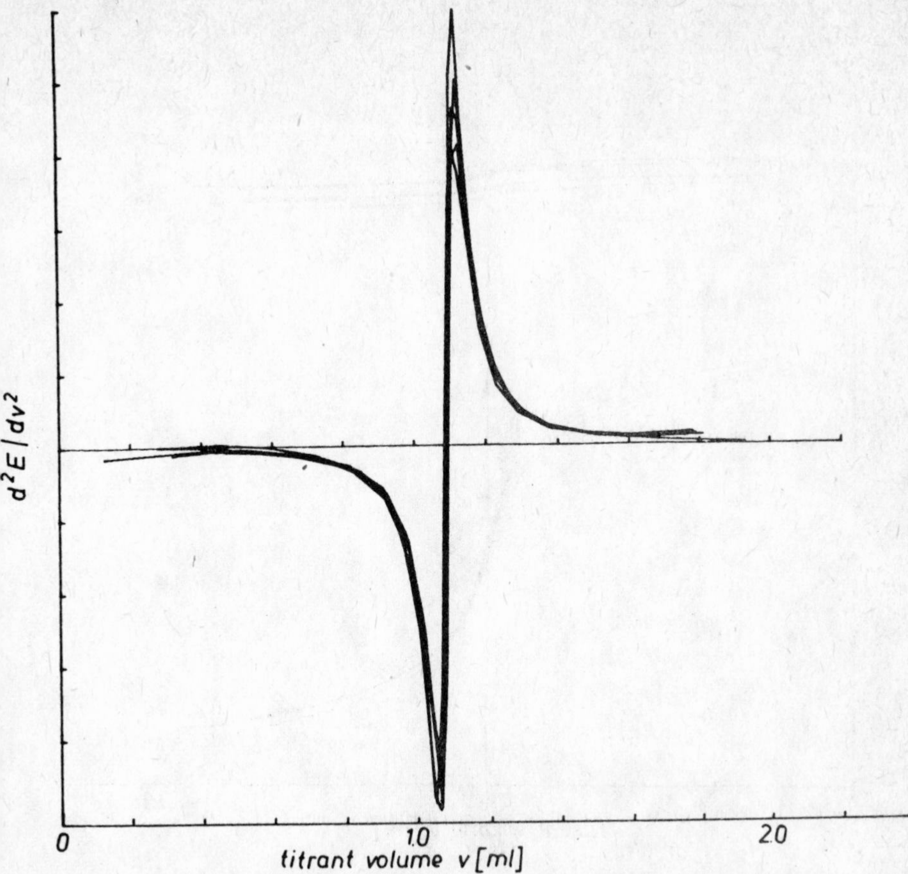

Fig.8 2. derivations of Cl^--titration curves of equal samples, using a potential-inconstant Cl^--ISE

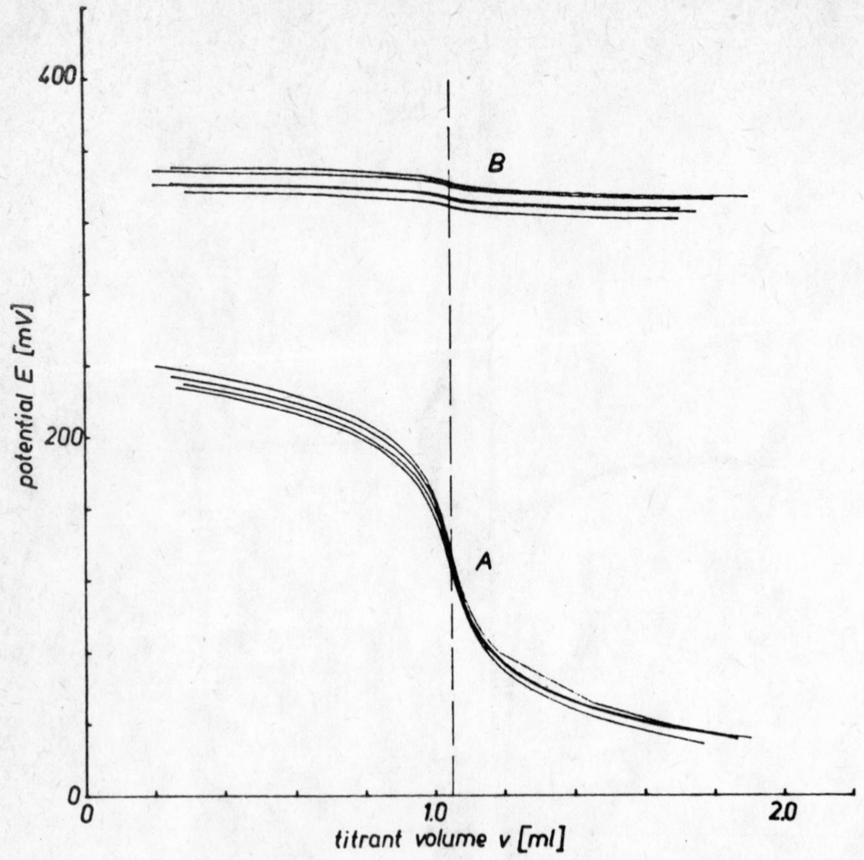

Results of computing inflection points with TITDAT:

	A		B
A:	1.047 ml	B:	1.044 ml
	1.044 "		1.049 "
	1.040 "		1.044 "
	1.046 "		1.043 "
			1.043 "
$\bar{v}$ =	1.044 ml	$\bar{v}$ =	1.045 ml

Fig.9 Titration of equal Cl^--samples using a normal sensitive /A/ and a very small sensitive /B/ Cl^--ISE

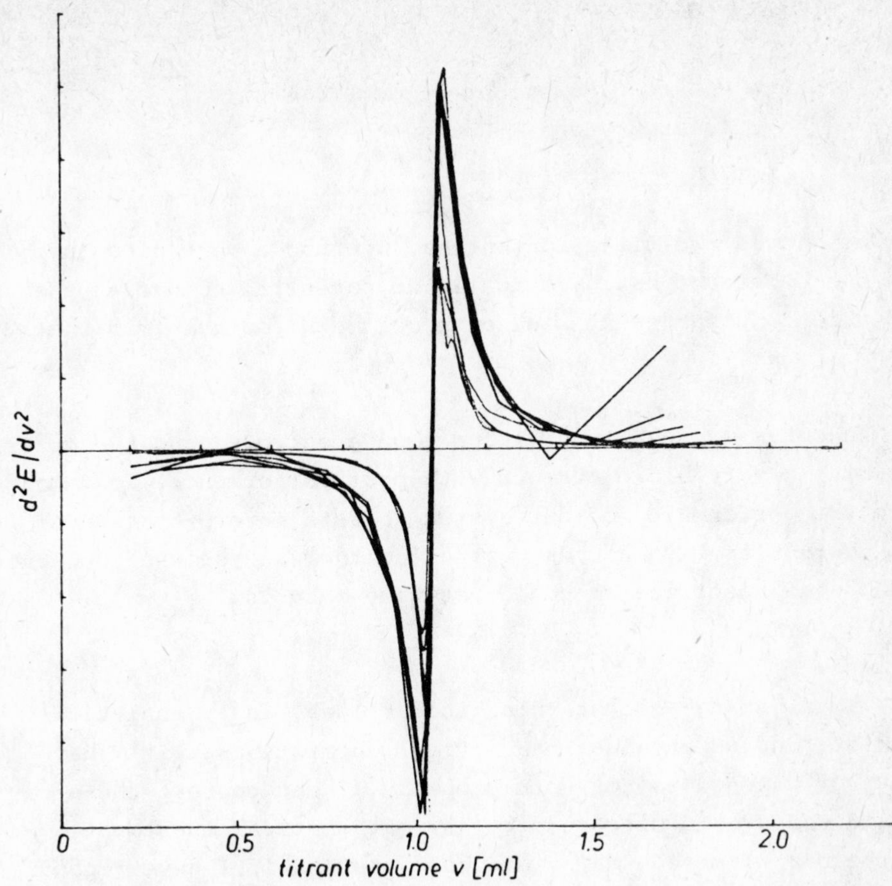

Fig.10 2.derivations of Cl^--titration curves of equal samples, using a normal sensitive and a very small sensitive Cl^--ISE

QUESTIONS AND COMMENTS

Participants of the discussion: P.Becker, G.Horvai, R.Kucharkowski, K.Nagy and E.Pungor

Question:

You showed two titration curves. One had a very nice jump, and the other one had practically no potential change at the equivalence point. What kind of result did you get from that ?

Answer:

One curve was obtained with a good electrode, and the other one with a very old electrode with poor performance. To demonstrate the performance of our system in data interpretation, I took both titration curves, evaluated them and showed that even with a bad electrode we could get some results.

Comment:

I am fighting against this kind of idea, as an analytical chemist, and as an electroanalytical chemist, that if I have a sensor which does not work properly, I can correct the error introduced by the electrode by any mathematical treatment, that a bad electrode may yield good results if I have a good mathematical treatment of the results.

Answer:

I completely agree. The results which we received are not excellent, I only chose this example to demonstrate that even in cases where the titration curve has the shape I have shown, it is possible to get an inflection point.

Question:

What kind of mathematical treatment did you use to calculate the inflection point ?

Answer:

We do not know the titration function. We approximate our values by a polynom and fit our titration curve segment by segment to a polynom. But for the exact determination of the equivalence point it is possible to fit the experimental values to the titration function. This is possible for a simple titration, e.g. neutralization, but in practical use for complex systems it is very difficult to derive the titration function. So we use a general concept for the interpretation of our titration curves.

Comment:

The problem with this general method seems to be that you cannot prove your accuracy or precision, because you do not know whether the equivalence point is the inflection point or not. And for complicated sytems they do not coincide. So I would not agree with you that for complicated systems this is the method of choice.

Answer:

By using such a program the results can be as good as are the chemical conditions.

For complicated titrations there could be a discrepancy between the equivalence point and inflection point. But I do not know a single method to calculate the equivalence point for such titrations. All other methods calculate the inflection point.

The only method to receive the exact equivalence point is fitting to the titration function. In some cases we investigated such problems, and calculated the difference between the inflection point and the equivalence point, and estimated the systematic error of the titration due to this difference. The other method is the calibration with solutions of similar concentration.

Question:

Did you carry out a systematic error analysis ?

Answer:

No.

4th Symposium on Ion-Selective Electrodes
Mátrafüred, 1984

DETERMINATION OF HALIDES AND THIOCYANATE BY NON-SUPPRESSED ION CHROMATOGRAPHY WITH ION-SELECTIVE FLOW-THROUGH ELECTRODES AS DETECTOR

H. MÜLLER and R. SCHOLZ

Section of Chemistry, Technical University "Carl Schorlemmer" Leuna-Merseburg, GDR

ABSTRACT

A non-suppressed ion chromatographic method has been developed for the determination of chloride, bromide, iodide and thiocyanate using an ion-selective flow-through electrode as detector. Trace amounts of iodide in sodium chloride, of bromide in sea water and in common salts were determined.

INTRODUCTION

Developments in ion chromatography suggest rapid and effective methods for the determination of ion concentrations in aqueous solutions. Since its public introduction in 1975 /1/, ion chromatography has gained increasingly wide acceptance as the method of choice for the analysis of many inorganic ions. Ion chromatography systems must use a detector which responds to inorganic ions. The most useful detector for IC is a conductivity detector. However, IC mobile phases are highly conductive, so some type of background suppression is required /2/. Suppressed IC has several disadvantages, for example, the regeneration of the suppressor column. Non--suppressed IC is performed using conventional LC instrumentation which can also be used for other tasks. There is no suppressor column required for this technique.

For the flow-injection analysis and the IC a new flow through detector has been developed /3, 4/. In this paper, the results obtained by a combination of an ion-exchange column with this flow-type detector are presented.

EXPERIMENTAL AND RESULTS

E.m.f. measurements were made with a precision pH-meter MV 87 /VEB Präcitronic, Dresden, DDR/ in connection with a recorder type "Endim" /VEB Meßapparatewerke, Schlotheim, DDR/. The sample loop size was 30 μl and the eluent flow rate was 1 ml/min. As an anion exchanger SB 10 /Machery and Nagel, Düren, FRG/ was employed. Other conditions for the chromatographic separation are given in Figure 3.

Figure 1 shows the construction of the flow-through electrode. The detector consists of a combination of two ion--selective solid-state electrodes of identical membrane material, both with direct internal contact. This arrangement enables one to measure the potential difference between the membranes which in turn is the basis of the differential potentiometric method. Until now, detectors were constructed for halides, thiocyanate, cyanide, sulphide and for silver, copper, cadmium and lead.

This type of electrode combined with liquid chromatographic separation offers the possibility for the determination of several ions with a differential potentiometric indication. We have investigated ion-selective electrodes of the silver sulphide basis as detector for halides and thiocyanate. The response of these electrodes to these ions under chromatographic conditions is shown in Figure 2 . The manifold used for the separation and determination is outlined in Figure 3.

Figure 4 show the influence of the eluent on the separation process. As eluents have been investigated: sodium acetate, sodium salicylate, potassium nitrate and sodium perchlorate. It was found out that potassium nitrate solution possesses high efficiency with respect to halides and thiocyanate. All further experiments were carried out with this eluent. By the method described, the simultaneous quantitative analysis of anions can be accomplished in less than 18 minutes.

<u>Determination of bromide in baltic sea water and of iodide and bromide in common salts</u>

The natural range of bromide in baltic sea water is about 2×10^{-4} M Br^- /Maritimes Observatorium, Zingst, DDR/.
The ion chromatogram of baltic sea water is shown in Figure 5.
As content of bromide in the sample was determined
$1.9 \pm 0.08 \times 10^{-4}$ M /f=2, P=0.95/.
An ion chromatogram of sodium chloride containing iodide is given in Figure 6. The results which were obtained with this technique are given in Table 1.
Figure 7 show other applications. In these cases bromide was determined in sodium chloride samples /common salt/.
The results are shown in Table 2.

DISCUSSION

The suggested method is a fast and relatively uncomplicated procedure for the simultaneous quantitative analysis of anions like Cl^-, Br^-, I^- and SCN^-.

Other ion-selective electrodes like Cd- and Cu-selective electrodes should be used successfully as detector for cations after preliminary separation by ion chromatography. Such applications of this type of detector will be reported later.

REFERENCES

1 H.Small, T.S.Stevens and W.C.Baumann, Anal.Chem., <u>47</u>, /1975/ 1801

2 H.Small, Anal.Chem., <u>55</u>, /1983/ 235 A

3 H.Müller et al., DDR-Patent WP G 01N /222869/1981/

Table 1. Determination of iodide in sodium chloride

Experimental technique	results /%/
IC:	
calibration curve	$/6.5\pm1.6/ \times 10^{-4}$ /f=2, P=0,95/
standard addition	$/6.4\pm1.5/ \times 10^{-4}$ /f=2, P=0.95/
reference method:	
iodometric method	$/11.5\pm1.2/ \times 10^{-4}$
producer	$/19.0\pm3.8/ \times 10^{-4}$

Table 2. Determination of bromide in sodium chloride /common salt/

Sample	bromide /%/	producer
sodium chloride	0.014	-
sodium chloride containing I^-	0.014	-
sodium chloride "Sanisal"[R]	0.17	0.23

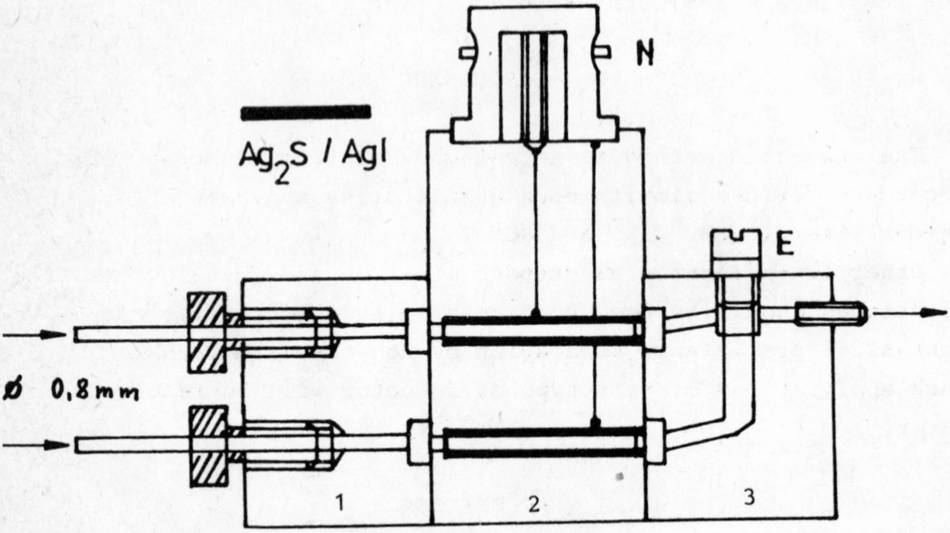

Fig. 1. Flow-through electrode/electrode block
1: cell inlet, 2: ISE, 3: cell outlet,
N: plug, E: electrolytic connection

Fig. 2. Response of the electrodes under chromatographic conditions
v_S = 30 µl, v = 1 ml/min, c = 3×10^{-1} M KNO_3
electrodes: A: Ag_2S/AgCl B: Ag_2S/AgBr C: Ag_2S/AgI
D: Ag_2S O Cl^- ● Br^- Δ I^- ◘ SCN^-

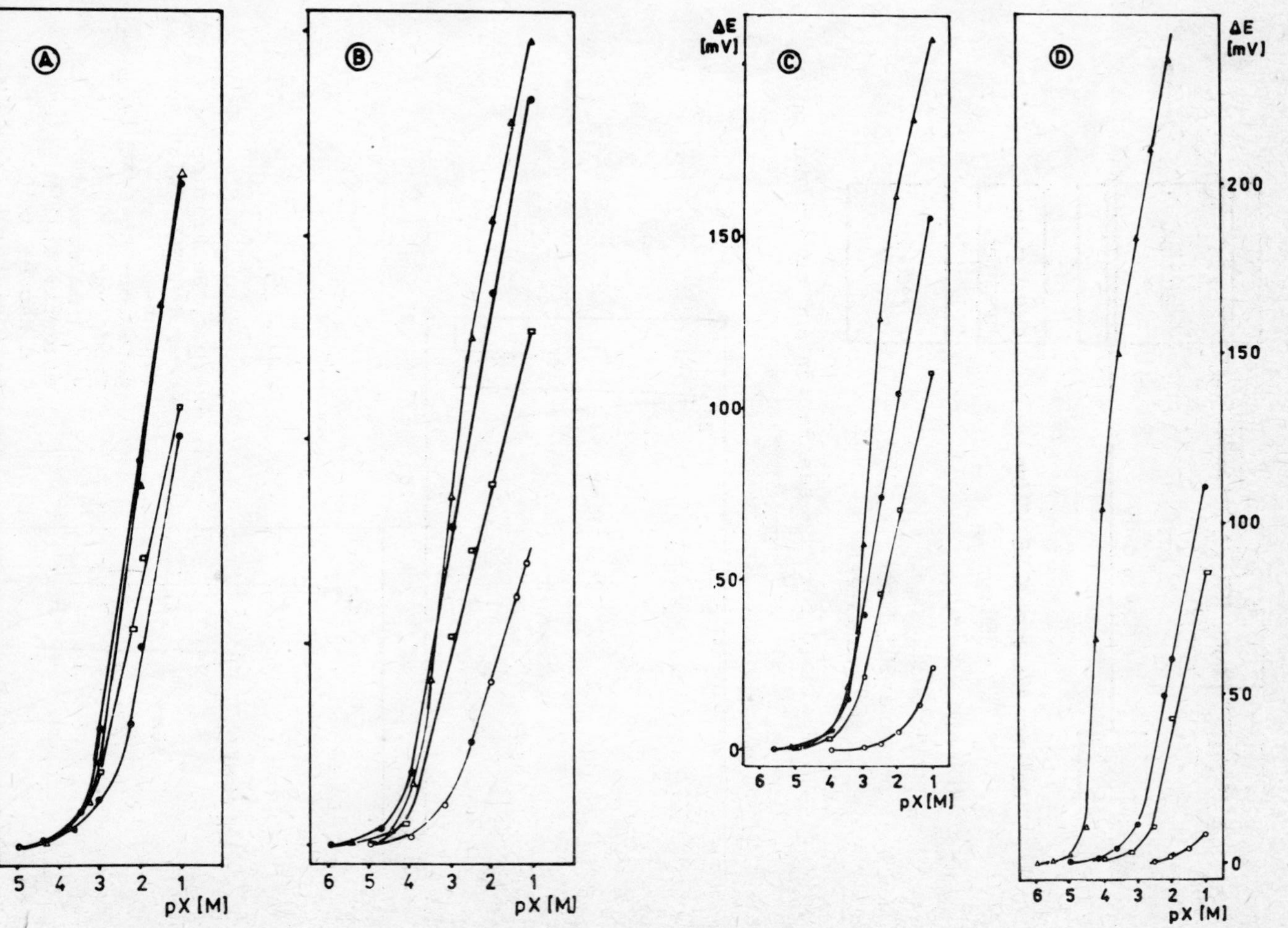
A
B
C
D
ΔE [mV]
pX [M]

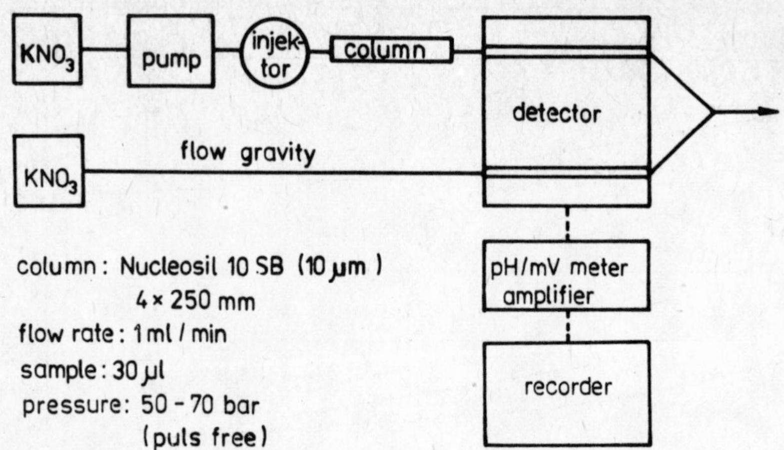

Fig. 3 Manifold for the separation and determination of Cl^-, Br^-, I^-, SCN^-

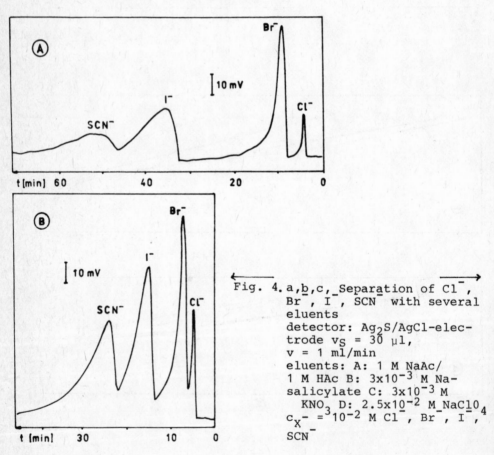

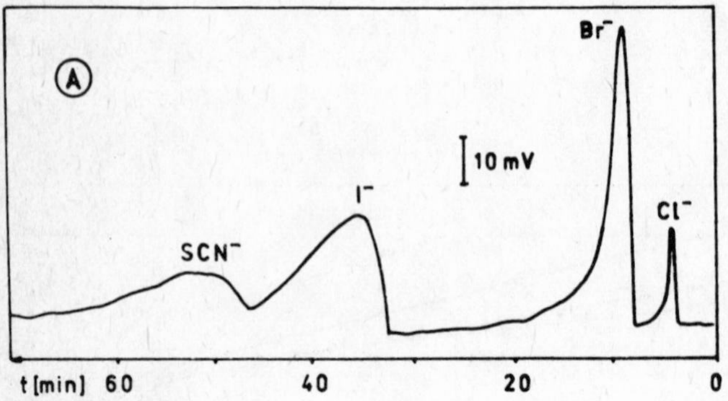

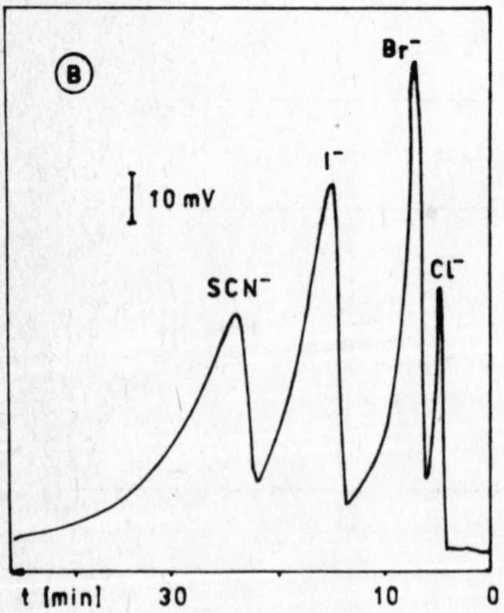

Fig. 4. a,b,c, Separation of Cl^-, Br^-, I^-, SCN^- with several eluents
detector: $Ag_2S/AgCl$-electrode v_S = 30 µl, v = 1 ml/min
eluents: A: 1 M NaAc/1 M HAc B: $3x10^{-3}$ M Na-salicylate C: $3x10^{-3}$ M KNO_3 D: $2.5x10^{-2}$ M $NaClO_4$
$c_{X^-} = 10^{-2}$ M Cl^-, Br^-, I^-, SCN^-

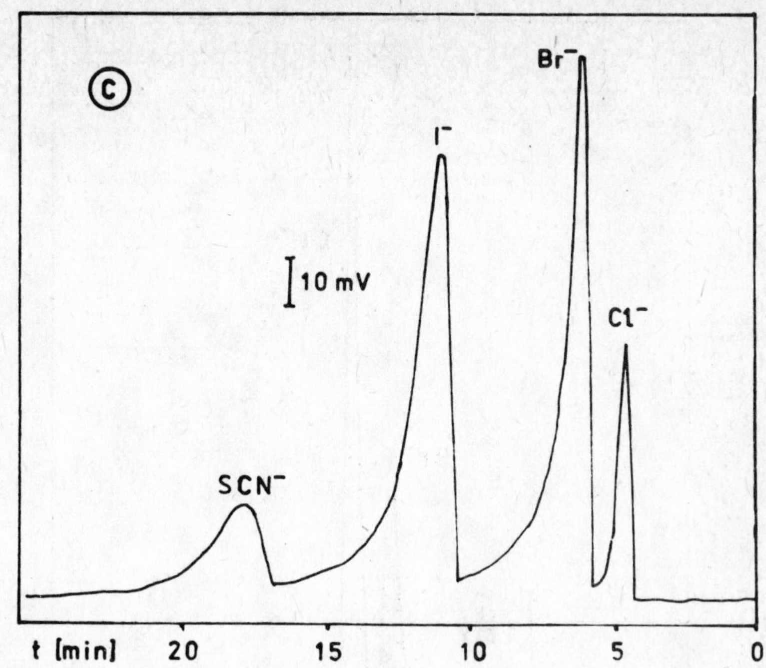
C
Br⁻
I⁻
10 mV
Cl⁻
SCN⁻
t [min]
20
15
10
5
0

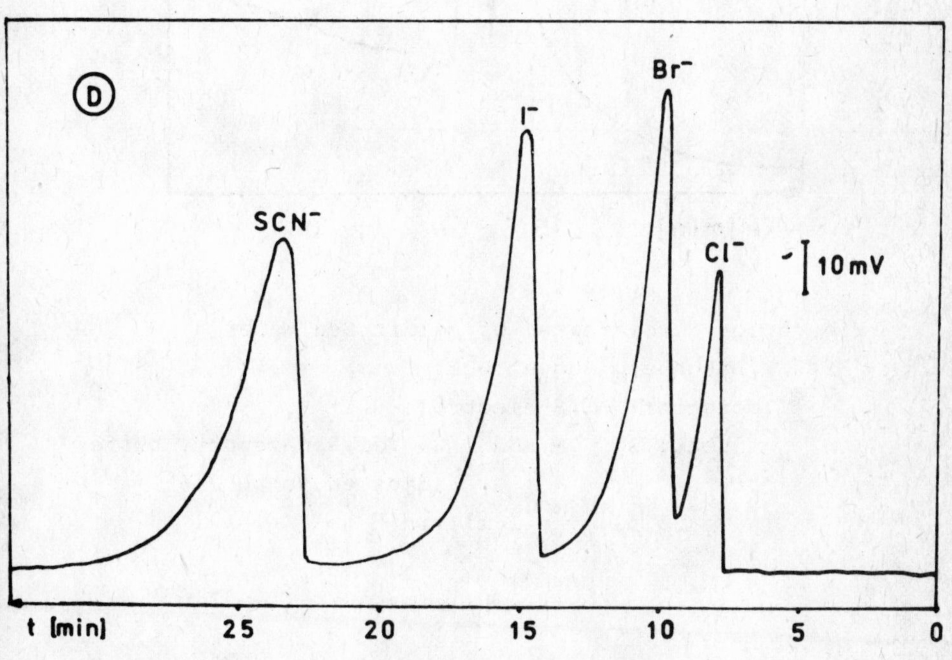
D
Br⁻
I⁻
SCN⁻
Cl⁻
10 mV
t [min]
25
20
15
10
5
0

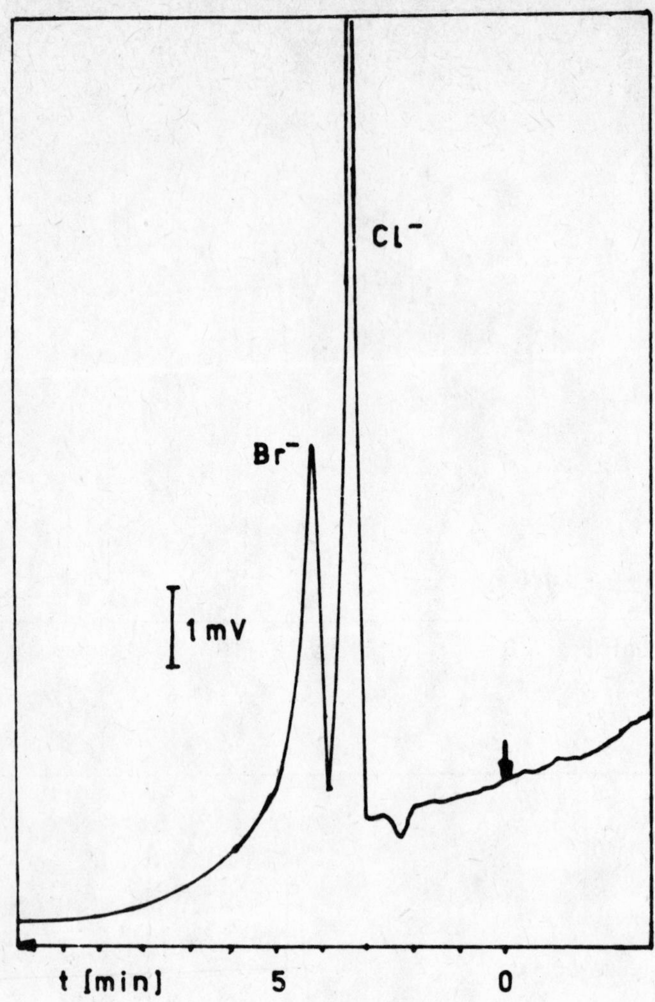

Fig. 5 Ion chromatogram of Baltic Sea Water /Bakenberg/Rügen, DDR/
detector: Ag_2S-electrode
eluent: 10^{-1} M KNO_3//25 Vol.% phosphate buffer adjusted to pH 7,6
v_S = 30 µl, v = 1 ml/min

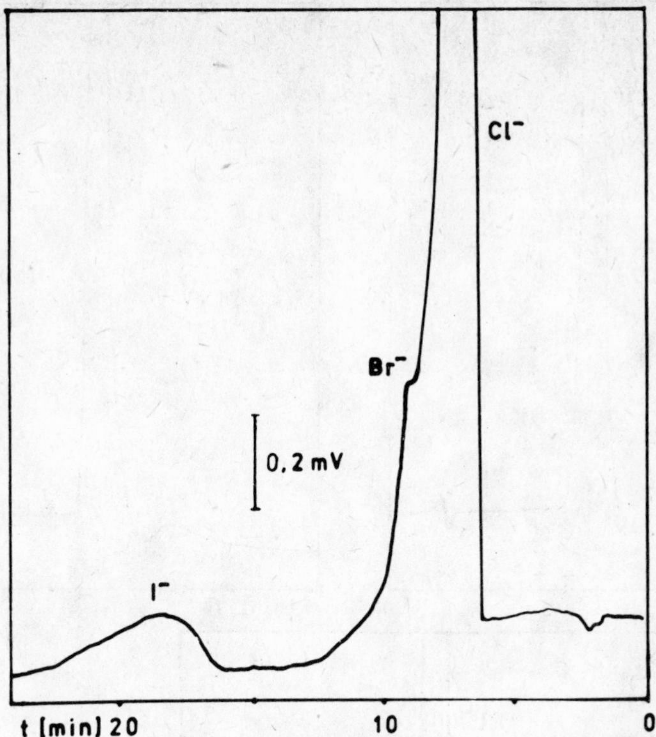

Fig. 6 Ion chromatogram of NaCl containing I^-
Detector: Ag_2S-electrode
eluent: $5x10^{-2}$ M KNO_3
v_S = 30 µl, v = 1 ml/min

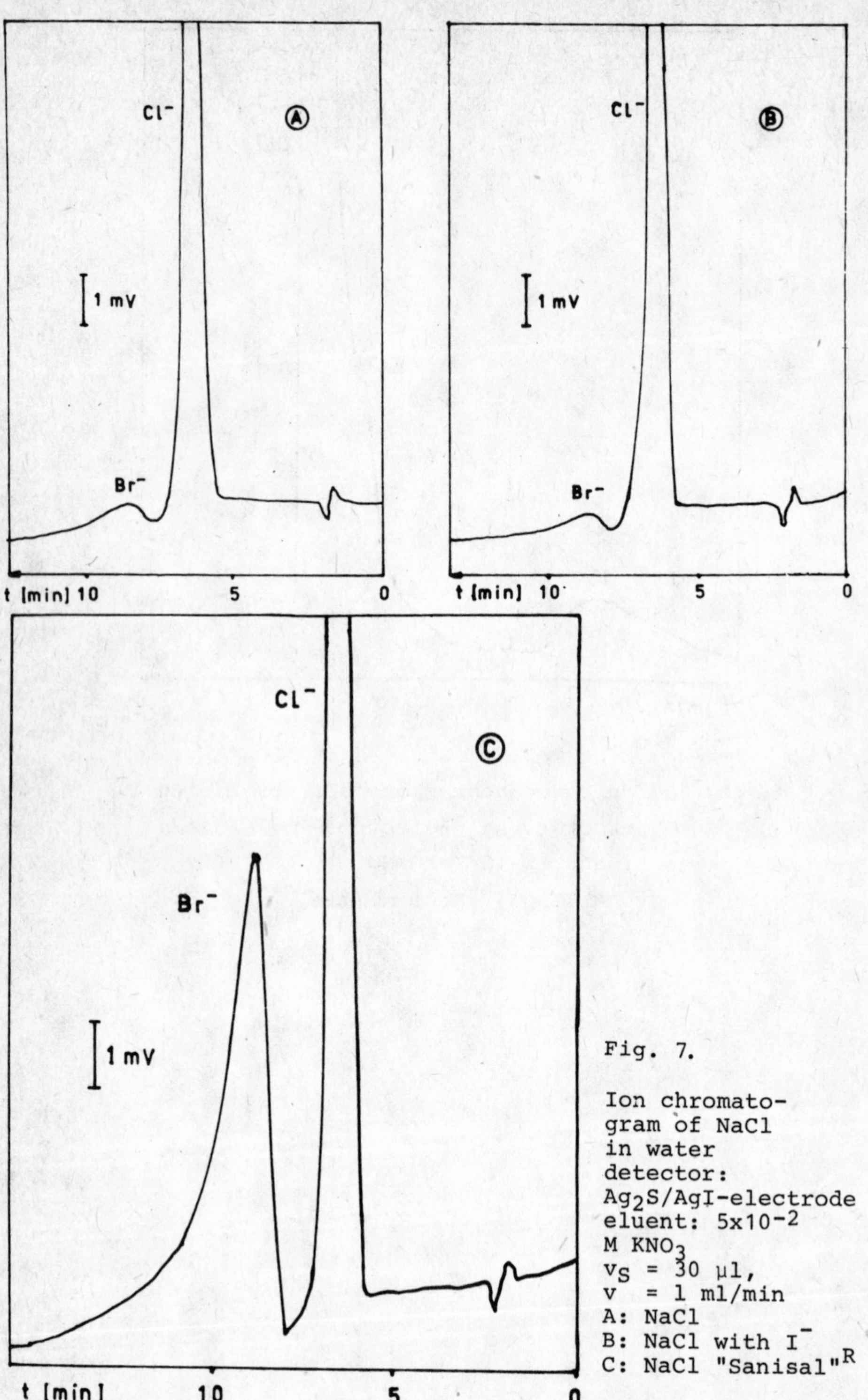

Fig. 7.

Ion chromatogram of NaCl in water
detector: Ag_2S/AgI-electrode
eluent: 5×10^{-2} M KNO_3
v_S = 30 µl,
v = 1 ml/min
A: NaCl
B: NaCl with I^-
C: NaCl "Sanisal"[R]

QUESTIONS AND COMMENTS

Participants of the discussion: G.Horvai, N.Ishibashi, H.Müller, K.Nagy, M.Neshkova, E.Pungor and J.D.R.Thomas

Question:

You have shown some calibration graphs in which the limit of the linear range was 10^{-3} mol/l. What was the reason for this ?

Answer:

This was under the conditions of flow injection analysis.

Comment:

We had lower limits for the linear range with flowing systems than with batch method. You may have some problems with the electrode.

Answer:

What I indicated, is the concentration of the solution injected. However, the concentration decreases significantly due to dispersion.

Question:

Could you give the concentration in the detector cell ?

Answer:

It is very difficult because even if we know the sample volume and flow rate, it is difficult to get an accurate value for the dispersion.

Comment:

When you have small sample plugs and the sample concentration is low, there is a tendency to get this shorter linearity range in the flow injection mode.

Answer:

The reason for these curves is the dispersion, I think.

In the batch method we had calibration curves linear down to 10^{-4}-10^{-5} mol/l. Here, in our system we have the ion-exchange column and the flow injection system.

Comment:

Even so, the dilution seems to be too high. And I think it would be necessary to determine the degree of dilution.

Comment:

I might add that if the flow rate is very high and the sample plug very small, the residence time will not allow to get signal wich may be evaluated.

Question:

You showed values of ΔE. What was the solution passing through the reference channel when you did the calibration ?

Answer:

We used potassium nitrate, without the potential determining ion.

Comment:

You use one type of ISE for measuring several ions. The response may be very good for one of them, but not so good for the others. So, in this respect the conductivity detector might be better, even if you need a suppression. And now there are also many non-suppressed techniques.

Answer:

The aim of this presentation was to show that it is possible to choose the best electrode for any analytical problem. For example, when I have to determine bromide in Baltic sea water, it is better to work with an electrode the sensitivity of which is not so high for chloride, but sufficiently high for bromide.

Question:

You used one and the same detector for different ions.

The electrode is responding with different types of mechanism to these ions. Did you have some troubles with the reproducibility and response time ?

Answer:

No, we had no problem.

Question:

Where was the reference electrode placed ? The liquid junction potential may depend on the kind of solution you are measuring. We have measured a 19 mV difference in liquid junction potential in the same experiment.

Answer:

We have no liquid junction, we have two solid-state electrodes.

Question:

Did not you observe adsorption of species ?

Answer:

The tailing for iodide is significant, so it is impossible to determine iodide and thiocyanate simultaneously when the concentration of iodide is high. And this tailing is not caused by the column but by the electrode, even when flow--injection technique is used.

Question :

Did you consider using silver wire as a universal sensor ?

Answer:

No, we have not.

4th Symposium on Ion-Selective Electrodes
Mátrafüred, 1984

APPLICATION OF ION-SELECTIVE ELECTRODES IN COMPLEX *IN VIVO* MEASUREMENTS

G. NAGY, J. TARCALI, K. TÓTH, R.N. ADAMS* and E. PUNGOR

Institute for General and Analytical Chemistry
Technical University, Budapest, Hungary

*Department of Chemistry, Kansas University
Lawrence, Kansas, USA

ABSTRACT

Simultaneous measurements were made with micropipette type potassium ion-selective and voltammetric micro electrodes in different areas of rat brain. Two different kinds of computer controlled voltammetric measuring apparatus and continuous potentiometric recording were employed. Valinomycin based and new bis/crown ether/s-type ligand based potassium selective electrodes were used. In vivo recording obtained after veratrine alcaloid stimulation are shown.

INTRODUCTION

Ion-selective electrodes have been employed in physiology research since their early years of development. During this period the researchers in the field have succeeded to work out an outstanding miniaturised version of ion-selective electrodes, i.e. the micropipette type ion-selective electrodes. In connection with this the pioneering work of Orme/1/, Walker/2/, Lux/3/, Thomas/4/, Nicholson/5/ and many others should be mentioned. In our days electrodes with submicron diameter measuring tip enable us to obtain immediate information about intracellular ion-activities, while easy to fabricate electrodes with a few micrometer diameter tip size are frequently used in extracellular in vivo measurements. Recently extracellular potassium activity monitoring is gaining application in brain research /6/ employed together with other physiological characteristic observation.

In vivo voltammetric studies with micro-size carbon electrodes were introduced in brain research by Adams and co-workers /7/. These measurements intend to obtain information on local concentration changes of electroactive neurotransmitter materials and related molecules. In practice of in vivo voltammetry graphepoxy electrodes with 50-300 μ diameter /8/, carbon paste ones /9/ with 200-500 μ diameter or electrodes made of carbon fiber of 6-40 μm diameter /10,11/ have been most frequently used. Different groups employ more or less different voltammetric measuring methods and apparatus ranging from the single pulse chronoamperometry to the differential double pulse measurements.

As it is well known the neurotransmitter material release of a neuron can coincide with massive local ion activity changes. Thus in order to get more detailed information it was obvious to combine the in vivo voltammetric measurements with ion--selective electrode monitoring of brain area adjacent to the voltammetric electrode. In our work detailed studies have been started to solve the technical problems of this combined application and to investigate its advantages. This short report is about our results obtained so far.

EXPERIMENTAL

Electrodes

Graphepoxy and carbon fiber microelectrodes were prepared and used in these studies. The preparation of the graphepoxy electrodes is described elsewhere /8/. For the preparation of the other type of electrodes, carbon fibers with 40 μ diameter were inserted in glass micropipettes and cemented in with epoxy or Loctite adhesive /Loctite /Ireland/ Limited Kylemore Park North DUBLIN I.S.415 Cat No 41506/. The fibers were cut back at about 100 μm from the conic end of the glass pipettes.

For the preparation of the potassium ion-selective electrodes micropipettes were pulled from sodium glass tubes of 1,5-2 mm id. The tubes previously were sooked in

H_2O_2 /30%/: H_2SO_4 cc 1:1 for a day and after washed to neutral and dried. The tip of the pipettes were broken off to get a tip diameter of about 5-40 μm. Silinizing solution $//CH_3/_2SiCl_2/$ was introduced with capillary action into the tip and the pipettes were placed in covered petry dish and kept for an hour at 100 ^{o}C. The preparation of the potassium electrodes was completed by back filling the pipettes with internal filling solution /145 mM NaCl, 5 mM KCl/ and introducing the liquid ion-exchanger solution through the tip with capillary action and a light suction.

Two different kinds of liquid ion-exchanger solution were used. One was made by dissolving 75% dibutyl sebacate /Fluka/,15% PVC and 10% valinomycin /Sygma/ in tetrahydrofurane /THF/.

The other one used the bis/crown ether/s-type ligand, BME 44, tailored out by Pungor and coworkers /12/. The preparation of the later solution was very simple:

In about 200 μl THF 5 mg BME 44 active materiall,36 mg sodium tetraphenyl borate /Aldrich/, 80.6 mg o-nitrophenyloctyl ether /Fluka/, and 13 mg PVC was dissolved. When it required the solutions were diluted with a few drops of THF

For voltammetric measurement a silver wire auxiliary electrode /A in Fig.1/ and a chloridised silver wire reference electrode /VR/ were introduced into the brain of the experimental animal, or into the in vitro measuring cell while micropipette type silver,silver chloride reference electrode /R/ placed into the close vicinity of the ion-selective electrode /ISE, in Fig.1/ served for the potentiometric measurements. Saline solution was employed as internal filling solution. For potentiometric ground electrode /Gr/ a silver wire was used.

In some cases for the local drug administration drug solution filled micropipette connected to a pressure ejector was joined to the electrode assembly. The schematic drawing of this assembly together with the other part of in vivo experimental arrangements are shown in Fig.1.

Apparatus

An automatic SDK 85 system based four channel chronocoulometric measuring unit described by Adams and coworkers /13/

was used. For "stair-case" voltammetric measurements a table minicomputer based /EMG-666, 8K byte Hungary/ system was put together in our laboratory. The block diagram of the system is shown in Fig.2. The potentiostat and the i/o interface was built in our laboratory. EMG type 893 printer, integrating digital voltmeter EMG type TR-1659/2 and plotter Videoton type 2000-666 Hungary/ completed the system.

Simple home made analog potentiometric unit connected to strip chart recorder served for the micropipette ion-selective electrode based measurements. To avoid ground loop problems the potentiometric unit was powered by batteries. The two high impedance inputs of the unit and the belonging voltage followers /AD 515 based ones/ were placed in a separate small size insulated metal capsule. In this was the cabels between the potentiometric electrodes and the input could be reduced to a few centimeter length.

In vivo work

Male Sprague Dawley rats /200-250 g/ were anesthetized with chloralhydrate /400 mg/kp ip/ or urethane solution /1 g/kg ip/ and placed on stereotaxic device. Part of the scull and dura was removed for the introduction of the electrode assembly and separate holes were prepared for the ground electrode, for the voltammetric reference /VR/ and auxiliary wire /A/ electrodes as shown in Fig. 1. The voltammetric working electrode and the potassium electrode were in vitro calibrated before and after the measurements. These calibration curves were used for the evaluation, to construct the concentration-time curves.

The chronocoulometric measurements were taken at 0.5 V /vs Ag/AgCl/ with 1 sec pulse width and 500 ms integration time. Relaxation times of 80-120 s were employed in different studies.

Stair-case voltammgrams were taken in the range of 0-0,6 V employing 20-25 steps. The recording of one voltammogram took about 2-4 s.

Potentiometric cell voltage was continuously recorded.

RESULTS

The potentiometric recording of the potassium ion activities with the device used has not interfered at all with the two different kinds of voltammetric measurements.
The voltammetric measurements however owing to the electric field generated made dramatic marking on the potentiometric trace. Since the voltammetric measurements were taken for a relatively short time /1-4 s/ this artifact did not disturbed our measurement aiming to detect relatively slow concentration changes. The periodic marking was even helpful in syncronising the two concentration-time curves.

Fig. 3 shows a potentiometric recording carried out with BME-44 ligand based potassium microelectrode /5 μ measuring tip diameter/ in the somatic cortex of an anestetized rat. The oscillation of the potassium activity was induced by addition of a small veratrine crystal on the surface of the cortex adjacent to the electrodes. The marking of the periodic stair case voltammetric measurements can well be noticed and distinguished. The cell depolarising-, in this way the extracellular potassium concentration increasing - action of the different veratrine alcaloids is well known /14/. The repetitive action of a single veratrine alcaloid dose was observed in different objects by severals /15,16/.

In our work the changes of potentiometric and voltammetric signals have been followed parallely after the local, i.v. or i.p. addition of different drugs in different brain areas. The detailed discussion of the results obtained will be given elsewhere.

Fig. 4 just as an example shows parallely recorded potassium and electroactive neurotransmitter concentration-time curves. The curves were taken in the striatum of a rat with graphepoxy and valinomycin based potassium electrodes. The local injection of about 50 nl droplet of a veratridine solution /1 mg/ml/ induced the regular, simultaneous periodic changes of both species in the extracellular liquid. This very well reproducible effect was observed in individual rats as long as 2-3 hours. The electroactive signal can be assigned as dopamine or dihydroxyphenyl acetic acid /DOPAC/ concentration change considering the area of electrode implantation.

REFERENCES

1 F.W.Orme, Liquid ion exchanger microelectrodes.
In: Glass microelectrodes 376-395
Eds. M.Lavallée et al. New York: John Wiley and Sons, Inc. 1969

2 J.L.Walker, Anal.Chem., 43, /1971/ 90A

3 I.Dietzel, U.Heinemann, G.Hofmeier and H.D.Lux, Exp. Brain Res., 40, /1980/ 432

4 R.C.Thomas, Ion-sensitive Intracellular Microelectrodes, Academic Press, London, 1978

5 C.Nicholson, G.ten Bruggencate, H.Stöckle and R.Steinberg, J.Neurophysiol. 41, /1978/ 1027

6 E.Lenieger-Follert, J.Cerebral Blood Flow and Metab. 4, /1984/ 150

7 R.N.Adams, Anal.Chem. 48, /1976/ 1126A
Anal.Chem., 52, /1980/ 2445

8 J.C.Conti, E.Strope, R.N.Adams and C.A.Marsden, Life Sci. 23 /1978/ 2705

9 A.W.Strenson, R.Mc Creery, B.Feinberg and R.N.Adams, J.Electroanal.Chem. 46, /1973/ 313

10 J.L.Ponchon et al., Anal.Chem. 51, /1979/ 1483

11 J.Tarcali, G.Nagy, G.Juhász, K.Tóth, T.Kukorelli, E.Pungor, in preparation

12 L.Tőke, B.Ágai, J.Bitter, E.Pungor, T.Szepesváry, E.Lindner, M.Horváth, J.Havas, Europatent, Application

13 H.Y.Cheng, W.White and R.N.Adams, Anal.Chem. 52, /1980/ 2445

14 M.C.W.Minchin, J.Neurosci.Methods, 2, /1980/ 111

15 M.Danko, J.Cseri, E.Varga, Acta Phys.Acad.Sci.Hung., 58, /1981/ 103, 275

16 T.L.Török, Z.Salamon, T.T.Nguyen and K.Magyar, Quart.J. Exp.Physiol., 69, /1984/ 1

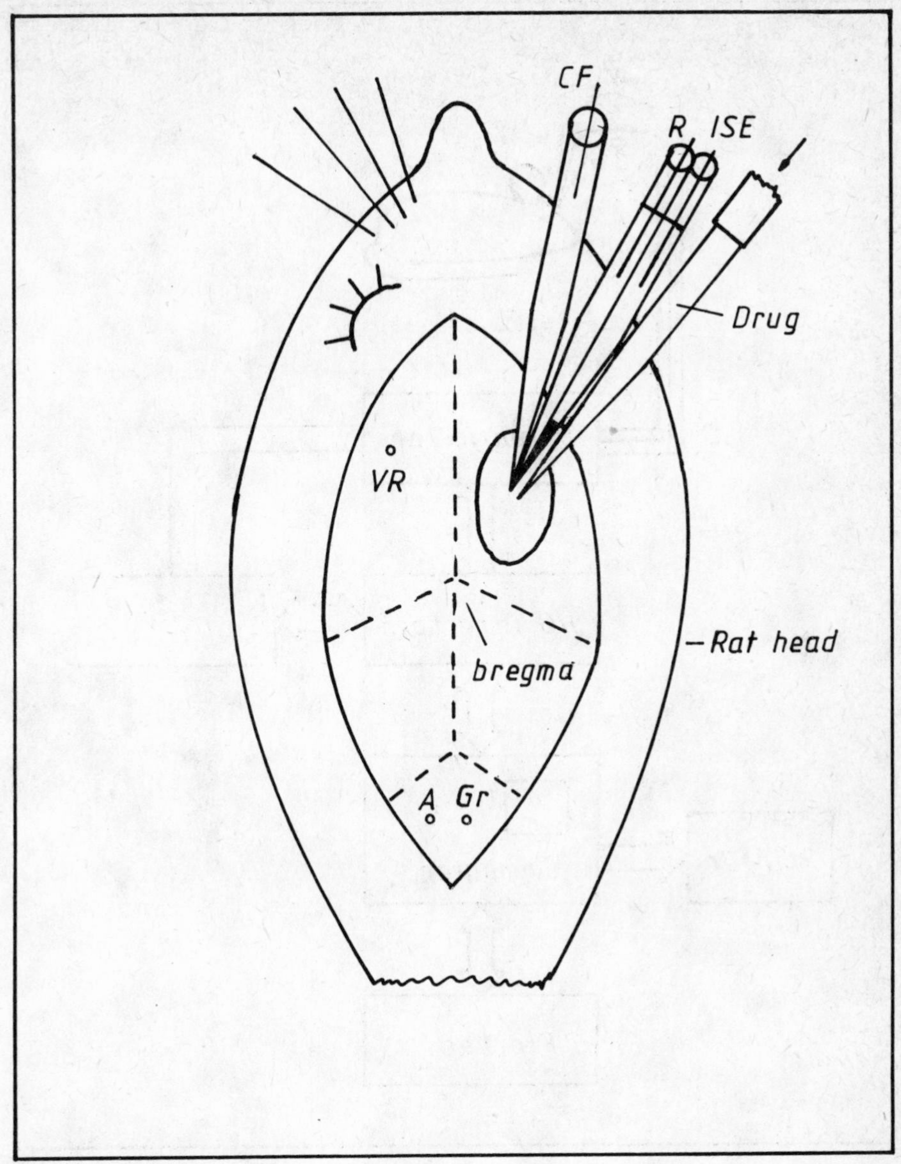

Fig. 1 Arrangement of the measuring assembly in the skull of the experimental animal

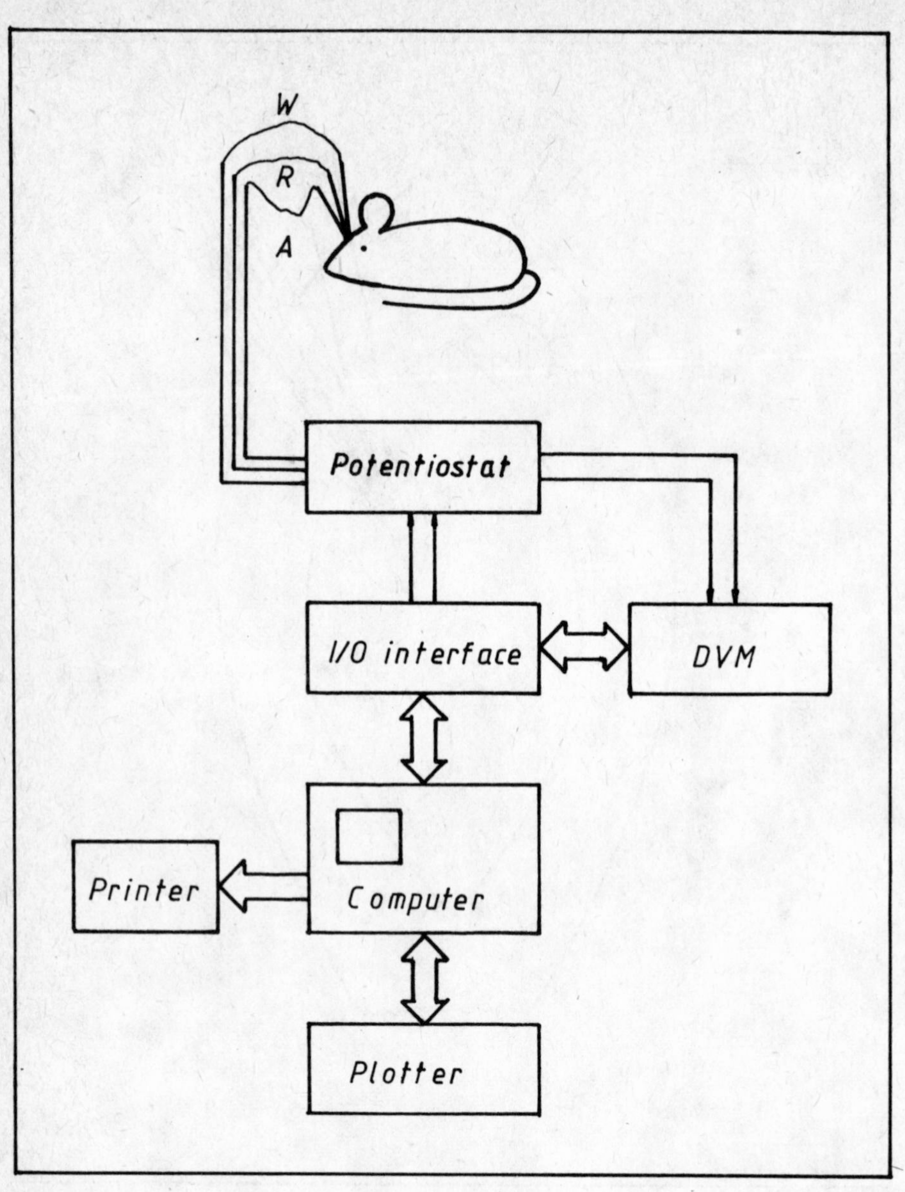

Fig. 2 Block diagram of the EMG /type 666/ computer controlled voltammetric measuring apparatus used in the experiment

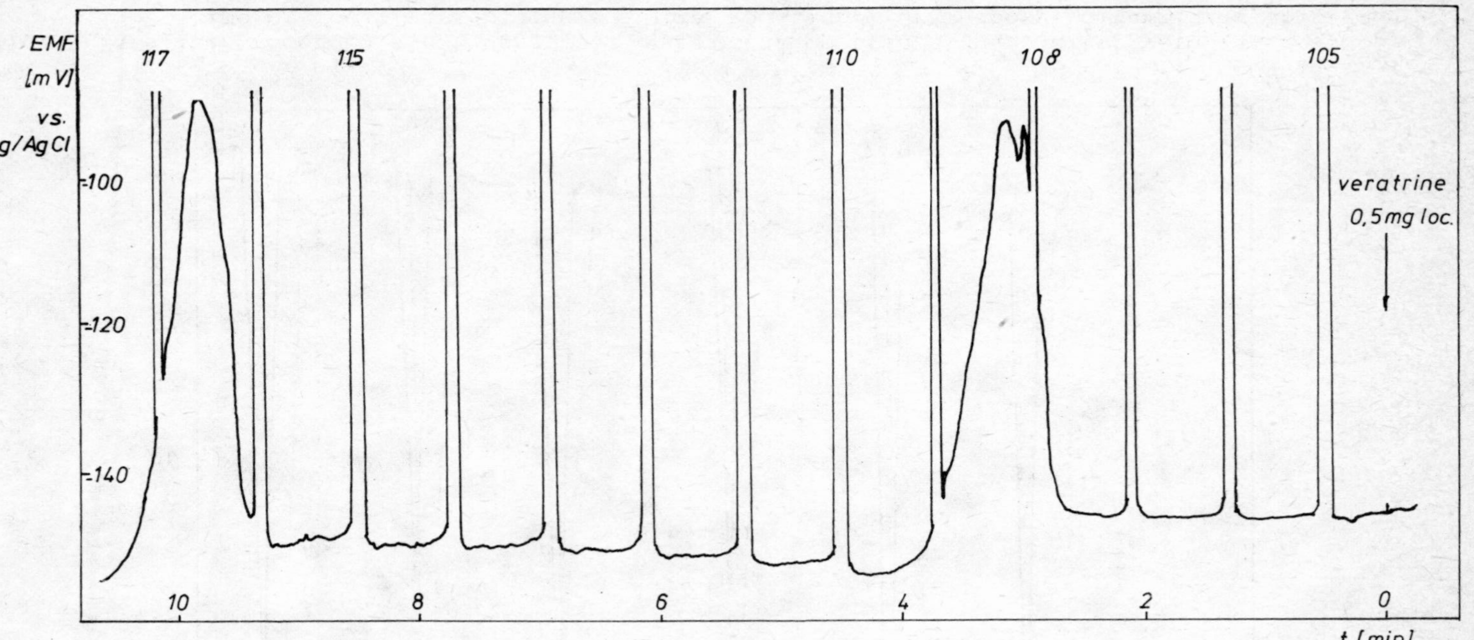

Fig. 3 Electrode potential-time trace recorded with potassium micropipette electrode in rat somato sensory cortex after local administration of veratrine crystal. /Ionophor-BME 44/

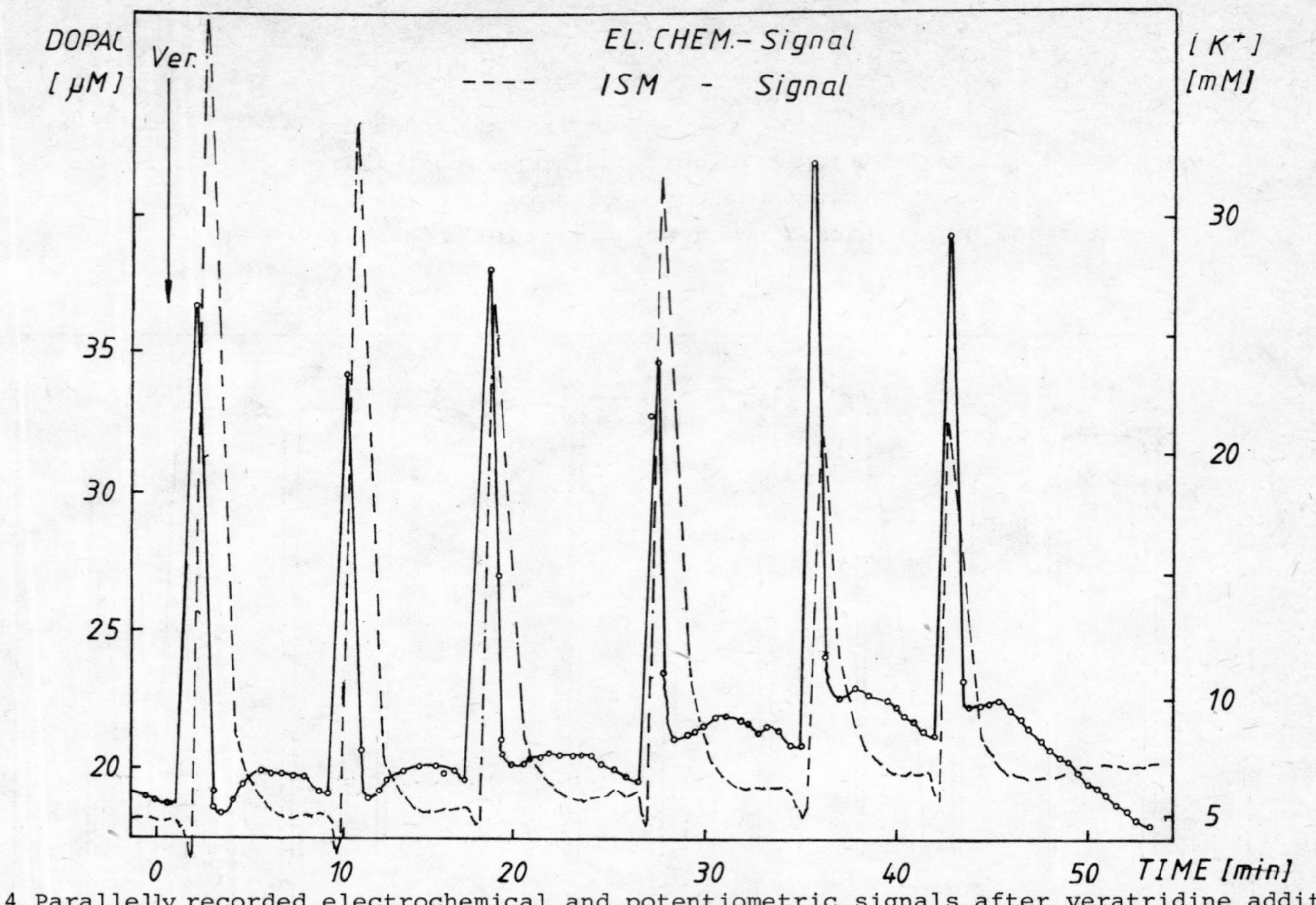

Fig. 4 Parallelly recorded electrochemical and potentiometric signals after veratridine addition. (Ver.-time of local veratridine solution addition.) Graphepoxy voltammetric electrode and valinomycin based ISE micropipette were used.

QUESTION

Participants of the discussion: M.Jänchen and G.Nagy

Question:

You have spoken about four electrodes in connection with brain research and I saw only 3 electrodes. How is this ?

Answer:

I said that Professor Adams' voltammetric apparatus can take four electrodes and make four voltammetric measurements at a time. Of course a reference and an auxiliary electrode is also needed, that is 4+2=6 electrodes altogether. In my measurements I generally used one voltammetric electrode, one reference and one auxiliary electrode. We also had a grounding electrode.

4th Symposium on Ion-Selective Electrodes
Mátrafüred, 1984

A NEW BROMIDE-SELECTIVE ELECTRODE FOR MONITORING OSCILLATING REACTIONS

Z. NOSZTICZIUS, M. WITTMANN and P. STIRLING

Institute of Physics, Department for Chemical Engineering
Technical University of Budapest, 1521 Budapest, Hungary

ABSTRACT

The highly corrosive media of the oscillatory Belousov-Zhabotinsky reaction and the fast concentration changes in the course of chemical oscillations present special requirements for bromide selective electrodes. Commercially available electrodes may show "memory" "stirring effects" and a relatively slow response. A simple new construction was developed to avoid the above problems.

INTRODUCTION

Bromide ions play a crucial role in the different mechanisms proposed for the oscillatory Belousov-Zhabotinsky (BZ) - type reactions (1 - 5). Thus it is understandable that bromide selective electrodes play an equally important role in the investigation of these oscillating reactions. In some recent works (6 - 8) we pointed out however that in the strong acidic media applied in the BZ reaction bromide selective electrodes give a potential response to different corrosive agents as well. Namely the hypobromous acid (which is an active intermediate of the BZ reaction causing a "fast" (8) corrosive process) and a considerable amount of bromate (which is an essential component of the BZ systems causing a "slow" (8) corrosive process) are the most important corrosive agents. It was shown that electrode potentials below the so called "solubility limit"

(7) are due to the presence of these corrosive agents and in these cases no bromide level can be calculated from the measured electrode potential. Nevertheless potentiometric traces recorded in the course of the BZ reaction are still very informative even below the solubility limit: changes in the hypobromous acid concentration can be detected in this region. Therefore a reliable bromide selective electrode can be one of the most important source of information for the reaction kineticist studying different BZ systems. Unfortunately the commercial electrodes are not designed for an extensive use in the highly corrosive media of the BZ reaction and that fact causes different problems. In this paper we discuss these problems and we propose a simple and inexpensive way to produce bromide-selective electrodes to study oscillatory reactions of the BZ type.

PROBLEMS OF THE DIFFERENT BROMIDE-SELECTIVE ELECTRODES

The commercial electrodes apply solid ionselective membranes containing silver bromide as an active material. Most of the membranes are pressed pellets made of pure silver bromide precipitate or silver bromide-silver sulfide coprecipitate or pellets made of molten silver bromide. Inside of the electrode the membrane is usually contacted directly or indirectly with silver metal. That is the ion-selective electrode resembles to the classical Ag/AgBr electrode but with an important advantage: there is no possibility for a direct contact between the inner silver metal and the solution tested by the electrode. This way a disturbing redox function can be avoided. This is especially important in the BZ reaction where different redox couples (Ce^{3+}/Ce^{4+} , Mn^{2+}/Mn^{3+} etc.) are applied as homogeneous catalysts. On the other hand the pellet-type membranes have some disadvantages too.

i) In the case of pressed pellets the appearing microcracks cause a major problem. These microcracks may develop as a result of the corrosion or they can appear spontaneously as a result of ageing. E.g. originally transparent AgBr pellets held in a dark dry container become cloudy at first and nontransparent later. These microcracks cause an electrode "memory" which is especially problematic in the monitoring of oscillating reactions where fast concentration changes have to be followed and a short response time is an important requirement. The memory is a result of adsorption in the microcracks and it can be disclosed by a characteristic "stirring effect". Namely if the electrode is dipped into a relatively concentrated (10^{-3} - 10^{-2}M) $AgNO_3$ or KBr solution then washed carefully and placed into a pure 1.5 M sulfuric acid solution (usual media of the BZ reaction) its potential will depend on stirring. The explanation of the phenomenon is the following. In the pure solution the adsorbed material (silver or bromide ions) diffuses out of the microcracks and if the solution is not stirred the contamination accumulates around the electrode pellet. That concentration buildup can be observed as a slowly creeping electrode potential. Now turning on a magnetic stirrer most of the contamination can be removed from the neighbourhood of the pellet and a sharp change in the potential (several tens of millivolts) can be recorded. The sign and the magnitude of that sudden jump (or fall) of the potential naturally depends on the pretreatment of the electrode (the circumstances and the duration of adsorption etc.). Anyway that memory effect slows down the response of the electrode just in a concentration range (10^{-5}M bromide or hypobromous acid and below) which is just the most important region in a BZ reaction. Especially the pressed pellets made of silver bromide-silver sulfide coprecipitate are prone to such error. Freshly pressed transparent silver bromide pellets can work satisfactorily for a while. Nevertheless the electrode memory has to be checked time to time because microcracks developed rather fast when the pellet is in contact with electrolyte solutions. In summary we may conclude that electrodes based on pressed pellets are not optimal to monitor oscillating reactions.

ii) Molten pellets are not prone to develop microcracks. In this case the main problem is a relatively high impedance of the thick pellets because that high impedance is usually accompanied by a sluggish response. Most probably some transient parasitic potentials can survive for a longer time due to the high impedance.

A SIMPLE ELECTRODE DESIGN FOR OSCILLATING REACTIONS

All the problems of the different pellets can be avoided by applying melt-coated wire type electrodes. Naturally a thin coating of low impedance does not provide a long life span in a corrosive media. However the tip of the electrode containing the coated wire is easy to change. (See the construction of the electrode: Figure 1a/ and 1b.) Also the melt-coated wire is easy to produce or to renew as it is depicted in Figure 2.

Preparation of the coated wire

i) Silver bromide was precipitated by adding 190 cm^3 of 0.1 M KBr solution to a continuously stirred 0.1 M $AgNO_3$ solution. (Volume: 200 cm^3). The latter also contained nitric acid in 1 M concentration. Silver nitrate was applied in an excess to avoid bromide ion adsorption on the precipitate. The AgBr precipitate was washed by decantation with distilled water. When the washing water was acid free the precipitate was mixed with 1 M HNO_3 and subsequently it was washed with distilled water again. At the end the precipitate was filtered and dried at $\sim 100^{\circ}C$ in a desiccator.

ii) Silver wire were freed of any contamination by etching them in a 3 M HNO_3 solution for some seconds. They were washed in distilled water and in acetone afterwards. A $\sim$20 mm long and $\sim$0.3 mm thick coating was produced on the dry pretreated silver wires by dipping them into a relatively cool ($t \sim 500°C$) molten AgBr bath (See Figure 2a/). Then the end of the electrode was dipped into a more hot AgBr melt ($t \sim 600°C$). This way at the end of the wire the thickness of the coating can be decreased to $\sim$0.02 mm along the last 5 mm (Figure 2b/).

MONITORING OSCILLATING REACTIONS WITH THE NEW ELECTRODE

A proposed measuring configuration is depicted in Figure 3. The rather complicated chain of salt bridges is applied to decrease the otherwise high diffusion potential and to avoid any contamination of the tested solution.

The electrode has to be shilded from direct sunshine or from other strong light sources. Otherwise a slow bromine evolution takes place from the electrode due to its light induced decomposition. A hydrolysis of that bromine produces bromide ions and consequently a potential response can be observed which depends on the stirring.

ACKNOWLEDGEMENT

The authors thank Prof.H.D.Försterling of University of Marburg and Prof.J.Sheppard of University of Zimbabwe for valuable discussions.

REFERENCES

1. R.J.Field, E.Kőrös and R.M.Noyes, J.Am.Chem.Soc. 94,(1972) 8649
2. Z.Noszticzius, J.A.Chem.Soc. 101,(1979) 3660
3. Z.Noszticzius, H.Farkas and Z.A.Schelly J.Chem.Phys. 80,(1984) 6062
4. R.M.Noyes, J.Chem.Phys. 80,(1984) 6071
5. Z.Noszticzius, V.Gáspár and H.D.Försterling J.Am.Chem.Soc. submitted
6. Z.Noszticzius, Acta Chim.Acad.Sci.Hung. 106,(1981) 317
7. Z.Noszticzius, E.Noszticzius and Z.A.Schelly J.Am.Chem.Soc. 104,(1982) 6194
8. Z.Noszticzius, E.Noszticzius and Z.A.Schelly J.Phys.Chem. 87,(1983) 510

Fig. 1. Construction and the main parts of the electrode →
a/ Structure of the tip containing the coated wire
b/ Cross-section of the bromide-selective electrode

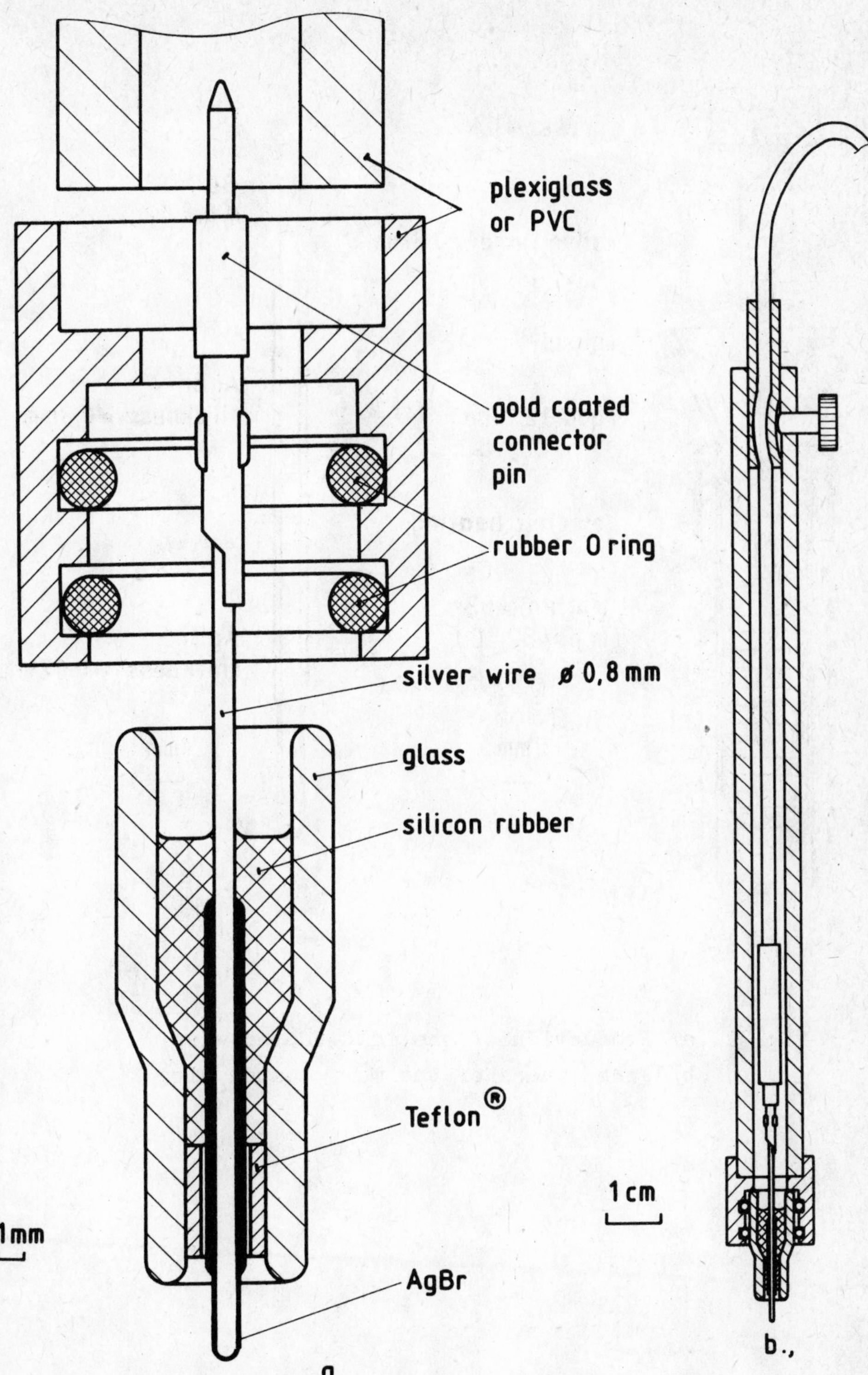
plexiglass
or PVC
gold coated
connector
pin
rubber O ring
silver wire ø 0,8 mm
glass
silicon rubber
Teflon®
AgBr
1 mm
a.,
1 cm
b.,

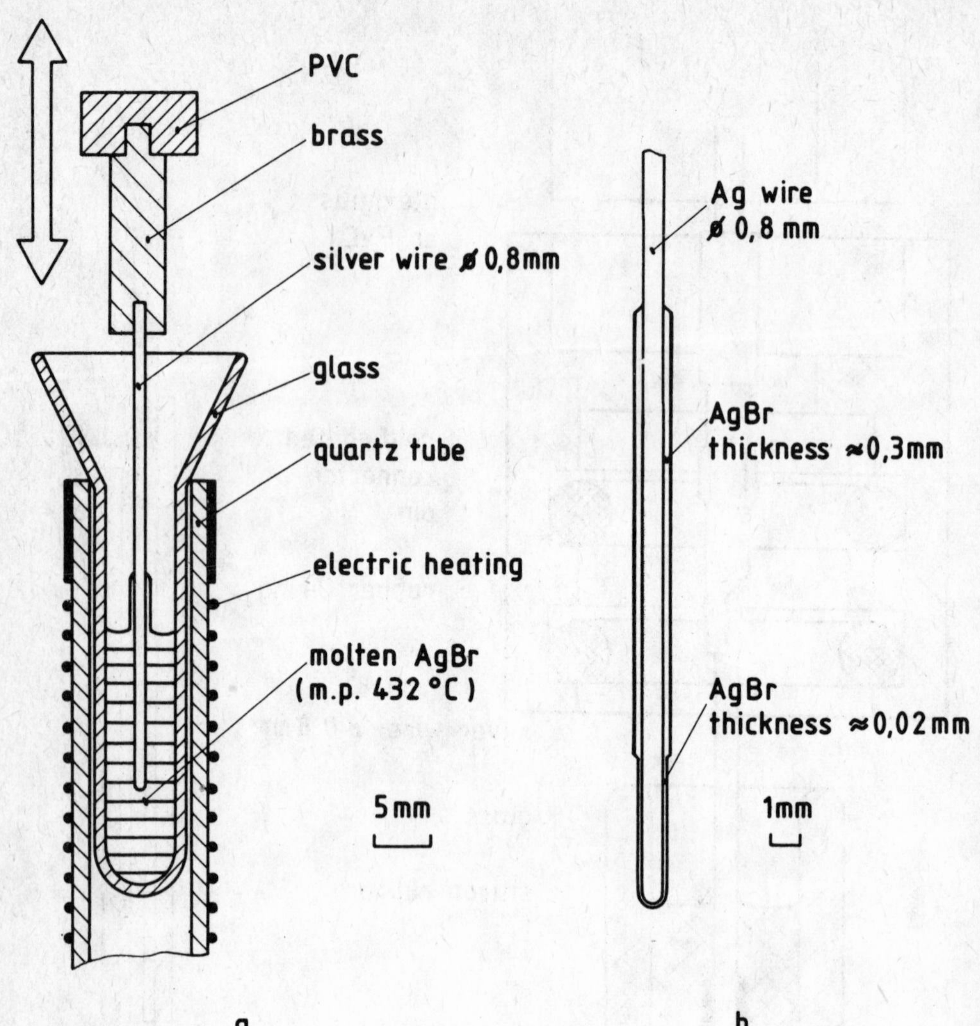

Fig. 2. a/ Production of the melt-coated wire
b/ Final shape of the melt coated wire

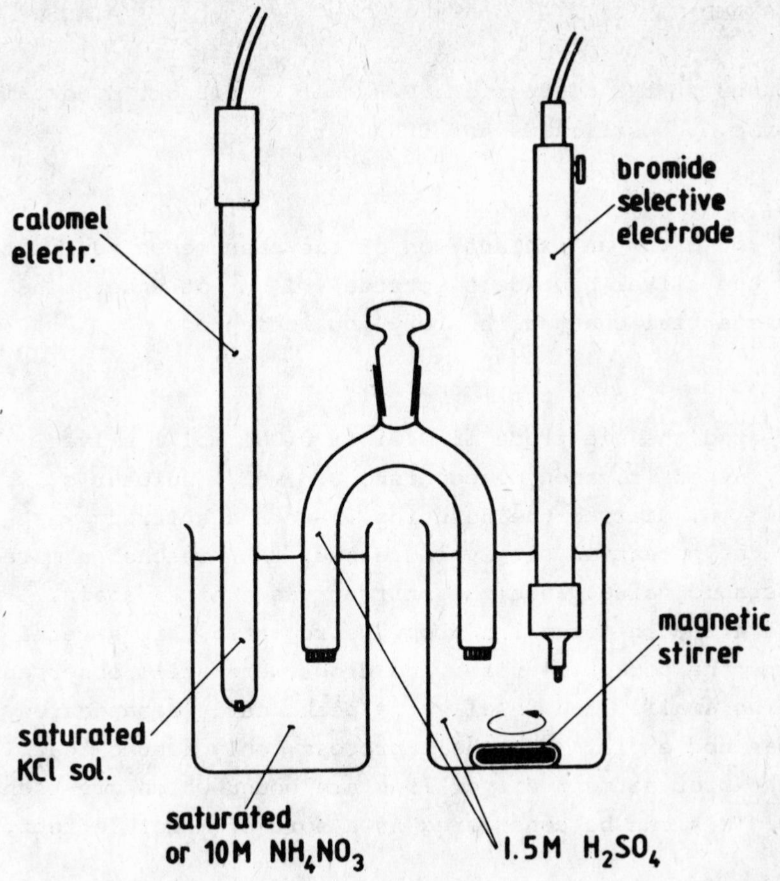

Fig. 3. Galvanic cell for potentiometric measurements

QUESTIONS AND COMMENTS

Participants of the discussion: P.Becker, R.P.Buck, G.Horvai, M.Jovanović, Z.Noszticzius and E.Pungor

Question:

Could you offer an explanation of the phenomenon you mentioned for the silver bromide electrode ? I do not understand why the potential changed in a way you indicated.

Answer:

We treated the electrode in 0.01 or 0.001 mol/l silver nitrate, washed it, then placed into 0.5 mol/l sulphuric acid solution, stirred the solution. When the stirring was stopped, the potential slowly increased, then reached a more or less stable value. When the stirrer was started again, there was a sudden potential drop. We repeated this several times, and the potential rises and drops were still observed, but getting smaller and finally the peak nearly disappeared. But we can not explain the whole process, only assume that during the pretreatment silver ions are bound which are then released. This can be considered as a sort of memory effect.

Comment:

I think if you have cracks, you have a relatively large surface area and the electrode responds to what is in the cracks.

Answer:

Maybe we have a mixed potential. The concentration in the cracks can be very high, whereas it is much lower at the electrode surface. And if silver ions come out of the cracks, the concentration decreases there, whereas it increases at the surface. This may result in a change in the mixed potential.

Comment:

One thing that came up some years ago is the question what is the effect of adsorbed ions on the potential. And it turned out that when the ion exchange process is really fast between the crystal and the solution, it does not make any difference if there are adsorbed ions because they do not contribute to the potential, and the ion exchange process goes right through the adsorbate and makes the same potential. But in the kinetics you see the adsorption because the adsorption is a sink-sam guessing - and all the ions that go in, adsorb, and do not change the potential. But when you let them come back out, the potential starts to drift. This is just one possibility, it may have nothing to do with the truth.

Comment:

The leaching of ions from the electrode should be taken into consideration.

Comment:

I would suggest to repeat the experiment with dilute sulphuric acid or potassium nitrate solution, because it has been shown in the case of the copper electrode that sulphuric acid was very efficient in cleaning the electrode. So, also in this case, it might attack the electrode.

Question:

Why has the electrode you prepared a part with a thick and a part with a thin layer ?

Answer:

The thin layer was applied to get a low impedance, and the thick layer was used to insulate the electrode, because here the impedance was high. This method proved to be good, the electrode performance was good.

Comment:

It would also be interesting to find out the redox potential of oxygen in 0.5 mol/l H_2SO_4, and also of the Br^-/Br_2 couple, to see whether processes other than adsorption and desorption may take place.

4th Symposium on Ion-Selective Electrodes
Mátrafüred, 1984

THE ELECTROCHEMICAL CHARACTERISTICS OF THE NON-STOICHIOMETRIC COPPER(I) SULPHIDE ELECTRODE IN SOLUTIONS OF COPPER(II) IONS

P. NOWAK and A. POMIANOWSKI

Institute of Catalysis and Surface Chemistry
Polish Academy of Sciences
ul. Niezapominajek 1,30-239 Cracow, Poland

ABSTRACT

Small amplitude cyclic voltammetry on stationary and rotating disc electrodes was used to evaluate the kinetic parameters of the reaction of anodic dissolution of cuprous sulphide and cathodic reduction of cupric ions at the sulphide electrode surface. It was stated that the reaction proceeds in two steps with the cuprous ion as an intermediate and the reaction $Cu^{+}_{aq} \rightleftharpoons Cu^{2+}_{aq} + e^{-}$ as the rate-determining step.

INTRODUCTION

Cuprous sulphide is frequently used for preparation of ion--selective electrodes, either as a pure substance or in mixtures with other sulphides. Despite the fact that many works concerning the electrochemical properties of cuprous sulphide have appeared in the last few years, little attention was paid to the mechanism of the reaction:

$$Cu_{2-x}S \rightleftharpoons Cu_{2-x-y}S + yCu^{2+}_{aq} + 2ye^{-} \quad /1/$$

which is a potential-determining process for the $Cu_{2-x}S$ electrode in solutions of Cu^{2+} ions.

Habashi and Torrez-Acuna /1/ suggested, that Cu^{+} ion should be an intermediate in reaction /1/, however Hepel /2/ and

Hepel /3/ stated on the basis of their measurements that this reaction proceeds in one two-electron step. On the other hand our measurements /4/ suggest that the Cu^+ ion should be an intermediate product in reaction /1/. To resolve this discrepancy we have studied the kinetics of the reaction /1/ using the method of small amplitude cyclic voltammetry /5/ for estimation of the exchange current densities and the transfer coefficients /α_a and α_c/ for the process investigated.

EXPERIMENTAL

The non-stoichiometric cuprous sulphide of the composition $Cu_{1.87}S$ was synthesized by melting appropriate amounts of copper and sulphur. X-ray diffraction analysis showed that it was a mixture of djurleite and digenite. The electrodes in the form of a cylinder were machined from the large lumps of sulphide and fastened in teflon or plexiglass holders. The measurements were made in the three-electrode system using a typical electrochemical set-up.

The triangular signals of overpotential of amplitudes: 4, 10, 20, 40, 100 mV and frequencies in the range 0.16 - 0.0025 Hz were used. Solutions containing $CuSO_4$ of concentrations in the range 1.10^{-4} - 1.10^{-1} $mol.dm^{-3}$ and background electrolyte /Na_2SO_4, KNO_3, H_2SO_4/ of such a concentration that the total ionic strength was equal to 1 $mol.dm^{-3}$ Na_2SO_4, were used.

The measurements were performed in acidic solutions /from pH=3 up to 1 $mol.dm^{-3}$ H_2SO_4/. All experiments were carried out at a temperature of 25 ± 0.1 ^oC. Solutions were deaerated before the measurements by bubbling for several hours with argon, from which oxygen was removed up to the level of 20 ppm. Before the experiments the surface of the electrode was polished on emery papers /ending with the 800 grade/, degreased with ethyl alcohol and finally polished on filter paper. Rotation speeds of the disc electrode in the range 0-35 Hz were used.

RESULTS AND DISCUSSION

The measurements were performed in such a range of frequencies that the contribution of the double layer capacitance to the overall interface impedance was negligible. The Warburg impedance was diminished using high rotation speed of the disc electrode. In Figure 1 is shown as an example the curve obtained in a solution of $CuSO_4$ of concentration 5.10^{-3} $mol.dm^{-3}$ at zero rotation speed. The hysteresis of the current observed after the reversal of the direction of the potential sweep is caused by the capacitive part of the Warburg impedance. As the Warburg impedance goes to zero when the frequency goes to infinity, one should expect that the slope of the current - potential curve at zero overpotential extrapolated to the infinitely large frequency should approximate the resistance of the faradaic reaction at the interface. Also the values of the current at the points of potential reversal as well as their ratio, from which the transfer coefficient may be calculated /6/, extrapolated to infinitely large frequency should approximate the respective quantities for the process under pure activation control. It was proved that all the quantities: Rea, Rec,Iamacx, Icmax and Iamax/Icmax gave linear plots versus reciprocial of the square root of frequency/for the explanation of the above symbols /see Figure 1/.

It can be concluded that the $Cu_{2-x}S$ electrode is very unstable in solutions of Cu^{2+} ions. After introduction of the electrode into the solution, the exchange current falls and even during several days does not attain a stable value. This is presented in Figure 2, where the changes of the impedance of the faradaic reaction at the interface /which is inversely proportional to the exchange current density/ in time are plotted.

Taking into consideration the correction for the ohmic drop of the potential in the solution it is to be seen that the exchange current falls almost two orders of magnitudes in a few days. In solutions of lower concentrations these changes are even faster. After several weeks the exchange current stabilizes - these electrodes /called further aged electrodes/ are not, however, stable enough and after the change of the solution concentration the exchange current changes further. It was proved that the main reason for this instability of the exchange current value was the residual oxygen, present in the solution. When a copper foil was introduced into the solution, the electrode potential became stable as long as the redox potential /measured with a Pt electrode/ was more negative than the redox potential for the Cu^{2+}/Cu^{+} redox couple. When the copper foil was withdrawn from the solution and the redox potential became higher than the potential of $Cu_{2-x}S$ electrode, the exchange current started to fall again.

It is very interesting that the most intense changes occur at very low oxygen concentrations - at higher oxygen concentrations the exchange current even rises, but it was impossible to study the kinetics of reaction /1/ in solutions containing oxygen because the interfering corrosion of the surface of the electrode covers up the results of measurements. The results of measurements conducted on both the "freshly prepared" /obtained by extrapolation of the parameters of the electrode reaction to zero time/ and "aged" electrodes are summarized in Table 1.

The anodic transfer coefficient α_a for both the aged and freshly prepared electrodes takes the value close to 0.75 /calculated from the slope of the dependence of the exchange current on the concentration or obtained for individual experiments by extrapolation of the value Iamax/Icmax to infinitely large frequencies/. This suggests that the process proceeds in two steps:

$$Cu_{2-x}S \rightleftharpoons Cu_{2-x-y}S + yCu^{+}_{aq} + ye^{-} \qquad /2/$$

$$Cu^{+} \rightleftharpoons Cu^{2+}_{aq} + e^{-} \qquad /3/$$

with the Cu^{+} ion as an intermediate and the second step much slower than the first one.

However it is to be noted that the difference between the exchange currents of the first and the second steps seems not to be as high as to fully fulfil the assumptions of the so--called rate-determining step concept. The asymmetry of the small amplitude cyclic voltammetry curves, especially for low frequencies, was much more significant than may be expected on the basis of the values of transfer coefficients obtained. The most probable explanation is the loss of the soluble intermediate /Cu^+ ion/ due to diffusion into the bulk of the solution. This problem will be discussed in more detail elsewhere /7/. The electrode may interact with the redox couple in the solution /the Cu^{2+}/Cu^+ systems/. When the redox potential of the solution is more negative than the potential of the $Cu_{2-x}S$ electrode, the electrode is continuously reduced /the quantity x in the formula $Cu_{2-x}S$ becomes lower/. On the other hand, very intense leaching of the electrode may proceed when the redox potential of the solution becomes higher than the potential of the $Cu_{2-x}S$ electrode. However, as long as the electrode is composed of two different phases /djurleite and chalcocite for example/ the rest potential of the electrode does not change significantly and the changes of the exchange current are relatively small.

In Figure 3. are presented the results of the experiment in which the electrode was continuously reduced by short-circuiting with the copper electrode in a solution of $CuSO_4$. The electrode had such a shape that the surface area to volume ratio of the electrode was very large and not only the composition of the surface of the electrode was changed but also of the whole volume of the electrode. During this experiment the composition of the electrode changed from $Cu_{1.87}S$ to $Cu_{1.94}S$. Except the initial period, when the results of the process of aging of the electrode were reversed, the rest potential did not change and the exchange current changed moderately.

It should be noted that during the experiment presented in Figure 3. all the time two sulphide phases were present in the solid body. The exchange current density for such reduced electrode is of the same order of magnitude as for the freshly polished electrode. This situation changes, however, significantly when only one sulphide phase is present. In that case both the rest potential and exchange current are strongly dependent on the composition of the electrode.

REFERENCES

1. F.Habashi,N.Torrez-Acuna, Trans AIME, 242 /1968/ 780
2. M.Hepel, J.Electroanal.Chem., 74, /1976/ 37
3. M.Hepel, T.Hepel, J.Electroanal.Chem., 81, /1977/ 161
4. P.Nowak, W.Barzyk, A.Pomianowski, J.Electroanal.Chem., 171, /1984/ 355
5. D.D.Macdonald, J.Electrochem.Soc., 125, /1978/ 1443
6. S.Barnartt, Electrochim.Acta, 15, /1970/ 1313
7. P.Nowak, E.Krauss, A.Pomianowski, submitted to J.Electroanal. Chem.

Table 1. The kinetic parameters for a non-stoichiometric cuprous sulphide electrode in reaction:
$Cu_{2-x}S \rightleftarrows Cu_{2-x-y}S + yCu^{2+} + 2ye^-$

Freshly polished electrode*			Aged electrode	
Concentration of Cu^{2+} mol.dm^{-3}	i_o mA.cm^{-2}	α_a	i_o mA.cm^{-2}	α_a
10^{-4}	0.023	.78	0.0015	.77
10^{-3}	0.22	.76	0.0069	.75
10^{-2}	2.1	.69	0.038	.70
10^{-1}	11.0	.66	0.12	.67

*The values of i_o for the freshly polished electrode were obtained by extrapolation to zero time, therefore they should be taken with limited confidence.

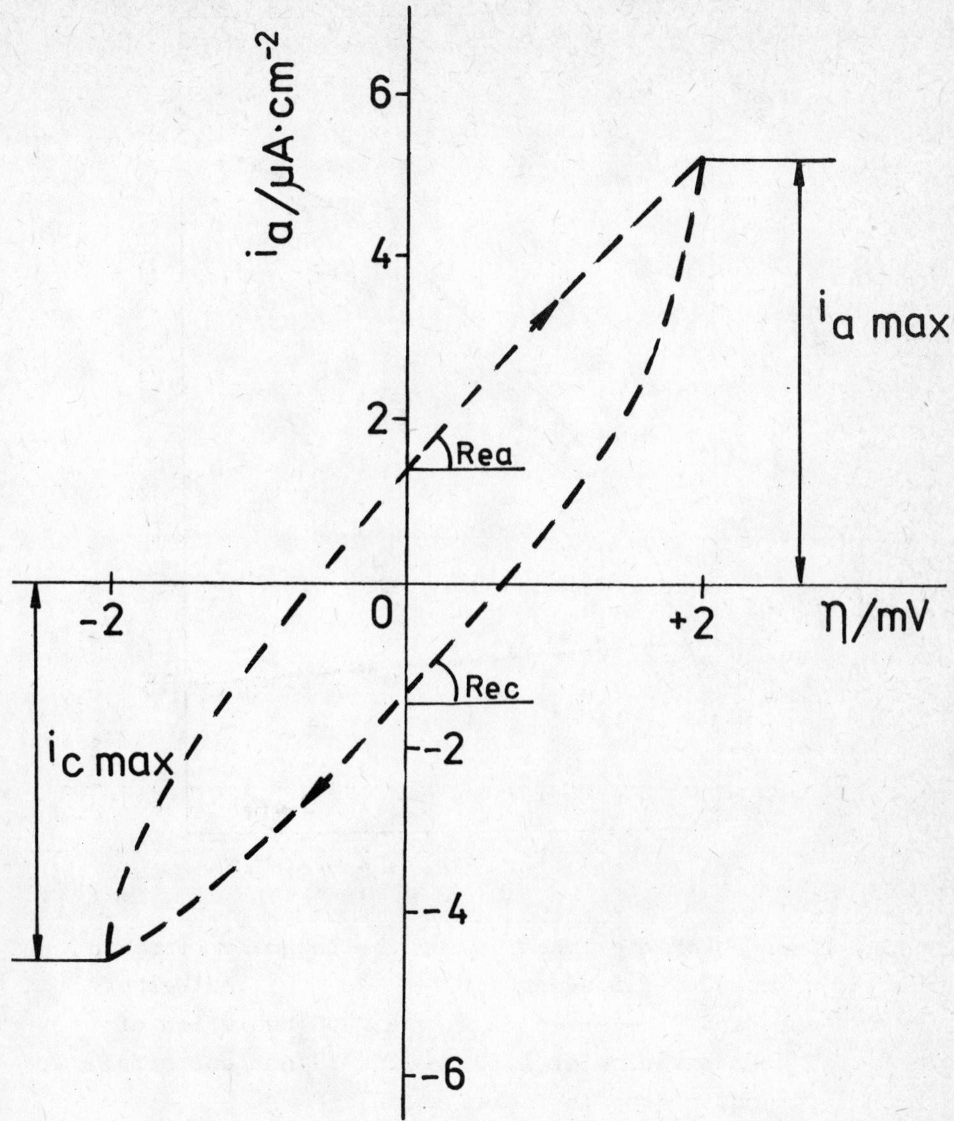

Fig. 1 Example of small amplitude cyclic voltammetry curve for $Cu_{1.87}S$ electrode in 5.10^{-3} $mol.dm^{-3}$ solution of $CuSO_4$, f=0.005 Hz.

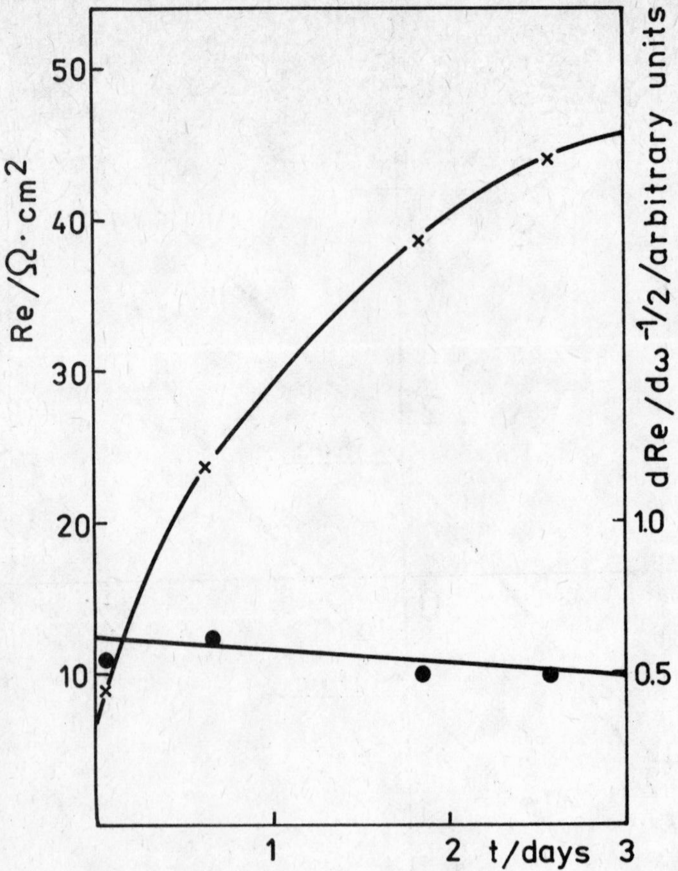

Fig. 2 Changes of the impedance of the faradaic reaction at the $Cu_{1.87}S$ electrode /——x——/ and Warburg constant /——•——/ in the $CuSO_4$ solution of a concentration of $1.10^{-1} mol.dm^{-3}$/ not corrected for ohmic drop in the solution/.

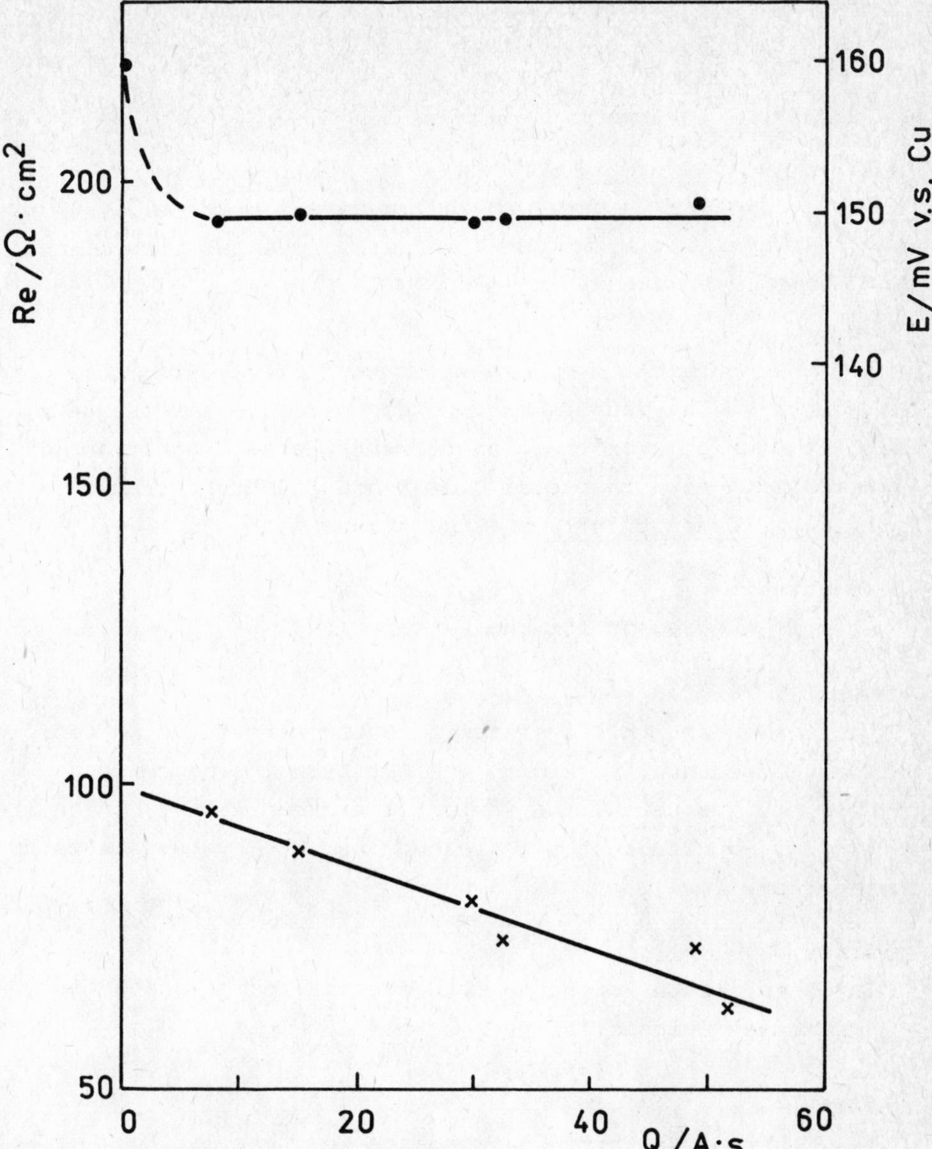

Fig. 3 Changes of the impedance of the faradaic reaction at the $Cu_{1.87}S$ electrode /——x——/ and of the rest potential /——•——/ of the electrode during the continuous reduction in 1.10^{-3} $mol.dm^{-3}$ $CuSO_4$ solution /not corrected for the ohmic drop in the solution/.

QUESTIONS AND COMMENTS

Participants of the discussion: R.P.Buck, T.Garai, J.Janata, M.Neshkova, P.Nowak and A. Zhukov

Question:

On the hysteresis curve which you have on the small amplitude cyclic voltammogram, where did you get the value of the double layer capacitance from?

Answer:

I cannot obtain the double layer capacitance from this curve because the measurement was done at such low frequency that the double layer charging current was negligibly small in comparison with the total current and the double layer capacitance did not influence the curve.

Question:

What is the reason for the hysteresis ?

Answer:

The reason for this hysteresis is the capacitive part of the Wartburg impedance. As known, the Wartburg impedance is composed of two parts, the resistive and capacitive, which are of similar magnitude at all frequencies. The hysteresis is due to the capacitive part.

Question:

Is there also a Wartburg impedance inside the space charge of the copper sulphide as well ?

Answer:

Yes, there is a Wartburg impedance in the solid body as well as in the solution, but that in the solid body is very small and can only be observed at high concentrations of copper in the solution.

Question:

How was the Wartburg impedance depending on the frequency in the frequency range studied ?

Answer:

I measured the frequency dependence of the impedance generally in the range 0.2-0.001 Hz.
The slope of the curves I have shown, as well as the reciprocal of the current of the point of reversal depends linearly on the reciprocal of the square root of the frequency. It is not surprising because the Wartburg impedance goes to zero when the frequency goes to infinity.

Question:

Why did not you use a system where you could plot the real and imaginary impedances against each other and the Wartburg ? The theory of this is very well worked out and it would be much clearer and easier to believe and understand if you gave us a plot of impedance vs. frequency, or impedance vs. impedance.

Answer:

You are right, but an impedance analysis is not so easy, you can do it with a linear system. However, this system is not linear because the transfer coefficient is 0.75, and thus there is a significant Faradaic rectification in the system and it was not easy to make a classical analysis in the frequency domain.

Comment:

I think you had too high an amplitude then.

Answer:

Even at such small amplitudes as e.g. 4 mV, the Faradaic rectification was significant. It came not only from the fact that the transfer coefficient was not 0.5 but also from a complex reaction with two consecutive steps. Part of the intermediate may diffuse in the solution. And this phenomenon makes the observed value even more asymmetric than it may be expected on the basis of the transfer coefficient. So, it was impossible to apply the classical impedance analysis in this case.

Question:

You polarized your electrode. Did not this cause a change in the structure and operation of the electrode ?

Answer:

The electrode function did not change after the treatment, because all these changes proceed in such a range of composition of the solid body that the two different sulphide phases are still present. Thus the electrode potential did not change and the slope remained Nernstian. However, the exchange current density changed significantly, and this may influence the behaviour of the electrode, since the selectivity depends on the exchange current density.

Question:

Does the standard potential of the electrode depend on the composition of the membrane ?

Answer:

Yes, it changes with composition.

Comment:

If you polarize the electrode, the process is a non--equilibrium one, and we cannot expect a Nernstian equilibrium potential. Could you comment on this ?

Answer:

The experiment was done as follows: The electrode was polarized in a solution of copper ion by means of short-circuiting with metallic copper electrode. Then the electrode was withdrawn from the solution, left for several days, then the exchange current was measured. Then the electrode was polarized again in solution, and the whole procedure was repeated several times. So, at the moment of the measurement of the exchange current the electrode was in equilibrium.

4th Symposium on Ion-Selective Electrodes
Mátrafüred, 1984

ION-SELECTIVE ELECTRODES FOR DETERMINATION OF CHARGED METAL COMPLEXES

O.M. PETRUKHIN, Yu.P. KHOLMOVOI, E.N. AVDEYEVA and S.L. ROGATINSKAYA

D.I. Mendeleyev Institute of Chemical Technology, Moscow, USSR

ABSTRACT

The characteristics of ion-selective electrodes based on ionic associates with the metallocomplex ions to be determined are discussed. Potentiometry can be used for quantitative determination of the total metal content in the solution under conditions optimal for the formation of metal-complex ion.

INTRODUCTION

The principal feature of potentiometry that is the dependence of the analytical signal on the activity of the analyzed ion makes it impossible to use the electrodes sensitive to metal ions to determine the complex metal forms. However, many industrially important solutions contain metals in the form of charged complexes. Various approaches can be employed to determine metals in complex forms. Hydrophobic stable complexes can be effectively determined with the help of ion-selective electrodes based on electrode-active compounds containing the complex ion to be determined.

For example we have developed ion-selective electrodes for determining the complexes $M/CN/^{-}_{2}$/1-3/, $M/Thio/^{+}_{n}$/4/ /M=Ag/I/, Au/I/, Thio is thiourea/.

To analyze solutions containing hydrophilic complexes another procedure can be used. A reagent which forms sufficiently stable hydrophobic charged complex with the metal to be determined is introduced into the solution. This approach has been employed for the determination of Cu/I/ in the presence of Cu/II/. The authors developed and used plasticized electrode containing $Cu/DMPhen/^{+}_{2}$.Pic as an electrode-active compound /DMPhen is 2,9-dimethyl-1,10-phenanthroline, Pic - picrate-ion//5/.

EXPERIMENTAL

We have studied the electrode characteristics of membranes based on ionic associates formed by the complex ions $Au/CN/^{-}_{2}$, $Ag/CN/^{-}_{2}$, $Au/Thio/^{+}_{2}$, $Ag/Thio/^{+}_{3}$, $Cu/DMPhen/^{+}_{2}$ and organophilic counterions such as tetradecylphosphonium, tetraphenylarsonium, picric acid, tetranitrodiaminocobaltate, sodium-tetraphenylborate. Chloroform, 1,2-dichloroethane and nitrobenzene were used as solvents. The electrodes reversible towards auridicyanide, argenticyanide and copper/I/ dimethylphenanthrolinate were prepared as polyvinylchloride films, with dibutyl phthalate as a plasticizing agent.

The membranes were prepared by substoichiometric extraction, and the metal concentration was an order of magnitude higher than that of the reagent. The content of electrode-active compound in the membrane was 10^{-2}-10^{-3}M. The solution of $KM/CN/_2$

at a concentration of 10^{-3} M in 0.45 M Na_2SO_4 served as a reference inner solution for $M/CN/^-_2$-selective electrode, 10^{-3} mol/l solution of $[/M/Thio/^+_n/]_2 \cdot SO_4$ in 0.15 M Na_2SO_4 performed the same function for $M/Thio/^+_n$-SE, and 10^{-4}M solution of $[/Cu/DMPhen/^+_2/]_2 \cdot SO_4^{2-}$ in 0.15 M Na_2SO_4 - for $Cu/DMPhen/^+_2$ - SE. Silver-silver chloride electrode for gold and silver and mercurous sulphate electrode for copper/I/ were used as reference electrodes. The ionic strength and pH of the solutions were kep constant, the e.m.f. of the electrochemical cell was measured by the compensation scheme using the pH-meter /model 121 and 340/ or ionomer model 74 types as high-resistance zero--indicators.

The limits of detection of metal complex ions were taken in accordance with IUPAC recommendations /6/. Electrode selectivity coefficients K_{ij}^{Pot} were determined graphically using the method of mixed solutions /7/.

RESULTS AND DISCUSSION

The comparison between the extraction data and the electrochemical properties of the system based on ionic associates containing the complex ions of gold and silver /1-4/ testify the fact that the extraction process results in the distributive character of equilibrium electrode potential. The electrochemical and extraction properties of this type of membranes depend on the stability and distribution parameters of the complex ion and the entire ionic associate. The conditions optimal for these electrodes coincide with the optimal extraction conditions /maximal excess of the extracted form in the solution, pH range suitable for potential-determining ion/ and provide high selectivity of electrodes and linear character of the electrode function in a broad range of concentrations of the metal determined /Table 1/.

The distribution coefficients of complex metal ions increase in the order chloroform > 1,2-dichloroethane > nitrobenzene, which results in a shift of the determination limits in the same sequence. Under similar conditions the limit of gold determination is lower than that for silver, which is in agreement with the values of distribution coefficients for the corresponding complex ions. The nature of the organic counter-ion which forms ionic associate with the potential-determining complex ion in the membrane phase affects the determination limit in a manner antibate to its extractability. The decrease in the concentration of the electrode-active compound in the membrane phase lowers the determination limits. However, the fall of these concentrations below 5.10^{-4} mol/l results in longer response time due to higher ohmic membrane resistance.

The operating pH range for ion-selective electrodes based on the complex ions $M/CN/_2^-$ and $M/Thio/_n^+$ is limited by the pH values for which the potential-determining form prevails in the analyzed solution, and by the conditions of maximal extractability of this form. The higher stability of complex ions of gold /$\lg\beta_{Au/CN/_2^-} = 56.0$; $\lg\beta_{Au/Thio/_2^+} = 25.3$/, as compared to similar complex ions of silver: /$\lg\beta_{Ag/CN/_2^-} = 21.1$; $\lg\beta_{Ag/Thio/_3^+} = 13.1$/ accounts for the wider pH range for the electrodes reversible towards the complex ions of gold and explains why this range becomes wider for higher concentrations of cyanide ion and thiourea in the analyzed solution.

The electrochemical properties of the ion-selective electrode used for determination of Cu/I/ in the form of phenanthrolinate complex obtained by preliminary conditioning of the analyzed solution are similar to those described above. The electrode function is linear in the range of Cu/I/ concentrations from 10^{-4} to 10^{-6} M. The lower limit of determination depends on the distribution of the electrode-active compound between the membrane and aqueous phase, which results in the concentration of potential-determining ions in the aqueous near-electrode layer exceeding that for the conditions when linear function is observed. The upper limit is due to the limited water solubility of $[Cu/DMPhen/_2^+]_2 \cdot SO_4^{2-}$ /when Na_2SO_4 is used as a background solution/.

The interval of linear dependence of electrode potential on the concentration of the potential-determining ion is markedly influenced by the competing reactions in the analyzed solution which shifts the equilibrium of the reaction yielding the potential-determining form. This form, however, is stabilized by the excess of DMPhen.

The operating pH interval for acid solutions is controlled by the protolytic properties of the reagent and by Cu/I/ concentrations in the alkaline pH region. The lower the concentration of potential-determining ion in the solution and the content of electrode-active compound in the membrane, the narrower is the operating pH range.

All these electrodes exhibit high selectivity /Table 2/. The metal ions which form stable complex ions similar to the potential-determining ones and can be extracted into the membrane organic phase with high distribution coefficients, usually interfere with the determined one. When there is no excess present in the analyzed solution, we cannot exclude competing complexing as in the case with the solutions containing Cu/II/ together with Cu/I/:

$$Cu/DMPhen/_2^+ + Cu^{2+} \rightleftharpoons Cu/DMPhen/_2^{2+} + Cu^+$$

For the electrodes described the sequences of electrode selectivity were shown to correspond to the extraction sequences and the interfering effect decreases along with the extractability of the complexes. Using some other solvent we can get different numerical values for the selectivity coefficients, while the selectivity sequences remain the same. The nature of counterion for the electrode-active compound and the concentration of this compound do not affect the electrode selectivity.

Analytical and performance characteristics of plasticized and liquid electrodes are actually the same.

These electrodes were used for analyzing the electrolytes of galvanic baths and industrial technological solutions, and sometimes as sensors in automated systems.

REFERENCES

1. A.S.Bitchkov, O.M.Petrukhin, V.A.Zarinski, Yu.A.Zolotov, L.V.Bakhtinova, G.G.Shanina, Zh. analit. khimii, 31, /1976/ 2114 /in Russian/

2. E.N.Avdeyeva, Yu.V.Shavnja, O.M.Petrukhin, Zh. analit. khimii, 37, /1982/, 1434 /in Russian/
3. V.P.Yankauskas, E.N.Avdeyeva, R.M.Kozlauskas, O.M.Petrukhin, Zh. analit. khimii, 38, /1983/ 636 /in Russian/
4. O.M.Petrukhin, Yu.V.Shavnja, A.S.Bobrova, Yu.M.Chikin, Ion-selective electrodes for determination of gold /I/ and silver in thiourea solutions, In: 3-rd Symposium on Ion--Selective Electrodes. Budapest, Akadémiai Kiadó, 1980, p.305
5. Yu.P.Kholmovoi, O.M.Petrukhin, L.M.Rub, V.D.Annapolski, Zav.Lab. 49, /1983/ /in Russian/
6. Recommendations for nomenclature of ion-selective electrodes /recommendation 1975/, IUPAC Analytical Chemistry Division, Pure Appl.Chem., 1976, v. 48, N.I.
7. K.Srinivasan, G.A.Rechnitz, Anal.Chem., 41, /1969/, 1203

Table 1. Characteristics of ISE, sensitive to metal-containing ions

Determined ion	Detection limit, M	pH-range	Selectivity series
$Au/CN/_2^-$	1.10^{-5}	2,5-11,5	$Au/CN/_2^- > Ag/CN/_2^- >$ $Hg/CN/_4^{2-} > Cd/CN/_4^{2-} >$ $Zn/CN/_4^{2-} > CN^- > Pd/CN/_4^{2-}$ $Ni/CN/_4^{2-} > Cu/CN/_3^{2-} >$ $Fe/CN/_6^{3-}$
$Ag/CN/_2^-$	3.10^{-6}	6-11,5	"
$Au/Thio/_2^+$	1.10^{-5}	$8NH_2SO_4$- -pH 4	$Au/Thio/_2^+ > Ag/Thio/_3^+ >$ $Cu/Thio/_4^+ > Fe/Thio/_2^{3+} >$ $Zn/Thio/_2^{2+} > Ni/Thio/_4^{2+}$
$Ag/Thio/_3^+$	7.10^{-5}	$2NH_2SO_4$- pH 3	"
$/DMPhen/_2^+$	1.10^{-6}	4-6	Cu/I/>Cu/II/>Ni/II/> Co/II/>Zn/II/>Mn/II/

Table 2. The selectivity coefficients of electrodes, reversible to metal-containing ions

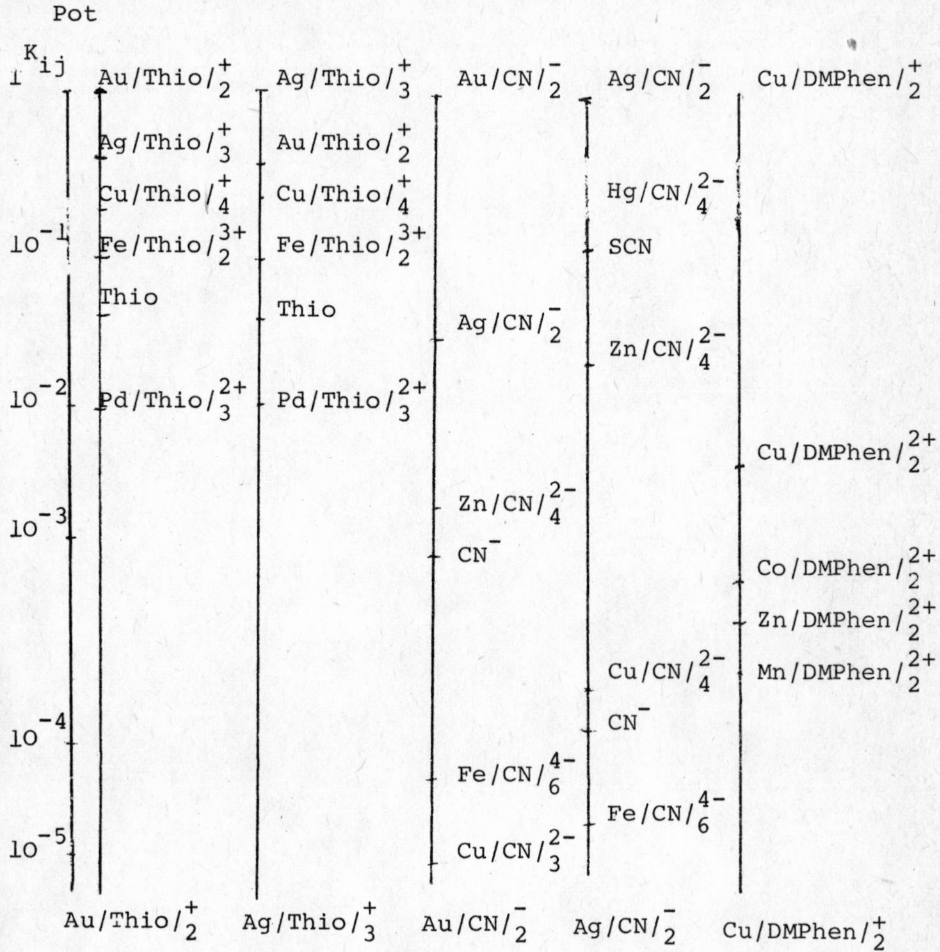

4th Symposium on Ion-Selective Electrodes
Mátrafüred, 1984

ION-SENSITIVE MEMBRANES APPLICABLE TO ION-SENSITIVE FIELD EFFECT TRANSISTORS MANUFACTURED BY ION IMPLANTATION

M.T. PHAM and W. HOFFMANN

Academy of Science of the GDR
Central Institute for Nuclear Research, Rossendorf
8051 Dresden, PF 19, GDR

ABSTRACT

Ion implantation may be a helpful tool in manufacturing ion sensitive membranes for ion sensitive field effect transistors (ISFET's). Model experiments with electrolyte-insulator-semiconductor (EIS) structures were conducted to this end. Alkali sensitive membranes could be prepared by implantation of Al and K into Si_3N_4 followed by an annealing step.

INTRODUCTION

To meet the requirements of ion sensitive membranes applicable to ISFET's (besides properties common to all ISE's like ion-sensitivity, selectivity, stability these are specially for ISFET's (i) good adhesion to the gate insulator surface, (ii) micro-structurability, (iii) semiconductor technology compatible manufacturing) we have investigated the doping of the gate insulator surface region itself by ion implantation. Recently we could show the variation of the ion-sensing properties of SiO_2 by implantation of groupe III elements /1/. It is the purpose of the present investigation to increase the alkali sensitivity of such membranes by implantation of additional ions. Basing on the theory of Eisenman /2/ accepted for glass electrodes this should be attainable by introducing group I elements.

EXPERIMENTAL

The fundamental ion sensing properties were studied with the help of the simple electrolyte-insulator-semiconductor structure /3/. The general sample preparation for this structure as well as the measurement of its quasistatic capacitance-voltage (CV) characteristics have been outlined in detail elsewhere /1/. In the present experiments (i) thermal SiO_2 (200nm thick) and (ii) Si_3N_4 (100nm thick), deposited by a CVD-process over SiO_2 (100nm thick) were applied as gate insulator materials on semiconductor silicon (⟨100⟩ p-type Si, 6-10 Ωcm resistivity). The implanted species were Al, Na und K.

RESULTS AND DISCUSSION

Implantation into SiO_2

Examination of the respective quasistatic CV-curves (Fig.1) leads to the following qualitative results:

- The implantation damage in SiO_2 may be annealed by "soft" annealing conditions as indicated by the pronounced minimum in the CV-curve nearby the SCE-potential around 0 V after annealing of Al implanted samples (compare curves 1 and 2).
- Na implantation followed by the same annealing gives no evidence of an electrolyte-insulator-semiconductor structure applicable to ISFET's. Na ions seem to have diffused to the SiO_2/Si interface causing a shift of the flatband voltage beyond the range of the measurements (curve 3).

Implantation into Si_3N_4

Si_3N_4 was layered on top of the SiO_2 because of its (i)high chemical stability and (ii) function as an alkali diffusion barriere as known from MNOSFET technique /4/. Na as the implanted species was substituted by K which has a reduced diffusion coefficient /5/.

Implantation of both Al and K to a high dose into Si_3N_4 results after annealing in a pronounced minimum in the CV-curves with only a slight shift to positive voltages compared to Si_3N_4 not implanted (Fig.2, curves 1-3), thus confirming the good barriere function of Si_3N_4 or the reduced alkali diffusion resp.

The structure of the crystal lattice is assumed to be influenced by the annealing conditions. Annealing at higher temperatures, longer times and in O_2 atmosphere results however in a strong shifting of the flatband voltage (Fig. 2, curve 4), caused by the high K-concentration as can be observed by reducing the implanted K-dose (Fig.2, curve 5). Annealing can be improved by increasing the temperature and reducing the annealing time as indicated by curve 6, which is brought back closer to the curve 1 of the original not implanted Si_3N_4. The further investigations were carried out at this annealing conditions.

The ion-sensitivity derivable from the shifting of the CV-curves measured with different electrolytes shows a significant dependence from the implantation dose. The K^+ sensitivity of Si_3N_4 can be increased by implantation of Al and by implantation of K with growing dose (Fig.3). It should be noticed, that increasing the K-dose leads not only to steeper curves with near Nernst'ian slope but also to an extended range of linearity. An analogous behaviour can be observed for the Na^+ sensitivity at the same sample preparation. However, for pNa>2 sensitivity for Na ions remains very low. At pNa = 3 the selectivity coefficient K_{KNa} = 0,5 was determined by the seperate solution method.

The H^+ sensitivity is changed in a quite different manner: At low implanted K-dose and with Al implanted alone the H^+ sensitivity changes continuously throughout all the pH range measured. At high implanted K-dose (5×10^{16} ions/cm^2) however, the sensitivity decreases sharply at pH 6 (Fig.4).

These experimental findings seem to confirm qualitatively the ion exchange theory for glass electrodes by Eisenman /2/. Doping of the Si_3N_4 layers with Al may result in the formation of $(Al-O-Si)^-$sites, where oxygen is supplied by O-traces present during Si_3N_4-CVD process /6/ or, even more probable, during implantation damage annealing in oxygen atmosphere. Implantation of additional alkali ions reduces further the field strength of the anionic ion exchange sites by "screening". For this case a further preference for cationic exchange over H^+ is expected by the theory.

A rough theoretical estimation of the surface composition after Al and K implantation for best K^+ sensitivity provides a proportional composition of K : Al : Si = 1 : 0,5 : 10. This is far from K : Al : Si = 5,3 : 1,3 : 10, which is given by Eisenman /2/ for K-selective glass electrodes KAS (20-5). This fact may account for the low K selectivity over Na of the implanted samples.

REFERENCES

1 M.T. Pham and W.Hoffmann, Sensors and Actuators, 5 /1984/ 217.

2 G.Eisenman, "Glass Electrodes for Hydrogen and other Cations" Dekker, New York 1967.

3 N.F.de Rooij and P.Bergveld, in "The Physics of SiO_2 and its Interfaces", S.T.Pantelides, Ed. Pergamon Press, New York 1978.

4 W.A.Pliskin and R.A.Gdula, "Passivation and Insulation", in "Handbook on Semiconductors", T.S.Moss, Ed. North Holland publishing Co. 1980 Vol.III Chapt.11.

5 J.S.Logan and D.R.Kerr; 1965 IEEE Solid State Device Res. Conf. Princeton NJ.

6 M.Esashi and T.Matsuo, IEEE Trans.BME-25 /1978/ 184.

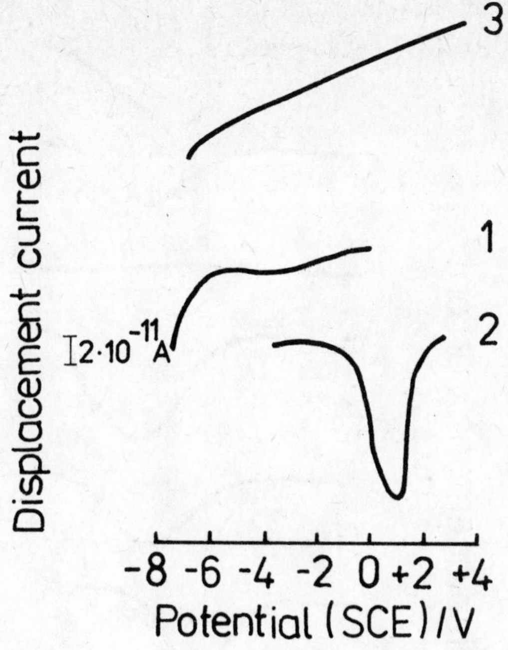

Fig.1 Quasistatic CV-curves measured at EIS-Structures showing the effect of ion implantation into SiO_2 and annealing

1 - Al-implanted (15keV, 10^{16} ions/cm^2)

2 - Al-implanted and annealed (N_2, 450 °C, 10 min)

3 - Na-implanted (15keV, 10^{15} ions/cm^2) and annealed (N_2, 450 °C, 10 min)

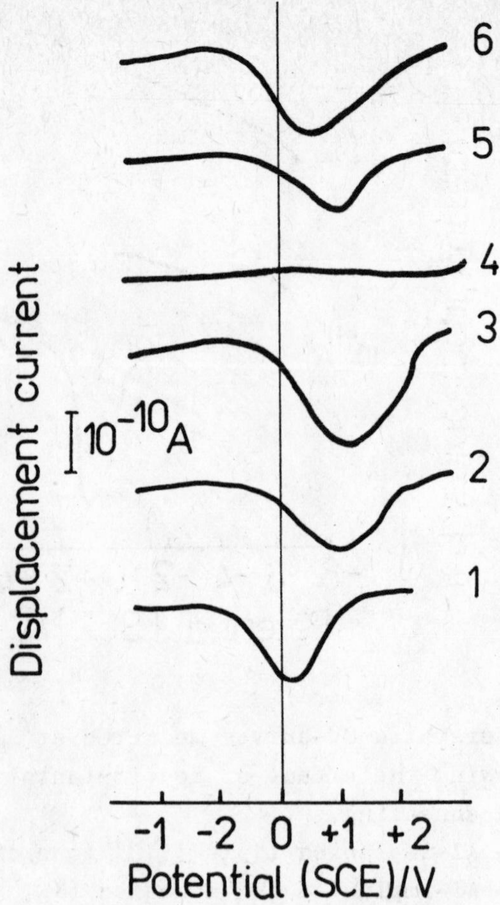

Fig.2 Quasistatic CV-curves showing the effect of ion implantation into Si_3N_4 and annealing

1 - not implanted

2 - Al-implanted (20keV, $5x10^{16}$ions/cm^2) and annealed (N_2, 450 °C, 10 min)

3 - Al-implanted like sample 2, K-implanted (30keV, 10^{17}ions/cm^2) and annealed (N_2, 450 °C, 10 min)

4 - Al- and K-implanted like sample 3 and annealed (O_2, 600 °C, 30 min)

5 - Al-implanted like sample 3, K-implanted (30keV, 10^{16}ions/cm^2) and annealed like sample 4

6 - Al- and K-implanted like sample 3 and annealed (O_2, 900 °C, 100 s)

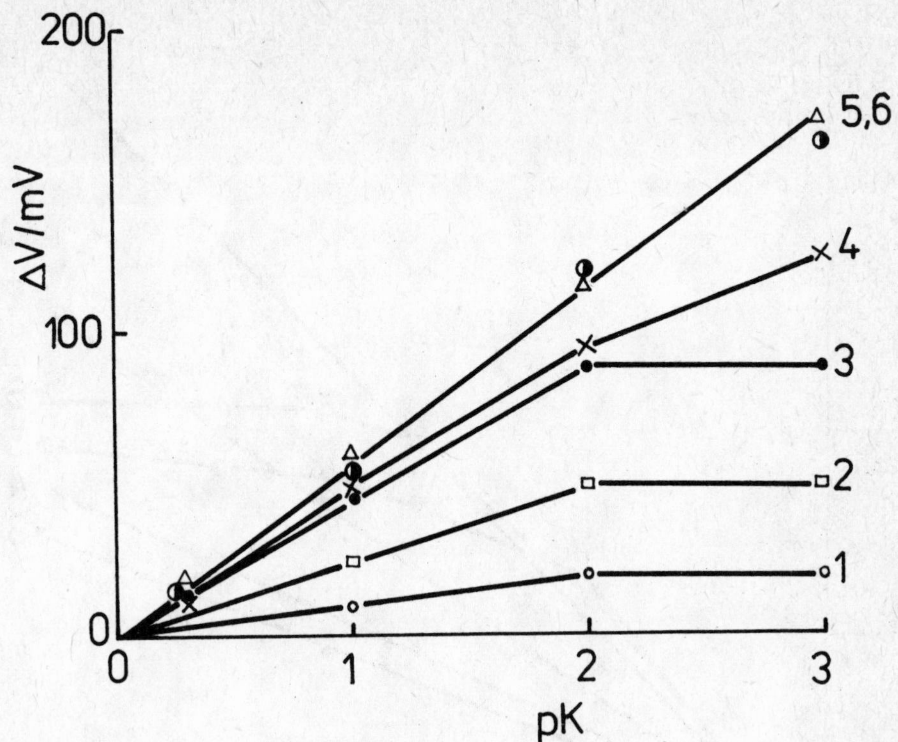

Fig.3 K^+ Sensitivity at pH = 8 of Si_3N_4, implanted with Al (20keV, $5x10^{16}$ions/cm^2), K (30 keV) and annealed (O_2, 900 ^oC, 100 s)

1 - not implanted

2 - without K-implantation

3 - K-dose $5x10^{15}$ ions/cm^2

4 - K-dose $1x10^{16}$ ions/cm^2

5 - K-dose $5x10^{16}$ ions/cm^2

6 - K-dose $1x10^{17}$ ions/cm^2

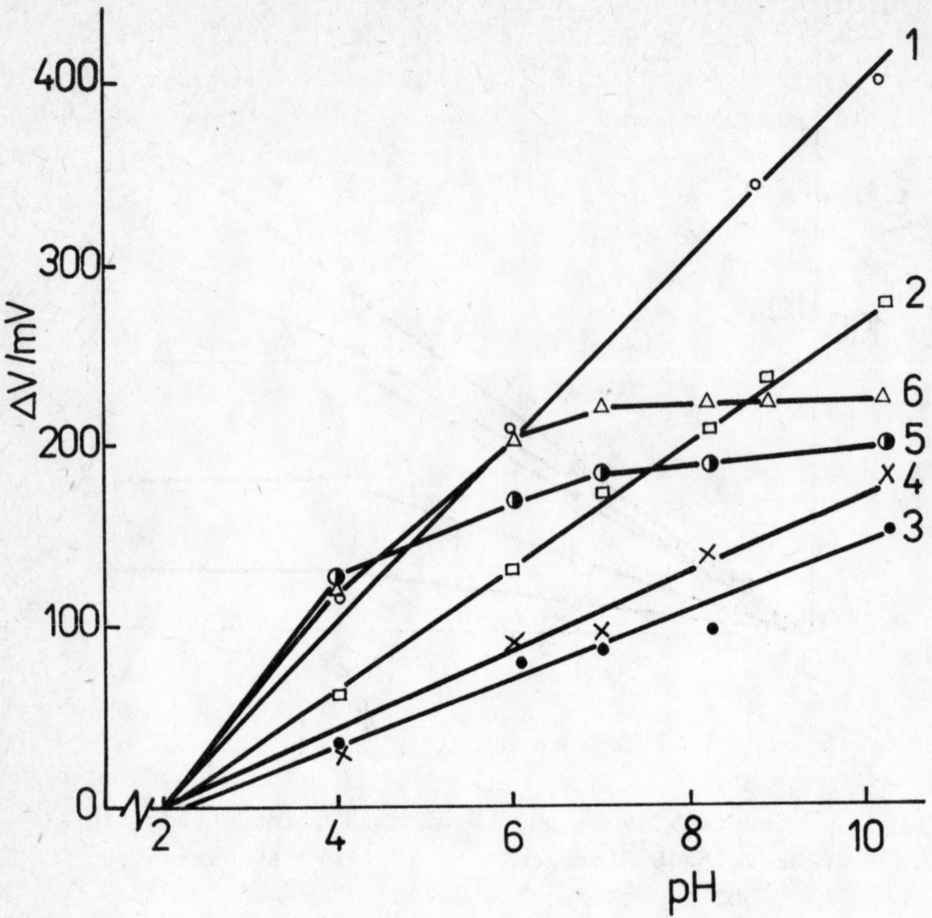

Fig.4 H^+ sensitivity of Si_3N_4, implanted and annealed - sample specification and curve marking like Fig.3

QUESTIONS

Participants of the discussion: W.Hoffmann, J.Janata and E.Pungor

Question:

You have shown some calibration diagrams. Why was the linear range so narrow in some cases ?

Answer:

In some cases we did not measure it in a wider range.

Question:

What is the depth of implantation?

Answer:

The maximum concentration has to be at a depth of 17 nm.

Question:

How many ions per cm^3 did you implant ?

Answer:

The ion dose was 10^{17} ions/cm^2.

Question:

Did you try to make a fluoride electrode ?

Answer:

No, not yet.

4th Symposium on Ion-Selective Electrodes
Mátrafüred, 1984

BEHAVIOUR OF THE SILVER SULPHIDE SELECTIVE ELECTRODE AS SILVER SENSOR IN ALKALINE MEDIUM IN THE PRESENCE OF COMPLEXING AGENTS

J. SIMÕES REDINHA and M. LUÍSA P. LEITÃO

Chemistry Department, University of Coimbra
3000 Coimbra, Portugal

ABSTRACT

A study of the silver sulphide electrode as silver sensor in alkaline media, in the presence of aminocarboxylic acid ligands, is made. For each case the concentration range within which the electrode response agrees with theoretical prediction is defined. Some data on the electrokinetic properties of the Ag_2S/solution phase boundary are presented in order to interpret the effect of ligands on the behaviour of the electrode.

INTRODUCTION

The study of complexes in solutions is one of the most important physicochemical applications of ion-selective electrodes. For many systems these electrodes presente characteristics which make them more advantageous than conventional electrodes in the measurement of low ionic activities, which is very important in the study of complexation reactions[1,2].

A large number of metallic complexes have been studied with reliable results. The most used electrode in this area has been lanthanum fluoride, although other electrodes have also been employed[3].

Nevertheless interferences in the response of the ion-selective electrodes have been referred. For example, the copper-selective electrode does not give correct results in the presence of an excess of aminopolycarboxilic ligands like EDTA[4-7]. Thus the utilisation of selective electrodes in the measurement of ionic activities in the presence of strongly complexing species is still, today, an open question and of great interest to elucidate.

We have been studying the complexation reactions of silver ion with aminocarboxylic acids, using the silver sulphide selective electrode to measure the silver ion concentration in solution. In this communication, data are presented on the behaviour of the electrode as an Ag^+ sensor in the presence of an excess of glycine or EDTA. Since the measurements were made at pH of approximately 12, so that the ligand would be entirely in its basic form, a study of the electrode response in alkaline media was also conducted.

EXPERIMENTAL

The electrode response was determined by stepwise addition of silver nitrate with a precision microsyringe to 50 cm^3 of a solution containing a known quantity of ligand. The pH and ionic strength of the ligand solution were adjusted to the values required with potassium hydroxide and potassium nitrate respectively. All potentiometric measurements were made in solutions of 0.1 M ionic strength, the cell emf being registered after each addition of Ag^+ solution. The electrodes were an ORION Ag_2S 94-16A electrode and an ORION 90-02-00 double junction Ag/AgCl electrode; data were recorded on a Philips PW9416 Ion Selective Analyser. Before each experiment the surface of the membrane of the selective electrode was polished. All measurements were carried out in a Teflon cell, thermostated at 25± ±0.02°C, with stirring and under a nitrogen atmosphere. The selective electrode was calibrated under the experimental conditions described above down to 10^{-4}M by means of silver nitrate solutions of 0.1M ionic strength and in the lower concentration region by addition of silver ion to a 0.1M potassium nitrate solution.

Zeta potentials were measured by electroomosis. The silver sulphide being prepared by mixing a silver nitrate solution with an excess of sodium sulphide solution.

RESULTS

The calibration curve for the Ag_2S electrode is presented in Fig.1. The electrode response is Nernstian down to about 10^{-5}M Ag^+ and the detection limit is 2×10^{-6}M.

Emf values as a function of the concentration of added silver ion, for various pH values are shown in Fig.2. It can be seen that the electrode potential is independent of pH for pH$\leq$11. For pH>11, owing to the formation of complexes, the potential decreases with increase in pH and a certain instability in the experimental readings is observed. The electrode response time to an increase in the Ag^+ concentration in the presence of hydroxide is 5 to 15 minutes, after which variations are less than 0.05 mV min^{-1}. The higher the Ag^+ concentration in solution, the more stable and reproducible are the values of the potential. The curves in Fig.2 show that the reproducibility of the electrode response is reasonable, especially for higher silver concentrations (>10^{-6}M).

In the case where there is formation of mononuclear complexes only, the relation between the concentrations of free silver ion and added silver ion is given by the following equation

$$|Ag|_T = |Ag|\ (1 + \Sigma\beta_n\ |L|^n) \qquad (1)$$

where $|Ag|_T$ is the total silver concentration, $|L|$ the concentration of free ligand and β_n the stability constants of the complexes formed. (Charges have been omitted for the sake of simplicity). Under these circunstances the electrode potential, E, is given by

$$E = E^{\ominus} - \frac{RT}{F}\ \ln\ (1 + \Sigma\beta_n\ |L|^n) + \frac{RT}{F}\ \ln\ |Ag|_T \qquad (2)$$

Equation (2) shows that E varies linearly with log $|Ag|_T$. The slope being 59.2 mV per concentration decade at 25°C, as long as the ligand concentration remains constant.

The linear region between 10^{-6}M and the start of silver hydroxide precipitation has a slope of ~70 mV per decade, significantly higher than that predicted by eqn. (2). The difficulty of reaching equilibrium and the non-Nernstian response of the electrode lead us to conclude that hydroxyl interferes in the functioning of the electrode.

In the presence of glycine or EDTA the electrode response time is much slower than in the absence of ligands, but is, however, quicker than for hydroxyl ion, one obtaining stable value after 2 to 5 minutes, depending on the Ag^+ concentration. Emf *vs* log $|Ag|_T$ curves for these two ligands are presented in Fig.3.

The slope of the linear portion of these curves is approximately 59 mV per decade.

For the linear regions of the curves (Fig. 2,3) the concentrations of free Ag^+ were calculated from the emf values and the equation for the electrode calibration curve. Thence, from the total silver ion, free silver ions and total ligand concentrations, and from the pH, the stability constants of the complexes formed were calculated using the computer program MINIQUAD[8]. Results are shown in Table 1.

The solubility constant of silver hydroxide was calculated from the concentration of Ag^+ in equilibrium with the silver hydroxide precipitate (upper plateau in Fig.2 curves); values obtained were $(3.1\pm0.2)\times10^{-8}$. These are concentration constants for an ionic strength of 0.1M.

The results obtained for the stability constants of the Ag^+ complexes with glycine and EDTA and for the solubility product of silver hydroxide are within the range of values obtained by other methods[9].

Data in the literature for complexes with OH^- are scarce and discordant. Nevertheless our results agree, in the species formed and in the values of the constants, with the more recent data obtained by Biedermann *et al.*[10]

DISCUSSION

Our results lead us to conclude that in the presence of silver complexing ligands the response of the Ag_2S electrode is slower than in their absence. For aminocarboxilic compounds the electrode response agrees with prediction, whereas with OH^- the response is affected.

It is known that OH^- interferes in activity measurements of fluoride ion using the LaF_3 electrode[11]. The interpretation given for this interference is that it is due to substitution of F^- by OH^- on the membrane surface. This explanation cannot be adapted to Ag_2S given the diferences in charge and ionic radius of OH^- and S^{2-} as well as the affinity of the two anions for Ag^+.

An important contribution to interpret the interferences produced by ligands, can come from the knowledge of the electrical double layer structure at the sensor membrane/Ag^+ solution interface. From adsorption studies, De Bruyn *et al.*[12] obtained data for the surface charge and double layer capacity in the interval pH 4.7-9.2. It is of interest to complement this data with zeta potential values in order to obtain information about the compact layer of the electrical double layer. Some results have already been obtained and are presented in Table 2 . The point of zero charge found for silver sulphide is pAg=10, which agrees with the value of De Bruyn.

Although the values obtained for ζ are as yet incomplete, they indicate that in the absence of ligands the isoelectric point and the point of zero charge coincide and that in the presence of ligands the former is lowered. The difference between the ζ values for solutions at pH=12 with or without glycine shows that glycine is adsorbed preferentially to hydroxyl giving rise to a lower charge density owing to its larger molecular dimensions.

Specific adsorption of anions will affect the surface charge and the rate of ion exchange between the surface of the solid and the solution. The experimental data do not allow conclusions to be drawn regarding the effect of adsorption on the thermodynamic potentials, but general observations of kinetic nature may be made. Effectively, the layer of adsorbed ligand appears to inhibit the diffusion of ions towards and from the surface, retarding the approach to equilibrium.

For glycine and EDTA, equilibrium is reached within several minutes as long as the total silver ion concentration is greater than 10^{-6} and $3x10^{-6}$ M respectively. In the case of OH^- the reading of potential obtained at the end of 5-10 minutes may not correspond exactly to the equilibrium values, but the differences are not significant , since they lead to reasonable values for the stability constants of the complexes formed. The difference between the value obtained for the potential and its real value will diminish with increasing silver ion concentration. This would explain the electrode super--Nernstian response in the presence of hydroxyl ion.

REFERENCES

1. J.H. Woodson and H.A. Liebhafsky, Anal. Chem. 41, 1895 (1969)
2. K. Cammann and G.A. Rechnitz, Anal. Chem. 48, 856 (1976)
3. G.J. Moody and J.D.R. Thomas in "Ion-Selective Electrodes in Analytical Chemistry" H. Freiser ed., Plenum Press, New York 1978 Vol. 1 Ch. 6 and references therein
4. G. Nakagawa, H. Wada and T. Hayakawa, Bull. Chem. Soc. Jpn. 48, 424(1975)
5. M.F. Taras and E. Pungor, Anal. Chim. Acta 82, 285 (1976)
6. I. Sekerka and J.F. Lechner, Anal. Letters A11, 415 (1978)
7. G.J.M. Heijne and W.E. Linden, Anal. Chim. Acta 96, 13 (1978)
8. A. Sabatini, A. Vacca and P. Gans, Talanta 21, 53 (1974)
9. "Stability Constants of Metal-Ion Complexes", The Chemical Society, London, Special Publ. 17, (1964) and 25 (1975)
10. G. Biederman and L.G. Sillen, Acta Chem. Scand. 14, 717 (1960)
11. G.Vesely and K. Stulik, Anal. Chim. Acta 73, 157 (1974)
12. I. Iwasaki and P.L. De Bruyn, J. Phys. Chem. 62, 594 (1958)

Table 1. Stability constants of silver complexes. μ=0.1M, t=25°C

Ligand	Stability constants	
	log β_1	log β_2
Hydroxyl	2.2 $\pm$0.1	3.9 $\pm$0.1
Glycine	3.81$\pm$0.03	6.75$\pm$0.02
EDTA	7.193$\pm$0.008	——

Table 2. Zeta potentials for Ag_2S/Ag^+ interface. t=25°C

pAg	pH	μ	ζ/mV
5	6.5	0.001	54
6	6.5	0.001	44
8	6.5	0.001	40
13	9.2	0.001	-53
8	12.2	0.01	-90
8	12.2	0.01 |gly|=10^{-3}M	0.1

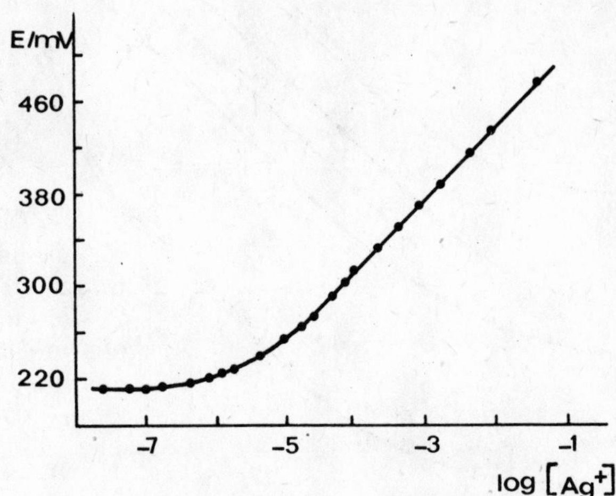

Fig.1 Calibration curve of Ag_2S ion-selective electrode μ=0.1M t=25°C. Equation for linear region E=551.4+59.18 log $|Ag^+|$

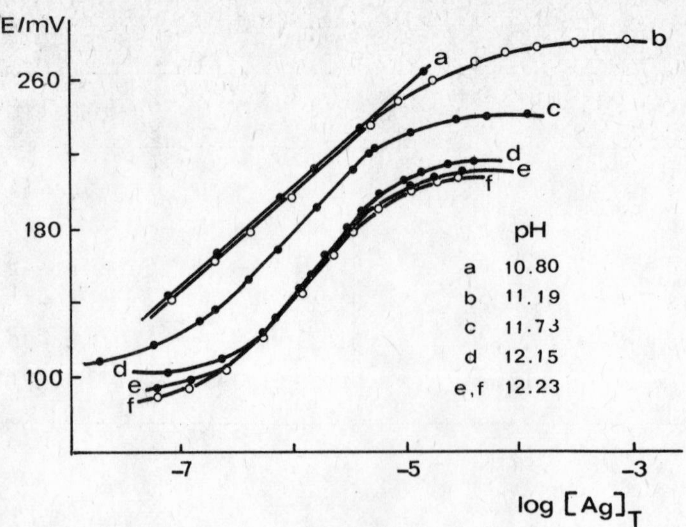

Fig.2 Emf *vs* log $|Ag|_T$ for different values of pH μ=0.1, t=25°C

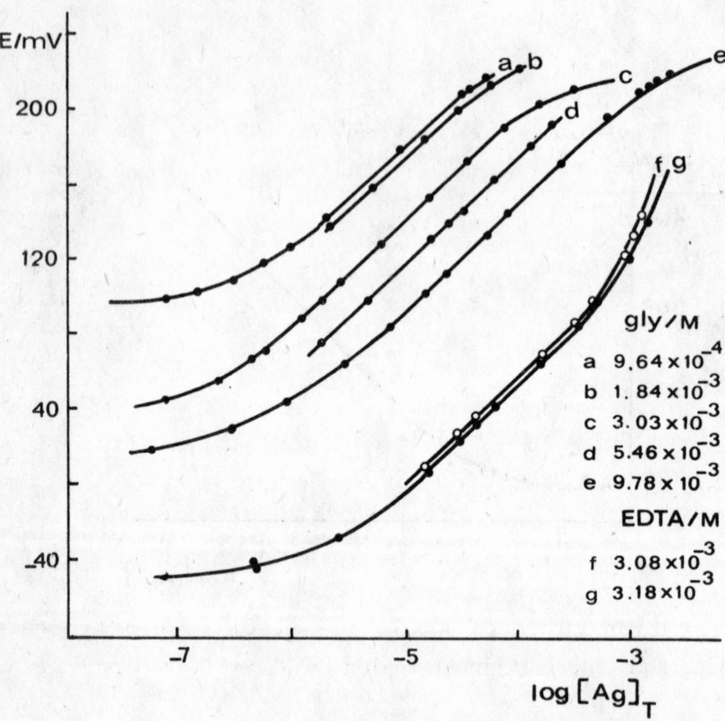

Fig.3 Emf *vs* log $|Ag|_T$ for different concentration of glycine or EDTA. μ=0.1M, t=25°C, pH=12.

QUESTIONS AND COMMENTS

Participants of the discussion: J.S.Redinha, J.D.R.Thomas and Zhang Zong-Rang

Question:

In the first part of your paper you showed us the calibration curve for silver ions with a cut-off at about 10^{-4} to 10^{-5} molar. It is our experience that it is possible to go down to lower levels. Adsorption can cause problems with silver ion determination. Could you tell us something about the vessels you employed ?

Answer:

The cut-off is at 10^{-5} mol/l. I was very careful about the sorption. I used teflon vessels.

Comment:

In calibrating ISEs in general, we found that it is sometimes useful to attack the problem from the other end. With more concentrated solutions, i.e. down to 10^{-4} to 10^{-5} mol/l it is possible to carry out serial dilution. But for calibration below that it is best to add small aliquots of a more concentrated solution to the background electrolyte. This way usually you can check the calibration to much lower concentrations.

Answer:

I did extrapolation, I was very careful about getting straight lines for the calibration, I was not so much pressed to go down to lower concentrations. As I was measuring complexes, the concentrations are out of the range where I calibrate, a precise slope value is very important.

Question:

You have mentioned that you measured zeta potentials. How did you do that ?

Answer:

We have many ways to measure them. One way is to have the particles moving and the solution stagnant. We have chosen this way, using a special cell built for this purpose.

4th Symposium on Ion-Selective Electrodes
Mátrafüred, 1984

THE USE OF CADMIUM ION-SELECTIVE ELECTRODE AND VOLTAMMETRIC TECHNIQUES IN THE STUDY OF CADMIUM COMPLEXES WITH INORGANIC LIGANDS

R. STELLA, M.T. GANZERLI VALENTINI and P.A. BORRONI

Department of General Chemistry - University of Pavia
Viale Taramelli 12 - 27100 Pavia, Italy

ABSTRACT

Cadmium complexes with inorganic ligands OH^-, CO_3^{2-}, SO_4^{2-} and Cl^-, at concentration levels higher than those found in natural waters, were studied by Ion-Selective Electrode potentiometry. Leden method was applied to calculate formation constants which were refined using an adequate computer programme and compared, when possible, with corresponding values obtained through voltammetric measurements.

INTRODUCTION

The study of labile complexes of heavy metals in natural waters is very important for understanding the transport and the biological interactions of the metals themselves and represents a problem extremely challenging.

Cadmium complexes with inorganic ligands are particularly interesting and a lot of investigations have been undertaken on this subject: a comprehensive review of the necessary literature is presented by Raspor (1). Total cadmium content in fresh water is in the range of 0.01 - 1 $\mu g\ l^{-1}$ (2) and consequently the aquatic environmental chemistry of cadmium belongs mainly to the field of trace or even ultratrace chemistry. Differential Pulse Anodic Stripping Voltammetry (DPASV) and, in some cases, Anodic Stripping Voltammetry

(ASV) are among the few techniques sensitive enough to allow the determination of cadmium at such low levels (3, 4). Moreover complexation produces a shift in peak potential E_p that, under conditions of electrochemical reversibility, may be related to the overall complex formation constant (5) and therefore exploited in speciation studies.

Cadmium Ion-Selective Electrode (Cd-ISE) is also an important tool in cadmium speciation studies as it allows to measure the free ion concentration. According to Gardiner (6) for Cd-ISE the linear relationship between the free ion concentration (or activity) and the electrode response extends not further than 100 μg Cd l^{-1} (10^{-6}M), though extrapolation to lower concentrations can be made with reasonable certainty. Therefore sensitivity is insufficient for direct measurements in fresh water systems. More recently Shephard et al. (7) performed analysis of cadmium speciation on water samples from a metal contaminated lake and showed that Cd-ISE exhibits Nernstian behaviour in the working range of 10^{-5} - 10^{-7} M Cd^{2+} concentrations.

Therefore in order to exploit the advantages offered by both techniques we performed voltammetric and Cd-ISE measurements in solutions having cadmium content higher than natural level and concentration of the natural ligands OH^-, CO_3^{2-}, HCO_3^-, Cl^- and SO_4^{2-} comparable with those found in fresh waters.

The principal aim of this work was to determine some complex formation constants of cadmium. The latter are very important because cadmium tends to stay in natural waters in the ionic form (predominantly as labile inorganic complexes) rather than becoming bound with the organic ligands (humic and fulvic acids).

EXPERIMENTAL

Reagents

A 10^{-2} M cadmium stock solution was prepared by dissolving with diluted "Suprapure" nitric acid, in a Teflon beaker, a weighed amount of "Ultrapure" cadmium wire (Alpha Products). The excess of nitric acid was then evaporated on a hot plate and the residue was taken up with water.

Perchloric acid 0.1 M was prepared from the 70% commercial solution (Merck) and standardized with Na_2CO_3 "primary standard" grade. Carbonate free sodium hydroxide 0.1 N was prepared from 50% w solution and potentiometrically titrated againts sulphammic acid.

Other chemical used were "reagent grade". Deionized and terdistilled water was used throughout all experiments.

Instrumentation and Cd-ISE calibration

Cd-ISE ORION S1 94-48 and single junction reference electrode ORION 90-01 were used connected to the high impendence ORION 701 A digital pH/mV meter. The electrode was carefully calibrated at room temperature (20.0 $\pm$ 1 °C) after adjusting ionic strength to 5.10^{-2} M with KNO_3 and pH value to 5 in the standard Cd solution. Diluted standard Cd solutions were prepared by delivering appropriate volumes of stock or more diluted solutions from an automatic microdispenser Dosimat E 655. The mV response plotted in a semilog scale versus the free ion concentration was a straight line in the range of 10^{-4} to 10^{-7} M Cd^{2+} concentration, provided a period of at least 20 minutes was allowed for the electrode system to reach equilibrium. The slope was 28.5 mV for each 10-fold increase in concentration and began to decrease below 10^{-7} M Cd^{2+}. Therefore we found, as already stated by Shephard et al. (7), that the electrode

Nernstian behaviour may be extended, as illustrated in Fig. 1, to 10^{-7} M Cd^{2+}, i.e. further than claimed by Gardiner (6), provided that enough time is allowed to reach equilibrium and also carefully cleaned plastic vessels are used. Also an ionic strength higher than 2.10^{-3} M, though unlikely in natural waters, improves electrode working features at very low cadmium levels. Care was taken in avoiding trace copper contaminations in cadmium solutions because it represents a serious interference as already mentioned by Gardiner (6); in a few cases little Cu^{2+} was detected with the corresponding Cu-ISE and the interference was eliminated by adding sodium iminodiacetate (IDA) as illustrated in a previous work (8).

The voltammetric measurements were run with the AMEL mod. 472 polarograph equipped with the capillary electrode of 30 to 40 sec dropping time.

An ORION microprocessor pH meter 811 equipped with an high precision glass electrode was used for pH measurements: calibration was made following the procedure of Rajan and Martell (9) using a pK_w value of 13.83 to convert pH in concentration units $[H^+]$.

Junction potential contribute to pH readings was considered irrelevant.

Cd-ISE measurement procedure

Free cadmium ion Cd^{2+} and pH were simultaneously measured in cadmium - ligand solutions of different composition.

In the case of Cd^{2+} - OH^- system 50 ml aliquots of Cd^{2+} solutions, whose concentration ranged from 4.04×10^{-4} to 5.80×10^{-7} M, were brought to pH 11.0 with carbonate free NaOH. Stepwise additions of 0.1 M $HClO_4$ were made and Cd-ISE response and pH were correspondingly recorded. After converting pH to $-\log[H^+]$ units plots such as those reported in

Fig. 2 were obtained.

In the Cd^{2+} - CO_3^{2-} system the concentration of the ligands (CO_3^{2-} and HCO_3^-) was similarly varied through pH changes (Fig.3) and in the Cd^{2+} - Cl^- and Cd^{2+} - SO_4^{2-} systems by stepwise addition of NaCl and K_2SO_4 respectively at fixed pH = 4.7.

Voltammetric experiment procedure

Same systems underwent Anodic Stripping Voltammetry (ASV) measurements with hanging mercury drop electrode (HMDE) by adopting experimental conditions proposed by Bilinski et al. (5). Ligand concentrations were stepwise changed as already described and E_p correspondingly recorded. Peak voltage shifts ΔE_p reported on the ordinate axis of Fig. 4 are related, after Bilinski et al. (5), to β values through the following relationship:

$$\Delta E_p = (0.059/n) \log \beta_j - (j\ 0.059/n) \log [L])$$

were n is the number of electrons involved in the redox process, β_j is the overall stability constant ($\beta_j = [ML_j]/[M][L]^j$), j is the number of ligands involved in the complex. The formula finds an evident application for graphic evaluation of β_j.

RESULTS AND DISCUSSION

In order to evaluate, through Cd-ISE measurements the stability constants, which rule the investigated systems, the graphic method suggested by Leden (10) was applied. It consists in calculating the function $f(L) = ([M_T] - [M])/[M][L]$, where $[M_T]$ represents the sum of the concentrations of the free metal ion M and of total metal complexed by the examined ligand. If the number of complexes formed is 2, as expected in these experiments, it becomes $f(L) = \beta_1 + \beta_2 [L]$: the graphic application for obtaining β_1 and β_2 is evident.

The Leden method was applied to the data points reported

in Fig. 2 over the nonlinear section of all curves. In the - log $[H^+]$ range 6 to 8.5 very low concentrations of cadmium-hydroxo species is in fact expected, whilst along the straight common tail of three upper curves precipitation of $Cd(OH)_2$ very likely happens and correspondingly solubility product K_{so} may be calculated.

Even more substantial corrections had to be applied to the experimental data before applying Leden method to Cd^{2+} - CO_3^{2-} system. The contributes of $CdOH^+$ and $Cd(OH)_2^o$ soluble species were evaluated from previously calculated constants and substracted from total Cd concentration for a correct evaluation of $[M_T]$. Improved fitting of the experimental data was obtained by considering the presence also of the hydrocarbonato complex Cd HCO_3^+: over the - log $[H^+]$ range 6 to 7.5, where it predominates, a log K = 1.93 was calculated for its formation constant.

Carbonate ion concentration $[CO_3^{2-}]$ was calculated from experimental pH values by using the formula:

$$[CO_3^{2-}] = C_T K_1 K_2 / P$$

where $P = [H^+]^2 + K_1 [H^+] + K_1 K_2$, K_1 and K_2 are the two dissociation constants of carbonic acid at ionic strenght 5 x 10^{-2} M ($K_1 = 10^{-6.21}$ and $K_2 = 10^{-10.02}$) and C_T is total carbonate concentration. Calculations for Cd^{2+} - Cl^- and Cd^{2+} - SO_4^{2-} systems resulted somewhat simplified as f(L) calculated in correspondence with the non linear section of the experimental curves resulted rather constant and equal to $10^{1.57}$ and $10^{1.59}$ respectively: therefore it must be concluded that experimental data are suitable for calculating just first formation constant.

An attemp was also made to refine β values graphically obtained, which are listed in the first column of Table 1.

For this purpose we found advantageous to employ the

computer programme by Meites and Meites (11), written in PolyBasic, that we modified to run on an Olivetti M 40 computer. Such a least-squares curve-fitting programme is able to fit an equation involving five adjustable parameters and that actually was the case in dealing with cadmium - carbonate system; for the latter the following equation was inserted into the programme:

$$[Cd^{2+}] = Cd_T/(1 + AK_w/[H^+] + B(K_w/[H^+])^2 + CK_1K_2C_T/P + D(K_1K_2C_T/P)^2 + EK_1[H^+]C_T/P)$$

where Cd_T and C_T are total concentrations of cadmium and carbonate respectively, K_1, K_2 and P were already defined and A,B,C, D,E are the five parameters corresponding to the overall formation constants of $CdOH^+$, $Cd(OH)_2^\circ$, $CdCO_3^\circ$, $Cd(CO_3)_2^{2-}$ and $CdHCO_3^+$ respectively. Hydroxo-complex data required an analogous but more simple function. Computer refined results are reported in the second column of Table 1. Results of ASV measurements were generally in good agreement, as shown in the third column of Table 1, with those based on Cd--ISE data: they also gave no evidence, in the adopted experimental conditions, of multiligand complex formation in the Cd^{2+}- Cl^- and Cd^{2+} - SO_4^{2-} systems.

Constants reported in Table 1 are mean values of five replicates; those listed in the first column, or in the second if available, were used to calculate the chemical form distribution which one should expect after adding cadmium to a water of known composition (Fig. 5).

REFERENCES

1 B. Raspor "Distribution and speciation of cadmium in natural waters". In J.O. Nriagu, Ed., Cadmium in the environment. J. Wiley and Sons, New York, /1980/ pp. 147-236.

2 H.W. Nürnberg, Proc. Anal. Div. Chem. Soc.(London),15/1978/ 275.

3 P.Valenta, L. Mart and H. Rützel, J. Electroanal.Chem.,82/1977/ 327.

4 H.W. Nürnberg, Sci. Total Environ., 12/1979/35.

5 H. Bilinski, R. Huston and W. Stumm, Anal.Chim.Acta, 84 /1976/ 157.

6 J. Gardiner, Water Res., 8 /1974/ 23.

7 B.K. Shephard, A.W. McIntosh, G.J. Atchison and D.W. Nelson, Water Res., 14, /1980/ 1061.

8 R. Stella and M.T. Ganzerli Valentini, Anal.Chim.Acta, 152, /1983/ 191.

9 K.S. Rajan and A.E. Martell, J.Inorg.Nucl.Chem.,26,/1964/ 789.

10 F.R. Hartley, C.Burgess and R.M. Alcock, Solution Equilibria, Horwood, Chichester, 1980.

11 T. Meites and L. Meites, Talanta, 19, /1972/ 1131.

12 D. Dyrssen and D. Lumme, Acta Chem.Scand., 16, /1962/ 1875.

13 I.M. Korenmann and V.N. Burova, Trudy po Khimi. Khim TeKK, 2, /1956/ 366.

14 P.T. Long and E.E. Angino, Geoch.Cosm.Acta, 41, /1977/ 1183

15 R. Pottel, Ber.Bunsengesellschafts Phys.Chem., 69, /1965/ 363.

16 W. Freitknecht and R. Reinmann, Helv.Chim.Acta, 34, /1951/ 2255.

17 V.B. Spivakovskii and L.P. Moisa, Russ. J. Inorg. Chem., 9, /1964/ 1239.

Table 1. Formation constants of cadmium complexes ($I = 5.0 \times 10^{-2}$ M; $T = 20 \pm 1$°C)

Reaction	Logarithmic constant			
	This work: graphic	computer	ASV	Literature value(°)
$Cd^{2+} + OH^- \rightleftharpoons CdOH^+$	4.40 ± 0.04	4.35 ± 0.03	4.51 ± 0.04	4.30 (12)
$Cd^{2+} + 2OH^- \rightleftharpoons Cd(OH)_2^{\circ}$	6.21 ± 0.09	6.74 ± 0.08	6.47 ± 0.08	10.62 (13)
$Cd^{2+} + CO_3^{2-} \rightleftharpoons CdCO_3^{\circ}$	3.55 ± 0.05	3.49 ± 0.04	3.86 ± 0.04	3.50 (5), 4.02 (6)
$Cd^{2+} + CO_3^{2-} \rightleftharpoons Cd(CO_3)_2^{2-}$	6.28 ± 0.10	6.37 ± 0.10	6.50 ± 0.09	
$Cd^{2+} + HCO_3^- \rightleftharpoons CdHCO_3^+$	1.93 ± 0.01	2.02 ± 0.01		2.10 (14)
$Cd^{2+} + Cl^- \rightleftharpoons CdCl^+$	1.57 ± 0.01			1.68 (6)
$Cd^{2+} + SO_4^{2} \rightleftharpoons CdSO_4^{\circ}$	1.59 ± 0.02			2.34 (6), 2.01 (15)
$Cd^{2+} + 2OH^- \rightleftharpoons Cd(OH)_2(s)$	-13.50 ± 0.12			-13.66 (16), -14.09 (17)
	$(-14.39 \pm 0.13)^{(°°)}$			

(°) different experimental conditions

(°°) aged precipitate

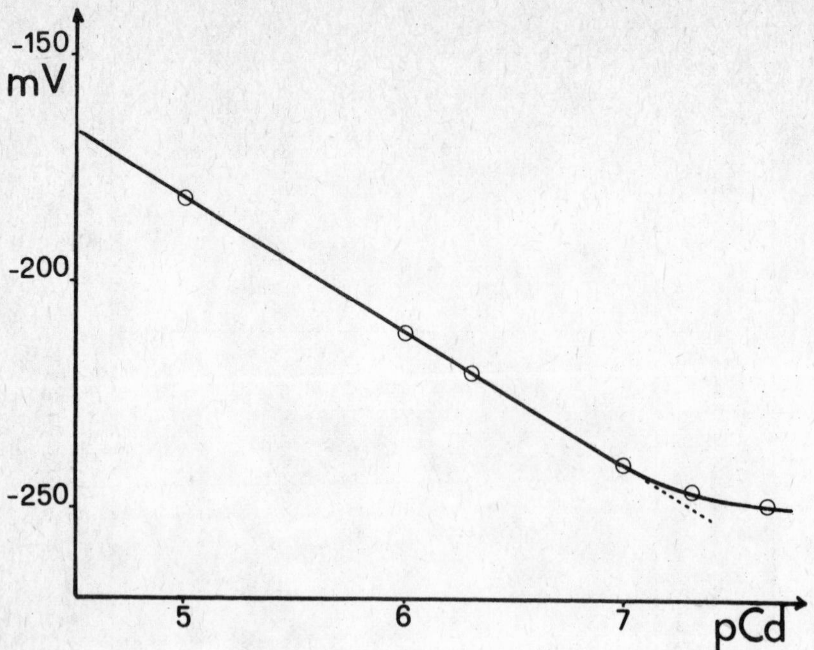

Fig. 1. Calibration curve of Cd-ISE. ($I = 5.0 \times 10^{-2}$ M).

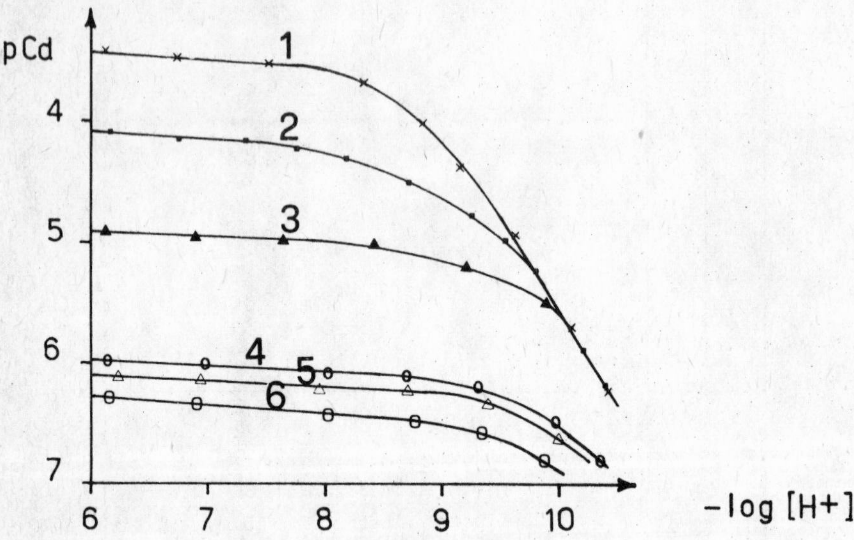

Fig. 2. Cadmium ion concentration (p Cd) as a $-\log [H^+]$ function in the $Cd^{2+} - OH^-$ system. ($Cd_T = 4.04 \times 10^{-4}$ M (1), 9.00×10^{-5} M (2), 1.19×10^{-5} M (3), 1.19×10^{-6} M (4), 9.00×10^{-7} M (5), 5.80×10^{-7} M (6)).

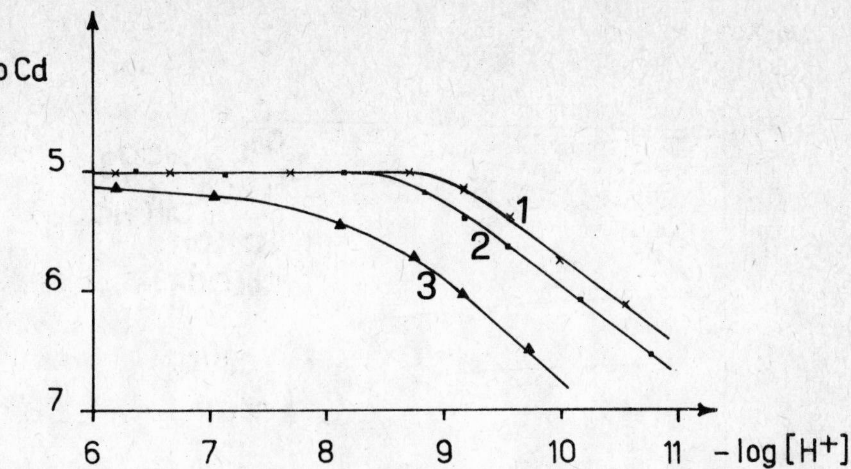

Fig. 3. Cadmium ion concentration (p Cd) as a -log $[H^+]$ function in the Cd^{2+} - CO_3^{2-} system. ($Cd_T = 1.00 \times 10^{-5}$ M; $C_T = 6.25 \times 10^{-4}$ M (1), 2.50×10^{-3} M (2), 1.25×10^{-2} M (3).

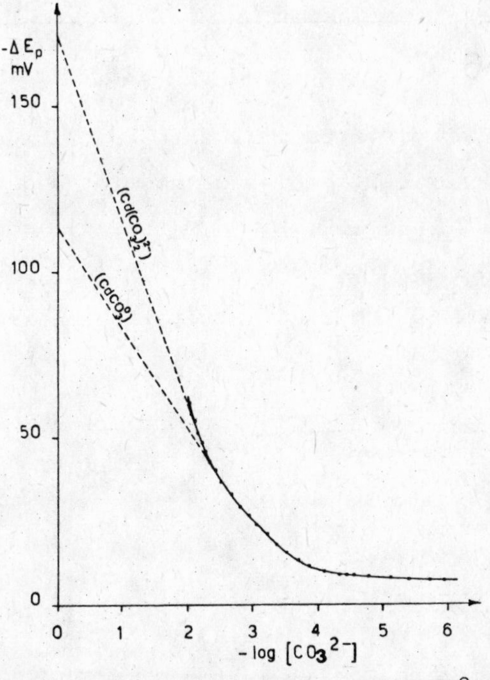

Fig. 4. Plot of $-\Delta E_p$ (ASV) versus $-\log[CO_3^{2-}]$ for Cd^{2+} - CO_3^{2-} system. ($Cd_T = 1.60 \times 10^{-4}$ M).

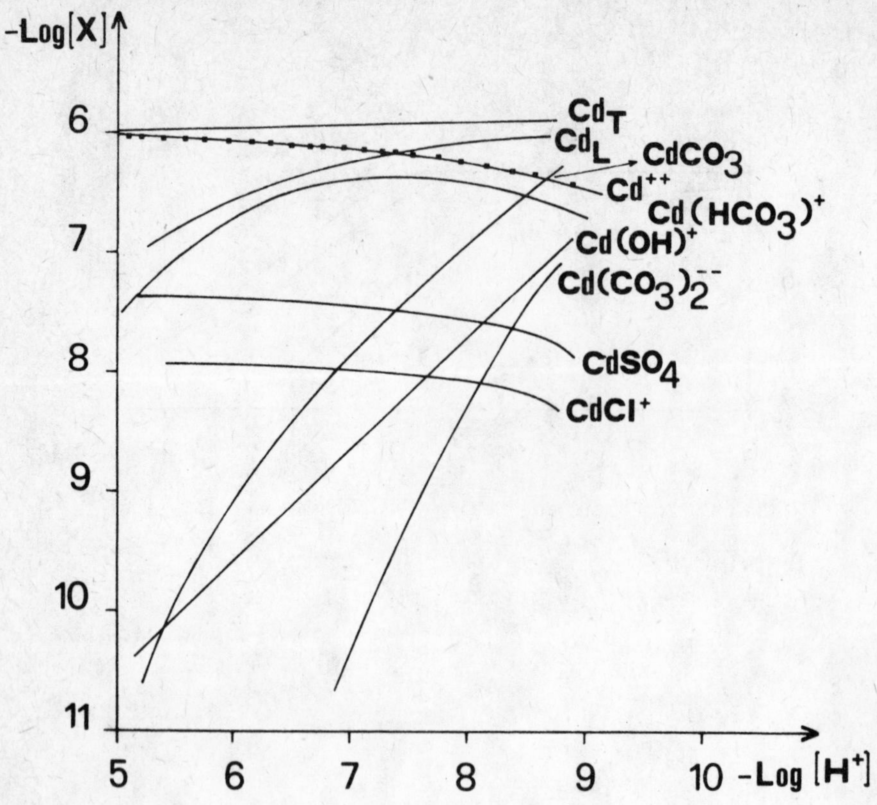

Fig. 5. Calculated distribution, as a function of -log $[H^+]$, of the chemical species of cadmium (added to a mineral water). (Cd_T = 1.09 x 10^{-6} M (added); $[Cl^-]$ = 3.72 x x 10^{-4} M; $[SO_4^{2-}]$ = 7.84 x 10^{-4} M; $[HCO_3^-]$ = 8.06 x 10^{-3}M; I = 5.0 x 10^{-2} M). Cd_L = total complexed cadmium.

QUESTIONS AND COMMENTS

Participants of the discussion: E.Pungor and R.Stella

Question:

You have shown a nice calibration curve which was linear down to 10^{-7} mol/l. Did you find many electrodes to work as well as this electrode, or was it an exceptionally good one ?

Answer:

I do not know what position the Cd ISE occupies in the kingdom of ISEs. It is not a king, not even a queen, but we wanted to nobilitate it. We think that one of the shortcomings in using the Cd ISE which leads to less favourable sensitivity, is the presence of copper, a quite common airborne contaminant. I think if you take care to mask it, you can improve the sensitivity of the Cd ISE very much. I do not have experience with any other electrode, just the Orion.

Question:

How was the 10^{-7} mol/l solution prepared ?

Answer:

By serial dilution and I did not check it. I think that at such low concentration it is an illusion to find an analytical method that is more precise than the dilution. However, I am not so sure that the 10^{-7} mol/l is really 10^{-7} mol/l.

Comment:

If you prepare such a solution in a glass vessel, the concentration will not be the same after 20 minutes.

Answer:

As I already said, I am not very sure about the range 10^{-6}-10^{-7}mol/l. That is why in the actual measurements we stayed over 10^{-6} mol/l.

4th Symposium on Ion-Selective Electrodes
Mátrafüred, 1984

MEASUREMENTS BY ION-SENSITIVE FIELD EFFECT TRANSISTORS

V. TIMÁR-HORVÁTH and G. VÉGH

Department of Electronic Devices
Technical University, Budapest, Hungary

ABSTRACT

A class of devices known as ion-sensitive field effect transistors is of particular interest for biomedical and industrial applications. Research efforts have been devoted to improving the preparation and packaging of an "n" channel pX sensor ISFET device as well as the chemical testing and characterization of the devices in different solutions. Several ion-sensitive layers /dielectrics/ were deposited on the gate /e.g. Si_3N_4 , Al_2O_3, SiO_2/ and selectivity measurements were carried out. Recently a new electrical measuring system has been under development for the parallel testing of several /e.g. twenty or more/ devices in the same solution.

INTRODUCTION

Numerous papers on the developing and studying of ion-sensitive field effect transistors were published [1-3], dealing in particular with the clarification of the sensing mechanism of the gate material. As it is well known, the ISFET sensor is basically identical with a conventional MOS transistor, only the gate metal is omitted and replaced by other dielectrics or membranes. The reference electrode and the ionic activity of the solution to be measured acts as a "solution gate" and the electrochemical potential developed at the interface between the gate membrane and the solution controls the drain current of the transistor. The theoretical characteristic of the ISFET device is expected to be linear function between the drain current and the ionic activity of the solution [4] assuming that the sensing layer on top of the gate dielectric is a conventional membrane material. In the case of thin dielectric films used as membranes /the evaluation of such films is compatible with the semiconductor technology/ the membrane resistance is exceptionally high /$> 10^{13} \Omega$/ and their irregular operation yields a function diverging from the linear [5]. A great number of papers on the behaviour of different thin dielectric membranes is available [6-7].

EXPERIMENTAL

The solid state processes for the fabrication of the ISFET devices are similar to those employed in standard MOS technology.

Two "n" channel devices: an ISFET and a MISFET were placed on one chip, the MISFET for testing the chip /Fig. 1./. Some chips were packaged in conventional TO18 transistor headers for the testing and physical characterization of the devices. The whole header and the chip with the gold bondings were coated with silicon rubber /Elastosil E 43/ except the gate area, to protect the metallic surfaces from getting damaged by the electrolyte. Other chips were also packaged experimentally in medical S1 injection needles. The arrangement was covered with silicon rubber in the same way. The sensor doesn't contain the micro-reference electrode yet. The structure is shown in Fig. 2.

Different thin dielectric films were deposited on top of a thin /50 nm/ obligatory SiO_2 layer: Si_3N_4 by the plasma enhanced low pressure chemical vapour deposition /PELPCVD/ method and Al_2O_3, Si_3N_4 and SiO_2 layers deposited by radio frequency /RF/ sputtering from dielectric targets. Some dielectric films were deposited not immediately on the SiO_2 layer, but without the removal of the protecting Al films on top of the latter /similarly to the extended gate structure [8]/. Some films were doped with Al and Na subsequently.

MEASURING SYSTEM

The pX measurements were carried out by the null-balancing method i.e. the drain current was held constant and at different ionic activities the external gate voltage connected in series with the reference electrode was varied automatically to null changes in the drain current. Reading the change of the applied gate voltage shows directly the double-layer potential.

The pX response of the ISFET sensors was measured by using solutions containing no common interfering cations. The solutions were prepared from a.g. HCl and NH_4OH diluted by bidistilled water. The solutions containing Na^+ and K^+ were diluted from a.g. NaCl and KCl. All measurements were performed at constant room temperature and lighting.

RESULTS

Some pH, pNa and pK characteristics are given in Figs.(3-6). The Si_3N_4 layer prepared by the PE LPCVD method became Al doped after the removal of the annealed protective Al layer. Fig.3. shows its better sensitivity to sodium than to hydrogen ions. The RF sputtered Si_3N_4 layer /Fig. 4./ is more selective to H^+, but it doesn't reach the theoretical slope probably due to the oxygen contamination in it. The sputtered Al_2O_3 layer shows a similar characteristic /Fig. 5./ All these films need apparently some additional heat treatment to change their stoichiometric ratio and site density.

The sputtered SiO_2 layer on top of an Al film shows a better behaviour than the bare thermal SiO_2, but its sensitivity fails to reach the theoretical slope nevertheless.

This layer which is doped with Na subsequently became alcali sensing /Fig. 6./. These preliminarily reported results need additional structural analysis.
Finally Fig. 7. shows a new electrical measuring system for the parallel testing of twenty /or more/ pX sensors in the same solution. In this arrangement the electrical parameters of ISFETs don't interfere with the measured potential, because the drain current, the reference voltage and the drain-source voltage are held constant and only the source potential is varying. All the devices under testing have own independent measuring cards.

SUMMARY

On evaluating our work we may state that these dielectric films are suited for sensing H^+, Na^+ and K^+ ions. Valuable results were obtained during the investigation of Al, Na addit-ions introduced into the sensing layer. This fact shows that it would be worthwhile to experiment in the following also with other substances /e.g. B, P etc./ eventually with their common application. The aluminium layer could be replaced favourably also by polysilicon.

A further task to be realized and pursued already is marking out the sensing area in a photolitographic way, instead of by free hand, for ensuring a uniform gate area and a higher productivity in the preparation.

Another development nearing to final stage is the elaboration of a multichannel measuring system, permitting a great number of various measurements.

REFERENCES

[1] P. Bergveld, IEEE Trans. Biomed. Eng. vol. BME-19, 1972.
[2.] W.M. Siu, M.S. Cobbold, IEEE Trans. Electron Dev. Vol. ED-26, No. 12. 1979.
[3.] A.G. Revesz, Thin Solid Films vol. 41, 1977.

[4.] P. Bergveld, Sensors and Actuators, vol. 1., 1981.
[5.] R.I. Lauks, IEEE Trans. Biomed. Eng. vol. ED-26, No.12, 1979.
[6.] J. Janata, R.J. Huber, Ion-Selective Electrode Rev. Vol. 1. 1979.
[7.] R.P. Buck, A. Fog, Sensors and Actuators, vol. 5. 1984.
[8.] J.V.D. Spiegel, I. Lauks, P. Chan, D. Babic, Sensors and Actuators, Vol. 4, 1983.

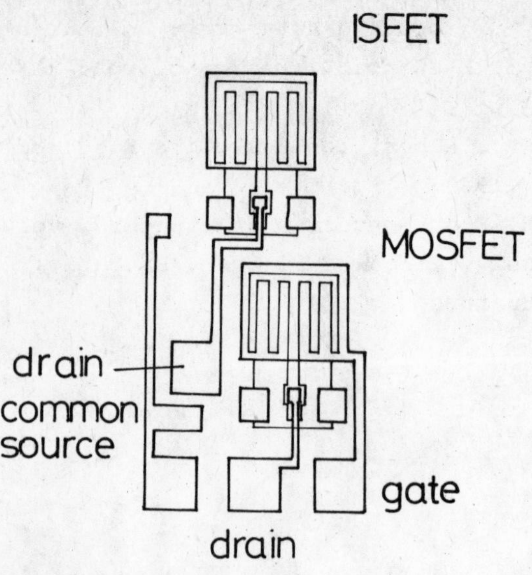

Fig.1. The overlay of the chip

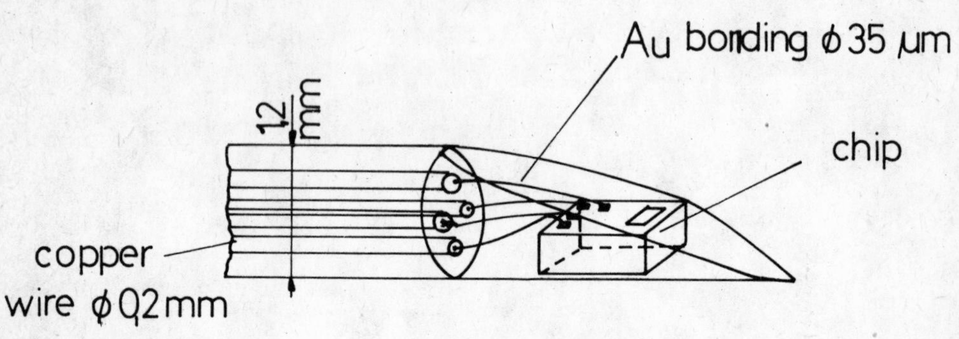

Fig.2. The view of the sensor structure

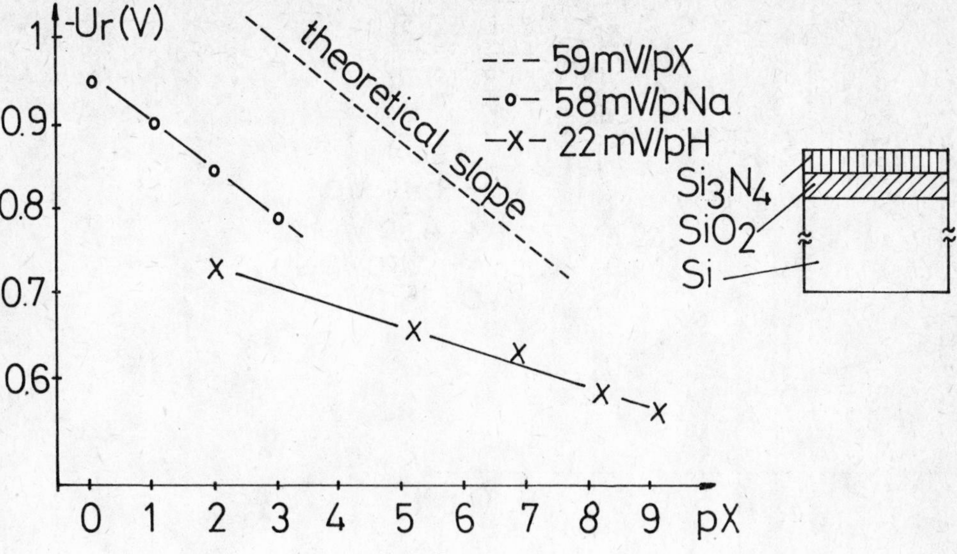

Fig.3. Device characteristics. The gate material is Si_3N_4 doped with Al

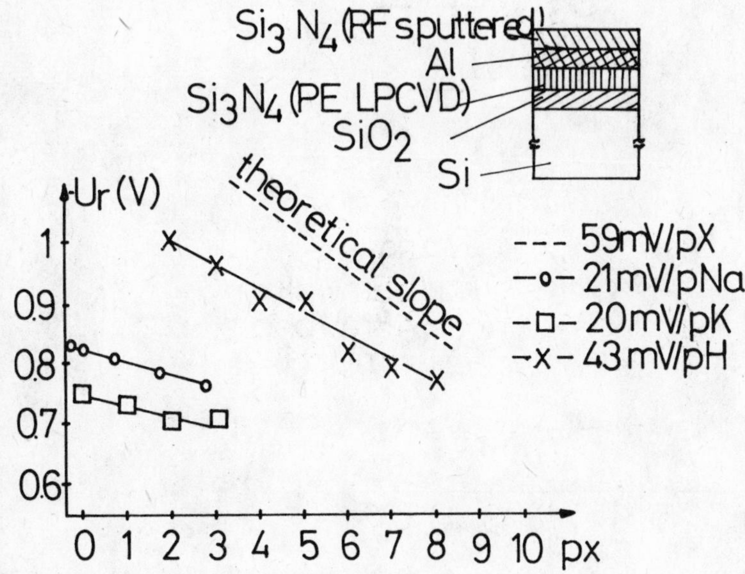

Fig.4. Device characteristics. The gate material is sputtered Si_3N_4

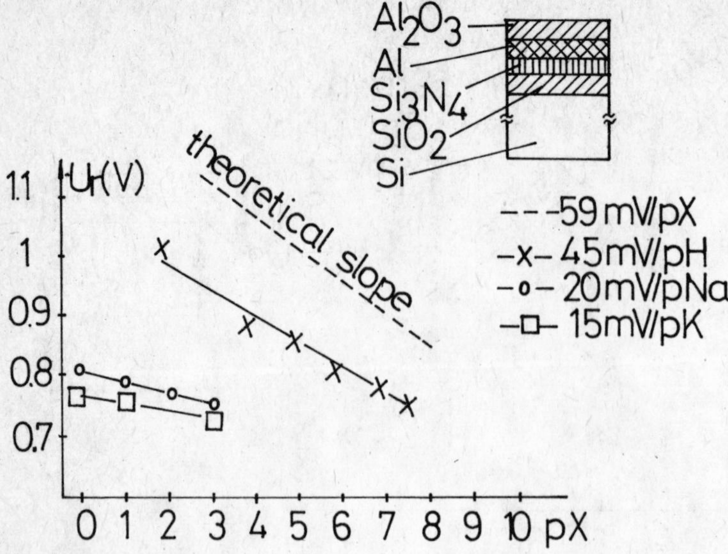

Fig.5. Device characteristics. The gate material is sputtered Al_2O_3

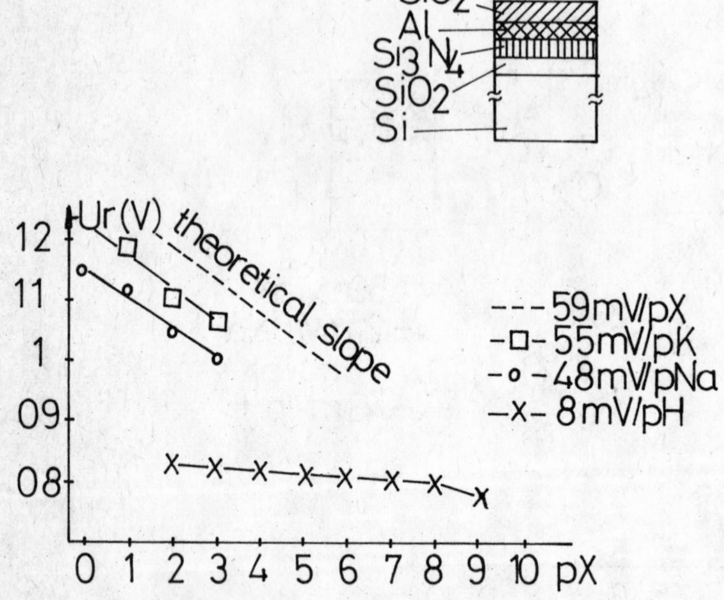

Fig.6. Device characteristics. The gate material is sputtered SiO_2 doped with Na

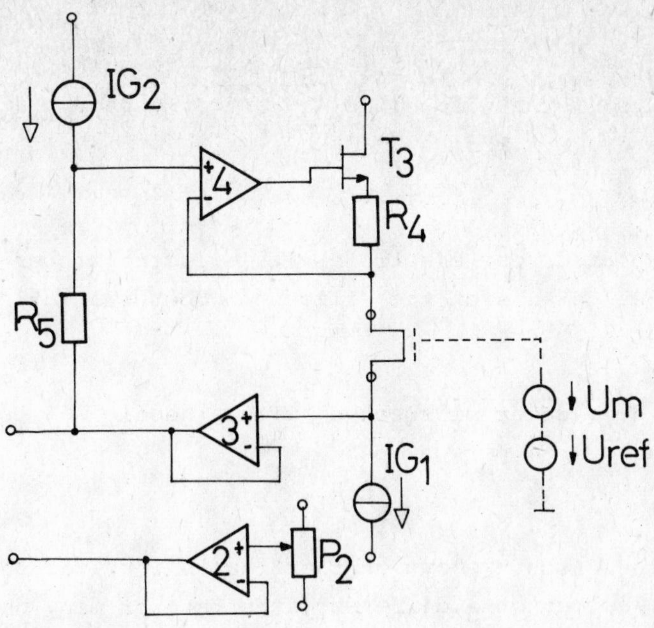

Fig.7. Electrical testing circuit. The operational amplifiers are µA 741

QUESTIONS

Participants of the discussion: J.Janata, G.Nagy, V.Timár-Horváth and J.D.R.Thomas

Question:
You insulated your electrode with silicone rubber. Could we have some details on the silicone rubber insulation ?

Answer:
It was done under microscope, with a needle. Of course, it requires a skilled person.

Question:
About the circuit shown on the final slide. Was this of different devices on a different substrate or many devices on the same substrate ?

Answer:
Different devices on different chips. We measure these devices parallel, with the help of one reference electrode.

Question:
You mentioned that you use very stable apparatus. What kind of apparatus were you using ?

Answer:
We have developed a measuring system ourselves for our purposes.

4th Symposium on Ion-Selective Electrodes
Mátrafüred, 1984

NEW LEAD ION-SELECTIVE CHALCOGENIDE GLASS ELECTRODES

Yu.G. VLASOV, E.A. BYCHKOV and A.V. LEGIN

Department of Chemistry, Leningrad
University, Leningrad 199164, USSR

ABSTRACT

Chalcogenide glasses containing lead iodide or lead sulfide were used as ion-selective electrode membranes. Such electrodes revealed high sensitivity and potential stability and short response time. Higher selectivity (5-20 times) to Cd^{2+} and lower sensitivity (10-100 times) for active oxidizing agents are advantages of chalcogenide glass electrodes compared to conventional polycrystalline ones. Closely interconnected electrochemical behaviour, transport properties and short-range order of chalcogenide glasses have been investigated. X-ray photoelectron and scanning Auger electron spectroscopy allowed to study chalcogenide glass membrane surface and to carry out depth-profiling analyses both before and after soaking in different aqueous solutions.

INTRODUCTION

Crystalline sensors based on $PbS-Ag_2S$ membranes are the most widespread lead ion-selective electrodes (see, e.g. /1/). However, easy oxidation of crystalline PbS leads to the deterioration of the electrode perfomance and sluggish response. It is the principle drawback of these electrodes /2,3/. Selectivity coefficient to Cd^{2+} according to many authors is about 0.1-1.0 /4/ and so accurate determination of Pb^{2+} ions is impossible when comparable quantities of lead and cadmium are present.

Chalcogenide glasses are promising membrane materials for solid-state ion-selective electrodes (ISE) /5-9/. Selectivity and stability in strong acid media of silver, copper, cadmium ion-selective chalcogenide glass sensors are higher than those of appropriate crystalline ones and sensitivity for active oxidizing agents is lower /10-12/. Some lead ISEs based on chalcogenide glasses have also been reported /7,13-15/. The electrochemical properties of arsenic selenide glasses containing lead were described by Owen /7/. Bohnke et al. /13,14/ studied the $0.6AgAsS_2$-$0.4PbI_2$ glass. The present paper describes the investigation of new lead ISEs based on PbS-Ag_2S-As_2S_3 and PbI_2-Ag_2S-As_2S_3 chalcogenide glasses, the wide composition range of solid-state membranes having been studied. Some preliminary results of this study have been reported earlier /15/.

EXPERIMENTAL

All glasses were prepared according to /16,17/ in evacuated quartz ampoules at 1000 K for 5-10 hours. Discs 0.1-0.4 cm thick and 0.5-1.0 cm in diameter were cut from the alloys obtained. Each disc was polished on both sides and sealed with epoxy resin into a plastic tube. The potentiometric measurements were carried out in the cell

$$\mathrm{Ag,AgCl} \left| \begin{matrix} \mathrm{KCl} \\ \mathrm{sat.} \end{matrix} \right\vdots\, \mathrm{1M\ KNO_3} \,\vdots\, \begin{matrix} \mathrm{test} \\ \mathrm{solution} \end{matrix} \left| \mathrm{glass} \right| \begin{matrix} \mathrm{10mM\ Pb(NO_3)_2} \\ \mathrm{10mM\ AgNO_3} \end{matrix} \left| \mathrm{Ag} \right. \quad (1)$$

Calibration curves were fitted by the least-squares method to get the slope S of the lineary region and its doubled standard deviation 2 σ_S. The electrode standard potential E_o and its deviation 2 σ_{E_o} were determined for 95 per cent confidence limit. Other details are described in /16,17/.

To compare the properties of different membrane materials all electrochemical measurements were made with two types of Pb ISEs - chalcogenide glass electrodes prepared in this research and crystalline sensors. The latter were as follows: (i) a commercial lead-selective electrode Crytur 82-17 with PbS-Ag_2S membrane obtained by solid-state synthesis /18/, (ii) an ISE with the same membrane composition prepared, however, by Ag_2S and PbS coprecipitation from solution /19/, and (iii) a sensor with single-crystal PbS membrane.

Hewlett Packard 5950A Esca spectrometer and Riber ASC 2000 scanning Auger microprobe were used to study the influence of various solutions on the surface composition of chalcogenide glass membranes.

RESULTS AND DISCUSSION

Glass-forming ability

The glass-forming regions are shown in Fig.1 and 2. Glasses containing up to 50 mol.% PbI_2 can be obtained in the iodide system PbI_2-Ag_2S-As_2S_3. In the sulfide system PbS-Ag_2S-As_2S_3 glasses enriched with silver sulfide (the ratio $R = [Ag_2S]/[As_2S_3] \geq 1$) may contain up to 25-30 mol.% PbS, while even 40-45 mol.% PbS can be incorporated in the alloys with low Ag_2S concentration. Glasses containing more PbS or PbI_2 (closed circles in Fig.1 and 2) are partly crystalline. It should be noted that glasses containing small amounts of silver sulfide can be phase-separated in both systems. Futher experiments are on this way now.

Optimum glass composition for ISE membranes

All the glasses investigated are sensitive to Pb^{2+} ions in solution. However, optimum glass composition regions in concentration triangles are of great interest.

Determination of glasses which do display the best electrode characteristics was carried out on the basis of statistical comparative analysis of alloy's ionic sensitivity throughout the glass-forming regions. More than 50 electrodes in the iodide system, and about 40 in the sulfide one were investigated. According to preliminary study, glassy/crystalline membranes demonstrated worse sensitivity, poor slope of the calibration curves, unsatisfactory potential reproducibility. On the other hand the detection limit, the Nernstian range, the response time, and the selectivity of the majority of glassy sensors are very close to each other. However, compositional dependence of the slope magnitude S and stability and reproducibility parameters can be determined. The slope S can vary from 24.0 to 29.2 mV/pPb, the standard potential stability during two weeks' continuous measurements (2 σ_{E_o}) makes up 2.7-17.0 mV, the reproducibility of repeated tenfold potential readings in 1 mM $Pb(NO_3)_2$ (2 $\sigma_{E(3)}$) is 0.9-5.0 mV.

To have a criterion for membrane composition choice we introduce an empirical parameter which characterises the membrane quality Q

$$Q = |S - S_o| + 2\sigma_S + 2\sigma_{E(3)} \tag{2}$$

The first term ($S - S_o$) reflects the real slope S deviation from the theoretical value S_o (S_o = 2.3RT/2F = 29.6 mV/pPb at 298 K) according to the Nernst equation

$$E = E_o + RT/2F \ln a_{Pb^{2+}} \tag{3}$$

the second term 2 σ_S deals with stability of response because $\sigma_{E_o} \approx 3.3\,\sigma_S$. The electrode characteristics reproducibility is taken into account by the latter term 2 $\sigma_{E(3)}$. The less is the sum Q the better are the total electrode properties.

The optimum glass composition areas were found with the help of Q values according to the Equation (2). Within the glass-forming regions in Fig.1 and 2 iso-Q lines are drawn. Having come into being by experimental data interpolation, these lines obviously show the optimum glass composition areas location. Thus, the best characteristics can provide the alloys containing 22-27 mol.% PbI_2 and comparable quantities of silver and arsenic sulfides (R = 1.0-1.5) in the iodide system, and the glasses containing 20-30 mol.% PbS (R = 1.0) in the sulfide one. Q value increase for the Ag_2S enriched glasses may be linked with the persistent tendency to crystallization. The worst ionic sensitivity of the materials containing a small amount of silver sulfide (R = 2/3-1/4) can be caused by: (i) the possible phase-separation or (ii) structural and chemical surface unstability. According to X-ray photoelectron (XPS) and Auger electron spectroscopy (AES) data obtained by the authors /23/, Ag_2S is the basic component of the modified surface layer of the glasses investigated, and so the lack of silver sulfide in the bulk for alloys with R < 1, gives rise to its deficiency at the surface, that can ruin its chemical stability.

The dependence of Q value on PbS or PbI_2 membrane concentration in the two systems is different. Both the lack and the excess of PbI_2 in the iodide glasses lead to deterioration of electrode response Q value varying less with composition than in the sulfide system. Perhaps, lead iodide is not a suitable material for doping chalcogenide glasses used as lead-selective electrode membranes. Some Q decrease with PbI_2 concentration increase is likely to be due to appearance of lead surrounded by sulfur as a result of the exchange type reaction in the melt

$$PbI_2 + Ag_2S \rightleftharpoons PbS + 2\,AgI \qquad (4)$$

The assumption concerning the exchange reaction (4) was put forward on the basis of ionic conductivity analysis by Bohnke et al. /14/. Vlasov and coworkers /20/ obtained direct evidence of the fact by XPS.

On the contrary, electrode characteristics of the sulfide alloys improve monotonously with PbS concentration increase, the best membrane material $30PbS\cdot35Ag_2S\cdot35As_2S_3$ (Q = 3.4) being just at the glass-forming region boundary. It should be pointed out that the glasses containing 40-45 mol.% PbS (R=2/3) have rather poor electrode characteristics (Q = 8-11) because of their low conductivity ($\sigma_{298\,K} = 10^{-8}- 10^{-9} ohm^{-1} cm^{-1}$).

Electrode response

Typical calibration curves are shown in Fig.3 for chalcogenide glass electrodes, polycrystalline $PbS-Ag_2S$ sensor and single-crystal PbS electrode. In the case of chalcogenide glass electrodes the detection limit is about 10^{-7} M, the Nernstian range lies within $10^{-1}- 10^{-6}$ M $Pb(NO_3)_2$, the slope is equal to the theoretical value S_o for the sensors with membranes which have the optimum glass composition. The stability and reproducibility of vitreous sensors are in no way inferior to those of crystalline ISEs. Q value of the latter is 3.5-4.5.

The analytical response time t_{95} of chalcogenide glass electrodes is as short as that of crystalline ones. No dependence of response time on lead nitrate solutions concentration is observed, t_{95} being a few seconds in 10^{-2} - 10^{-5} M concentration range.

Graphic comparison of the selectivity coefficients of chalcogenide glass sensors and crystalline ISEs is shown in Fig.4. In general, chalcogenide glass electrode selectivity is about that of crystalline sensors and literature data /4,18,19/, however, the former are 5-20 times more selective to Cd^{2+} (Fig.5(a)). Such high selectivity for cadmium ions only is rather surprising. The glass stability and selectivity in $Fe(NO_3)_3$ solutions are greater as well (Fig.5(b)). Constant potential drift in positive direction is typical for crystalline sensors when $c_{Fe(NO_3)_3} > 5 \cdot 10^{-5}$ M. The potential of chalcogenide glass electrodes was steady or its drift was 1-2 order of magnitude less than that of crystalline ones. Lower sensitivity of chalcogenide glass electrodes for oxidation results in easily reached low detection limit (10^{-7} M) and in high long-term potential stability. Chalcogenide glass sensors do not need any surface treatment during approximately 3 months of action, while crystalline ISEs under the same conditions must be polished at least once in two weeks.

Oxidizing agents influence

It is convenient to simulate the ageing of ISE membrane by oxidizing agent treatment. The calibration curves of chalcogenide glass electrodes of the iodide system and crystalline sensor after their oxidation by 1 mM $KMnO_4$ are shown in Fig.6 and 7. Soaking of glasses in $KMnO_4$ causes some shift of the electrode response to the positive potential region, loss of sensitivity in low $Pb(NO_3)_2$ concentrations and, finally, complete degradation of the electrode response after a few hours of soaking. As to crystalline ISEs, a five-minute $KMnO_4$ treatment is enough to ruin their sensitivity totally.

The sulfide glasses are highly stable to oxidation as well. The dependence of the detection limit on the soaking time in the 3% hydrogen peroxide solution is shown in Fig.8. The glasses are preserving their ionic sensitivity 10-20 times longer and their detection limit is 10-30 times lower compared to crystalline sensors.

High stability of chalcogenide glass electrodes to oxidizers is stipulated by some peculiarities of oxidation of glasses. Pungor et al. /21/ were the first to show by XPS that crystalline ISE response deterioration occurs due to the presence of the oxidized sulfur, particularly sulfates, on the membrane surface. Earlier Thompson and Rechnitz /22/ found that lead-selective sensor membranes, electrode performance of which was unsatisfactory, contain $PbSO_4$. However, our XPS studies of the surface of the glasses showed that the influence of ferric ions, $KMnO_4$, H_2O_2 sulfur is present on the membrane surface in the sulfide form only. Fig.9 displays S 2p-photoelectron spectra of several glasses after oxidation. The chemical shift of unresolved spin-orbit doublets of sulfur corresponds to S^{2-} ions. The oxidizers essentially change the glass surface composition (Fig.10), but oxygen appearance on the membrane is due to oxidation of lead and arsenic.

Potential generating processes at the chalcogenide glass/ /solution interface

The most common point of view on silver-based solid-state ISE sensing mechanism is as follows. The principle process, which determines ionic sensitivity of these sensors is silver ion exchange between solution and membrane. The activity of silver ions near membrane surface depends on Pb^{2+} concentration in solution and solubility products of PbS and Ag_2S (see, e.g. /1/). Earlier we put forward an assumption that potential generating process is direct Pb^{2+} exchange between solution and modified surface layer of the glass /15/. This idea is confirmed by the evidence obtained in the present study.

The first is ionic sensitivity of single-crystal PbS electrode (Fig.3), its selectivity is close to that of the other lead-selective sensors investigated, containing silver in their membranes (Fig.4).

Secondly, Ag_2S-As_2S_3 binary glasses are sensitive to Pb^{2+} though they contain no lead in the original composition. Such situation resembles oxide glass electrode performance - their response to H^+ ions is a result of soaking in acid solutions and alkali cation exchange for hydrogen ions in the modified surface layer /24/.

Thirdly, we found by XPS and AES that a modified surface layer does exist after chalcogenide glasses come in contact with solutions (Fig. 11).

Mechanism study has been only initiated, so it is too early to discuss the details. However, some important results of radioactive tracer study of ion exchange /25/ should be pointed out. We have found using ^{82}Br that if the direct Br^- ion exchange between $AgBr$-Ag_2S-As_2S_3 glass membrane and solution takes place, the electrode is sensitive to bromine anions. If no exchange is observed, no ionic sensitivity can be found either.

ACKNOWLEDGEMENTS

We should like to thank Dr. Yu.P.Kostikov and V.S.Strykanov for the XPS measurements, Dr. V.Ivanov for the AES measurements, Dr. Yu.E.Ermolenko for PbS-Ag_2S electrode preparation.

REFERENCES

1. J.Koryta, Anal.Chim.Acta, 61, (1972) 329; 91, (1977) 1; 111, (1979) 1; 139, (1982) 1; 159, (1984) 1.
2. S.N.K.Chaudhari, F.C.Chang, K.L.Cheng et al., Anal.Chem., 53, (1981) 2048.
3. E.Pungor, K.Tóth, G.Nagy et al., Anal.Chim.Acta, 147, (1983) 23.
4. E.Pungor, K.Tóth and A.Hrabéczy-Páll, Pure Appl.Chem., 51, (1979) 1913.
5. C.T.Baker and I.Trachtenberg, J.Electrochem.Soc., 118, (1971) 571.
6. R.Jasinski, I.Trachtenberg and G.Rice, J.Electrochem.Soc., 121, (1974) 363.
7. A.E.Owen, J.Non-Cryst.Solids, 35&36 (1980) 999.
8. Yu.G.Vlasov and E.A.Bychkov, Hung.Sci.Instrum., 53, (1982) 35.

9. N.Tohge, T.Minami and M.Tanaka, Yogyo-Kyokai-Shi, 91, (1983) 32.
10. Yu.G.Vlasov, E.A.Bychkov, E.A.Kazakova et al., Zh.Anal. Khim., 39, (1984) 452.
11. Yu.G.Vlasov, E.A.Bychkov and A.M.Medvedev, Zh.Anal.Khim., in press.
12. Yu.G.Vlasov, E.A.Bychkov, A.D.Safarov et al., Zh.Anal. Khim., in press.
13. C.Bohnke, A.Saida and G.Robert, C.R.Acad.Sc.Paris, C290, (1980) 97.
14. C.Bohnke, J.P.Malugani, A.Saida et al., Electrochim.Acta, 26, (1981) 1137.
15. Yu.G.Vlasov, E.A.Bychkov and A.V.Legin, in Ionoselectivnye elektrody i ionnyi transport (Ion-Selective Electrodes and Ion Transport). Nauka, Leningrad, 1982, p.133.
16. Yu.G.Vlasov, E.A.Bychkov and A.V.Legin, Elektrokhimia, in press (part 2).
17. Yu.G.Vlasov, E.A.Bychkov and A.V.Legin, Zh.Anal.Khim., in press.
18. M.Semler and B.Manek, in Ion-Selective Electrodes,2. Ed. by E.Pungor and I.Buzás. Akadémiai Kiadó, Budapest, 1978, p.529.
19. Yu.G.Vlasov, Yu.E.Ermolenko and O.A.Iskhakova, Zh.Anal. Khim., 34, (1979) 1522.
20. Yu.G.Vlasov, E.A.Bychkov and A.V.Legin, Elektrokhimia, in press (part 1).
21. E.Pungor, K.Tóth, M.K.Pápay et al., Anal.Chim.Acta, 109, (1979) 279.
22. H.Thompson and G.A.Rechnitz, Chem.Instrum., 4, (1972) 239.
23. Yu.G.Vlasov, E.A.Bychkov and A.V.Legin, to be published.
24. Glass Electrodes for Hydrogen and Other Cations. Principles and Practice. Ed.by G.Eisenman. Marcel Dekker, New York, 1967.
25. Yu.G.Vlasov, E.A.Bychkov and D.V.Golikov, Elektrokhimia, in press.

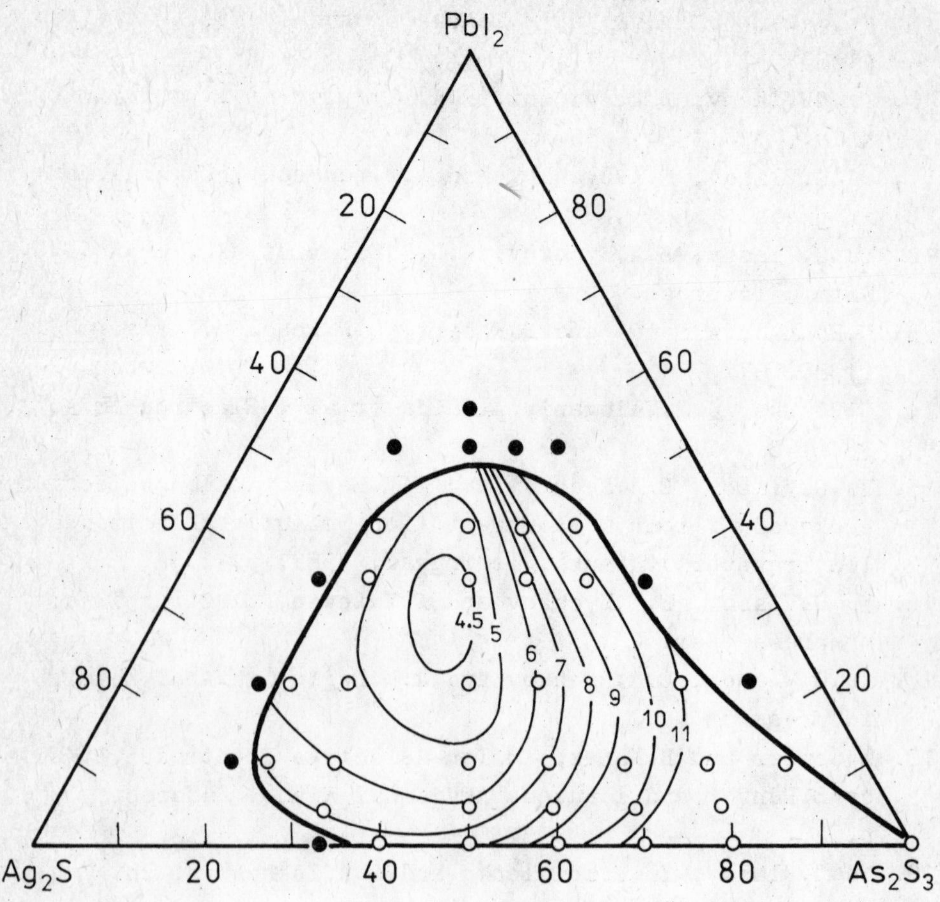

Fig.1 Glass-forming region in PbI_2-Ag_2S-As_2S_3 system. (o) glasses, (•) glassy/crystalline alloys. Iso-Q lines are drawn within the glass-forming region. Numbers correspond to Q values. The less is Q the better are the total electrode properties.

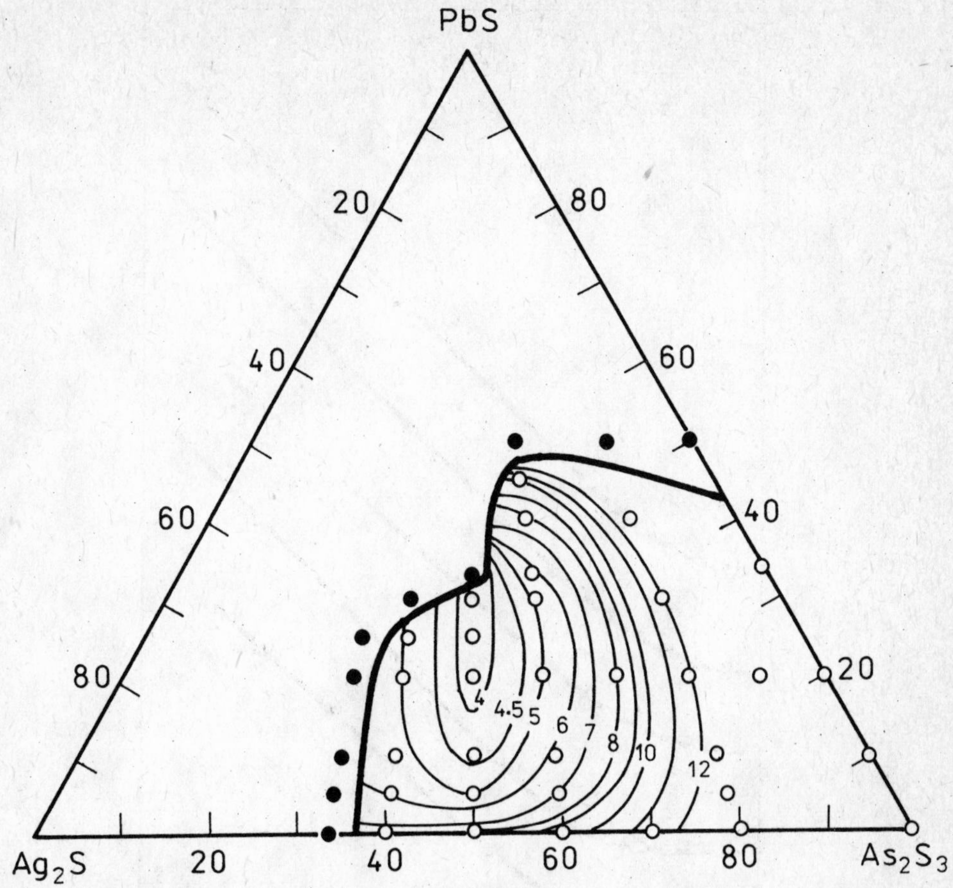

Fig.2 Glass-forming region in PbS-Ag_2S-As_2S_3 system. For details see Fig.1.

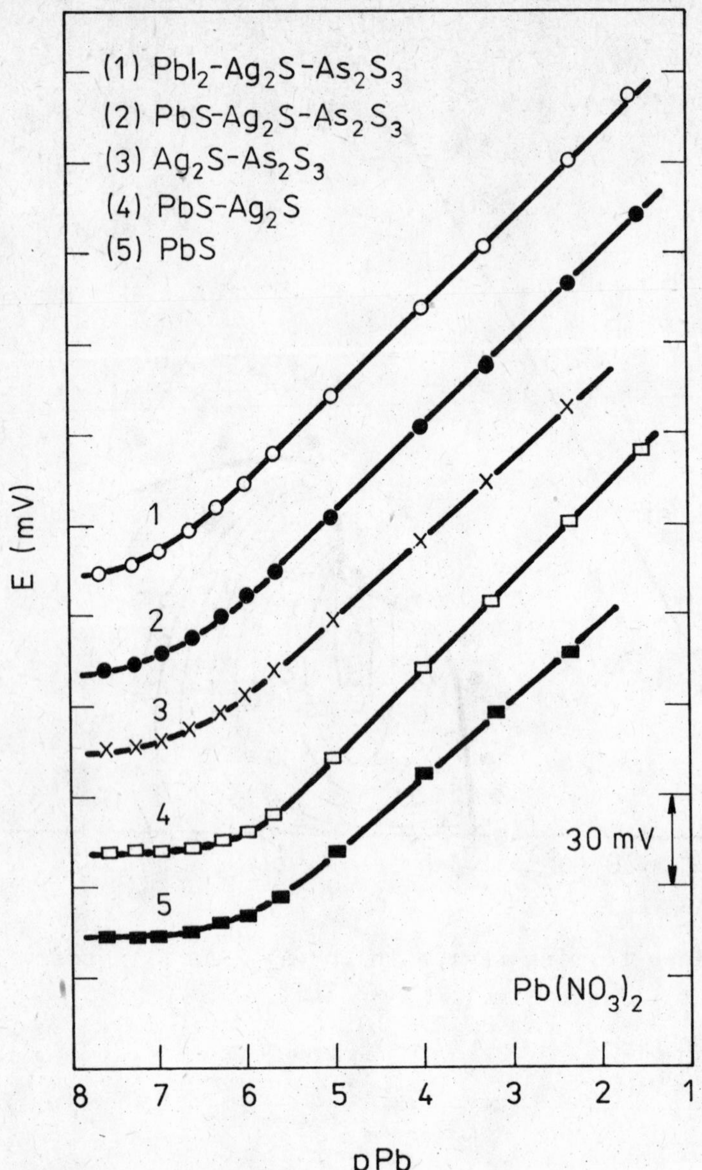

Fig.3 Typical response of lead ISEs in $Pb(NO_3)_2$ solutions. Membrane compositions: 1 - PbI_2-Ag_2S-As_2S_3 (glass), 2 - PbS-Ag_2S-As_2S_3 (glass), 3 - Ag_2S-As_2S_3 (glass), 4 - PbS-Ag_2S (polycrystalline pressed-pellet), 5 - PbS (single-crystal).

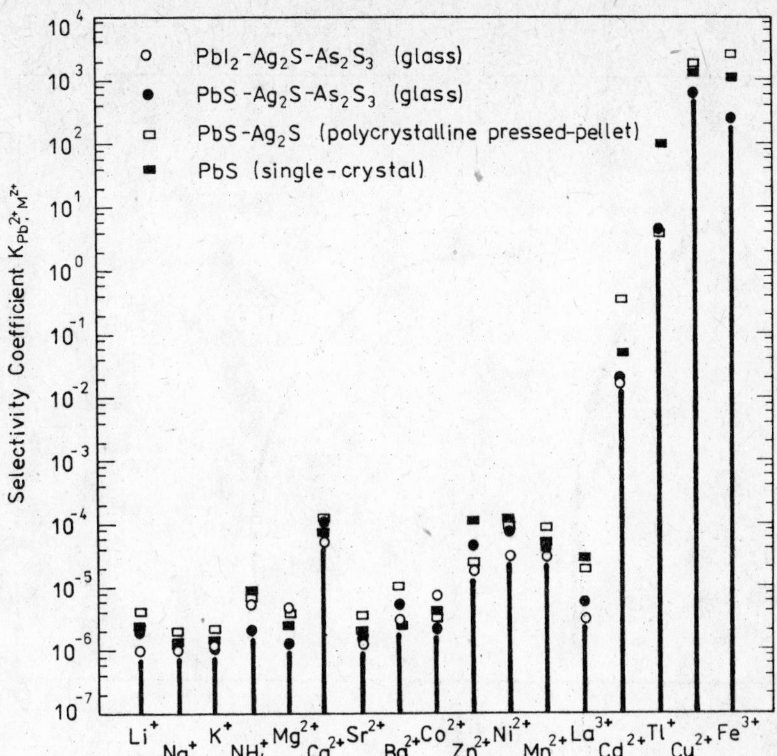

Fig.4 Selectivity coefficients of chalcogenide glass electrodes and crystalline sensors. Membrane compositions: (○) PbI_2-Ag_2S-As_2S_3, (●) PbS-Ag_2S-As_2S_3, (□) PbS-Ag_2S, (■) PbS.

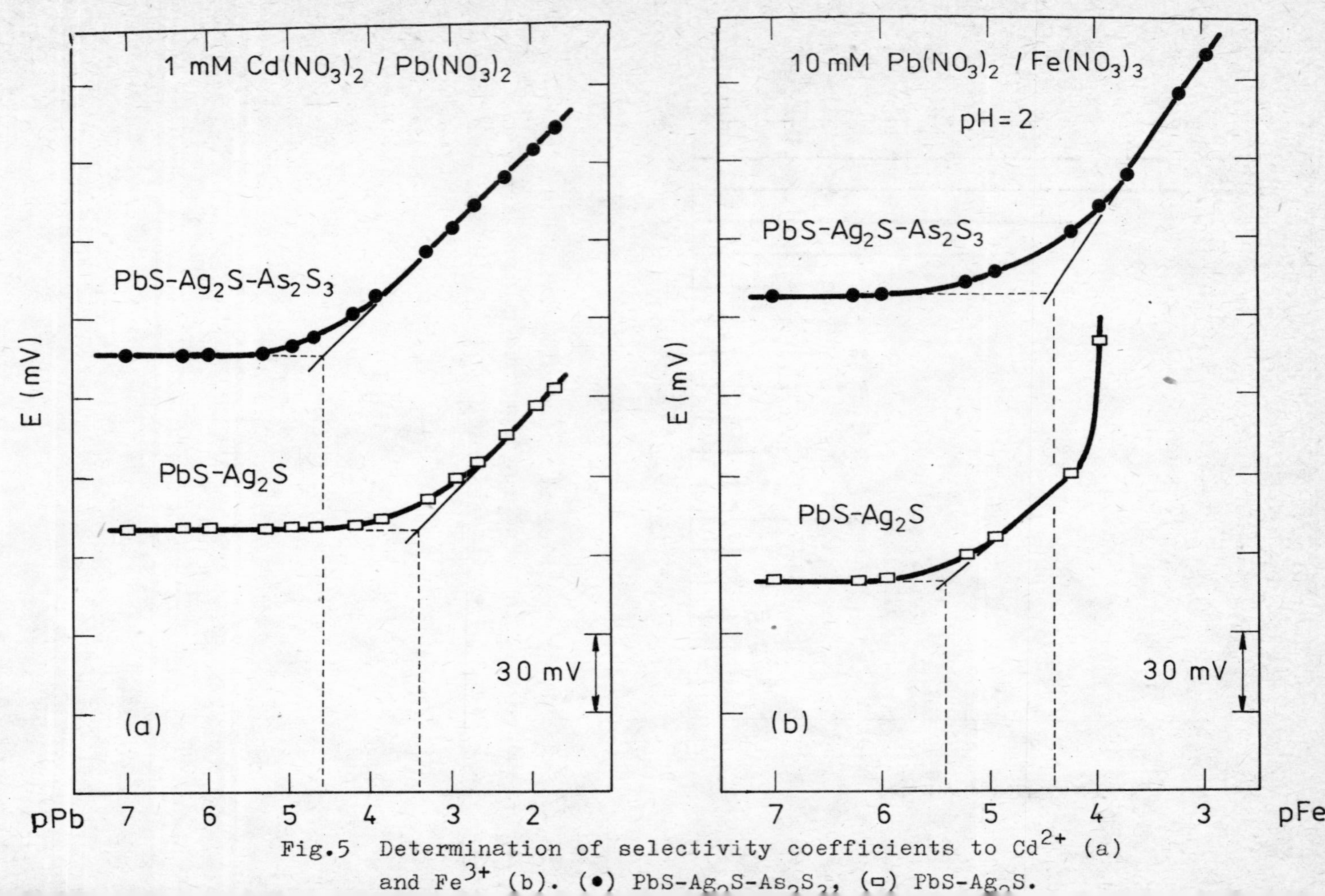

Fig.5 Determination of selectivity coefficients to Cd^{2+} (a) and Fe^{3+} (b). (●) PbS-Ag_2S-As_2S_3, (□) PbS-Ag_2S.

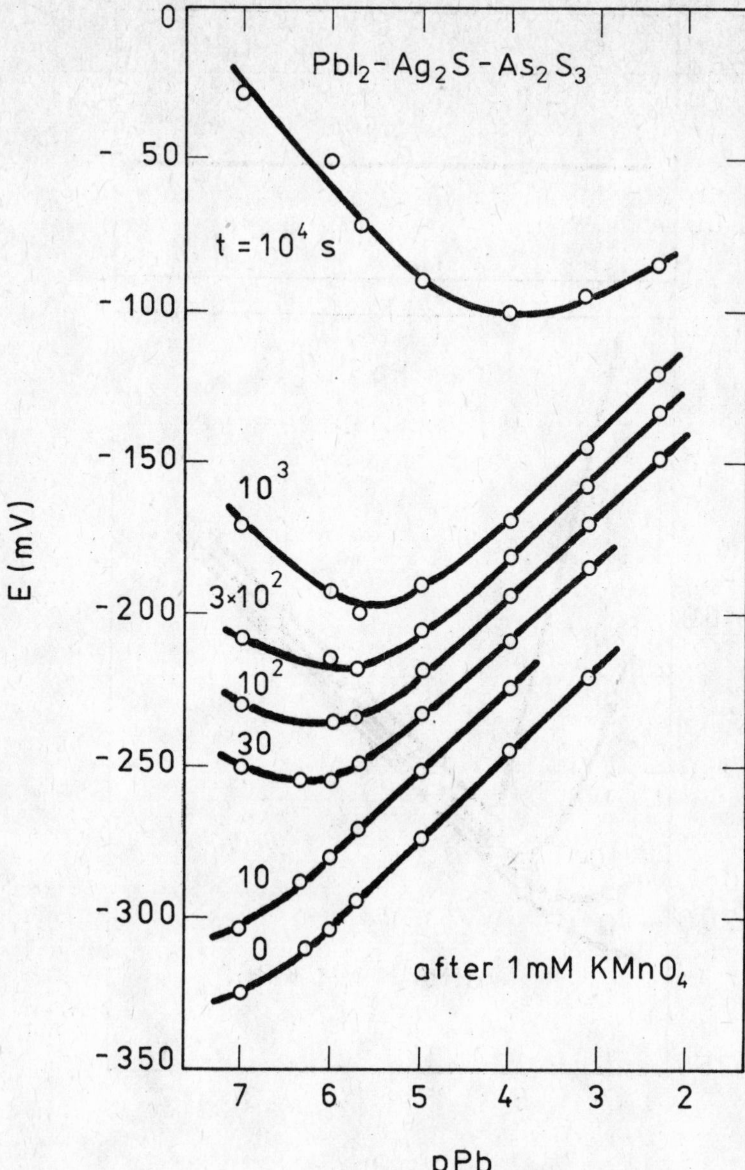

Fig.6 Response of $PbI_2-Ag_2S-As_2S_3$ chalcogenide glass electrode after oxidation in 1 mM $KMnO_4$. Numbers near calibration curves show treatment time in seconds.

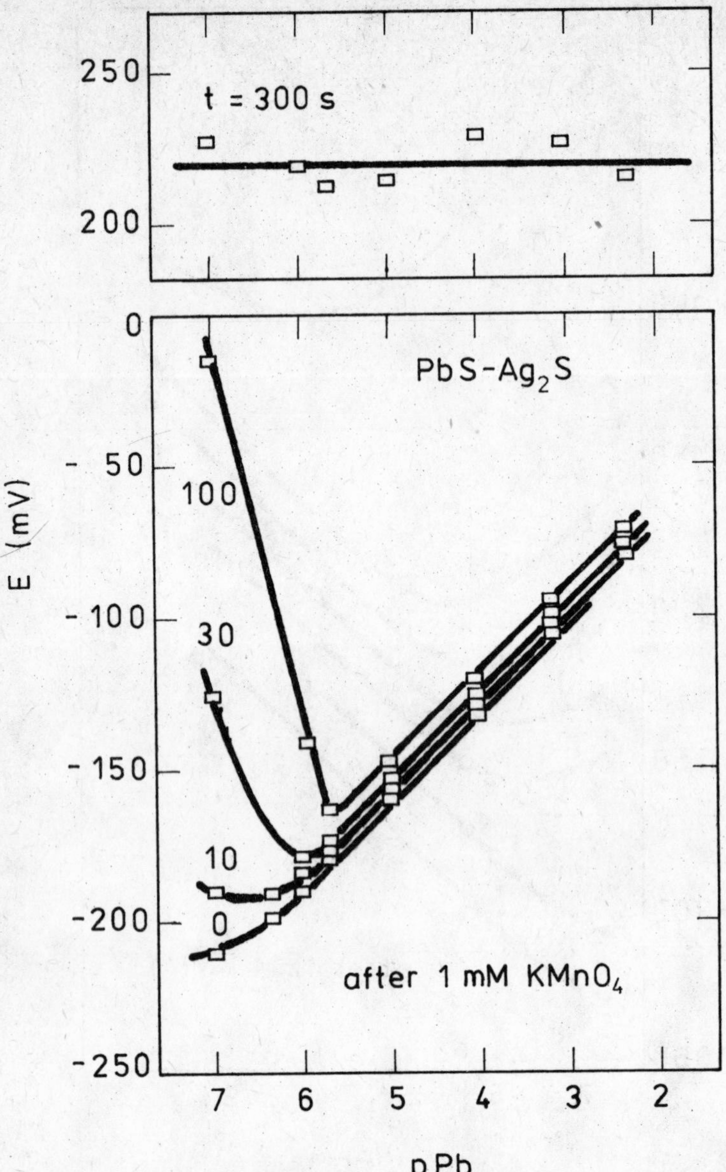

Fig.7 Response of PbS-Ag_2S electrode after oxidation in 1 mM $KMnO_4$. For details see Fig.6.

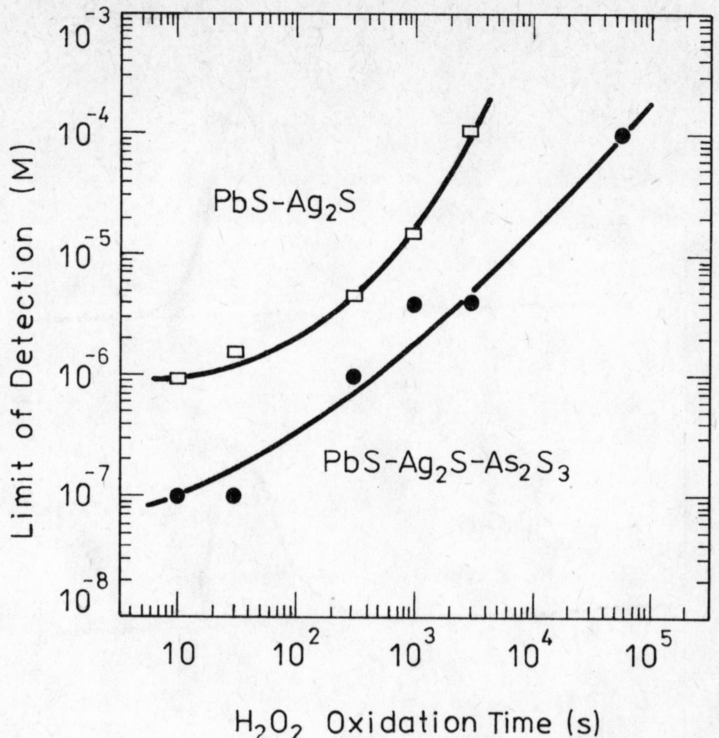

Fig.8 Detection limit of $PbS-Ag_2S$ electrode (□) and $PbS-Ag_2S-As_2S_3$ sensor (●) after oxidation in 3% H_2O_2.

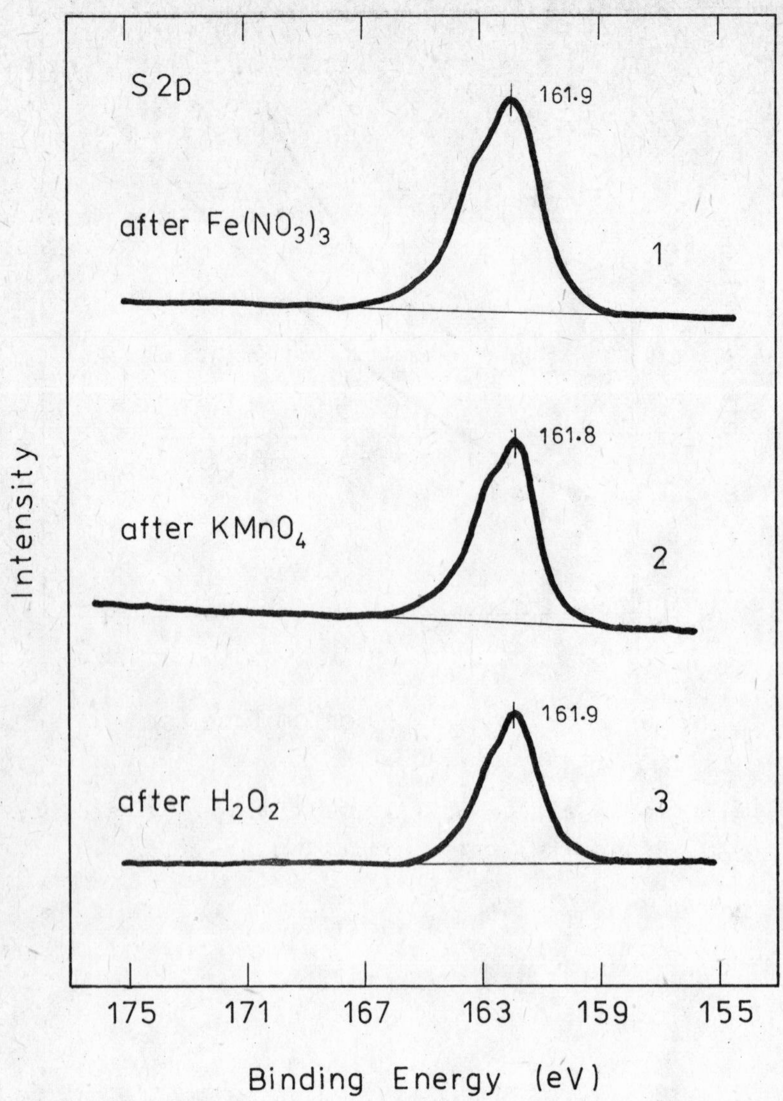

Fig.9 S 2p - photoelectron spectra of chalcogenide glasses after oxidation in 0.1 M $Fe(NO_3)_3$ (1), 1 mM $KMnO_4$ (2) and 3% H_2O_2 (3) solutions.

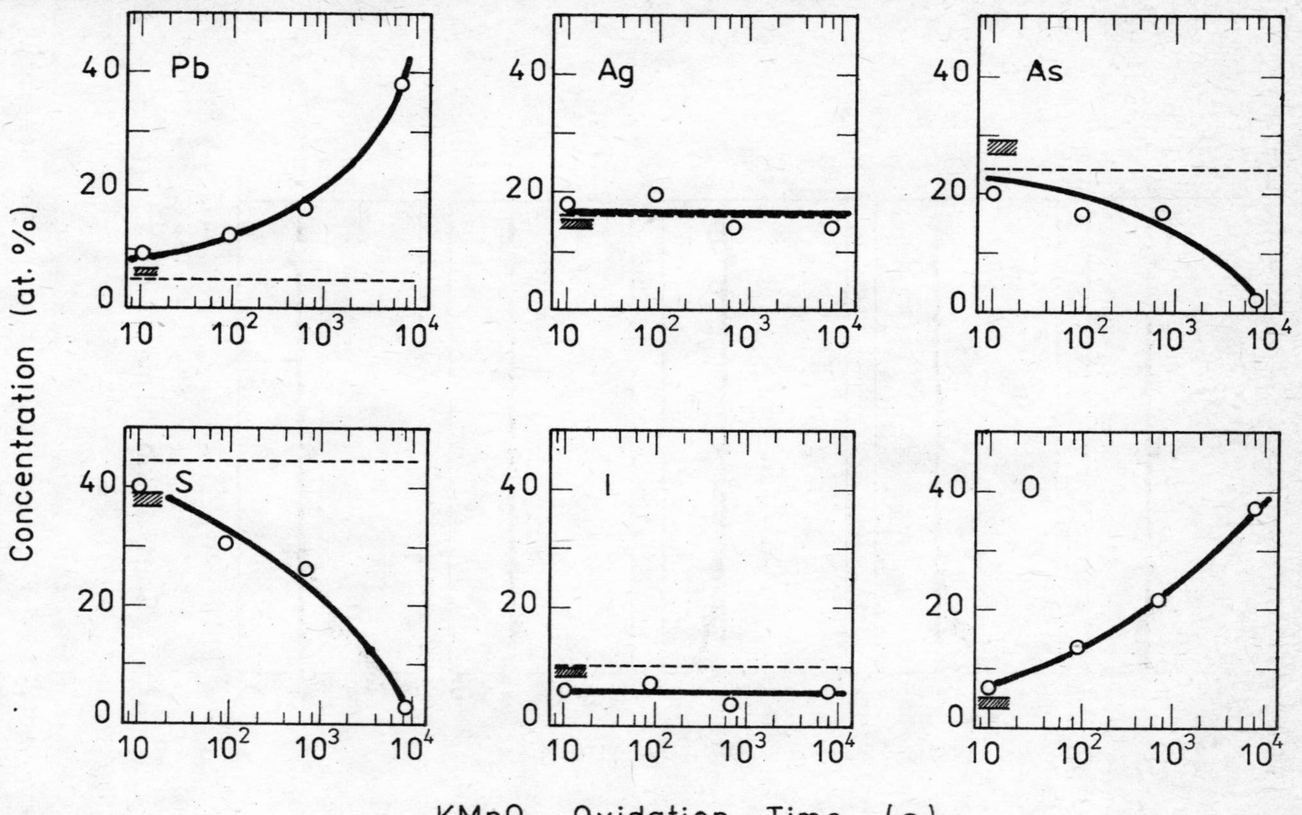

Fig.10 Surface composition of PbI_2-Ag_2S-As_2S_3 glass after 1 mM $KMnO_4$ treatment. (------) bulk concentration, (▨) original surface concentration.

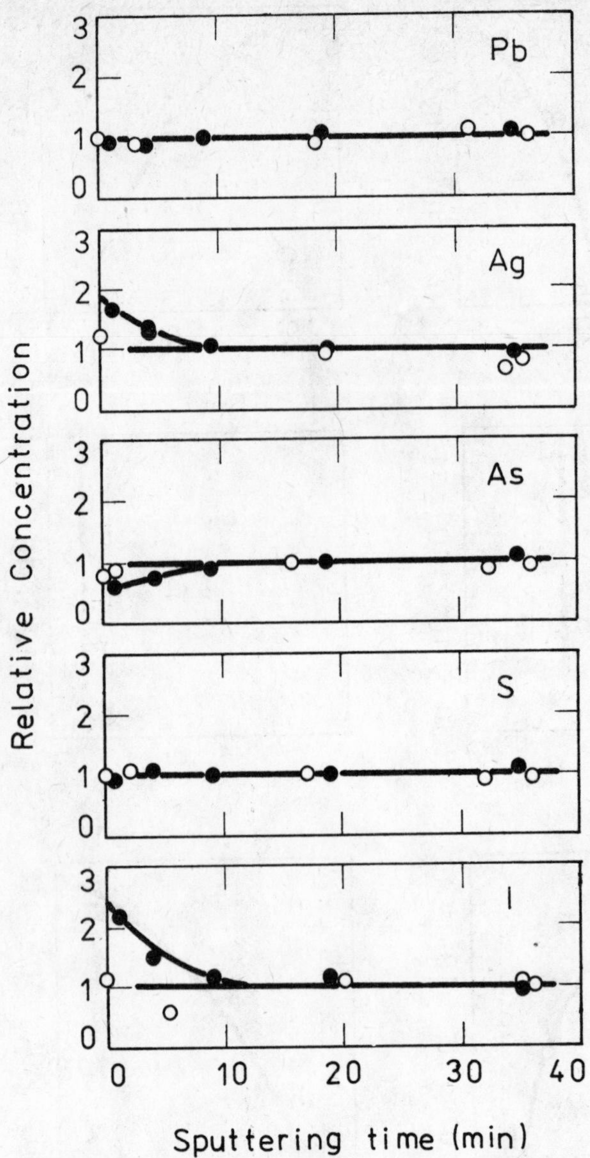

Fig.11 Depth profile of PbI_2-Ag_2S-As_2S_3 glass after soaking in 10 mM $Pb(NO_3)_2$. (O) original glass, (●) the glass after soaking for half an hour.

QUESTIONS AND COMMENTS

Participants of the discussion: R.P.Buck, M.Jovanović, E.Pungor, J.D.R.Thomas and Yu.G.Vlasov

Question:

You mentioned the effect of treatment with H_2O_2 on the detection limit of the electrode. What was the concentration of the H_2O_2 employed ?

Answer:

It was 3%

Question:

Why do you call your glasses chalcogenide glasses, although you have only S, and not the other members of the group ? Why do you not call it sulphide glass ?

Answer:

You are right, this is just a member of the big class of chalcogenide glasses.

Question:

I was very interested in the result that if you oxidize the surface, you still have sulphide. Does this mean that you are injecting holes into the electrode? Or is there something else oxidized ? It was suggested some years ago that chalcogenide glasses have holes, moving all over the surface.

Comment:

On investigating lead sulphide electrodes, we have detected lead sulphate on the surface after oxidation with H_2O_2.

Answer:

We know about the result you mentioned. We used spectroscopic methods to study the surface and only sulphide could be identified at the surface. We do not know the explanation yet, maybe there are holes in the membrane, but we only know the experimental fact.

4th Symposium on Ion-Selective Electrodes
Mátrafüred, 1984

APPLICATION OF Ag^+-SELECTIVE ELECTRODE FOR STUDY OF COMPLEX FORMATION OF SILVER IONS WITH GLYCINE, HISTIDINE AND DIPEPTIDES FORMED BY THESE AMINO ACIDS

Yu.G. VLASOV, V.V. PALCHEVSKY and V.I. SCHERBAKOVA

Department of Chemistry, Leningrad University
Leningrad 199164, USSR

ABSTRACT

Ag^+-selective electrode was applied for investigation of complex formation of silver ions with Glycine, Glycylglycine, L-Histidine, Glycyl-L-Histidine and L-Histidylglycine. Dissociation constants of aminoacids and dipeptides mentioned above were determined by potentiometric acidimetric titration. Using the method of particular derivatives the compositions of complexes existing in aqueous solutions were found. Stability constants of the complexes were calculated.

INTRODUCTION

The present paper is a part of our investigation of complex compounds formed by metal ions and fragments of protein /1,2/. The purpose of this study is to determine the influence of the nature of the central atom and ligands on composition and properties of complexes. Glycine (Gly), histidine (His), glycylglycine (Gly-Gly), glycylhistidine (Gly-His) and histidylglycine (His-Gly) were taken as ligands and silver - as a central atom. The investigation was carried out potentiometrically, using silver-selective electrodes in a wide range of pH, at the temperature 25°C and constant ionic strength I=1.0.

The complex formation of silver ions with histidine was investigated only at pH=11 /3/. There is no information concerning stability of complex compounds of silver with histidylglycine and glycylhistidine in literature. Protolytic properties of glycine NH_2-CH_2-COOH and glycylglycine $NH_2CH_2CONHCH_2$-COOH are due to the presence of acidic (carboxylic) and basic (amino) groups. In addition to these groups there is one more proton-donor group - an imidazole group - in the molecules of histidine, glycylhistidine and histidylglycine (Fig.1). Gly, His and dipeptides based on them exist as different ions, depending on solution pH. The composition of these ions changes with the increase of pH from $H_iL^{(i-1)+}$ to L^- where i=2 in the case of Gly and Gly-Gly and i=3 in the case of His, Gly-His and His-Gly. For example, the histidine molecule in acid solutions exists as the cation H_3L^{2+}. Hydrogen ions splitting out takes place with the increase of pH as follows:

$$\underset{H_3L^{2+}}{^{+}HN\text{-}C(=CH\text{-}NH\text{-}CH=)\text{-}CH_2\text{-}CH(NH_3^+)\text{-}COOH} \xrightleftharpoons{pK^{(a)}_{COOH}} \underset{H_2L^{+}}{^{+}HN\text{-}C\text{-}CH_2\text{-}CH(NH_3^+)\text{-}COO^-} \xrightleftharpoons{pK^{(a)}_{ImH^+}} \underset{HL^{\pm}}{N\text{-}C\text{-}CH_2\text{-}CH(NH_3^+)\text{-}COO^-} \rightleftharpoons$$

$$\xrightleftharpoons{pK^{(a)}_{NH_3^+}} \underset{L^-}{N\text{-}C\text{-}CH_2\text{-}CH(NH_2)\text{-}COO^-} \qquad\qquad HL^{\pm} \rightleftharpoons \underset{HL}{N\text{-}C\text{-}CH_2\text{-}CH(NH_2)\text{-}COOH}$$

Processes of protolytic dissociation of His-Gly and Gly-His may be shown by similar schemes.

To bring about necessary calculations one must find the values of dissociation constants $K_i^{(a)}$ for the aminoacids and the dipeptides under experimental conditions.

EXPERIMENTAL

A silver-selective all-solid-state electrode with polycrystalline Ag_2S membrane /4/ and a metallic Ag electrode, prepared by electrodeposition of metallic silver on a platinum wire were used as electrodes reversible to silver ions.

Calibration of a membrane electrode in $AgClO_4$ solutions of various concentrations (Ag^+ = 1.0 10^{-5}- 1.0 10^{-2} M) showed that the slope of the dependence of electrode potential on silver ion concentration was close to the theoretical one and equal to 58.1 mV/decade. Calibration of metallic Ag electrode in $AgClO_4$ solutions showed that its readings varied from the readings of ion-selective electrode by a constant value (2 - 3 mV). To determine pH a glass electrode calibrated in $HClO_4$ and HNO_3 solutions of known concentration was used. Ag/AgCl electrode was used as reference electrode. Potentiometric measurements were carried out in the following electrochemical cells:

Ag | Ag_2S | studied solution ¦ $NaNO_3$ sat. ¦ KCl sat. | AgCl | Ag (1)

Ag | studied solution ¦ $NaNO_3$ sat. ¦ KCl sat. | AgCl | Ag (2)

glass electrode | studied solution ¦ $NaNO_3$ sat. ¦ KCl sat. | AgCl | Ag (3)

To study the complex formation of silver with Gly and Gly-Gly $AgClO_4$ was used. $AgClO_4$ was prepared by dissolving Ag_2O in 1M solution of $HClO_4$ /5/. Chemically pure Gly, His and Gly-Gly, recrystallized from water, were used. His-Gly and Gly-His were synthesized in this investigation by the technique suggested in /6,7/.

The purity of prepared dipeptides was tested by high-voltage paper electroforesis and by thin-layer silica gel chromatography. Histidine and dipeptides containing histidine were L-isomers.

RESULTS AND DISCUSSION

Dissociation constants values

Values of dissociation constants were found for the case when concentration of a particular aminoacid or dipeptide in solution was 0.1 M, I=1.0. Determination of the dissociation constants and investigation of complex formation were carried out in solutions containing $NaClO_4$ (in the case of Gly and Gly-Gly) and $NaNO_3$ (in the case of His, Gly-His and His-Gly) as a supporting electrolyte. The potentiometric acidimetric titration in argon atmosphere was used for determination of dissociation constants /8/. To find out the values of $K^{(a)}_{COOH}$

for His, Gly-His and His-Gly, a solution containing 0.10 M of the substance and 0.10 M of HNO_3 was titrated by 1.0 M solution of HNO_3 so that the dilution during titration should be negligible. $K^{(a)}_{COOH}$ values for Gly and Gly-Gly as well as $K^{(a)}_{ImH^+}$ for His, Gly-His and His-Gly were determined by titration of 0.10 M solution of the substance by solutions of $HClO_4$ (Gly and Gly-Gly) and HNO_3 (His, Gly-His, His-Gly). $K^{(a)}_{NH_3^+}$ values for the aminoacids and the dipeptides were determined by titration 0.10 M solution of the substance by NaOH solution free from carbonates. Concentrations of HNO_3, $HClO_4$ and NaOH used for titration were also 1.0 M. Table 1 represents the values of dissociation constants which were found by this method.

Table 1. Dissociation constants at 25°C and I=1.0

Substance	$pK^{(a)}_{COOH}$	$pK^{(a)}_{ImH^+}$	$pK^{(a)}_{NH_3^+}$
Gly	2.37 ± 0.03	-	9.27 ± 0.04
L-His	1.74 ± 0.03	5.99 ± 0.01	8.92 ± 0.04
Gly-Gly	3.17 ± 0.03	-	8.12 ± 0.02
Gly-L-His	2.2 ± 0.1	6.6 ± 0.1	8.24 ± 0.04
L-His-Gly	2.61 ± 0.04	5.91 ± 0.04	7.60 ± 0.04

Composition and stability constants of complex compounds

Complex compounds with general composition

$$Ag_qL_xH_k(OH)_y^{(q-k-x-y)^+} \tag{1}$$

may be formed in aqueous solutions of the investigated aminoacids and dipeptides containing silver ions (L is anion of the aminoacid or dipeptide).

The total stability constant β_{qxky} of the complex will be as follows (ion charges are omitted):

$$\beta_{qxky} = \frac{[Ag_qL_xH_k(OH)_y]}{[Ag^+]^q[L^-]^x\,[H^+]^k[OH^-]^y} \tag{2}$$

In the present research compositions of silver complexes formed in solutions, i.e. the set of q, x, k, y in Eq.(1) were determined by the method of partial derivatives of the electrode potentials /9/. This method is based on the construction and analysis of plots of ion-selective electrode potentials vs. each concentration parameter, other parameters being constant. In general Ag^+-selective electrode potential may be expressed by the equation:

$$E = E_0 + \theta \lg C_{Ag^+}^{tot} - \theta \lg \left(1 + \sum_{q=0}^{Q} \sum_{x=0}^{X} \sum_{k=0}^{K} \sum_{y=0}^{Y} q \beta_{qxky} [Ag^+]^{q-1} [L^-]^{x} h^{k-y}\right) \qquad (3)$$

The partial dependences were obtained in the following way: two solutions were prepared (a solution being titrated and a titrant) that differed in concentration of one component only, concentrations of all the other components being kept constant. Thus, for example, in obtaining experimental dependences of the electrode potential E on $pC^{(a)}$ ($pC^{(a)}$ total concentration of aminoacid or dipeptide) in the case of histidine, the solution being titrated and the titrant differed only in histidine concentration (pH, total concentration of silver ions and ionic strength of both solutions were the same). According to the method of partial derivatives, the value q in Eq.1 of complex predominating in solution may be determined by the slope of the linear parts of the dependence of an electrode potential E on pC_{Ag^+}:

$$\left(\frac{\partial E}{\partial pC_{Ag^+}}\right)_{pH, pC^{(a)}} = -\frac{\theta}{q} \qquad (4)$$

where $\theta = 2.3RT/F$.

If the number q is known, the number x of anions L^- coordinated in the complexes is determined by the analysis of the dependence of membrane electrode potential on the total concentration of aminoacid or dipeptide in solution:

$$\left(\frac{\partial E}{\partial pC^{(a)}}\right)_{pH, pAg^+} = \theta \frac{x}{q} \qquad (5)$$

Interpretation of the partial derivative E - pH obtained at constant pC_{Ag^+} and $pC^{(a)}$ is difficult, because the slopes of linear parts of this dependence are affected by the process of dissociation of aminoacid (dipeptide). Therefore the experimental dependence E - pH in most cases cannot be used for direct determination (i.e. the determination by the slope of linear parts of this dependence) of the number of protons k or hydroxyl groups y. The difference (k-y) can be found by the successive approximation method for every point of the experimental dependence E - pH, using appropriate equations /9/.

Determination of the compositions of the complexes predominating in solution will be shown for silver complexes with Gly-Gly.

Experimental dependences of Ag^+-selective electrode potential on pC_{Ag^+} (Fig.2) are straight lines with the slope θ. According to the method of partial derivatives /Eq.4/, it means that only mononuclear complexes of silver (q=1) are formed in 0.10 M solutions of Gly-Gly in the concentration range of silver 10^{-4}- 10^{-2} M and pH=2-10. The number of Gly-Gly anions x in complex is determined from the dependence of membrane electrode potential on $pC^{(a)}$. Using Eq.5 for the analysis of experimental dependences E - $pC^{(a)}$ (Fig.3) it can be shown, that complexes of silver and Gly-Gly with the ratio metal/ligand 1:1 and 1:2 are formed (some curves have linear parts with the slope θ or 2θ, respectively). Silver complexes with Gly-Gly may contain $HL^{\pm}$ and L^- particles as ligands. Determination of the type of coordination particles ($HL^{\pm}$ or L^-) was carried out using experimental dependence E - pH (Fig.4) and the following expression which was deduced from equation (3)

$$\left(\frac{\partial E}{\partial pH}\right)_{pC_{Ag^+},\, pC^{(a)}} = \qquad (6)$$

$$= \theta\left(\frac{k-y}{q} - \frac{x}{q}\,\frac{2h^2 + K^{(a)}_{COOH}\cdot h}{h^2 + K^{(a)}_{COOH}\cdot h + K^{(a)}_{COOH}\cdot K^{(a)}_{NH_3^+}}\right)$$

where h - concentration of hydrogen ions.

The types of coordination particles were determined as follows. It is known from the experimental data that e.g. at pH=7 the composition of silver complex with Gly-Gly is 1:2 (q=1, x=2). Taking into account that graphically found value of the derivative $(\partial E/\partial pH)$(constant $pC_{Ag^+}, pC^{(a)}$) at pH=7 is equal to -2θ and substituting the values $K^{(a)}_{COOH}=6.76\ 10^{-4}$; $K^{(a)}_{NH_3^+} = 6.166\ 10^{-9}$; $h=1.0\ 10^{-7}$, q=1 and x=2 into Eq.(6), we have:

$$\left(\frac{\partial E}{\partial pH}\right)_{pC_{Ag^+},pC^{(a)}} = \theta\left(k-y - \frac{2.0\cdot 10^{-14}+6.76\cdot 10^{-4}\cdot 1.0\cdot 10^{-7}}{1.0\cdot 10^{-14}+6.76\cdot 10^{-4}1.0\cdot 10^{-7}+ +6.76\ 10^{-4}6.166\cdot 10^{-9}}\right) = \theta(k-y - 2\ 0.94) = -2\theta \quad (7)$$

As it is unlikely that H^+ and OH^- particles should exist in Gly-Gly complex with silver simultaneously, one can conclude from Eq.(7) that k=y=0 at pH=7 and that, consequently deprotonated complex of silver with Gly-Gly of the composition AgL_2^- exists in solution.

Thus, the analysis of the experimental data obtained and application of Eq.(6) enable one to draw a conclusion that complex compounds of the composition AgL and AgL_2^- are formed in solutions containing 0.10 M of Gly-Gly and $1.0\cdot 10^{-3}$ M $AgClO_4$ in pH range from 2 to 10. In this case the equation for silver electrode potential will be:

$$E = E_0 + \theta \lg C_{Ag^+}^{tot} - \theta \lg(\beta_{1100}\cdot[L^-] + \beta_{1200}\cdot[L^-]^2 + 1) \quad (8)$$

where E_0 - standard potential, $C_{Ag^+}^{tot}$ is total concentration of silver ions in solution. Experimental dependence E - pH and Eq.(8) were used to calculate total stability constants of Gly-Gly complexes with silver. Stability constant β_{1100} of the complex AgA was the first to be determined. According to Eq.(8) in the region where the first complex predominantly exists, we have:

$$\beta_{1100} = \frac{10^{\frac{E_0-\theta \lg C_{Ag^+}^{tot} - E_i}{\theta}} - 1}{[L^-]} \quad (9)$$

Stability constant values β_{1200} for the complex AgA_2^- were calculated by the equation:

$$\beta_{1200} = \frac{10^{\frac{E_0-\theta \lg C_{Ag^+}^{tot} - E_i}{\theta}} - 1 - \beta_{1100}\cdot[L^-]}{[L^-]^2} \quad (10)$$

Average values of stability constants for Gly-Gly complexes with silver calculated by Eq.(9) and (10) for each point of experimental dependence E - pH are given in Table 2. Compositions and stability constants of silver complexes with other ligands are shown in Table 2 also, these data being found in the same way. Distribution diagram clearly shows the state of silver ions in 0.10 M solution of Gly-Gly (Fig.5). It can be seen from the Table 2 that in Gly and Gly-Gly solutions silver ions form deprotonated complexes only. On the other hand, in solutions of histidine and dipeptides protonated complex compounds exist along with deprotonated ones.

REFERENCES

1. Yu.G.Vlasov, V.I.Scherbakova and I.V.Mikhailova, Zh.Prikl. Khim., 55, (1982) 304.
2. Yu.G.Vlasov, V.V.Palchevsky and V.I.Scherbakova, Zh.Prikl. Khim., 56, (1983) 559.
3. I.Vallasdas-Dubois, Compt.rend., 237, (1953) 1408.
4. Yu.G.Vlasov, Yu.E.Ermolenko and O.A.Iskhakova, Zh.Anal. Khim., 34, (1979) 1522.
5. Handbuch der Präparativen Anorganischen Chemie. Herausgegeben von G.Brauer, Ferdinand Enke Verlag, Stuttgart, 1954.
6. N.C.Davis, J.Biol.Chem., 223, (1956) 935.
7. A.Yokoyama, H.Aiba and H.Tanaka, Bull.Chem.Soc. of Japan, 47, (1974) 112.
8. A.Albert, and E.Sergeant, Ionization Constants of Acids and Bases. Barnes and Noble, New York, 1962.
9. Redoximetry (in Russian). Ed.by B.P.Nikolsky, V.V.Palchevsky. Khimia, Leningrad, 1975.

Table 2. Stability Constants β_{qxky} for Aminoacids and Dipeptides Complexes with Silver

Stability Constants β_{qxky}	lg β_{qxky}				
	Gly	Gly-Gly	His	Gly-His	His-Gly
$\beta_{1110} = \frac{[Ag(HL^{\pm})]}{[Ag^{+}][HL^{\pm}]}$	–	–	3.5 ± 0.1 pH 3.3÷4.2	3.8 ± 0.1 pH 3.6÷4.5	3.22±0.04 pH 3.1÷3.8
$\beta_{1100} = \frac{[AgL]}{[Ag^{+}][L^{-}]}$	3.3 ± 0.1 pH 6.0÷7.5	3.1 ± 0.1 pH 5.5÷6.8	–	–	–
$\beta_{1220} = \frac{[Ag(HL^{\pm})_2]}{[Ag^{+}][HL^{\pm}]^2}$	–	–	6.77± 0.03 pH 4.8÷6.8	7.07 ± 0.03 pH 5.1÷6.6	6.02±0.03 pH 4.6÷5.5
$\beta_{1210} = \frac{[Ag(HL^{\pm})L]}{[Ag^{+}][HL^{\pm}][L^{-}]}$	–	–	8.1 ± 0.1 pH 6.8÷8.5	8.0 ± 0.1 pH 7.6÷8.5	7.52± 0.04 pH 6.4÷7.3
$\beta_{1200} = \frac{[AgL_2]}{[Ag^{+}][L^{-}]^2}$	6.9 ± 0.1 pH 7.8÷10	5.8 ± 0.1 pH 7.2÷9.5	8.41±0.04 pH 8.9÷9.9	7.7 ± 0.2 pH 8.6÷11	7.7± 0.1 pH 7.8÷11

Gly NH_2-CH_2-COOH

Gly-Gly $NH_2-CH_2-CO\ NH-CH_2-COOH$

His

```
              H
              |
N - C - CH2 - C - COOH
‖   ‖         |
HC  CH        NH2
  \ /
   N
   |
   H
```

Gly-His

```
                       H
                       |
NH2-CH2-CONH -  C - CH2 - C - N
                |         ‖   ‖
               COOH       HC  CH
                            \ /
                             N
                             |
                             H
```

His-Gly

```
              H
              |
N - C - CH2 - C - CONH-CH2-COOH
‖   ‖         |
HC  CH        NH2
  \ /
   N
   |
   H
```

Fig.1 Substances under investigation.

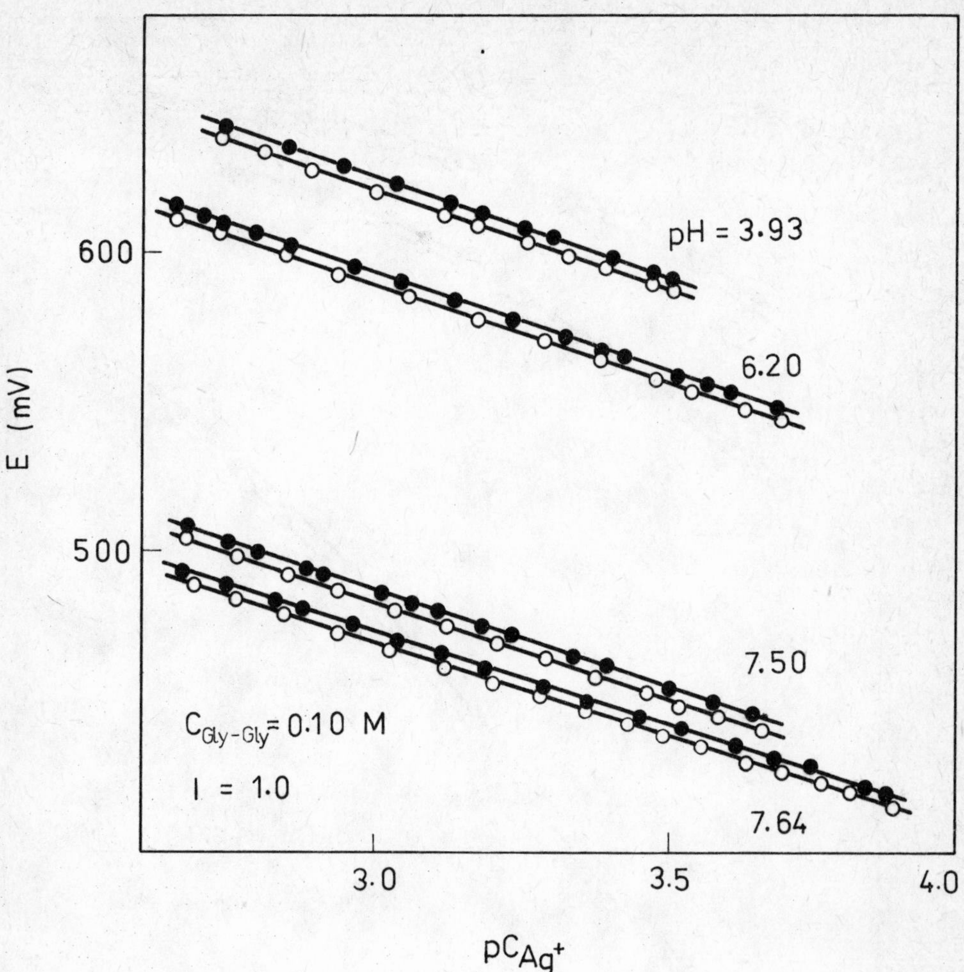

Fig.2 Dependence of silver-selective membrane (•) and silver electrode (o) potentials on the total concentration of silver ions. I=1.0; $C_{Gly-Gly}$=0.10 M; pH: 1 - 3.93, 2 - 6.20, 3 - 7.50, 4 - 7.64.

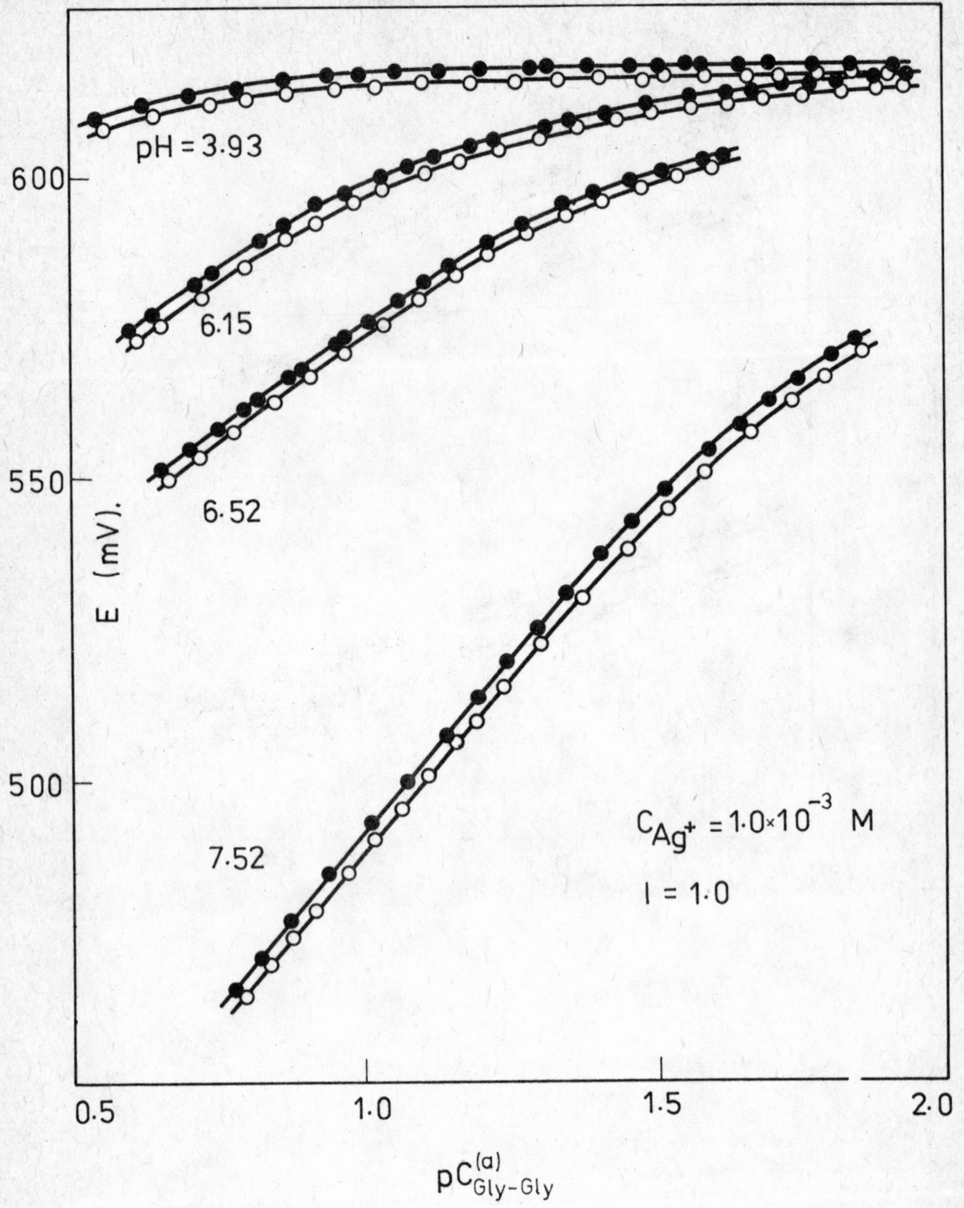

Fig.3 Dependence of silver-selective membrane (•) and silver electrode (o) potentials on the total concentration of Gly-Gly. I=1.0; $C^{tot}_{Ag^+}$=1.0 10^{-3} M; pH: 1 - 3.93, 2 - 6.15, 3 - 6.52, 4 - 7.52.

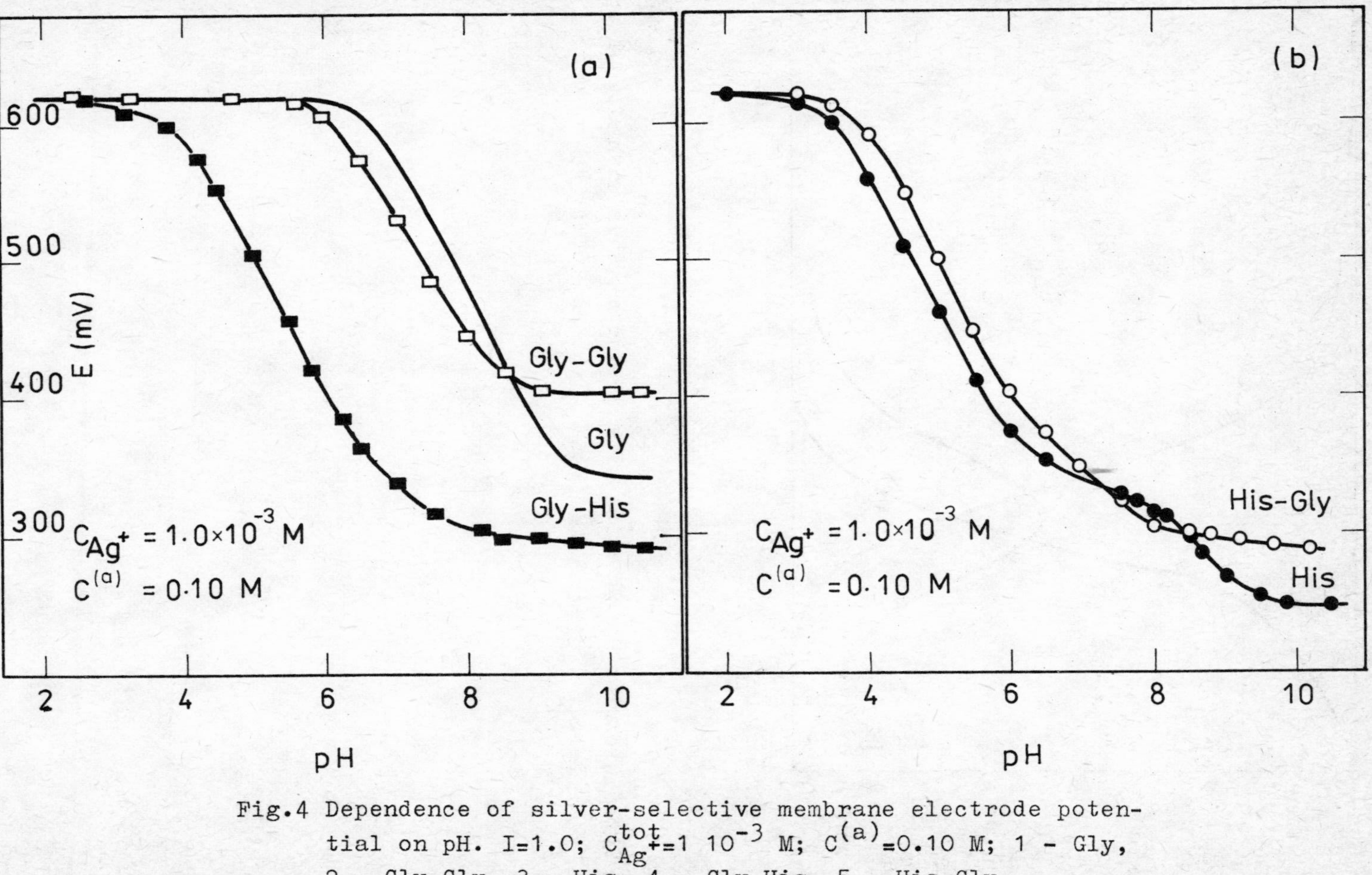

Fig.4 Dependence of silver-selective membrane electrode potential on pH. I=1.0; $C^{tot}_{Ag^+}=1\ 10^{-3}$ M; $C^{(a)}=0.10$ M; 1 - Gly, 2 - Gly-Gly, 3 - His, 4 - Gly-His, 5 - His-Gly.

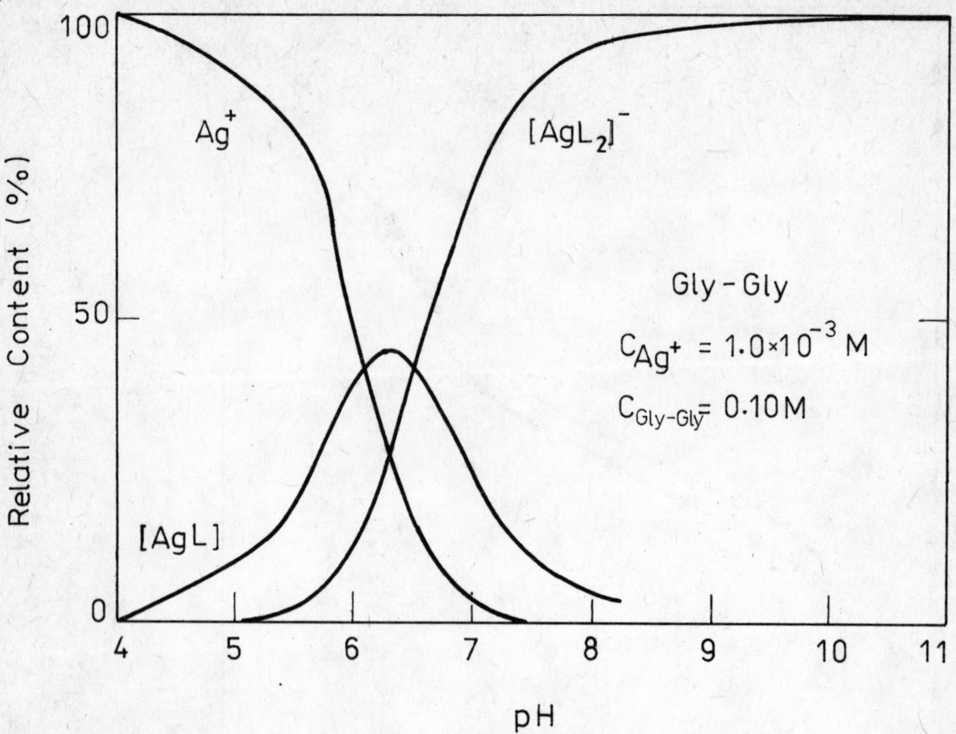

Fig.5 Distribution diagram of silver complexes with Gly-Gly. I=1.0; $C^{tot}_{Ag^+}$=1.0 10^{-3} M; $C_{Gly\text{-}Gly}$=0.10 M;

QUESTIONS AND COMMENTS

Participants of the discussion: J.S.Redinha and Yu.G.Vlasov

Comment:

I was glad to see that the Ag_2S electrode can yield reliable results in studies on complexes. We have done similar studies, but with a different technique: we preferred to take the basic form and add the silver to the solution. You used the ligand in the protonated form, added the silver and studied the pH range 2-10.

Question:

You are using two electrodes: a pH and a silver electrode. In what region did you get better results using the one or the other electrode ?

Answer:

We also used a third electrode, silver, for comparison. The electrodes were found to work well in a wide range of concentration without any problem, and the method gave us reproducible results.

Question:

What is the concentration of glycine in your studies ?

Answer:

0.1 mol/l

4th Symposium on Ion-Selective Electrodes
Mátrafüred, 1984

RECENT PROGRESS OF ION-SELECTIVE ELECTRODES IN CHINA

ZHANG ZONG-RANG

Department of Chemistry
Shanghai Teachers' College
Shanghai, China

ABSTRACT

There are many works in Ion-Selective Electrodes in China now, both in the development of new electrodes and applications to a wide variety of practical and theoretical problems. The author has collected more than 500 papers which have appeared in special Chinese journals and proceedings in recent three years, and attempts give a brief review on this field in China.

INTRODUCTION

In recent years, there have been some interesting developments in this field in China. It is very difficult to give a complete survey on it. We try to show the trends between 1981 to 1984.

We have had several conferences on this topic, the total number of papers exceeded 800. The number of papers published in Chinese journals /1981-1983/ was about 400. The special journal, "Comunications of Ion-Selective Electrodes", was founded in 1981, now it is a quarterly journal.

Chinese factories have produced more than 20 types of commercial electrodes. Some of these have very good performance such as the fluoride electrode and pH electrode.

Several reviews have been published about the trends of ISE research in China, the titles of papers published in Chinese from 1972-1983 were listed in the above reviews.

SOME INTERESTING NEW TRENDS

Design of new electrodes

The design of new electrodes was the subject of about 100 papers. Two types were long-chain tetraalkyl-ammonium and basic--dye based electrodes. Some organic chemists have synthesized many kinds of long-chain tetraalkyl-ammonium as the electroactive materials, they were sensitive to NO_3^-, I^-, ClO_4^-, BF_4^- , TaF_6^- etc. Some chemists prefer to use basic-dyes as the electroactive material , such as green blue, methyl violet, basic red, crystal violet, etc. sensitive to SCN^-, $SbCl_6^-$, $TlCl_4^-$, $AuCl_4^-$, ReO_4^-. Based on their experience, some specialists have proposed some empirical or semi-empirical rules of design and mechanism of response, and some quantum chemical computation or thermodynamic derivations.

Development of drug or organic substance electrodes was another exciting trend in this field. In this year's two meetings, there were about 50 papers on this topic, as presented in Table 1 below.

A special kind of electrode was the so called rare-earth electrode, it can be used for the determination of rare-earth elements, which occur in China.

Work has also been done in China on the development of microelectrodes and ISFETs. The size of microelectrodes was small for in-vivo detection, mainly for Ca^{2+}, K^+, Na^+ and pH determination. Chemists, cooperating with experts of electronics developed ISFETs, not only sensitive to one ion, but also multi-channel systems combined with reference electrode.

Development of application technique and expansion of application fields

ISE application in organic and biological systems is one of the most exciting fields for Chinese specialists now, not only to the detection of inorganic species in organic substances and biological systems. Efforts have been made to study bio-active materials, such as amino acids, choline, vitamins, antibiotics, and to monitor the metabolism and heart action of animals. Some reviews have been published on this topic in Chinese journals.

Dynamic techniques using ion-selective electrodes have been developed to extend the application areas,e.g. for the determination of Mn^{2+}, I^-, F^-, Hg^{2+}, Se^{2+}.

Some papers focused on the development of data processing methods, including both mathematical model design and computation. Most of them were devoted to multi-addition methods.

Some electrodes have been used to check aqueous standards, such as those for F^-, NO_3^-, CN^-, NO_2^-, Cl^-, NH_3, Cd^{2+}, Ag^+, etc. and have yielded good correlation with the results of other methods.

Theoretical works on ISEs

In recent years, the theoretical works on ISEs have become an important area in China. Chinese specialists have applied some new techniques as follows.

1. Electrochemical methods, such as voltammetry, AC impedance, wide range polarization for studying the transport behaviour of charge carriers.
2. Surface analysis techniques, such as X-ray diffration ESCA, etc. have been used for silver and lead electrodes.
3. Radiotracer techniques. This method has given some information about the transference phenomena of Ba^{2+} and TaF_6^- electrodes.
4. Studies on the phase boundary between two immiscible liquids using polarization method, such as cyclic voltammetry, chronopotentiometry, current scan voltammetry, etc.
5. Modified electrodes.
6. Determination of thermodynamic parameters.

Design and production of computer combined instruments

Except for the production of common type pH meter and ion-analyzer, we have designed three types of ionanalyzer and titrator combined with computer. It has high accuracy and resolution, and is easy to control.

SUMMARY

Chinese experts and analysts have made successful efforts to cover a wide area, both development and research. Some foreign specialists visited China, they have got deep impression on it. We have exchanged many informations in both directions.

ACKNOWLEDGEMENT

This work was partly supported by the colleagues in my laboratory and other institutes.

REFERENCES

1. Wang He-Ji, Anal.Chem. /China/ 7, /1979/ 343
2. Zhang Zong-Rang, J.Shanghai Teachers' College /Science Edition/ /1/, 125 /1981/
3. Zhang Zong-Rang, Communication of ISEs, 2, /4/ 1 /1982/
4. Zhang Zong-Rang, Plenary Lecture on ISEs Symposium, September, 1984, /in Press/
5. Proceedings of following conferences held in China:
   a/ Symposia of Electrochemistry
      1. August, 1981, Shanghai
      2. September, 1984, Xiamen
   b/ Symposia of Electroanalytical Chemistry
      1. October, 1981, Kunmin
      2. July, 1984, Changchun

c/ Symposia of Ion-Selective Electrodes

1. January, 1977, Fuzhou
2. December, 1979, Taixien
3. September, 1984 Yangzhou

d/ Symposia of Electroanalytical Instruments

1. December, 1980, Suzhou
2. July, 1982, Xiamen

Table 1. Number of published papers in Chinese Journals /1981-1983/

"Analytical Chemistry"	58
"Communication of ISEs"	251
"Physico-Chemical Testing"	21
Others	ca 70
Total	400

Table 2. Kinds of commercial electrodes

F^-, Cl^-, Br^-, I^-, NO_3^-, CN^-, ClO_4^-,
Ag^+, Na^+, K^+, Ca^+, Cu^{2+}, Pb^{2+}, Hg^{2+}, Cd^{2+},
Hardness, BF_4^-, TaF_6^-,
NH_3, CO_2,
Surfactant, Saccharin

Table 3, Electrodes for drugs and organic compounds

Atropine	Diphenhydramine
Procaine	Adrenalinum
Chlorpromatine	Promethazine
Salicylic acid	Bendroflumethiazidum
Choline	Amino acid
Picrate	Tri-chloro acetic acid
Dinitro phenol	

Table 4. Ionic association type electrodes

1. Tetraalkyl ammonium and tetraalkyl phosphonium

   NO_3^-, I^-, ClO_4^-, BF_4^-, TaF_6^-,

2. Basic dye system

   SCN^-, $SbCl_6^-$, $TlCl_4^-$, $AuCl_4^-$, ReO_4^-, BF_4^-,

   $Zn/SCN/_4^{2-}$, $In/SCN/_4^{2-}$, $InBr_3^-$,

3. $UO_2/C_6H_5COO/_3^-$, $GaCl_4^-$, IO_4^- , ClO_4^- , HgI_3^-

QUESTION

Participants of the discussion: J.D.R.Thomas and Zhang Zong--Rang

Question:

In studying the effect of Tiron on various metals, in one case you had two orders of magnitude change in the metal concentration, but the change in emf was fairly small, whereas I would have expected a bigger emf change.

Answer:

The figure was on the effect of copper. If we use 10^{-1} mol/l copper, its effect is very big whereas at 10^{-3} mol/l the emf value is very close to the original one.

4th Symposium on Ion-Selective Electrodes
Mátrafüred, 1984

ENVIRONMENTAL APPLICATIONS OF CADMIUM ELECTRODES

ZHANG ZONG-RANG, WU XA-QING and LU HONG-YU

Department of Chemistry
Shanghai Teachers' College
Shanghai, China

ABSTRACT

Cadmium electrodes have been applied in environmental control, e.g. in the analysis of waste water from metal plating and TV tube plants. This special method has provided precise results and rapid procedures.

The authors have found an effective TISAB for masking and buffering. The effective TISAB consisted of suitable proportions of tiron, formalin, zinc ions and pH buffer. The lower detection limit of cadmium ions was as low as 0.05 ppm. The standard deviation and recovery was good.

INTRODUCTION

Cadmium ions are an important source of water pollution. The determination of cadmium pollutant in natural and waste waters is the subject of the present study. There are some papers published on the design and applications of cadmium electrodes. For the elimination of the interference by heavy metal ions and complexing reagents, such as Pb/II/, Fe/III/, Ag/I/, Hg/II/, NTA,CN/I/ and F/I/, commonly, extraction and masking methods are applied. We used the commercial cadmium electrode designed in our laboratory and a special method was developed for the monitoring of cadmium ions in waste water from metal plating and TV tube plants or in natural water.

Several factories in Shanghai and China have used the method developed by us.

EXPERIMENTAL

Interference and masking of heavy metal ions

Tiron is a good complexing agent for Cu/II/, Fe/III/, Pb/II/, Zn/II/ and tiron addition to the TISAB has little effect on the response of the electrode. Hence, it can be used as a masking reagent for the above ions /Figs. 1-3/.

Formalin is a reducing reagent for Ag/I/ and Hg/II/ and when it is used together with tiron, Ag/I/ and Hg/II/ are easily reduced at room temperature.

Interference by NTA and elimination of interference

In plating baths, some complexing agents are also present such as CN^- or NTA. A significant effect of CN^- and NTA on the electrode potential was observed. There is only little effect on the membrane of the Cd electrode itself, but the concentration of free Cd/II/ ions decreases due to the formation of $/CdNTA/^-$.

Zinc ions prefer to combine with NTA to form $/ZnNTA/^-$, which is more stable than $/CdNTA/^-$. By adding zinc ions to the sample solution, the effect of NTA could be reduced.

Procedure of cadmium determination /NTA contained/

Some tiron was added to the sample solution /1:100/, then some Zn^{2+} solution and formalin solution. The pH of solution was adjusted to 7. Potentiometric measurements were carried out by using the standard addition method. Tables 1,2 and 5 show the results obtained by this method.

Determination of trace cadmium ions in plating bath /cianide contained/

In China, cyanide containing waste water was treated with NaClO. Residual NaClO,however, may change the results of measurements. It was found that the addition of Na_2SO_3 effevtively reduced this effect. Table 3 and 4 show the high accuracy and the good correlation between different methods.

Determination of cadmium ions in the waste water of TV cathodic tube factory

Fluoride is present in the waste water of TV tube factory in a large amount. Sometimes, it may be more than 100 ppm, and a large amount of silicate gel may also be present. It may have some effect on the determination. This problem could be solved by addition of boric acid and alum. The procedure used was as follows: 0.5 g of alum was added to 100 ml of sample solution, it was heated and the pH adjusted to 7,5, filtered to yield a clear solution. Then some boric acid was added to the clear solution, to convert interfering F^- to BF_4^-.

Tables 6, 7 and 8 show the results of this method. The detection limit was 0.05 ppm, standard deviation and variation coefficient were both good.

RESULTS AND DISCUSSION

1. The authors have proposed effective TISAB solution for use in measurements with cadmium ion-selective electrodes. One of them was a Tiron-formalin-Zn^{2+}-buffer, the other solutions had other components for use in other matrices.

2. The detection limit in these systems may be down to 0.05 ppm under common situations.

REFERENCES

1. J.Koryta, Anal.Chim.Acta 91, /1977/ 35
2. W.E.Lindon and R.Dostervink, Anal.Chim.Acta 108, /1979/ 169
3. D.D.Perin, "Masking and Demasking of Chemical Reactions", /1970/, John Wiley Press
4. S.L.Grassino et al, J.Inorg. Nucl. Chem. 33, /1971/ 421
5. Zhang Zong-Rang et al, Proceedings of the Meeting of Environmental Protection of Chinese University, 1978.
6. Zhang Zong-Rang et al, Proceedings of ISE Conference of China, 1980.
7. Wu Xia-Qing et al, J.Shanghai Teachers' College /Science Edition/, 2, /1982/ 85
8. Lu Hong-Yu et al, J.Shanghai Teachers' College /Science Edition/, 1, /1981/ 95
9. Wu Xia-Qing et al, J.Shanghai Teachers' College /Science Edition/, 1, /1983/ 57

Table 1. Recovery of cadmium at different concentrations

No.	Cd^{2+}	Observed value	Recovery /%/
1	22.4 ppm	22.5 ppm	101%
2	11.2 ppm	11.9 ppm	106%
3	1.1 ppm	1.2 ppm	109%
4	0.6 ppm	0.6 ppm	100%

Table 2. Recovery of synthetic sample of plating bath /Cyanide contained/*
Cd^{2+} = 0.1 ppm

1	2	3	4	5	6	7	8
100%	97%	97%	91%	100%	96%	100%	96%

*0.01M Na_2SO_3 was added

Table 3. Results of waste water of metal plating plant /NTA contained/

No.	Observed value /electrode/	Observed value /AAS/
1	39 ppm	35 ppm
2	32 ppm	29 ppm
3	5.4 ppm	5.1 ppm
4	0.2 ppm	0.2 ppm
5	0.2 ppm	0.2 ppm

Table 4. Comparison between electrode method and AAS

No	Observed value /electrode/	Observed value /AAS/
1	0.07 ppm	0.08 ppm
2	0.20 ppm	0.16 ppm
3	0.24 ppm	0.24 ppm
4	0.05 ppm	$<$0.05 ppm
5	0.05 ppm	$<$0.1 ppm
6	0.03 ppm	$<$0.05 ppm

Table 5. Standard deviation of test

No.	1	2	3	4	5	6	7	$\bar{X}$	S
1	0.07	0.07	0.07	0.06	0.06	0.06	0.06	0.064	0.0053
2	0.06	0.06	0.05	0.05	0.05	0.05	0.06	0.054	0.0053

Table 6. Results for TV tube waste water

No	1	2	3	4	5	6
1	0.035	0.035	0.035	0.035	0.039	0.039
2	0.035	0.031	0.031	0.031	0.035	0.031

No	7	8	9	$\bar{X}$	S	P_s
1	0.035	0.039	0.039	0.037	0.0020	5.3%
2	0.031	0.031	0.031	0.032	0.0019	5.9%

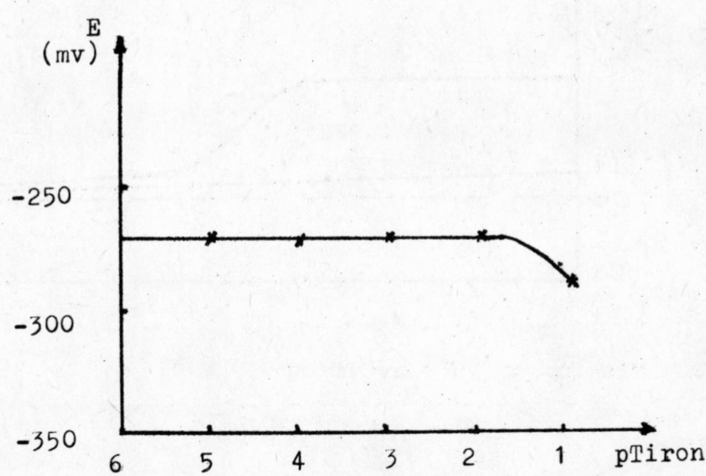

Fig.1 Potential change with the concentration of Tiron

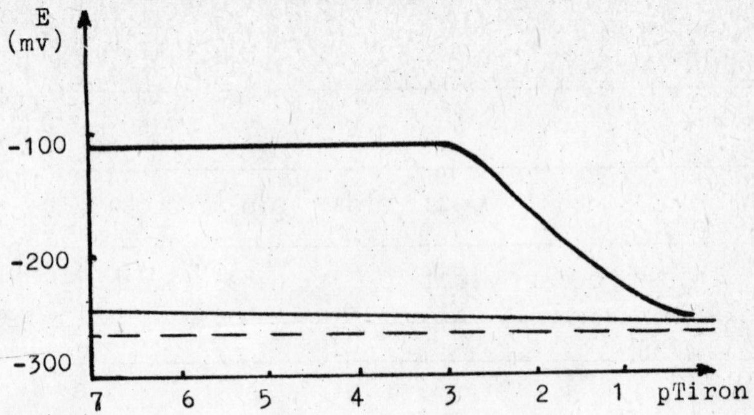

Fig.2 Masking of Cu^{2+} by Tiron

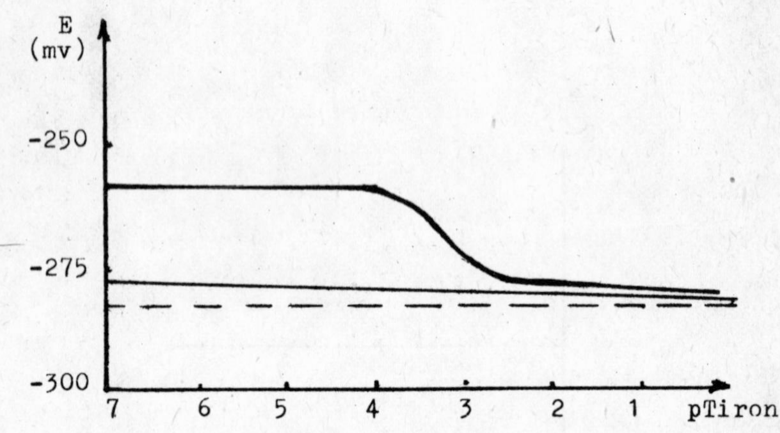

Fig.3 Masking of Pb^{2+} by Tiron

4th Symposium on Ion-Selective Electrodes
Mátrafüred, 1984

NEUTRAL BIDENTATE ORGANOPHOSPHORUS REAGENTS AS ELECTRODE-ACTIVE COMPOUNDS IN ION-SELECTIVE ELECTRODES

A.F. ZHUKOV, A.B. KOLDAYEV, N.P. NESTEROVA, T.YA. MEDVED' and M.I. KABATCNIK

D.I. Mendeleyev Institute of Chemistry and Chemical Technology, Moscow, USSR

ABSTRACT

The electrode characteristics of plasticized electrodes in relation to alkaline and alkaline-earth elements and magnesium were studied. The selectivity behaviour of the electrodes was analyzed for different substituents at phosphoryl and carbonyl groups. The selectivity characteristics of the electrodes based on the compounds studied have been determined for plasticizing agents with various dielectric constants. The dependence of selectivity on the concentration of lipophilic anion in the membrane was examined.

INTRODUCTION

Organophosphorus neutral ligands are widely used in extraction owing to selective chelate formation and complexing properties at low pH values /1,2/. These ligands are as effective as their carboxyl-containing analogs which have found broad application in ionometry /3,4/. Organophosphorus compounds exhibit a number of specific properties resulting from a peculiar structure and stereochemistry of the phosphoryl group. π-bonds between oxygen and phosphorus atoms in the phosphorus-containing ligands are much weaker than those between carbon and oxygen atoms in carboxyl-groups. The polarizability of the P=O-bond in the phosphoryl group is higher than that of the C=O-bond in carboxyl, so phosphorus-

-containing ligands are marked by more significant electron effects.

In the synthesis of phosphorus-containing ligands with the desired properties we used the revealed correlation between the ligand structure and its properties by varying its dentate value, the nature of functional groups and the stereochemistry of the entire molecule.

EXPERIMENTAL

We studied the selectivity characteristics of electrodes based on neutral organophosphorus compounds of the formula

$$\begin{matrix} R_1 \\ R_1 \end{matrix} \!\!> \underset{\underset{O}{\|}}{P} - R - \underset{\underset{O}{\|}}{P} <\!\! \begin{matrix} R_2 \\ R_2 \end{matrix}$$

where R is CH_2 or C_6H_4 $/CH_2/_2$; R_1, R_2 - alkyl, phenyl or tolyl groups.

The compounds of various classes with different dielectric constants such as dioctylphthalate, dibutylphthalate, tributylphosphate, triethylhexylphosphate, o-nitrophenyloctyl and o--nitrophenylheptyl esters were used as plasticizing agents. The membranes were prepared as described in /5/. The electrode characteristics were studied using the following galvanic cell. Ag, AgCl/ KCl saturated // studied solution//membrane/10^{-1}M KCl + 10^{-1}M $CaCl_2$/AgCl, Ag.
The ionic strength of the solution was kept constant at 0.5 mol/l. The electromotive force of the electrochemical cell was measured by the compensation scheme with different types of potentiometers as high-resistance zero indicators.

The limit of detection was established in accordance with IUPAC recommendations, and selectivity coefficients were determined by the method of biionic potentials in 0.1 M solutions of metal chlorides at a constant concentration of an interfering ion /6/.

RESULTS AND DISCUSSION

The ligands studied fall into three groups:

I. Ligands containing phosphoryl and carbamoyl groups:

$$(R_1)_2P(=O)-CH_2-C(=O)-N(R_2)_2$$

II. Ligands containing two phosphoryl groups with alkylene or vinylene bridge:

$$(R_1)_2P(=O)-R-P(=O)(R_2)_2$$

III. Ligands containing two phosphoryl groups with xylene bridge:

$$C_6H_4[CH_2-P(=O)(R_1)_2]_2$$

The extraction selectivity for these compounds depends on the structure of the bridge between P=O-goups, types of substituents at phosphorus atoms in phosphoryl groups, chelate effects, and on steric factors. By varying the substituents at phosphorus atoms we can alter the electron density at the oxygen atoms of phosphoryl groups, thus changing the extraction ability of organophosphorus compounds with respect to various ions.

The selectivity characteristics of ligands with respect to calcium ions for o-nitrophenyloctyl ester are shown in Figs. 1-3. The lowest selectivity coefficients towards the ions of alkaline and alkaline-earth elements were obtained for the ligand No 16 /tetra-/p-tolyl/-o-xylylenediphosphine/. The high extraction potential of this compound is due to the interaction of two P=O groups with calcium ion followed by formation of chelate.

The selectivity of calcium-selective electrodes based on ligand No.16 was not influenced by plasticizing agents with different dielectric constants. A bulk para-chloro-tetraphenylborate anion which prevents the transfer of counterions from the solution into the organic phase did not change the electrode selectivity /Fig. 4/. Potassium tetraphenylborate present in the membrane at a concentration of more than 50% by weight markedly decreased the selectivity of a calcium-selective electrode, in the presence of a background electrolyte /KCl > 0.1 M/, indicating that potassium ion becomes the main potential-determining ion.

The study of the electrode characteristics of calcium selective electrode shows that a stable electrode function is exhibited only in solutions containing KCl as a background electrolyte, but not in pure calcium chloride solutions /Fig. 5/. Raising the concentration of the background solution from 0.01 M to 0.3 M KCl we can reduce the electrode potential of calcium-selective electrode at the expense of the solution counterions. The change in the electrode potential is most likely caused by a change in the composition of the complex ion at the membrane/solution interface. In this case the selectivity depends on the equilibrium and kinetic factors /7/.

$$K_{Co,M} = \frac{U^{*}_{CaL^{2+}}}{U^{*}_{iL^{2+}}} \; \frac{K^{ext}_{Ca}}{K^{ext}_{i}}$$

where $U^{*}_{CaL^{2+}}$ and $U^{*}_{iL^{2+}}$ are the mobilities of ligand complexes with the analyzed calcium ion and with interfering ion, respectively; K^{ext}_{Ca}/K^{ext}_{i} is the ratio between the extraction equilibrium constants of CaX_2 and iX_2 /where X is a univalent counterion/ into the membrane phase. The ratio K^{ext}_{Ca}/K^{ext}_{i} is formally identical to $K^{exch}{}_{Ca,i}$, the equilibrium constant of the exchange reactions;

$$Ca^{2+}/sol./+iL^{2+*}/mem./ \xrightleftharpoons{K^{exch}_{Cai}} CaL^{2+*}/mem./ + i^{2+}/sol./$$

$$\frac{K_{Ca}^{ext}}{K_i^{ext}} = \frac{K_{CaL^{2+}}}{k_{iL^{2+}}} \cdot \frac{\beta_{CaL^{2+}}}{\beta_{iL^{2+}}}$$

$K_{CaL^{2+}}$ and $k_{iL^{2+}}$ are the partition coefficients of complex cations between aqueous and organic phases; $\beta_{CaL^{2+}}$ and $\beta_{iL^{2+}}$ are the stability constants for the complexes CaL^{2+} and iL^{2+}.

For the cations of the same charge and for complex cations of the same type the ratio $k_{CaL^{2+}} / k_{iL^{2+}}$ is close to unity and independent of the solvent nature. The mobility ratio for these complex cations in the organic phase is also close to unity. Hence, the selectivity of this system will depend only on the chemical properties reflected by the values of stability constants of complex cations in the aqueous phase.

Calcium and organophosphorus ligand can interact according to the following scheme:

$$Ca^{2+} + L \underset{}{\overset{\beta_1}{\rightleftharpoons}} [CaL]^{2+} \qquad /1/$$

$$Ca^{2+} + 2\,L \underset{}{\overset{\beta_2}{\rightleftharpoons}} [CaL_2]^{2+} \qquad /2/$$

yielding positively charged complexes CaL^{2+} and CaL_2^{2+}. β_1 and β_2 are the stability constants of the corresponding complex cations. The complexes CaL^{2+} and CaL_2^{2+} can further interact with the counterion Cl^-:

$$CaL^{2+} + 2\,Cl^- \underset{}{\overset{\beta'_1}{\rightleftharpoons}} CaLCl_2 \qquad /3/$$

$$CaL^{2+} + 2\,Cl^- \underset{}{\overset{\beta'_2}{\rightleftharpoons}} CaL_2Cl_2 \qquad /4/$$

where β'_1 and β'_2 are the stability constants of the corresponding complexes. Since the electrode characteristics of calcium-selective electrodes are influenced by the concentration of the counterion in the membrane phase, complexing at the membrane/solution interface can be assumed to follow the pathways /3/ or /4/.

The selectivity coefficients depend not only on complexing processes and the mobility of the complex with the ion to be determined in the membrane phase, but also on the equilibrium and kinetic characteristics of the complexes of electrode active compounds with interfering ions. So, to explain the selectivity behaviour of the calcium selective electrode based on ligand No.16, calcium complex with this ligand should be studied in further detail.

REFERENCES

1. Lobana T.S., Sandhu S.S., Solvent extraction properties of ditertiaryphosphine dioxides. - Coord.Chem.,Rev., 1982, v.47, 283-300.
2. T.Ya.Medvedev, Yu.M.Polikarpov, L.E.Bertina, V.V.Kossih, K.S.Yadina, M.I.Kabachnik, Alkylenediphosphine dioxides as extractants and complexing agents. - Advances in Chemistry, 1975, v.44, 1003-1024 /in Russian/.
3. M.E.Morf. The Principles of Ion-Selective Electrodes and Membrane Transport, Budapest, Akadémiai Kiadó, 1981.
4. W.Simon, W.E.Morf. Ion-Selective Electrodes Based on Neutral Carriers, in: Ion-Selective Electrodes in Analytical Chemistry /H.Freiser/, Plenum Press, N.Y., 1978.
5. Camman K., Das Arbeiten mit Ionenselectiven Electroden, Springer Verlag, 1977.
6. Recommendations for the Nomenclature of Ion-Selective Electrodes /Recommendation 1975/, IUPAC, Analytical Chemistry Division, Pure Appl.Chem., 1976, v.48, N.1, p.127.
7. In: ref. 3, p.264.

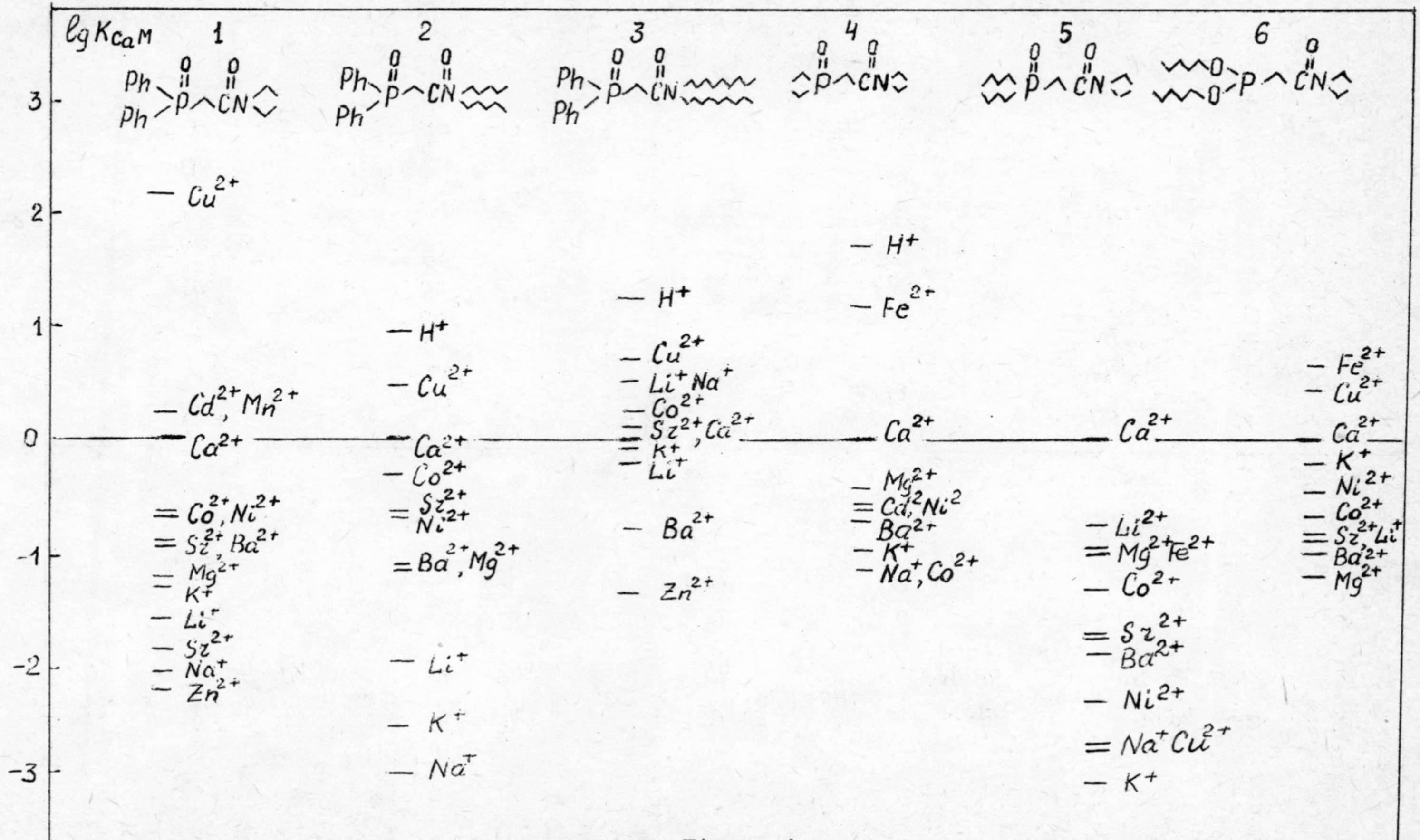

$lg K_{CaM}$

1 $Ph_2P(O)CH_2C(O)N$(cyclic): Cu^{2+}; Cd^{2+}, Mn^{2+}; Ca^{2+}; Co^{2+}, Ni^{2+}; Sr^{2+}, Ba^{2+}; Mg^{2+}; K^+; Li^+; Sr^{2+}; Na^+; Zn^{2+}

2 $Ph_2P(O)CH_2C(O)N$(alkyl)$_2$: H^+; Cu^{2+}; Ca^{2+}; Co^{2+}; Sr^{2+}, Ni^{2+}; Ba^{2+}, Mg^{2+}; Li^+; K^+; Na^+

3 $Ph_2P(O)CH_2C(O)N$(alkyl)$_2$: H^+; Cu^{2+}; Li^+ Na^+; Co^{2+}; Sr^{2+}, Ca^{2+}; K^+; Li^+; Ba^{2+}; Zn^{2+}

4: H^+; Fe^{2+}; Ca^{2+}; Mg^{2+}; Cd^{2}, Ni^{2}; Ba^{2+}; K^+; Na^+, Co^{2+}

5: Ca^{2+}; Li^{2+}; Mg^{2+} Fe^{2+}; Co^{2+}; Sr^{2+}; Ba^{2+}; Ni^{2+}; Na^+ Cu^{2+}; K^+

6 (RO)$_2P(O)CH_2C(O)N$(cyclic): Fe^{2+}; Cu^{2+}; Ca^{2+}; K^+; Ni^{2+}; Co^{2+}; Sr^{2+}, Li^+; Ba^{2+}; Mg^{2+}

Axis: 3, 2, 1, 0, -1, -2, -3, -4

Figure 1

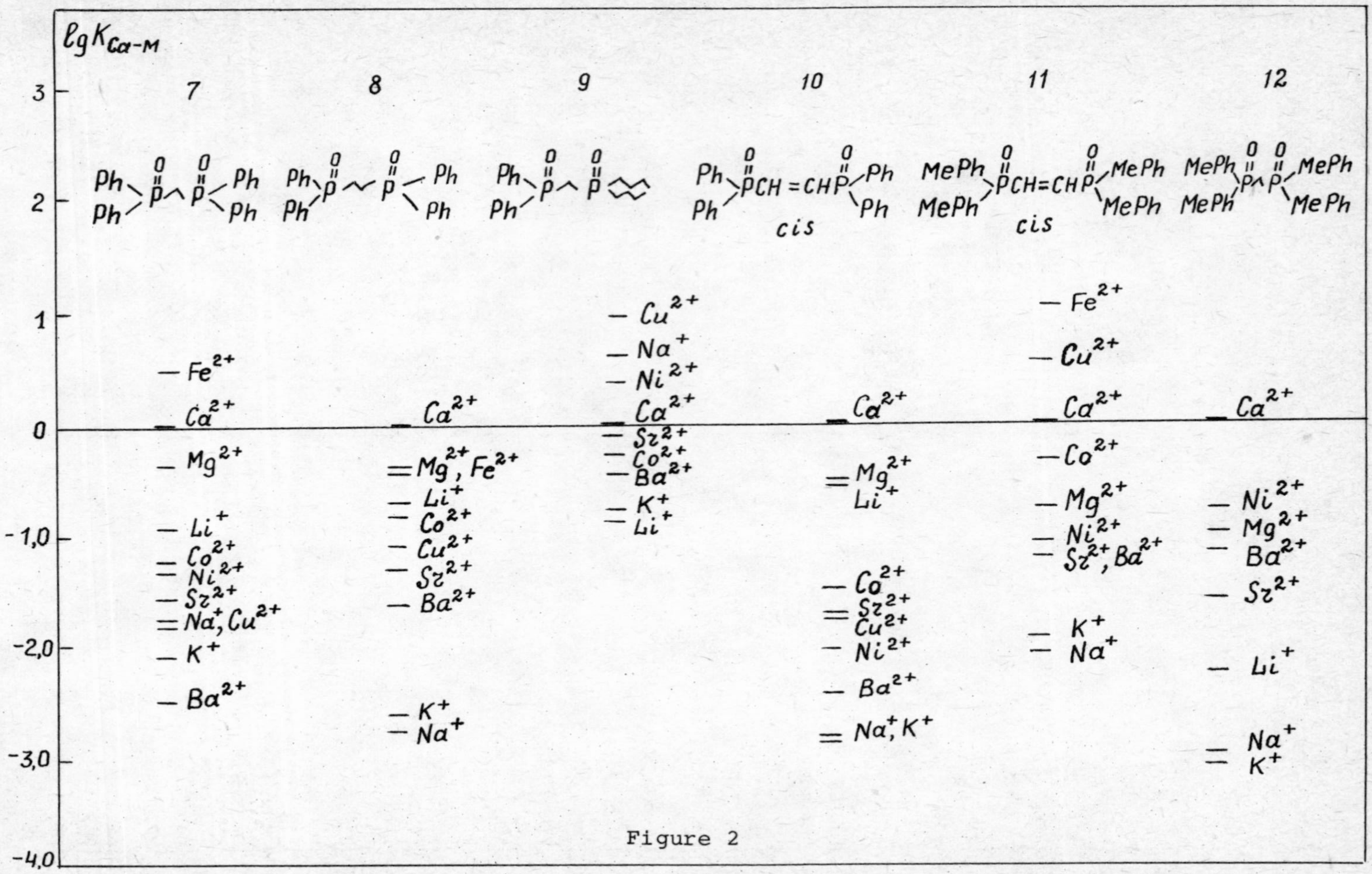

$lg K_{Ca-M}$

Axis: 3, 2, 1, 0, -1,0, -2,0, -3,0, -4,0

7 $Ph_2P(O)CH_2P(O)Ph_2$

Fe^{2+}, Ca^{2+}, Mg^{2+}, Li^{+}, Co^{2+}, Ni^{2+}, Sr^{2+}, Na^{+}, Cu^{2+}, K^{+}, Ba^{2+}

8 $Ph_2P(O)(CH_2)_nP(O)Ph_2$

Ca^{2+}, Mg^{2+}, Fe^{2+}, Li^{+}, Co^{2+}, Cu^{2+}, Sr^{2+}, Ba^{2+}, K^{+}, Na^{+}

9 $Ph_2P(O)CH_2P(O)$ (cyclic)

Cu^{2+}, Na^{+}, Ni^{2+}, Ca^{2+}, Sr^{2+}, Co^{2+}, Ba^{2+}, K^{+}, Li^{+}

10 $Ph_2P(O)CH{=}CHP(O)Ph_2$ cis

Ca^{2+}, Mg^{2+}, Li^{+}, Co^{2+}, Sr^{2+}, Cu^{2+}, Ni^{2+}, Ba^{2+}, Na^{+}, K^{+}

11 $(MePh)_2P(O)CH{=}CHP(O)(MePh)_2$ cis

Fe^{2+}, Cu^{2+}, Ca^{2+}, Co^{2+}, Mg^{2+}, Ni^{2+}, Sr^{2+}, Ba^{2+}, K^{+}, Na^{+}

12 $(MePh)_2P(O)CH_2P(O)(MePh)_2$

Ca^{2+}, Ni^{2+}, Mg^{2+}, Ba^{2+}, Sr^{2+}, Li^{+}, Na^{+}, K^{+}

Figure 2

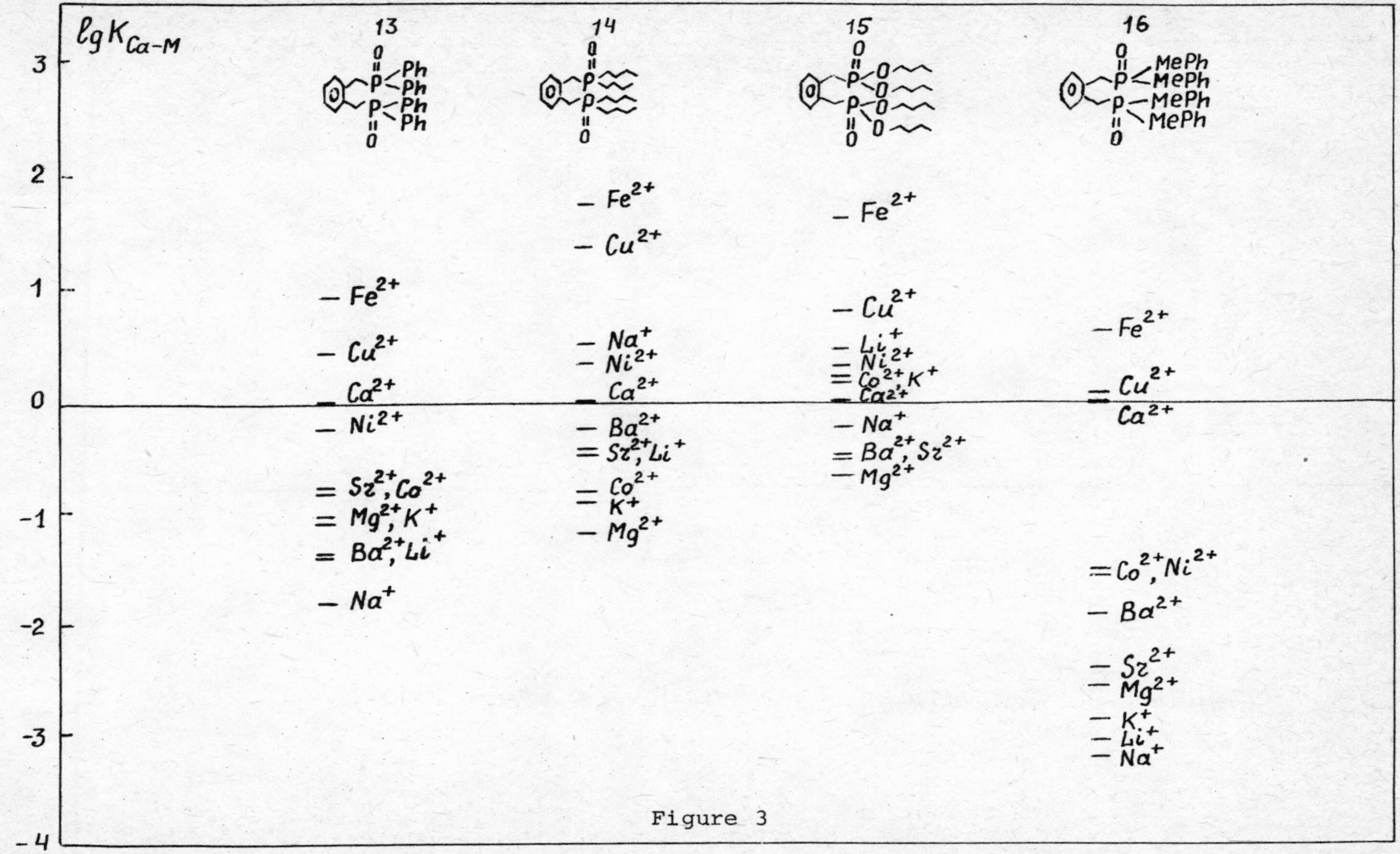

lg K_Ca-M
3
2
1
0
-1
-2
-3
-4
13
14
15
16
Ph
MePh
— Fe2+
— Cu2+
Ca2+
— Ni2+
= Sr2+, Co2+
= Mg2+, K+
= Ba2+, Li+
— Na+
— Fe2+
— Cu2+
— Na+
— Ni2+
Ca2+
— Ba2+
= Sr2+, Li+
= Co2+
K+
— Mg2+
— Fe2+
— Cu2+
— Li+
= Ni2+
= Co2+, K+
Ca2+
— Na+
= Ba2+, Sr2+
— Mg2+
— Fe2+
= Cu2+
Ca2+
= Co2+, Ni2+
— Ba2+
— Sr2+
— Mg2+
— K+
— Li+
— Na+

Figure 3

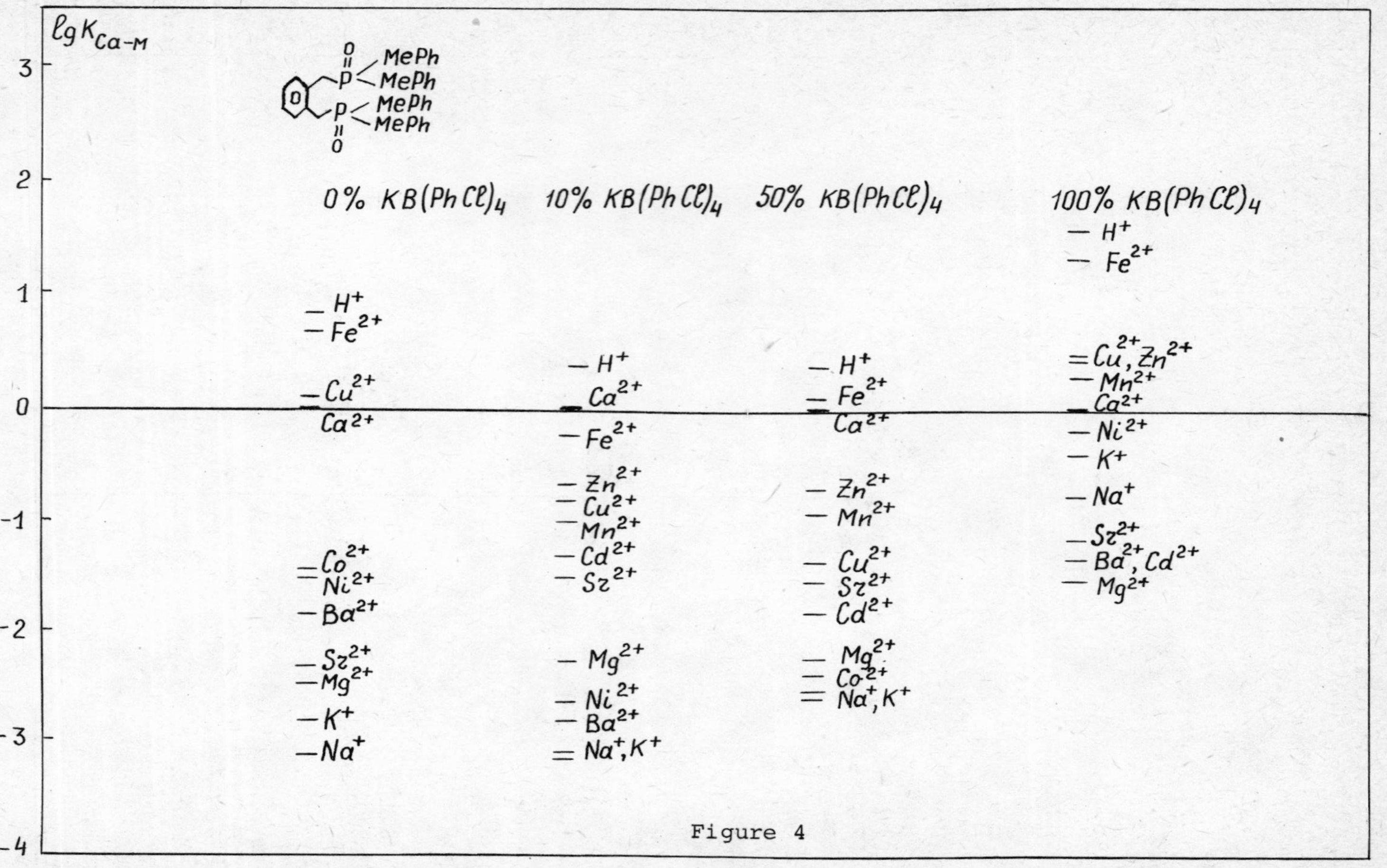
lg K_Ca-M
3
2
1
0
-1
-2
-3
-4
MePh
MePh
MePh
MePh
0% KB(PhCl)4
10% KB(PhCl)4
50% KB(PhCl)4
100% KB(PhCl)4
H+
Fe2+
Cu2+
Ca2+
Co2+
Ni2+
Ba2+
Sz2+
Mg2+
K+
Na+
H+
Ca2+
Fe2+
Zn2+
Cu2+
Mn2+
Cd2+
Sz2+
Mg2+
Ni2+
Ba2+
Na+, K+
H+
Fe2+
Ca2+
Zn2+
Mn2+
Cu2+
Sz2+
Cd2+
Mg2+
Co2+
Na+, K+
H+
Fe2+
Cu2+, Zn2+
Mn2+
Ca2+
Ni2+
K+
Na+
Sz2+
Ba2+, Cd2+
Mg2+

Figure 4

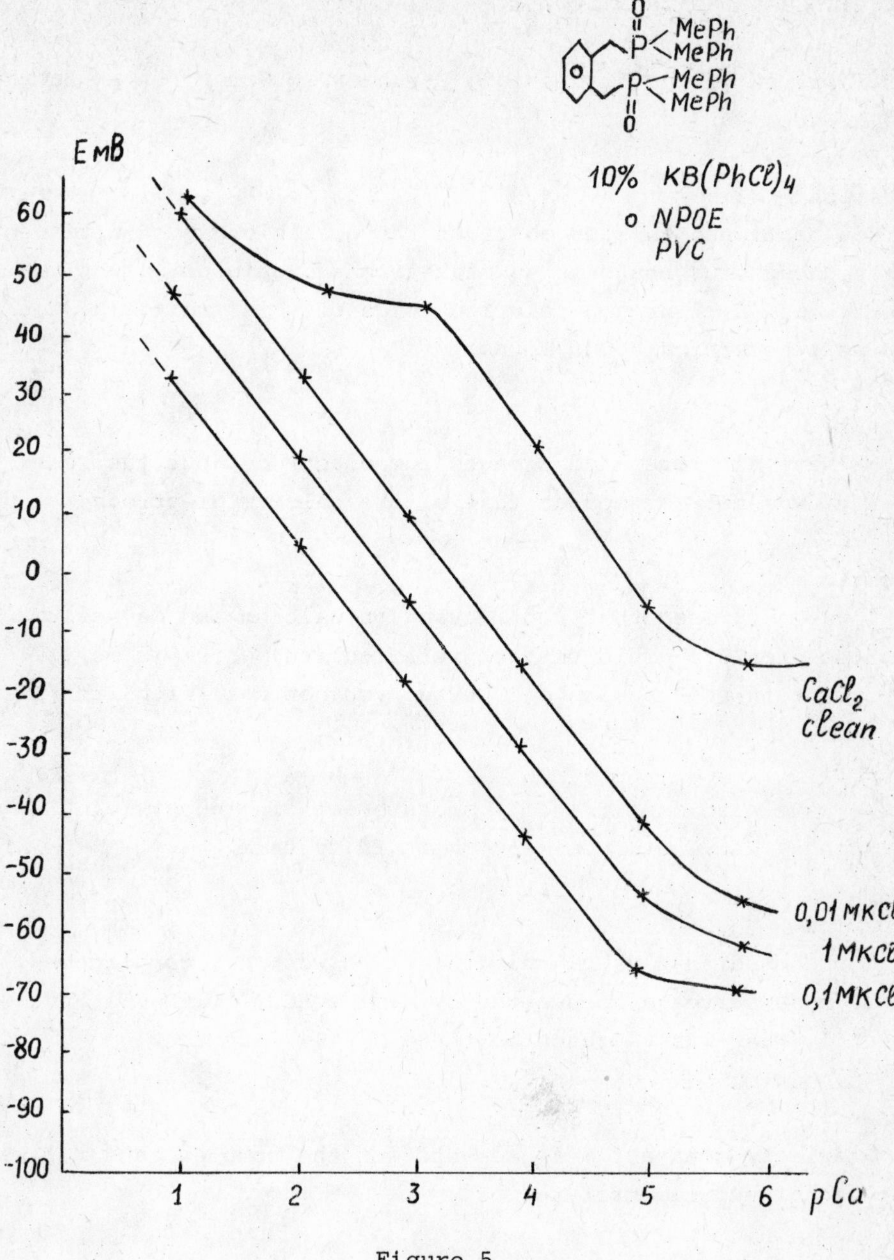
EмB
60
50
40
30
20
10
0
-10
-20
-30
-40
-50
-60
-70
-80
-90
-100
1
2
3
4
5
6
pCa
MePh
MePh
MePh
MePh
10% KB(PhCl)4
o NPOE
PVC
CaCl2
clean
0,01 MKCl
1 MKCl
0,1 MKCl

Figure 5

QUESTIONS AND COMMENTS

Participants of the discussion: E.Pungor, W.Simon, J.D.R.Thomas and A.Zhukov

Question:

You mentioned in your abstract the possible use of different organophosphorus reagents as magnesium and calcium selective electrodes. I think you referred to calcium in your talk. Could you tell something about magnesium ?

Answer:

We have also prepared a magnesium electrode, but the selectivity was not as good as that of the calcium electrode.

Question:

I was surprised that you focused on calcium and magnesium, because iron/II/ would be the preferred ion, which is very important in soil analysis ? Why do you not deal with iron/II/?

Answer:

We have also made iron/II/ measurements in industrial solutions, but I did not mention it in my talk.

Question:

You showed an E vs p_{Ca} diagram which was very complicated. There were curves also denoted by 0.01 M KCl, 0.1 M KCl and 1 M KCl. What was represented there ?

Answer:

Stability constant measurements for the complex, and KCl was the background electrolyte.

ROUND TABLE DISCUSSION ON THE PROBLEMS OF STANDARDS IN WORK WITH ION-SELECTIVE ELECTRODES

4th Symposium on Ion-Selective Electrodes
Mátrafüred, 1984

INTRODUCTORY LECTURES

FURTHER STUDIES OF IONIC ACTIVITY SCALES

ROGER G. BATES

Department of Chemistry, University of Florida,
Gainesville, Florida 32611 USA

ABSTRACT

Ion-selective electrodes are thermodynamic sensors and hence ion activity standards are essential in order to calibrate them most effectively. At low ionic strengths ($I<0.1$), activity coefficients can be estimated with sufficient precision by the Debye-Hückel theory. Specific ion-ion and ion-solvent interactions become important, however, in concentrated solutions. It was shown earlier that a model ascribing specific ionic properties to differing degrees of hydration can provide a reasonable basis for scales of ion activity at elevated ionic strengths. Three methods have now been used to determine hydration indexes (h) for unassociated electrolytes. Values of h for individual ionic species have been evaluated by assigning h=0 to chloride ion, and it is now suggested that bromide and iodide may also be regarded as essentially unhydrated. It is shown that an uncertainty of 1 in the hydration index alters pM for alkali chlorides MCl by less than 0.02 unit when the molality is 2 mol kg^{-1} and has a negligible effect on pM for alkaline earth chlorides MCl_2.

INTRODUCTION

There are three primary reasons that lend importance to the study of standards for ion-selective electrode (ISE) measurements. First of all, like other potentiometric methods, the determination of ions requires calibration of the ISE/reference electrode assembly to compensate for time variations in the base potentials of the components of the measuring system. Secondly, the potentials of many ISE's do not vary with ionic composition in accordance with the Nernst slope. This defect does not prevent useful data from being obtained but requires that the effective slope of the system be determined. For this purpose, at least two reference standards are required. Thirdly, and not the least important, is the fact that ISE's, as thermodynamic sensors, generate a potential response to some function of activities rather than concentrations.

It is evident then that activity standards are needed in most instances. In dilute solutions and in constant ionic media, it is nonetheless possible to standardize an ISE assembly with reference solutions of known ionic concentration. Under these conditions, experimental measurements of ion concentrations in "unknown" solutions may be derived from the difference of emf between a standard (E_S) and a test solution (E_X). In the ionic strength (I) range 0.01 to 0.1 mol kg^{-1}, where long-range ion-ion interactions predominate, the Debye-Hückel theory (with reasonable values of the ion-size parameter $\mathring{a}$) leads to useful ion activities. Even at I=0.16, there is little practical difference among values of pNa in NaCl solutions derived by applying several common conventions, as well as by measurements of cells with liquid junction (1,2). This conclusion may be drawn from the data in Table 1.

Metal-ion buffers offer a promising route to standards for some metal ions in the range of ionic molality (m) from 10^{-5} to 10^{-2}, and they deserve further study. With these, one might establish reference points at very low metal-ion activities without exceeding the range of ionic strengths in which activity coefficients can be estimated with some degree of reliability.

One cannot, however, buffer a solution with respect to the alkali ions, for example, in the high concentration range ($m > 1$ mol kg^{-1}). It is thus not possible to take full advantage of the useful response of ISE's without entering the concentration region where ion-ion and ion-solvent interactions are highly specific and activity coefficients are no longer a function of ionic strength alone. Recent efforts toward establishment of reasonable (though conventional) scales of ionic activity in concentrated solutions are the subject of this contribution.

EVOLUTION OF THE HYDRATION MODEL

In 1920, Bjerrum (3) recognized that hydration of ions in concentrated aqueous solutions leads to a reduction in the amount of "free" water and hence to a difference between the activity of hydrated and unhydrated ions. This difference is a function of $(p/p_o)^h$, where p and p_o are the partial pressures of the solution and pure water, respectively, and h is an average hydration number for the electrolyte. This approach was extended and refined first by Stokes and Robinson (4) and later by Glueckauf (5). In both instances, it was assumed that the activity coefficient of the unhydrated electrolyte (γ_{el}) could be calculated from the "extended" Debye-Hückel formula with ion size $\mathring{a}$. It was shown that the stoichiometric mean activity coefficient $\gamma_{\pm}$ could be expressed as a function of $\mathring{a}$ and h very satisfactorily even in rather concentrated solutions. Thus,

$$\log \gamma_{\pm} = \log \gamma_{el} + f(h) = f(\mathring{a}) + f(h) \qquad (1)$$

It was observed further that both $\log \gamma_{el}$ and f(h) varied in nearly linear fashion with m; this variation of $\log \gamma_{el}$ is in accord with the so-called Hückel form of the Debye-Hückel equation (6).

The two treatments differed in the manner by which the non-electrostatic or ion-solvent term f(h) was related to the hydration number. Whereas Stokes and Robinson used "mole-fraction statistics", Glueckauf's model was based on volume fractions, covolume effects being embodied in a quantity r, namely the ratio of the apparent molar volume V of the electrolyte to the molar volume of pure water.

At an earlier conference on ion-selective electrodes (Budapest, 1977), we reported on our development of a modified form of the hydration model for unassociated electrolytes (7). The 1969 equation of Glueckauf for γ_{el} was adopted, in view of his convincing demonstration (8) that this formula is superior to the Debye-Hückel equation up to $\kappa\mathring{a}=2.8$, or to an ionic strength of about 5 mol kg^{-1}. We based the term f(h) on volume fraction statistics. The complete equation for $\log \gamma_{\pm}$

in the modified version at 25°C thus becomes

$$\log \gamma_{\pm} = -A|z_+z_-|I_c^{\frac{1}{2}}\left[\frac{1+0.5B\mathring{a}I_c^{\frac{1}{2}}}{1+B\mathring{a}I_c^{\frac{1}{2}}}\right]^2 + \frac{0.018mr(r+h-\nu)}{2.303\nu(1+0.018mr)}$$

$$+ \frac{h-\nu}{\nu}\log(1+0.018mr) - \frac{h}{\nu}\log(1-0.018mh) \quad (2)$$

In this equation, I_c is the ionic strength on the molar concentration (c) basis, A and B are the appropriate Debye-Hückel constants, and ν is the number of ions per molecule of electrolyte.

In our earlier work (7), $\mathring{a}$ and h were determined from accepted values of the activity coefficients by a graphical method based on the assumed linearity with m of both $\log \gamma_{el}$, the first term on the right of Eq. (2), and f(h), the sum of the last three terms. The value of $\mathring{a}$ giving the best straight line for $\log \gamma_{\pm}-\log \gamma_{el}$ vs.m, and the slope β_1 of this line, were determined. Similarly, nearly straight lines (slopes β_2) were obtained when f(h) for various values of h was plotted as a function of m. The best value of h was taken to be that yielding $\beta_2=\beta_1$. A refinement of this method of calculation, described below, is termed Method A in the present work. It is perhaps appropriate to note here the opinion of Glueckauf (8) that factors other than hydration of ions (hard-sphere effects, etc.) doubtless enter into the value of h but that these are likely also to vary in linear fashion with the molality. We therefore prefer to term h the "hydration index" rather than the "hydration number". The values of h obtained by this method nonetheless agree well with the primary hydration numbers estimated by Bockris and Reddy (9).

Method A. In the present work, data for the activity coefficients were taken from the tables of Robinson and Stokes (10) and the slope-matching procedure was performed by a micro-computer. The program, written in BASIC language, calculated $\log \gamma_{el}$ for an initial estimate of $\mathring{a}$ and determined the slope β_1 of the straight-line plot of $\log (\gamma_{\pm}/\gamma_{el})$ vs. m, as well as the standard deviation for regression. The value of $\mathring{a}$ was then altered by a pre-assigned increment (usually 0.01Å) until a minimum in the standard deviation had been located. Next the value of h was varied in assigned increments until the best match with β_1 of the slope β_2 of f(h) vs. m was found. Finally, these best values of $\mathring{a}$ and h were substituted in Eq. (2) and the standard deviation of the calculated $\log \gamma_{\pm}$ printed. The value of r changes with the electrolyte concentration; in accordance with earlier suggestions, the value of r for 1:1 electrolytes was taken at c=1M and at c=0.7M for 2:1 electrolytes. The probable error thus introduced is estimated below.

Method B. Method A proved quite satisfactory when applied to 1:1 halides, the standard deviation of $\log \gamma_{\pm}$ (N = 17 to 19) ranging from 0.0003 to 0.0085. With most 2:1 halides, however, the deviations were too large (in the range 0.01 to 0.025) to be acceptable. This difficulty was clearly due to the fact that f(h) was not a linear function of m, as can be seen for $CaCl_2$ in Figure 1. The best matching of the slope β_2 (dotted lines) with β_1 (solid line) was found with h=10.49, but the standard deviation for $\log \gamma_{\pm}$(calc.) was 0.023.

In Method B, therefore, the requirement of linearity in m for both log $(\gamma_{\pm}/\gamma_{el})$ and f(h) was abandoned and values of å and h giving the best fit of Eq. (2) to the experimental data were found by nonlinear least squares methods. A modified Marquardt algorithm (11), programmed in BASIC by Dr. A. G. Dickson, was combined with Eq. (2) to yield å and h and the standard deviation of log $\gamma_{\pm}$(calc.)

Method C. By use of Method B, the standard deviations of log $\gamma_{\pm}$(calc.) were reduced to about one-half those of Method A for 2:1 electrolytes but remained larger than desired. The most likely cause of an inadequacy of Eq. (2) is the inherent assumption of a hydration index that does not vary with molality. A variation in h would be most evident with the highly hydrated 2:1 electrolytes. This point has been explored by Loewenthal and Marais (12) in their study of carbonate systems in seawater. These investigators found some improvement when h was assumed to vary with m in the following manner:

$$h = h^{o} - qm^{y} \qquad (3)$$

Their suggestion of y=2 did not prove suitable in our work, but a linear variation of h with m (y=1) provided a significant improvement. In Method C, therefore, Eq. (2) was modified by substitution of $(h^{o}-qm)$ for h and the three parameters å, h^{o}, and q determined by nonlinear least squares methods. By this means, standard deviations for the 2:1 chlorides were reduced to acceptable values in the range 0.001 to 0.003. The parameters for 1:1 and 2:1 chlorides resulting from the application of the three methods are collected in Tables 2 and 3. It may be noted that the present study is restricted to data at 25°C alone. Nevertheless, Deák (13) offered evidence in the 1980 Mátrafüred symposium that h for NaCl and KCl varies by only 0.3 to 0.4 between 10 and 60°C.

Figure 2 illustrates the deviations of the individual values of $[\log\gamma_{\pm}(\text{exp.}) - \log\gamma_{\pm}(\text{calc.})]$ for $CaCl_2$ as a function of m. The solid curve was calculated with $\mathring{a}$=5.32Å and h=10.49 (Method A results), while the dots represent $\mathring{a}$=5.44 and values of h varying according to h = 11.04-0.462m (Method C). Evidently, a decrease in h from 11.0 at m=0 to h=10.1 at m=2 mol kg^{-1} is implied. The physical significance of such a figure is a subject for speculation. A decrease of this magnitude in the time-average hydration number is not unreasonable, inasmuch as about 40% of the water in a 2M solution of an electrolyte with h about 11 would be bound by the ions and could no longer be regarded as free solvent.

Estimation of I_c and r. Values of V, the apparent molar volume, are needed in order to calculate both the molar concentrations (c) and the value of r. Apparent molar volumes of electrolytes in solution usually obey the relationship

$$V = V_o + S_v c^{\frac{1}{2}} \tag{4}$$

The constants V_o and S_v for 32 electrolytes in aqueous solution are listed by Harned and Owen (14). For other systems, V was calculated from d and d_o, the densities of the solution and of water, respectively, and M_2, the molar mass of the electrolyte, by

$$V = \frac{1000}{mdd_o}(d_o - d) + \frac{M_2}{d} \tag{5}$$

Concentrations (c) were derived by

$$c = md_o(1 - 0.001cV) \tag{6}$$

using an iterative procedure incorporated into the computer program. Values of r were obtained by

$$r = \frac{Vd_o}{M_1} \tag{7}$$

where M_1 is the molar mass of water (18.015 g mol^{-1}).

Errors in h resulting from fixing r at the nominal concentrations (c_r) of 1M (1:1 electrolytes) and 0.7M (2:1 electrolytes) can be estimated from values of S_v:

$$\frac{dr}{dc_r} = \frac{S_v d_o}{2M_1 c^{1/2}} = \frac{0.028 S_v}{c^{1/2}} \tag{8}$$

Evidently the errors increase with the magnitude of S_v, which varies over the range from 0.83 for HCl to 9.9 for $SrCl_2$ and 12.16 for Na_2SO_4, according to Harned and Owen (14). A change of c_r from 1M to 2M can be shown to lower h by less than 0.04 for the 1:1 electrolytes studied, while a change from 0.7M to 1.5M produces a decrease of 0.15 in h for $SrCl_2$ and 0.19 for Na_2SO_4.

INDIVIDUAL ION ACTIVITY COEFFICIENTS

In 1970, Bates, Staples, and Robinson (15) showed that hydration indexes for ions could characterize the specific ion-solvent interactions leading to a dependence of the activity coefficient of a particular ion not only on the ionic strength and $\mathring{a}$ but on the properties of the counter ions present in the solution as well. Assuming additivity of ionic hydration indexes, they showed that it was not unreasonable to take h=0 for chloride ion as a conventional basis for assigning hydration indexes to other ionic species. Further evidence for this convention was presented at the 1977 Budapest conference (7) in the form of a plot of h for 1:1 chlorides as a function of $1/r_+$, where r_+ is the crystal radius of the cation. A similar plot of the data derived from the present study is shown in Figure 3. This type of plot suggests that h_{Cl} is near zero, if it is accepted that hydration of very large cations is minimal.

Table 4 compares the values of h for chlorides , bromides, and iodides of 10 cations. In general, it may be said that the agreement is of the order of the precision in the determination of h by the three different methods. It seems to justify the conclusion, also reached by Loewenthal and Marais (12), that h is zero for all three halide anions. This convention leads to the following relations for the activity coefficient of the cation in solutions of 1:1 and 2:1 halides, MX and MX_2, respectively (15):

$$\log \gamma_{M^+} = \log \gamma_{\pm} + \frac{0.018}{\ln 10}(h_M - h_X)m\phi \qquad (9)$$

and

$$\log \gamma_{M^{2+}} = 2\log \gamma_{\pm} + \frac{0.018}{\ln 10}h_M m\phi + \log[1+0.018(3-h_M)m] \qquad (10)$$

where ϕ is the osmotic coefficient.

Provisional reference values for pM and pX in solutions of six unassociated electrolytes at molalities of 0.1, 1, and 2 mol kg^{-1} are given in Table 5. They were calculated with the aid of Eq. (9) and (10), using h values derived from the data for chlorides, which are thought to be more precise and internally consistent than for the bromides and iodides.

In this connection, it is important to inquire as to the error in pM and pX caused by the uncertainties in determining the values of h. These uncertainties are evident in the averages given in the last column of Table 4 ; they must be increased slightly in accordance with the uncertainty (0.05 to 0.15) resulting from fixing r at nominal values of 1M or 0.07M as discussed above. The values of d(pM)/dh given in Table 6 show that pM and pX are quite insensitive to errors in h. For the 1:1 chlorides, a change of 0.3 in h corresponds to less than 0.01 in pM at m=3, while an error of 1 unit in h for the 2:1 chlorides is almost without effect at m=2 and can easily be tolerated.

TEST OF THE HYDRATION CONVENTION

Conventions are not subject to experimental confirmation, but it is possible to assess the consistency of the hydration approach by making use of experimental data for cells with liquid junction. For this purpose, one must estimate liquid-junction potentials (E_j); for this, we have used the Henderson equation. The emf E of the following cell was measured at $25^{o}C$:

$$Hg;Hg_2Cl_2,KCl(satd.)||MCl(m)|AgCl;Ag \tag{11}$$

where M was Li, Na, K, Cs, NH_4; m varied from 0.1 to 3 mol kg^{-1} (1). Accepting the convention of pCl given by Eq. (9), one can write

$$\frac{E-E_j}{0.05916} + \log(m\gamma_\pm) - 0.00782(h_+-h_{Cl})m\phi = E^{o\prime} \equiv \Delta \tag{12}$$

the left side of which should be independent of m when the proper value of (h_+-h_{Cl}) is inserted, subject of course to the limitations in estimating E_j.

The results for solutions of NaCl are shown in Figure 4. It is evident that our data as well as those of Hurlen (2), shown as dots, suggest a value of $(h_{Na}-h_{Cl})$ near 2.5. Inasmuch as $(h_{Na}+h_{Cl})$ was found to be about 3.6, these measurements lead of h=0.6 for chloride ion. Data for four of the salts gave an average near h_{Cl}=1, with an average deviation of about 0.5. A value of h_{Cl}=0.9 was shown by Bagg and Rechnitz (16) to be consistent with their measurements of cells with liquid junction. As we have already indicated, 1) the estimation of E_j limits severely the accuracy of such a test, and 2) pM and pX are not strongly dependent on the value chosen for h. Thus it may be concluded that a convention assigning a zero hydration index to the chloride ion provides a reasonable approach to the establishment of reference standards for ion-selective electrodes at ionic strengths well above the range of the Debye-Hückel theory.

ACKNOWLEDGMENT

This work was supported in part by the National Science Foundation (U.S.A.) under Grant CHE81 20592.

REFERENCES

1 D. R. White, Jr., and R. G. Bates, Unpublished measurements.

2 T. Hurlen, Acta Chem. Scand., A33, (1979) 631.

3 N. Bjerrum, Z. Anorg. Allgem. Chem., 109, (1920) 275.

4 R. H. Stokes and R. A. Robinson, J. Am. Chem. Soc., 70, (1948) 1870.

5 E. Glueckaur, Trans. Faraday Soc., 51, (1955) 1235.

6 E. Hückel, Physik. Z., 26, (1925) 93.

7 R. G. Bates and R. A. Robinson, in Ion-Selective Electrodes (E. Pungor, ed.), Akadémiai Kiadó, Budapest, 1978, pp.3-19.

8 E. Glueckauf, Proc. Roy. Soc., A310, (1969) 449.

9 J. O'M. Bockris and A.K.N. Reddy, Modern Electrochemistry, Plenum Press, New York, 1970, Vol. I, Chap. 2.

10 R. A. Robinson and R. H. Stokes, Electrolyte Solutions, 2nd edition revised, Butterworths, London, 1970, appendix 8.10.

11 J. C. Nash, Numberical Methods for Computers. Linear Algebra and Function Minimization, Halsted Press, New York, 1979.

12 R. E. Loewenthal and G.v.R. Marais, Carbonate Chemistry of High Salinity Waters, Research Report No. W46, Department of Civil Engineering, University of Cape Town, 1983.

13 E. Deák, in Ion-Selective Electrodes, 3 (E. Pungor, ed.), Akadémiai Kiadó, Budapest, 1981, pp. 203-213.

14 H. S. Harned and B. B. Owen, The Physical Chemistry of Electrolyte Solutions, 3rd edition, Reinhold Publ. Corp., New York, 1958, pp. 358-361.

15 R. G. Bates, B. R. Staples, and R. A. Robinson, Anal. Chem., 42, (1970) 867.

16 J. Bagg and G. A. Rechnitz, Anal. Chem., 45, (1973) 1069.

Table 1. pNa in 0.16 molal NaCl at 25°C

Convention	pNa
MacInnes	0.913
Guggenheim	0.921
pH	0.918
Hydration	0.916
Emf, Henderson equation:	
White, Bates	0.914
Hurlen	0.918

Mean: pNa=0.917±0.003
pCl=0.923

Table 2. Parameters for 1:1 chlorides

	Method	Max. m	å/Å	h [a]	q	s(log $\gamma_{\pm}$) [b]
HCl	A	3	4.46	5.43		0.0085
	B	3	5.25	5.21		.0038
	C	4	4.40	5.65	0.117	.0009
LiCl	A	3	4.41	5.11		0.0066
	B	3	4.92	4.98		.0025
	C	4	4.45	5.26	0.089	.0011
NaCl	A	4	4.46	3.60		0.0015
	B	4	4.51	3.54		.0006
	C	3	4.42	3.61	0.021	.0004
KCl	A	4.5	4.02	2.46		0.0010
	B	4	4.04	2.46		.0006
	C	4	3.94	2.55	0.019	.0002
RbCl	A	5	3.57	2.19		0.0011
	B	4	3.53	2.23		.0005
	C	4	3.45	2.30	0.016	.0003
CsCl	A	5	2.74	1.86		0.0021
	B	4	2.67	1.91		.0015
	C	4	2.84	1.66	-0.045	.0013
NH_4Cl	A	4	4.15	2.11		0.0015
	B	4	4.11	2.13		.0013
	C	4	3.86	2.35	0.044	.0003

[a] For Method C, h^o [b] Standard deviation of fit

Table 3. Parameters for 2:1 chlorides

	Method	Max. m	å/Å	h [a]	q	s(log $\gamma_{\pm}$) [b]
$MgCl_2$	A	2	5.32	11.28		0.0252
	B	2	6.75	10.78		.0103
	C	2	5.48	11.97	0.523	.0018
$CaCl_2$	A	2	5.32	10.49		0.0227
	B	2	6.30	9.90		.0095
	C	3	5.44	11.04	0.462	.0017
$SrCl_2$	A	2	5.38	10.06		0.0199
	B	2	6.34	9.45		.0086
	C	3	5.41	10.45	0.433	.0029
$BaCl_2$	A	1.8	5.60	8.69		0.0119
	B	1.8	6.11	8.40		.0085
	C	1.8	5.05	9.88	0.686	.0016

[a] For Method C, h^o [b] Standard deviation of fit

Table 4. Hydration indexes for 1:1 and 2:1 halides derived by three methods [a]

Cation	Chloride	Bromide	Iodide	Mean
H^+	5.39	5.63	5.60	5.54±0.10
Li^+	5.09	5.23	5.15	5.16±0.05
Na^+	3.57	3.84	3.93	3.78±0.13
K^+	2.48	2.44	2.34	2.42±0.05
Rb^+	2.23	1.90	1.60	1.91±0.21
Cs^+	1.83	1.50	0.81	1.38±0.38
NH_4^+	2.20	-	-	-
Mg^{2+}	11.2	11.8	12.0	11.7 ±0.3
Ca^{2+}	10.4	10.9	11.1	10.8 ±0.3
Sr^{2+}	9.9	10.2	10.6	10.2 ±0.2
Ba^{2+}	8.8	9.1	10.1	9.3 ±0.5

[a] For 2:1 halides (Method C), h is given at m=0.7.

Table 5. Provisional reference values of pM and pX at 25°C

	$h_+ - h_-$	m=0.1		m=1.0		m=2.0	
		pM	pX	pM	pX	pM	pX
LiCl	5.1	1.099	1.106	0.071	0.152	-0.356	-0.174
NaCl	3.6	1.106	1.112	0.156	0.209	-0.181	-0.070
KCl	2.5	1.112	1.115	0.201	0.236	-0.095	-0.024
NH_4Cl	2.2	1.112	1.115	0.204	0.235	-0.088	-0.026
KF	0.6	1.110	1.111	0.186	0.195	-0.128	-0.110
$CaCl_2$	10.4	1.570	0.842	0.579	-0.139	-0.188	-0.507

Table 6. Estimation of errors in pM caused by uncertainties in hydration index

	d(pM)/dh at -			
	m=1	m=2	m=3	m=4
LiCl	-	-0.017	-0.030	-0.091
NaCl	-	-0.016	-0.025	-0.035
KCl	-0.007	-0.014	-	-0.030
CsCl	-0.007	-0.014	-0.021	-0.029
$MgCl_2$	0.000	-0.002		
$CaCl_2$	+0.001	0.000		
$BaCl_2$	+0.001	-0.004 [a]		

[a] At m = 1.8

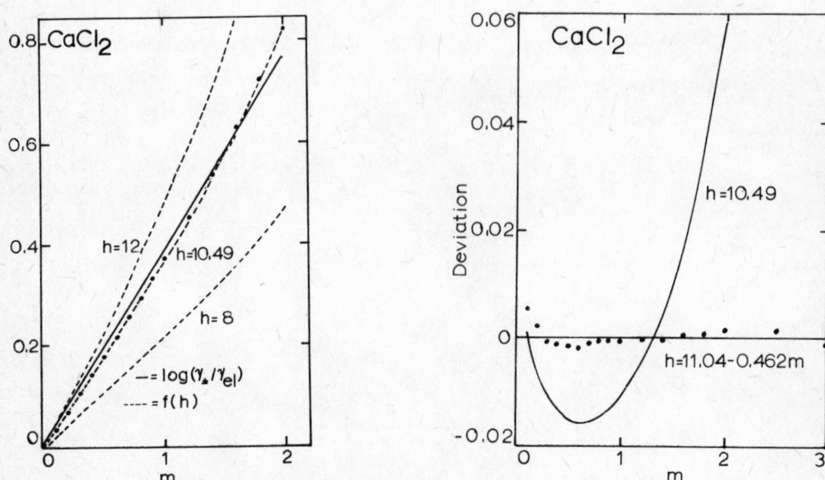

Fig. 1. Analysis of activity coefficient data for $CaCl_2$ by Method A

Fig. 2. Deviations of individual values of [log $\gamma_{\pm}$(exp.) - log $\gamma_{\pm}$(calc.)] from Method A (h=10.49) and Method C (h=11.04-0.462m)

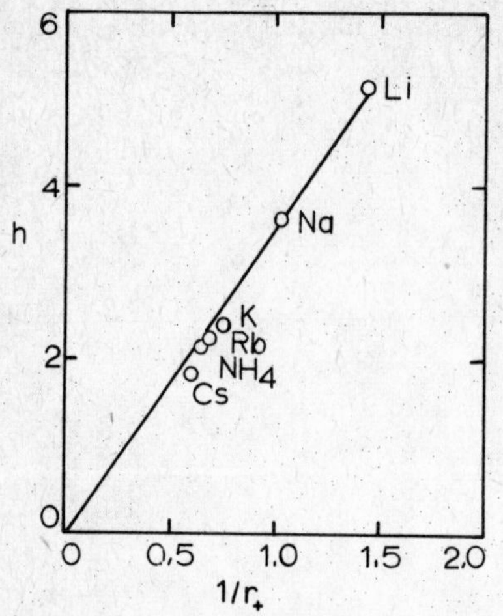

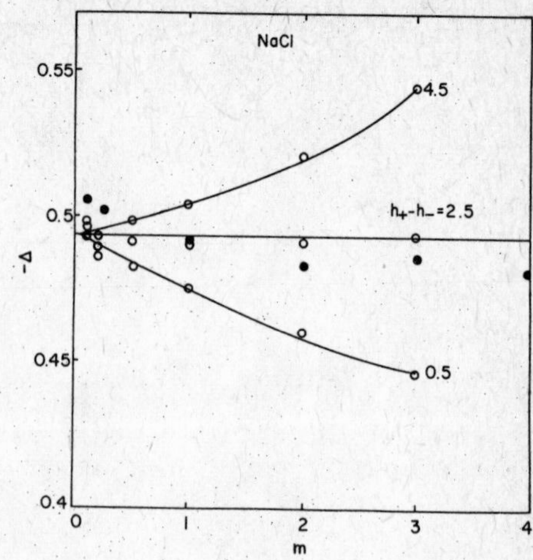

Fig. 3. Plot of the hydration index for 1:1 chlorides as a function of the reciprocal of the crystal radius of the cation.

Fig. 4. Estimation of $(h_+ - h_-)$ for NaCl from the emf of cells with liquid junction

4th Symposium on Ion-Selective Electrodes
Mátrafüred, 1984

DETERMINATION OF MEAN ACTIVITY COEFFICIENTS WITH ION-SELECTIVE ELECTRODES

E. LINDNER, M. GRATZL, A. HRABÉCZY-PÁLL and E. PUNGOR

Institute for General and Analytical Chemistry,
Technical University, Budapest, Gellért tér 4.
1111-Hungary

ABSTRACT

A computerized procedure has been developed for the determination of mean activity coefficients using ISEs, and the conditions of the applicability of the method were established by comparison of the results obtained with accepted literature data.

INTRODUCTION

As ion-selective electrodes respond to the activity of ions, they appear to be ideal sensors for the determination of activity coefficients at known concentrations. The range of electrolytes that might be studied by potentiometric technique has remarkably broadened by the introduction of an increasing variety of new sensors. However, the number of contributions dealing with the determination of activity coefficients by potentiometric detectors is rather small. In those studies mainly cation-selective glass, solid state crystal electrodes, and silver/silver chloride electrode were used /1-4/.

Mean activity coefficients of electrolyte solutions can be determined from the emf of galvanic cells without liquid junction /5/. The essence of one of the methods used for the determination is that first the emf is measured in the presence of a dilute solution with known activity coefficient, then in the presence of a solution of known concentration but unknown activity coefficient. The latter is then calculated from the emf difference, using an assumed theoretical or a previously determined slope value.

In another widely used method which does not presume any known $\gamma\pm$ value, but requires a very precise determination of E^o, the cell emf values are measured at relatively low molalities where the values of log $\gamma\pm$ can be estimated by the Debye--Hückel equation. The E^o value is obtained by extrapolation to zero molality /5/. Then, the mean activity coefficients in more concentrated solutions are determined from measured emf values, using the E^o value thus determined, and theoretical slope values. This, however, implies an extrapolation, assuming the same E^o and slope value for low and extremely high molatilies.

It has been shown /6/ that the extrapolation means the implicit determination of the constants of the linear and quadratic terms in the extended Debye-Hückel equation.

The aim of the present work was to study how and under what conditions different type ion-selective electrodes can be applied to determine mean activity coefficients.

The electrodes used for studying $NaNO_3$, KNO_3 and $Ca/NO_3/_2$ as model electrolytes were $Na^+$and K^+-sensitive glass electrodes /Corning and Beckman, respectively/, valinomycin-based K^+, and ETH 1001-based Ca^{++}-selective electrodes prepared with or without sodium tetraphenylborate additive, ion-exchanger based Ca^{++} and divalent cation electrodes, as well as Orion NO_3^--selective electrode /6/.

In this work the emf values of two cells with liquid junction containing the same reference electrode were measured, and the emf of the cell without transference was obtained by subtraction.

RESULTS AND DISCUSSION

To circumvent the problems posed by the two methods widely accepted in the literature a new method was developed for the determination of mean activity coefficients from measured emf data. This method /i/ does not require known activity coefficient values, /ii/ allows the contants in the Debye-Hückel equation as well as the E^{o} and the slope value of the cell to be determined explicitly at the same time, and /iii/ does not involve any extrapolation.

The basic idea of the new method is that a computerized least-squares minimization procedure is applied to all experimental points, those measured in concentrated as well as in dilute solutions. In this way the statistical reliability of the procedure could be enhanced, as the number of points considered is greater. However, by checking the new procedure using simulated emf data it could be clearly shown that the procedure yields mean activity coefficient data of adequate accuracy only if no systematic error /manifested e.g. in a change in the slope of E^{o} value of the cell within the relatively wide concentration range studied/ occurs and if the random errors do not exceed $\pm$ 0,5 mV /6/. In such cases the theoretical limit of the accuracy of $\gamma_{\pm}$ values determined by the minimization procedure is set by the suitability of the Debye-Hückel equation to describe $\gamma_{\pm}$ values as function of the ionic strength in a wide concentration range. This limit is in the range of 0.2-0.6% /in $\gamma_{\pm}$/ /6/.

However, it cannot be controlled in the case of an unknown system /where we do not have literature data for $\gamma_{\pm}$/. A further assumption involved in all potentiometric methods used for activity coefficient determination is the linearity of the cell calibration curve in the whole concentration range. This assumption, however, is not always justified in the case of ISEs, especially at high concentrations. Moreover, it cannot be checked experimentally when using the electrode for activity coefficient determinations. This explains why neutral carrier based potassium and calcium, as well as different types of ion-exchanger based calcium electrodes were not found appropriate for this type of measurement. When these electrode types are used at high molalities, the slope value of the calibration graph decreases due to anion interference /7,8,9/. The additives recommended resulted in some but insufficient improvement at molalities as high as 3-5 mol/kg.

Besides its advantages, the weak points of the new method reside in the strong correlation of the errors of the parameters determined for the Debye-Hückel equation /e.g. a and C, the ion size parameter and the constant in the linear term, respectively/, which means that significantly different values of a and C may result in the same apparent accuracy of the $\gamma_{\pm}$ values determined. This means that the conditions set are not suitable for their separate determination. A further deficiency of the method is due to the drastic shift in the minimum of the error function even at relatively small systematic errors, that is, there is practically no chance to obtain acceptable Debye-Hückel constants and $\gamma_{\pm}$ values.

Summing up it can be stated that the new method offers several advantages, as mentioned above, but does not eliminate all the problems encountered during the potentiometric determination of activity coefficients. It yields results of adequate accuracy /an error of 1-2% in $\gamma_{\pm}$/ only if

/i/ the cell calibration graph is strictly linear over the whole concentration range studied;

/ii/ the random errors do not exceed $\pm$ 0.5 mV. If the above conditions are fulfilled, our method yields acceptable results, as shown by some data in Table 1 for KNO_3 and NaCl.

REFERENCES

1. R.G.Bates, In "Ion Selective Electrodes 3", Pungor E., Ed., Akadémiai Kiadó, Budapest, 1981.
2. A.Shatkay, A.Lerman, Anal.Chem., 41, /1969/ 514.
3. A.Shatkay, Talanta 17, /1970/ 381.
4. A.Shatkay, Electrochim.Acta 15, /1970/ 1759.
5. R.A.Robinson, R.H.Stokes, Electrolyte solutions, Butterworths, London, 1955.
6. R.G.Bates, A.G.Dickson, M.Gratzl, A.Hrabéczy-Páll and E.Pungor, Anal.Chem.55, /1983/ 1275.
7. J.H.Boles, R.P.Buck, Anal.Chem. 45, /1973/, 2057.
8. A.Hulanicki, M.Trojanowicz, Z.Augustowska, In "Ion-Selective Electrodes 3" Pungor E. Ed.; Akadémiai Kiadó: Budapest, 1981.

9. W.E.Morf, D.Ammann, W.Simon, Chimia 28, /1974/ 65.

Table 1. Comparison of $S(\gamma_{\pm})$ data for full range calibration

Electrolyte studied	RMSD /mV/	$S\,\gamma_{\pm}$	Reference
KNO_3	2.09	0.023	Bates et al/6/
NaCl	0.57	0.012	Shatkay et al/2/

RMSD /Root Mean Square Deviation/

$$RMSD = \sqrt{\frac{\sum_{i=1}^{N} \left(E_{i/measd/} - E_{i/calcd/}\right)^2}{N - p}} \quad /see/6//$$

$$S\,\gamma_{\pm} = \frac{\sum_{i=1}^{N} \left[\gamma_{i/calcd/} - \gamma_{i/lit/}\right]^2}{N}$$

QUESTIONS AND COMMENTS

Participants of the discussion: R.P.Buck, R.A.Durst, G.Horvai, E.Pungor and J.D.R.Thomas

Comment:

We have already discussed the problems connected with the hydration theory at earlier meetings, also here at Mátrafüred. However, the problems are not completely solved, there are still open questions, that we can discuss during this session.

Question:

The equations used to describe the activity coefficients are growing in length. However, people still tend to use the shorter forms. It would be good to hear a comment from someone about the usefulness of the longer forms in real practice, e.g. in complicated clinical systems.

Comment:

In the case of blood the goal would be to have a standard which covers the range of sodium, potassium, calcium, chloride and bicarbonate. In order to do that, I would want to start with activity coefficients for each bicarbonate and each chloride. And then, in order to be sure of having made the most efficient use of the existing data, you have to use the equations that contain the terms that go far enough to give you the statistical certainty of the experiment. And I think that Prof.Bates is just about there.

There are plots that clearly show that log γ is not a linear function of molality, therefore you have to include the non--linear terms. So, if I were to do this, I would start by determining all the parameters from the existing data on all the bicarbonates and the chlorides, and then I would proceed to the mixtures. However, we may be years away from that, as we have to determine the hydration data in mixtures first, which is also a problem.

Comment:

In my opinion, there may only be a correlation between the hydration index and the hydration number, and if I think about it, I begin to doubt that there is a correlation between them at all.

We made measurements at the beginning of the 1960-s to find out what happens to halide ions if they are adsorbed at a surface. The surface was that of a silver halide, and the ions studied were the halides. The results of calculations suggested that the adsorbed anions are non-hydrated. Obviously, if you dissolve something, it has to solvate, as without solvation you cannot dissolve it. But it is very difficult to describe the solvation.

The equations we have seen are a very good help, but to get results closer to observed values, we have to increase the number of terms, the number of constants involved. However, one must be careful in doing so, because if constants without a physical meaning are chosen, one can describe the whole universe, it is only a fitting problem.

Comment:

I agree with everybody's comments so far in terms that in order to be able to describe electrolyte systems adequately, one has to keep adding terms to equations, but I agree that this cannot go on indefinitely. And I think Professor Bates is trying to achieve the most thermodynamically accurate description of the system, and he is probably one of the few people in the world who have the ability to do that. And the rest of the world is waiting for something to happen in terms of standard and may not be quite so desperate for an absolute thermodynamic definition of the activities.

At some point we are going to make decisions and come out with standards. We have had the activity standards for years for sodium, potassium, chloride and fluoride and these are based on Professor Bates' original hydration theory. Work is in progress for calcium, too.

However, when we are dealing with mixed electrolytes, then much of the theory that has been developed really breaks down as not being applicable. Here, I believe we will have to choose a strictly empirical type of development of standards. That is, make measurements to the best of our abilities, and just assign activities, activity coefficients to the various mixed electrolyte standards based on the measurements that we can do, and perhaps as already mentioned, try to relate these to the single systems as much as possible, and try to transfer the information from the simple electrolytes to the more complex mixed electrolyte systems. I also agree with what was said about the hydration indices, that these in some ways may be fragments of imagination as there is no way one can say definitely what the hydration number of an ion is.

Twentyfive years ago I was doing some measurements of hydration numbers, and I realized that the hydration number one obtains is very much technique dependent. That is, if you have a very sensitive technique, you will find a larger hydration number because it will measure more of the coordination sphere of an ion, whereas if you use a less sensitive technique, it will only measure the water molecules that are very closely associated with that same ion. With alkali ions, the variation in hydration numbers may be tremendous, and I guess what we have to do is more or less what Professor Bates has done, that is, take a best educated quess, in this case the hydration index of zero for the chloride ion, for example, which seems to be a good approximation.

Comment:

I agree with what has been said about the difficulties in getting reliable activity coefficient data. I wish to add that we made a zero-order approach to determine activity coefficient values experimentally about twenty years ago. We had to measure ionic constituents in the presence of large organic molecules, and we made the following approach, I call it a zero-order approach, as there remained unsolved problems. We have chosen an ion which was not present in the solution originally and for which we had a good electrode. We prepared a dilute aqueous solution of it, and measured the potential of the electrode in this solution. Then we added the same constituent to the sample solution in an amount to yield the same concentration as that of the pure aqueous solution, and measured the electrode potential in this solution. From the difference of the potentials measured we calculated a practical value of the activity coefficient in the sample solution. This is a zero-order approach, a practical one which we cannot describe theoretically. However, this way we could characterize the effect of the organic material on the activity of the ion measured.

Comment:

When designing scales for single ionic activities, we also need a convention for calculating liquid junction potentials. I believe the standardization problem begins with a general question: can we find algorithms to calculate from the compositions of two solutions and of the connecting salt bridge the single ionic activities of the species measured on either side of the cell and also the value of the liquid junction potential in such a way that the emf calculated from these values agrees well with the emf measured. If such algorithms can be found in general or for a suitably wide range of practically relevant compositions, the standardization problem will be solved. If not, we have to set up practical scales by experimental definitions. In the latter case the problem is to find a self-consistent scale. Since the problems arise mainly with solutions of high ionic strength, one might perhaps start by measuring the emf of cells with transference where on one side of the cell there is a solution of low ionic strength. Based on this emf and some agreed convention for calculating the liquid junction potentials, one could assign the single ionic activity to the species measured in the high ionic strength side of the cell. Such a scale would perhaps be self-consistent.

Comment:

It is quite right that you can set up these self-consistent

scales but the question is whether these scales will work for other people in other laboratories. You made two assumptions that probably could cause trouble. One is that the diffusion potential has a given value and that will depend on the construction and type of the liquid junction you are using. So if someone in another laboratory is using a different type of construction for the liquid junction, he could get significantly different results. The second assumption you made is that the response of the electrode is Nernstian over the entire range, and you would have to prove that in some way. But I think the key thing is that although you get internal consistency, unless everybody is going to make the measurements with the same type of electrodes and junctions, you do not have the international type of standard or approach that is really necessary.

Comment:

I would try to avoid the problem of non-Nernstian response by proving that the electrode works for measuring mean activity coefficients which are known from other measurements. The other thing you say is right, namely that if someone has a liquid junction with a different diffusion potential, the whole thing fails. However, I think that part of the standardization would be to standardize the liquid junction. There are several types of liquid junction which reportedly work in the same way in different laboratories. If this were not the case, I would not see any way for standardization.

Comment:

I think that if you have standards that are available and that everybody agrees on, that have certain values, then it does not matter what the liquid junction is like because in preparing your working curve, you are making all your measurements using this liquid junction.

Question:

about the practical activity scale suggested here by an earlier speaker. Is there any evidence that you make internally consistent measurements with this system ? Is it just theoretically true that the junction potentials are perfect derivatives, or is it experimentally true ? I suppose an assumption was made about the liquid junction being only dependent on the concentration ratios and the mobilities and therefore each one is calculatable from the other two. But is that experimentally true ?

Answer:

As far as I can remember there are situations where the junction potentials are perfect derivatives, for example in the case of concentration cells. However, in general they are not.

Comment:

In the clinical field we are developing reference methods so that all laboratories will be getting internally consistent measurement. We are trying to develop for example for the pH measurement of blood a reference method where everyone will use the same type of junction, and achieve consistency with these very difficult - to - measure systems by everybody using the same method. So if we could standardize the method this would also help in coming up with the appropriate standards.

Question:

in connection with nomenclature. What is the correct English name for the high input impedance mV-meters we are using with ISEs ? Someone suggested the word "potentiometer", but this word has also another meaning.

Answer:

You are correct that in the past "potentiometer" meant a device for producing a potential very accurately, and there is some confusion in terms of also using it as a device for measuring the potential. But if you look at the word, "potentiometer" is obviously a device for measuring the potential. So, the present usage of the term as a potential measuring device rather than a potential producing device is probably more correct, and now for the one that produces the potential "potentiostat" is the more common term.

Comment:

But I think that the word potentiometer does not really imply that you are using a high input impedance instrument. And one would probably not call a digital voltmeter potentiometer. This is why I think that this is not a really good word.

Comment:

I think a good many authors, when they use these high impedance measuring devices tend to use them in the sense of research, which means high precision. pH/mV meter and the word pH meter itself implies that it is a high impedance instrument. And then, if people modify their instrument by putting in the extra unit in order to cope with two high impedance electrodes, they usually specify that.

Question:

What term should be used then with an instrument like blood sodium analyzer, which has a high input impedance mV meter, but does not measure pH ?

Answer:

I think the tendency here is to specify the name of the actual instrument itself, such as Radiometer ICA 1. The manufacturer's specification then defines what is in the instrument. However, there may come a day when the Electroanalytical Commission of IUPAC will have to attend to this.

Comment:

When I teach this subject I teach that a potentiometer is a Poggendorf potentiometer, or a battery with a Kohlrausch slidewire and it applies a voltage which is variable. And this happens to be a low impedance system.
However, when you use electronic measuring devices, I think you should always call them voltmeters or mV meters. Electrometer may be an alternate term. It is a high impedance electronic device and I would never call a pH meter potentiometer.

Comment:

If the instrument does not have a pH scale on it, then the tendency is to call it high impedance mV meter.

QUESTION

Participants: R.A.Durst and E.Lindner

Question:

You showed the experiment where you made measurements of the cells with liquid junction with the silver chloride electrode vs. the ion-selective electrodes and you said you combine the values from these to give you the value for the cell without liquid junction. Is this done to avoid experimental problems with the cell without liquid junction where you would need two high impedance inputs ?

Answer:

Yes, we did this only because we did not have a pH meter with two high impedance inputs in Professor Bates' laboratory.

SUBJECT INDEX